Lessons Learned From Research on Mathematics Curriculum

A Volume in
Research in Mathematics Education

Series Editors

Denisse R. Thompson
University of South Florida

Mary Ann Huntley
Cornell University

Christine Suurtamm
University of Ottawa

Books in This Series

Lessons Learned From Research on Mathematics Curriculum (2024)
edited by Denisse R. Thompson, Mary Ann Huntley,
and Christine Suurtamm

International Perspectives on Mathematics Teacher Education (2023)
edited by Denisse R. Thompson, Christine Suurtamm,
and Mary Ann Huntley

Researching Pedagogy and Practice with Canadian Mathematics Teachers (2020)
edited by David A. Reid, Christine Suurtamm, Annie Savard, Elaine
Simmt, Dominic Manuel, Lisa Lunney Borden, and Richard Barwell

International Perspectives on Mathematics Curriculum (2018)
edited by Denisse R. Thompson, Mary Ann Huntley,
and Christine Suurtamm

Digital Curricula in School Mathematics (2016)
edited by Meg Bates and Zalman Usiskin

Approaches to Studying the Enacted Mathematics Curriculum (2012)
edited by Daniel J. Heck, Kathryn B. Chval,
Iris R. Weiss, and Steven W. Ziebarth

Variability is the Rule:
A Companion Analysis of K–8 State Mathematics Standards (2010)
edited by John P. Smith III

Language and Mathematics Education:
Multiple Perspectives and Directions for Research (2010)
edited by Judit N. Moschkovich

A Five-Year Study of the First Edition
of the Core-Plus Mathematics Curriculum (2010)
edited by Harold Schoen, Steven W. Ziebarth,
Christian R. Hirsch, and Allison BrckaLorenz

Future Curricular Trends in School Algebra and Geometry:
Proceedings of A Conference (2010)
edited by Zalman Usiskin, Kathleen Andersen, and Nicole Zotto

A Decade of Middle School Mathematics Curriculum Implementation:
Lessons Learned from the Show-Me Project (2008)
edited by Margaret R. Meyer and Cynthia W. Langrall (with associate
editors Fran Arbaugh, David C. Webb, and Murrel Brewer Hoover)

Series titles continued on next page

*Mathematics Curriculum in Pacific Rim Countries—
China, Japan, Korea, and Singapore: Proceedings of a Conference (2008)*
edited by Zalman Usiskin and Edwin Willmore

The History of the Geometry Curriculum in the United States (2008)
by Nathalie Sinclair

The Classification of Quadrilaterals: A Study of Definition (2008)
Prepared by Zalman Usiskin and Jennifer Griffin with the help of
David Witonsky and Edwin Willmore

*The Intended Mathematics Curriculum as Represented in
State-Level Curriculum Standards: Consensus or Confusion? (2006)*
edited by Barbara Reys

*Mathematics Curriculum in Pacific Rim Countries—
China, Japan, Korea, and Singapore: Proceedings of a Conference (2008)*
edited by Zalman Usiskin and Edwin Willmore

The History of the Geometry Curriculum in the United States (2008)
by Nathalie Sinclair

The Classification of Quadrilaterals: A Study of Definition (2008)
Prepared by Zalman Usiskin and Jennifer Griffin with the help of
David Witonsky and Edwin Willmore

*The Intended Mathematics Curriculum as Represented in
State-Level Curriculum Standards: Consensus or Confusion? (2006)*
edited by Barbara Reys

Lessons Learned From Research on Mathematics Curriculum

Editors

Denisse R. Thompson
University of South Florida

Mary Ann Huntley
Cornell University

Christine Suurtamm
University of Ottawa

INFORMATION AGE PUBLISHING, INC.
Charlotte, NC • www.infoagepub.com

Library of Congress Cataloging-in-Publication Data

CIP record for this book is available from the Library of Congress
http://www.loc.gov

ISBNs: 979-8-88730-709-1 (Paperback)

  979-8-88730-710-7 (Hardcover)

  979-8-88730-711-4 (ebook)

Copyright © 2024 Information Age Publishing Inc.

All rights reserved. No part of this publication may be reproduced, stored in a retrieval system, or transmitted, in any form or by any means, electronic, mechanical, photocopying, microfilming, recording or otherwise, without written permission from the publisher.

Printed in the United States of America

CONTENTS

Foreword
  Janine T. Remillard xi

Preface xxi

1. Introduction: Learning About Conducting Curriculum Research
   *Denisse R. Thompson, Mary Ann Huntley,
   and Christine Suurtamm* 1

SECTION I: DEVELOPMENT AND ANALYSIS OF THE OFFICIAL CURRICULUM

2. Towards the Development of a Primary Mathematics Curriculum in Ireland
   Elizabeth Dunphy and Thérèse Dooley 17

3. Combined Reading: A Methodological Approach for Documental Curricular Analysis
   Priscila Dias Corrêa and Leticia Rangel 37

4. Creating Space for a Topic in the Null Curriculum
   Lynn M. McGarvey, Jennifer S. Thom, and Josh Markle 57

5. Analyzing the Official and Intended Curricula: Methods and Techniques to Study Mathematics Standards and Textbooks
   *Shannon Dingman, Dawn Teuscher, Lisa Kasmer,
   and Travis Olson* 79

SECTION II: TEXTBOOK ANALYSIS

6. Choosing Samples in Mathematics Textbook Research: Textbooks, Chapters, Pages, and Tasks
   Dustin L. Jones 97

7. Lessons Learned From Cross-Cultural Curricular Research: The Case of Proving-Related Activities
   Hangil Kim, GwiSoo Na, and Eric Knuth 111

8. Lessons Learned in Creating Mosaics to Portray the Algebra Strand of U.S. High-School Textbooks
   Mary Ann Huntley, Maria S. Terrell, and Nicole L. Fonger 131

SECTION III: CURRICULUM DESIGN AND DEVELOPMENT

9. The Curriculum Research Framework: Supporting the Scientific Creation, Implementation, and Evaluation of Curricula
   Julie Sarama and Douglas H. Clements 157

10. Lessons Learned in Designing an Effective Early Algebra Curriculum for Grades K–5
    Maria Blanton, Angela Murphy Gardiner, Ana C. Stephens, Rena Stroud, Eric Knuth, and Despina Stylianou 189

11. Lessons Learned From Decades of Designing, Researching, and Revising a Kindergarten to Grade 6 Mathematics Program
    Karen C. Fuson 215

12. Lessons Learned for Developing and Enacting a Digital Collaborative Platform With an Embedded Problem-Based Mathematics Curriculum
    Alden J. Edson 245

13. Study of Technology-Based Inquiry Learning: Automatic Analysis of Online Personal Feedback Processes
    Raz Harel and Michal Yerushalmy 273

14. Narrowing the Gap Between School Mathematics Curricula and Contemporary Mathematics: Lessons Learned During a Four-Phase Curriculum R&D Study
    Nitsa Movshovitz-Hadar, Atara Shriki, Ruti Segal, and Boaz Silverman 291

SECTION IV: RESEARCH ON THE TEACHER INTENDED CURRICULUM

15. Broadening Conceptualizations of the Intended and Enacted Curriculum to Inform Research Design
    Emily Saxton, Nicole Rigelman, Corey Drake, and Kim Markworth 323

16. Our Journey Through Data Analysis to Develop a Model of Curricular Reasoning
    Dawn Teuscher, Shannon Dingman, Lisa Kasmer, and Travis Olson 347

17. Lessons Learned From Studying the Teacher Intended Curriculum: The Value of Capturing and Analyzing Attention
    Kelsey Quaisley and Lorraine M. Males 367

18. Opportunities and Challenges in Surveying and Interviewing Elementary Teachers About Their Mathematics Curricular Use
    Bima Sapkota, Kristin Doherty, Mona Baniahmadi, Nan Jiang, Laurel Hendrickson, Doris Fulwider, Amy M. Olson, Marcy B. Wood, and Jill A. Newton 385

19. Lessons Learned From Researching Online Mathematics Curriculum Supplementation
    Lara K. Dick and Amanda G. Sawyer 397

SECTION V: RESEARCH ON THE ENACTED CURRICULUM

20. From Participants to Partners: Lessons Learned From Enacted Curriculum Research in Early Numeracy Contexts
    Arielle Orsini, Julie Houle, Helena P. Osana, Alison Tellos, and Anne Lafay 419

21. Analyzing Enacted Curriculum: Seven Lessons Learned From Collecting, Coding, and Analyzing Large Sets of Qualitative Data
    Kristy Litster 441

22. Examining Curriculum Openings in Early Algebra: A Closer Look at Teachers' Interaction With Curricula
    Despina Stylianou, Ingrid Ristroph, Boram Lee, Eric Knuth, Ana C. Stephens, Maria Blanton, Rena Stroud, and Angela Murphy Gardiner 457

23. Lessons Learned in Studying the Impact of Teaching
    Mathematics in Context on Student Achievement
    Mary C. Shafer 481

24. Lessons Learned From a Multi-Site Study of an Algebra
    Support Curriculum
    Deborah Spencer, June Mark, Mary Beth Piecham,
    Julie K. Zeringue, Kelsey Klein, and Laura M. O'Dwyer 505

25. Studying the Enacted Secondary Mathematics Curriculum:
    Lessons Learned From the University of Chicago School
    Mathematics Project
    Denisse R. Thompson and Sharon L. Senk 529

SECTION VI: PERSONAL REFLECTIONS ON CURRICULUM DEVELOPMENT AND RESEARCH

26. Lessons Learned From the Development of Innovative
    Mathematics Materials
    Zalman Usiskin 563

27. Disentangling Mathematics Curriculum From Contextual
    Enmeshment: Lessons Learned Across Eight Studies
    Thomas E. Ricks 581

28. Everything I Really Need to Know About Curriculum,
    I Learned From CSMC
    Jill A. Newton 599

About the Editors and Authors 623

FOREWORD

Janine T. Remillard
University of Pennsylvania

In the early 1990s, mathematics education in the U.S. was undergoing one of its many reforms, this time in response to a visionary set of recommendations proposed by the National Council of Teachers of Mathematics (NCTM), named the *Curriculum and Evaluation Standards for School Mathematics* (NCTM, 1989). Commercial publishers updated their textbooks, adding problem-solving strategies and visual models to the standard fare, adorning the cover of the teacher's guide with a bold "NCTM Standards version" medallion. Just back from a visit to an elementary school, a colleague paused at my office door and asked, "Do these textbooks really meet the *Standards*?" Flustered by both the simplicity and complexity of the question, I stammered and then offered something like, "It's really teachers, not textbooks, who meet standards." I then clumsily tried to explain that the new textbooks most likely contained components that reflected the *Standards*, but we didn't have a clear picture of what a textbook that met the *Standards* looked like. This gap in knowledge was not only because the vision in the *Standards* was novel, at least to the majority of educators and textbook authors, but because staking out aspirational ideas for K–12 students' mathematics learning was an entirely different project than designing instructional pathways that could come close to achieving these aims.

Lessons Learned From Research on Mathematics Curriculum, pp. xi–xix
Copyright © 2024 by Information Age Publishing
www.infoagepub.com
All rights of reproduction in any form reserved.

That was in 1994. Since that time, U.S. educators have seen and studied a range of curriculum programs designed to meet the *Standards*, and then the *Common Core State Standards for Mathematics* (NGA & CCSSO, 2010) 20 years later. Developing curriculum materials that can influence the mathematics content taught *and* pedagogy used has become a foremost mission in the U.S. and to some extent around the world. And research has followed. We now understand something about the range of ways curriculum resources can reflect policy objectives (Remillard & Kim, 2020), and we can anticipate that teachers will make meaning and use these resources in very different ways, providing varied opportunities for student learning, and teachers' learning as well.

These understandings reflect over 30 years of research and theory building around mathematics curriculum design and use. This volume, *Lessons Learned From Research on Mathematics Curriculum*, represents both progress made and valuable lessons learned from research. It comes at an ideal time: the field of mathematics curriculum research is now firmly established, but it is also facing new questions and opportunities to think and work differently. New questions emerge from advancing digital technologies that have already shifted the way we source and interact with information, tools, and one another. These advances ask us to rethink what *curriculum materials* are, their organization and capacities, and how they might be accessed and by whom; they also ask us to reconsider what *the curriculum* is. What needs to be learned and for what purposes? These questions are deeply connected to pernicious social issues confronting educational practice, policy, and research that mathematics educators cannot and should not ignore, such as inequality, poverty, and the health of the planet.

By asking authors to speak to lessons learned from their particular corners of the curriculum-research map, the editors have gone beyond accumulating research findings. They have also collected insights and recommendations related to the research process itself. Some insights involve nuts and bolts suggestions for specific aspects of curriculum research, such as the importance of "determining a common unit of analysis when studying textbooks" (Dingman et al., Chapter 5, p. 88) or using a different font color to distinguish aspects of the original framework from new annotations (Huntley et al., Chapter 8). Others speak to the relationships one builds with collaborating teachers. Orsini and colleagues (Chapter 20) found that the research was more productive when the relationships were reciprocal and researchers and teachers believe they could learn from each other. Many reported learning the value of patience with the research process and a willingness to learn and make changes along the way. All insights—large or small, broadly or narrowly focused—are valuable, as they reflect the accumulated wisdom of experience in the field.

In keeping with the focus on lessons learned in the pages that follow, I synthesize lessons that crystallized as I read across the 28 chapters in this volume. These lessons, I found, resonated with my three decades of research on mathematics curriculum and even longer working with curriculum materials as a teacher and teacher educator.

Lesson 1: Curriculum is a complex, multi-faceted, and dynamic construct; in order to understand and make visible the complexity of curriculum, the field needs research that explores the topic from different perspectives and vantage points.

The conversation with my colleague, recounted earlier, points to my own struggles to name the complexity of both *curriculum* as an aspirational vision or policy statement and curriculum *materials* as a classroom resource, as well as the dynamic relationship between the two. Drawing on various authors over time, in the introductory chapter, the editors discuss a number of ways researchers and educators think about curriculum: It is both a thing and a process, the *what* of teaching and learning and the *path* by which it happens; it is aspirational and descriptive, both what is deemed desirable and an accounting of what is actually accomplished. In the piece that I coauthored with my colleague, Daniel Heck (Remillard & Heck, 2014), we used the term *experiences* to suggest that curriculum includes what students "actually experience to support their learning" (p. 707). We also present a model of what we call the "curriculum policy, design, and enactment system," which visualizes the way the meaning of *curriculum* varies depending on where it occurs in a complex system of design and enactment.

This model leads us to another important characteristic of curriculum: it is dynamic, unfolding in real time through interpretative and decision-making processes. The curriculum materials designed to follow the *Common Core State Standards*, for example, reflect the designers' interpretations of the standards and their own visions for teaching and learning, constrained of course, by the realities of any education system. Similarly, a teacher's planned curriculum, developed through reading and interpreting the teacher's guide, represents their interpretation through the lens of what is appropriate for their students. And that is all before the teacher begins to enact the curriculum in real time.

Perusing the chapters in the volume, I am encouraged to find that the complexity of curriculum as a construct is embraced and fully tapped. Not only have researchers honed in on different instantiations of curriculum, often linking their constructs to the Remillard and Heck (2014) model, but they have not shied away from complex representations of aspects of curriculum under study. For example, Corrêa and Rangel (Chapter 3) describe an approach to analyzing curriculum documents that they call "combined reading," a multi-phased process of critically examining and questioning multiple features of curriculum documents, including the mathematical topics in relation to one another, terms and expressions used to surface underlying values, principles, and beliefs, and conceptions of teacher and teacher knowledge.

I am especially heartened to discover that several chapters add depth to the responsive and improvisational nature of the operational curriculum, exploring in both conceptual and systematic ways teachers work as curriculum (en)acters. Stylianou and colleagues (Chapter 22), for example, build on the construct of "openings in the curriculum" (Remillard & Geist, 2002) to assess the impact of teacher take-up of emergent opportunities during instruction ("algebra moments") to deepen students' understanding of algebraic concepts. Saxton and colleagues (Chapter 15) explore similar questions from the perspective of curriculum material development. How can a curriculum resource be designed to support teachers to respond to students' ideas in the moment, making "productive adaptations" (DeBarger et al., 2013), while maintaining essential elements of the designed curriculum? Importantly, Saxton and colleagues argue that the essential elements are big ideas, characteristic features, and commitments, not specific lesson steps or teacher scripts, reminiscent of a distinction made by Brown and colleagues (2009) between literal fidelity and fidelity to ideas. In my view, the field will benefit from research that further conceptualizes the work of curriculum enactment and the complex relationship between curriculum design and the real, responsive labor teachers do in the classroom.

Finally, in a retrospective reflection on his research, Ricks (Chapter 27) deepens and extends our understanding of the complexity of curriculum by looking to complexity science (Davis & Simmt, 2003) as a guiding framework. From this perspective, Ricks argues, research on curriculum must represent "all aspects of education as interacting, nested, complex systems of agents, which are themselves other complex systems" (p. 588). These complex systems include culture, language, institutional structures and mandates, and social and global concerns.

Lesson 2: Frameworks are critical to research on mathematics curriculum because, if well developed, they can capture complex and dynamic phenomena within and beyond specific contexts. Yet they are partial representations, and there is always a tension between simplifying and complexifying that researchers must navigate.

Many authors of chapters in the volume introduce frameworks or visual models to represent the constructs they are studying or to guide their design work. As a fan of visual models, I am partial to such frameworks because of the way they have helped me capture and represent complex ideas and relationships that I seek to understand. I especially appreciate the spatial nature of such representations, which communicate in ways that linear text does not. Upon reviewing the chapters in this volume, I see that I am not alone. Developing, refining, and using visual models to guide research and practice is an instructive activity for many curriculum researchers. It is evident that the analytical process of model construction affords as much benefit as the resulting products. In reviewing these chapters, I am reminded of the following lessons about the possibilities and uses of frameworks and models.

A principal benefit of visual models is the amount and type of information they are able to convey. In addition to identifying core constructs of importance, visual models put them in relation to one another, showing influence or simply co-occurrence. This capability allows researchers to represent complex concepts along with underlying explanatory mechanisms. Importantly, developing these representations should be an evolving process and tool for thinking about one's data. Moreover, they need to communicate meaningfully. When working on models myself, I often channel Iris Weiss, former president of Horizon Research Inc. and a healthy skeptic of visual models. Drawings where every construct is connected to every other object in the same way, Iris would warn, communicate nothing. Model development requires embracing this skepticism, along with an interest in communicating meaningfully. It can also be a messy, time-intensive process. In their chapter, Teuscher and colleagues (Chapter 16) recount the iterative, messy process they underwent to develop a model of curricular reasoning.

Frameworks can also guide curriculum design work, as is illustrated by several chapters in this volume. As Blanton and colleagues (Chapter 10) emphasize: "Frameworks provided critical scaffolding around how we designed our curriculum.... [They] helped focus and stabilize our process and define more clearly the steps we needed to take" (p. 194). Similarly, Sarama and Clements (Chapter 9) and Saxton and colleagues (Chapter 15) provide two additional examples of how frameworks can guide curriculum design work. Importantly, not all frameworks are presented as visual models. A well-structured list of sequential steps or characteristic features to attend to can be sufficient to offer systematic guidance for action. The most useful frameworks are research based, make concepts and relationships visible, and offer an analytical approach to decision making.

Finally, chapters by Ricks (Chapter 27) and Shafer (Chapter 23) highlight ways that frameworks can support researchers to move back and forth between an analytical perspective, which focuses on key constructs and influential factors, and curriculum processes as they occur in the real world. As Ricks (Chapter 27) points out, frameworks can help researchers "disentangle" phenomena related to mathematics curriculum activity from contexts in which they are enmeshed. This disentangling facilitates the research process, allowing important constructs and influential mechanisms to surface above the hustle and bustle of contextual factors. Still, it is often the case that one goal of curriculum research is to understand how realities of the context influence curriculum enactment. In her chapter, Shafer (Chapter 23) shares a framework to guide the monitoring of curriculum enactment that takes into account many contextual factors, including teacher background, school context, support in the environment, and classroom events, to name a few. Looking at these two chapters together serves as a reminder that frameworks are important in research because of their ability to both simplify and complexify the phenomena of interest. It is up to researchers to know when either is desirable.

Lesson 3: New methodological approaches, often made possible by technology, can (and do) advance how we understand curriculum and what we know about its use.

The research discussed in this volume spans topics, research foci, and methodologies. Many researchers describe the use of new approaches to collecting, analyzing, or representing data, enabling them to understand new phenomena or explore new questions. Technological advances and access to digital tools are often, but not always, at the heart of these advances. In fact, one of the most important contributions of this volume is the extensive collection of new methodological innovations being introduced to the field. I briefly touch on two examples. In their chapter, Quaisley and Males (Chapter 17) describe their use of eye-tracking software to explore what teachers are attending to when planning with print curriculum guides or scrolling on a computer screen. In conjunction with more conventional methods of gathering data on teacher attention, Quaisley and Males used images and metrics generated by eye-tracking tools to document what teachers attended to on a page, in what order, and for how long. They then used innovative data representations (heat maps and gaze plots) to illustrate this otherwise hidden process of curriculum use. Also using digital technologies, Huntley and colleagues (Chapter 8) created visual maps, called mosaic timeline plots, to show the treatment of different topics in a mathematics curriculum program over time. These visual representations display the density, distribution, and sequencing of different types of content in a single textbook, allowing for comparative analysis across different curriculum programs.

Lesson 4: Digital environments can be harnessed to design curriculum tools that extend, rather than undermine, teacher capacity and expertise.

Digital curriculum resources are widely available and are being extensively used by teachers, leading to varied results. While many examples exist where technologies are used to support collaborative problem solving in the classroom, students in the U.S. are increasingly spending substantial amounts of time on digital platforms, often working individually to complete mathematics assignments. This reality is part of what van Dijck and Poell (2018) refer to as the platformization of education. In my research, I have found consistent use of *online student-facing platforms* (OSFPs) across elementary schools (Remillard et al., 2021). We use the term OSFPs to refer to an array of software programs, such as Dreambox, IXL, Prodigy, or Zearn, that have what Choppin et al. (2014) refer to as "individual learning designs." Our analysis suggests that most OSFPs engage students in low to moderate levels of mathematical rigor and provide limited opportunities for student agency. Further, students' use of OSFPs replaces interactions with a teacher or other students, and teachers have decreased access to information about student thinking or understanding (Condon et al., 2021; Kavanagh et al., 2024).

It is encouraging, then, to see examples of digital tools and environments that extend teacher capacity, rather than undermine it. Edson (Chapter 12) describes a masterful effort to design digital tools for students and teachers that foster student sensemaking and facilitate collaboration. The aim of this research team was not only to deepen student learning, but give teachers access to student thinking through the teacher dashboard and the capacity to generate "just-in-time supports" for specific students as needed. Like many others, this chapter documents the extensive design process, making evident the critical role that frameworks play. The other two chapters on digital curriculum designs (Harel & Yerushalmy, Chapter 13; Movshovitz-Hadar et al., Chapter 14) similarly describe efforts to leverage digital tools to supplement and enhance student learning, illustrating the possibilities of digital curriculum resources.

Lesson 5: The value of community cannot be overstated. Mathematics curriculum research has advanced considerably over the last quarter century, and these advances are a product of a strong, scholarly community that nurtures intellectual discourse and methodological pluralism, and welcomes new members into its ranks.

Reading the accounts of different projects and research approaches in this volume was inspiring and exciting. It was easy to glimpse markers of a robust community of researchers undertaking diverse and novel projects. This sense of community is confirmed by the reminiscent chapter by Jill Newton (Chapter 28), which details how the U.S. National Science Foundation-funded Center for the Study of Mathematics Curriculum (CSMC) nurtured her during her doctoral studies and launched her research career. Bringing together mathematics curriculum researchers and developers from multiple universities, along with their doctoral students, CSMC initiated research on mathematics curriculum, lifted up the research of others, hosted important and challenging conversations, and prioritized doctoral education. Twenty years after the center was funded, Jill reflects on being "raised" in the strong intellectual community and family that CSMC became.

Jill's reflection struck a chord for me, as I also benefited from being part of the CSMC extended family during my early research career. And I would be remiss if I did not acknowledge the critical role that Barbara Reys played in leading the Center throughout and beyond its funding period. All of us currently engaged in mathematics curriculum research do so, in part, because of Barb's vision and dedicated labor.

REFERENCES

Brown, S. A., Pitvorec, K., Ditto, C., & Kelso, K. R. (2009). Reconceiving fidelity of implementation: An investigation of elementary whole-number lessons. *Journal for Research in Mathematics Education*, *40*(4), 363–395.

Choppin, J., Carsons, C., Bory, Z., Cerosaletti, C., & Gillis, R. (2014). A typology for analyzing digital curricula in mathematics education. *International Journal of Education in Mathematics, Science, and Technology*, *2*(1), 11–25.

Condon, L., Machalow, R., Remillard, J., Van Steenbrugge, H., Krzywacki, H., & Koljonen, T. (2021). Doing math in the digital age: An analysis of online mathematics platforms. In D. Olanoff, K. Johnson, & K. Johnson (Eds.), *Proceedings of the 43rd annual meeting of the North American Chapter of the International Group for the Psychology of Mathematics Education* (pp. 1714–1722). Philadelphia, PA.

Davis, B., & Simmt, E. (2003). Understanding learning systems. *Journal for Research in Mathematics Education*, *34*, 137–167.

DeBarger, A. H., Choppin, J., Beauvineau, Y., & Moorthy, S. (2013). Designing for productive adaptations of curriculum interventions. *Teachers College Record*, *115*(14), 298–319.

Kavanagh, S. S., Bernhard, T., & Gibbons, L. K. (2024). 'Someone else in the universe is trying to teach you': Teachers' experiences with platformized instruction. *Learning, Media and Technology*, 1–17. https://doi.org/10.1080/17439884.2024.2337396

National Council of Teachers of Mathematics [NCTM]. (1989). *Curriculum and evaluation standards for school mathematics*.

National Governors Association Center for Best Practices & Council of Chief State School Officers [NGA & CCSSO]. (2010). *Common core state standards for school mathematics.* https://www.thecorestandards.org/Math/

Remillard, J. T., & Geist, P. (2002). Supporting teachers' professional learning through navigating openings in the curriculum. *Journal of Mathematics Teacher Education, 5*(1), 7–34.

Remillard, J. T., & Heck, D. (2014). Conceptualizing the curriculum enactment process in mathematics education. *ZDM Mathematics Education, 46*(5), 705–718. https://doi.org/10.1007/s11858-014-0600-4

Remillard, J. T., & Kim, O. K. (2020). *Elementary mathematics curriculum materials: Implications for teachers and teaching.* Springer.

Remillard, J. T., Van Steenbrugge, H., Machalow, R., Koljonen, T., Krzywacki, H., Condon, L., & Hemmi, K. (2021). Elementary teachers' reflections on their use of digital instructional resources in four educational contexts: Belgium, Finland, Sweden, and US. *ZDM Mathematics Education, 53*(6), 1331–1345. https://doi.org/10.1007/s11858-021-01295-6

van Dijck, J., & Poell, T. (2018). Social media platforms and education. In J. Burgess, A. Marwick, & T. Poell (Eds.), *The SAGE handbook of social media* (pp. 579–591). SAGE. https://doi.org/10.4135/9781473984066.n33

PREFACE

In *International Perspectives on Mathematics Curriculum*, a 2018 volume in this series *Research in Mathematics Education*, the focus was on the underlying educational structure of the curriculum in eight countries around the world. Some researchers have described curriculum as follows:

> ... a kind of underlying "skeleton" that gives characteristic shape and direction to mathematics instruction in educational systems around the world.... The plan that expresses these aims and intentions, which takes them from vision to implementation, and serves as the broad course that runs throughout formal schooling, is curriculum. (Schmidt et al., 1997, p. 4)

The curriculum described in those eight countries could be considered as the official curriculum. However, educators recognize that many variations occur as that official curriculum is instantiated in textbooks, in classroom implementation, and in student learning.

In this volume, we focus on research related to mathematics curriculum. But rather than focusing on *results* of research, we instead focus on *lessons learned about conducting research on curriculum*, whether about design and development, analysis of curriculum in the form of official standards or textbook instantiations, teacher intentions related to curriculum implementation, or actual classroom enactment. In other words, for new or experienced curriculum researchers interested in broadening their focus of curriculum research, what lessons about conducting curriculum research might be learned from the work of others? Could challenges and issues researchers faced and their solutions to address them be gathered so that others could build on their work?

The work in this volume helps fill the gap in the research literature described in the report of the latest study sponsored by the International Commission on Mathematical Instruction. The directors of that study noted "limited writings about appropriate research methodologies" in studying mathematics curriculum at different levels and related to different components of the curriculum (Shimizu & Vithal, 2023, p. 4). Although the current volume was conceived before this report was published, the chapters in our volume provide a response to this observation—they discuss lessons related to methodologies in studying different levels of the mathematics curriculum.

In addition, the international perspective provided by various authors of the chapters in this volume is evidence that curriculum issues cut across country boundaries. Sharing lessons from authors of different countries can strengthen the broader mathematics research community and provide insights that can help researchers make important strides forward.

OUR CALL FOR MANUSCRIPTS

As editors of this volume, we have individually focused on different aspects of curriculum research over several decades. Thompson has researched implementation of the secondary curriculum developed by the University of Chicago School Mathematics Project (UCSMP). Many lessons have been learned about studying enactment in a range of U.S. schools and districts and addressing challenges in comparing instruction and achievement of students and teachers using UCSMP materials to those using a somewhat comparable comparison curriculum (Thompson & Senk, 2010, 2012).[1]

Huntley has studied the secondary mathematics curriculum from various perspectives. One comparative study involved examining effects of the Core-Plus Mathematics Project (CPMP) curriculum and more conventional curricula on growth of student understanding, skill, and problem-solving ability in algebra (Huntley et al., 2000). Another study involved development of instruments to examine teacher enactment of two NCTM *Standards*-based curricula (Huntley & Heck, 2014), and another study involved a detailed content analysis of the algebra strand of six textbook series (Huntley & Terrell, 2014; Mayer et al., 2019). Each of these areas of curriculum research involved starts and adjustments in how to conduct, analyze, and report the research.

Suurtamm has been instrumental in curriculum design and development with her background research report informing the official provincial curriculum in Ontario (Suurtamm & McKie, 2019). She has also investigated teachers' pedagogical actions, helping teachers reflect on their own and other teachers' classroom implementations of curricula (e.g., Suurtamm & Quigley, 2020), with lessons learned about the use of video analysis to enhance classroom pedagogical practice.

Although the three of us have learned many lessons about studying our respective areas of curriculum, we believed the field would benefit from lessons learned by many other researchers who have also engaged in curriculum research. And, we believed there would be value in gathering many lessons about conducting curriculum research into one volume. We thus placed an open call for potential manuscripts on the Information Age Publishing website, and shared the call through various listserves and organizations (e.g., Association of Mathematics Teacher Educators, Special Interest Group on Research in Mathematics Education, Canadian Mathematics Education Study Group). The call is shown in the following text box.

Call for Chapter Proposals for the Volume

As evidenced by sessions on mathematics curriculum research at international conferences and meetings, researchers around the globe have increasingly been engaged in various aspects of mathematics curriculum research. This includes all levels of curriculum, as noted in the TIMSS framework (Valverde et al., 2002) and expanded on by Remillard and Heck (2014):

- The *official curriculum* consists of national, state, or local standards and goals for what students should learn and on what they should be assessed.
- The *intended curriculum* often consists of the curriculum identified in textbooks in many countries of the world.
- The *teacher intended curriculum* consists of those lessons and intentions of the teacher related to how they intend to implement the curriculum within their instructional environment.
- The *implemented* or *enacted curriculum* consists of what aspects of the curriculum actually occur in classrooms.
- The *assessed curriculum* consists of the elements of the curriculum that appear on local (i.e., classroom) or state or national assessments.
- The *learned curriculum* consists of what students actually master.

Researchers engaging in one or more of these aspects of curriculum research have typically had to develop their own methodologies, instruments, and analysis techniques. They have often faced difficulties or challenges when conducting their research and have had to find ways to address them. This volume is intended to be a compilation of the lessons learned from engaging in curriculum research that might be useful to current curriculum researchers as well as the next generation of curriculum researchers. Rather than reinventing the wheel, they can use lessons learned from experienced mathematics curriculum researchers. The purpose of the volume is not to present research per se, but rather to illustrate lessons learned or to discuss how challenges were overcome when conducting mathematics curriculum research.

Potential authors were encouraged to submit a brief outline of a proposed chapter. Interest in the topic led to the submission of 57 brief proposals. We recommended 42 of these authors submit a first draft of a chapter, with no guarantee of acceptance. Of the 38 drafts submitted, we accepted 27.

POTENTIAL AUDIENCE FOR THE BOOK

We believe the range of lessons learned across the levels of curriculum will be a valuable contribution to the profession and to curriculum researchers, whether novice researchers or those more experienced in conducting curriculum research. We hope the volume might generate conversations among those engaged in research at a particular level of curriculum and/or bring groups of curriculum researchers together. The volume might be used by university instructors teaching graduate level courses related to mathematics curriculum and/or curriculum research.

We believe this volume is a natural complement to several other volumes about curriculum published by Information Age Publishing that focus on issues in mathematics curriculum, as indicated below.

- *Researching Pedagogy and Practice with Canadian Mathematics Teachers* (Reid et al., 2020) explores the results from a study related to pedagogical practices in several provinces in Canada.
- *International Perspectives on Mathematics Curriculum* (Thompson et al., 2018) provides views on the official curriculum in eight countries around the globe.
- *Digital Curricula in School Mathematics* (Bates & Usiskin, 2016) considers how researchers in different countries are addressing challenges surrounding digital curriculum in the current educational and globalized environment.
- *Enacted Mathematics Curriculum: A Conceptual Framework and Research Needs* (Thompson & Usiskin, 2014) focuses on research related to the enacted curriculum and expands on discussions at a conference related to researching the enacted curriculum.
- *Approaches to Studying the Enacted Mathematics Curriculum* (Heck et al., 2012) identifies instruments used to study the enacted curriculum from researchers studying various curricula.
- *Variability Is the Rule: A Companion Analysis of K–8 Mathematics Standards* (Smith, 2011) analyzes U.S. state standards documents for geometry, measurement, probability, and statistics.

- *A Five-Year Study of the First Edition of the Core-Plus Mathematics Curriculum* (Schoen et al., 2010) shares results from a longitudinal study of the high-school Core-Plus curriculum.
- *Future Curricular Trends in School Algebra and Geometry: Proceedings of a Conference* (Usiskin et al., 2010) explores curricular issues in two specific content areas that are included in most school experiences around the world.
- *The Classification of Quadrilaterals: A Study of Definition* (Usiskin et al., 2008) studies the definition and classification of quadrilaterals from an analysis of many textbooks.
- *The History of the Geometry Curriculum in the United States* (Sinclair, 2008) explores the geometry curriculum from a study of textbooks and policy documents.
- *A Decade of Middle School Mathematics Curriculum Implementation: Lessons Learned from the Show-Me Project* (Meyer et al., 2008) shares stories of adoption and implementation of several *Standards*-based curriculum materials.
- *Mathematics Curriculum in Pacific Rim Countries—China, Japan, Korea, and Singapore: Proceedings of a Conference* (Usiskin & Willmore, 2008) investigates the mathematics frameworks used to design curriculum in each of the named countries.
- *The Intended Mathematics Curriculum as Represented in State-Level Curriculum Standards: Consensus or Confusion?* (Reys, 2006) explores the placement of mathematics topics in the intended curriculum of number and operations and algebra in grades K-8 in the United States.

STRUCTURE OF THE VOLUME

The volume consists of twenty-eight chapters. Chapter 1 provides an introduction to the volume. In addition to sharing several definitions of curriculum, we discuss different classifications or levels of curriculum, share information about research that has been conducted on these levels of curriculum, and discuss gaps in the research the volume aims to fill.

The remainder of the book, Chapters 2–28, consists of lessons learned in conducting curriculum research. We have grouped these chapters into six sections, with each section focused on the lessons learned in one of the levels of curriculum identified in the Introduction (Chapter 1). In Chapter 1, we highlight some of the lessons found in the section chapters, as well as some connections across chapters that transcend the curriculum levels.

- I. *Development and Analysis of the Official Curriculum*: Chapters 2–5
- II. *Textbook Analysis*: Chapters 6–8
- III. *Curriculum Design and Development*: Chapters 9–14
- IV. *Research on the Teacher Intended Curriculum*: Chapters 15–19
- V. *Research on the Enacted Curriculum*: Chapters 20–25
- VI. *Personal Reflections on Curriculum Development and Research*: Chapters 26–28.

The volume concludes with a brief biography of the authors and editors.

ACKNOWLEDGMENTS

We appreciate the dedication and hard work of the many authors in writing and revising their chapters, as well as responding to our many questions and edits. We also appreciate the willingness of George Johnson, President of Information Age Publishing, to publish the volume and to respond to our many queries about production issues in the preparation of this volume.

REFERENCES

Bates, M., & Usiskin, Z. (Eds.). (2016). *Digital curricula in school mathematics*. Information Age Publishing.

Heck, D. J., Chval, K. B., Weiss, I. R., & Ziebarth, S. W. (Eds.). (2012). *Approaches to studying the enacted mathematics curriculum*. Information Age Publishing.

Huntley, M. A., & Heck, D. J. (2014). Examining variations in enactment of a grade 7 mathematics lesson by a single teacher: Implications for future research on mathematics curriculum enactment. In D. R. Thompson & Z. Usiskin (Eds.), *Enacted mathematics curriculum: A conceptual framework and research needs* (pp. 21–45). Information Age Publishing.

Huntley, M. A., Rasmussen, C. L., Villarubi, R. S., Sangtong, J., & Fey, J. T. (2000). Effects of *Standards*-based mathematics education: A study of the Core-Plus Mathematics Project algebra and functions strand. *Journal for Research in Mathematics Education, 31*, 328–361.

Huntley, M. A., & Terrell, M. S. (2014). One-step and multi-step linear equations: A content analysis of five textbook series. *ZDM Mathematics Education, 46*(5), 751–766. https://doi.org/10.1007/s11858-014-0627-6

Mayer, J. M., Huntley, M. A., Fonger, N. L., & Terrell, M. S. (2019). Professional learning through teacher-researcher collaborations. *Mathematics Teacher, 112*, 382–385.

Meyer, M. R., Langrall, C. W., Arbaugh, F., Webb, D. C., & Hoover, M. B. (Eds.). (2008). *A decade of middle school mathematics curriculum implementation: Lessons learned from the Show-Me Project*. Information Age Publishing.
Reid, D. A., Suurtamm, C., Savard, A., Simmt, E., Manuel, D., Borden, L. L., & Barwell, R. (2020). *Researching pedagogy and practice with Canadian mathematics teachers*. Information Age Publishing.
Remillard, J. T., & Heck, D. J. (2014). Conceptualizing the curriculum enactment process in mathematics education. *ZDM Mathematics Education, 46*(5), 705–718. https://doi.org/10.1007/s11858-014-0600-4
Reys, B. J. (Ed.). (2006). *The intended mathematics curriculum as represented in state-level curriculum standards: Consensus or confusion?* Information Age Publishing.
Schmidt, W. H., McKnight, C. C., Valverde, G. A., Houang, R. T., & Wiley, D. E. (1997). *Many visions, many aims: A cross-national investigation of curricular intentions in school mathematics* (Volume 1). Kluwer.
Schoen, H. L., Ziebarth, S. W., Hirsch, C. R., & BrckaLorenz, A. (2010). *A five-year study of the first edition of the Core-Plus mathematics curriculum*. Information Age Publishing.
Shimizu, Y., & Vithal, R. (2023). School mathematics curriculum reforms: Widespread practice but under-researched in mathematics education. In Y. Shimizu & R. Vithal (Eds.), *Mathematics curriculum reforms around the world: The 24th ICMI Study* (pp. 3–21). Springer.
Sinclair, N. (2008). *The history of the geometry curriculum in the United States*. Information Age Publishing.
Smith, J. P., III. (2011). *Variability is the rule: A companion analysis of K–8 mathematics standards*. Information Age Publishing.
Suurtamm, C., & McKie, K. (2019). *Research to inform elementary mathematics curriculum revision*. Report presented to the Ontario Ministry of Education.
Suurtamm, C., & Quigley, B. (2020). Nested noticing: Valuing voice. In D. A. Reid, C. Suurtamm, A. Savard, E. Simmt, D. Manuel, L. L. Borden, & R. Barwell (Eds.), *Researching pedagogy and practice with Canadian mathematics teachers* (pp. 43–60). Information Age Publishing.
Thompson, D. R., Huntley, M. A., & Suurtamm, C. (Eds.). (2018). *International perspectives on mathematics curriculum*. Information Age Publishing.
Thompson, D. R., & Senk, S. L. (2010). Myths about curriculum implementation. In B. Reys, R. Reys, & R. Rubenstein (Eds.), *Mathematics curriculum: Issues, trends, and future directions* (pp. 249–263). National Council of Teachers of Mathematics.
Thompson, D. R., & Senk, S. L. (2012). Instruments used by the University of Chicago School Mathematics Project to study the enacted curriculum. In D. J. Heck, K. B. Chval, I. R. Weiss, & S. W. Ziebarth (Eds.), *Approaches to studying the enacted mathematics curriculum* (pp. 19–46). Information Age Publishing.
Thompson, D. R., & Usiskin, Z. (Eds.). (2014). *Enacted mathematics curriculum: A conceptual framework and research needs*. Information Age Publishing.
Usiskin, Z., Andersen, K., & Zotto, N. (Eds.). (2010). *Future curricular trends in school algebra and geometry: Proceedings of a conference*. Information Age Publishing.
Usiskin, Z., Griffin, J., Witonsky, D., & Willmore, E. (Eds.). (2008). *The classification of quadrilaterals: A study of definition*. Information Age Publishing.

Usiskin, Z., & Willmore, E. (Eds.). (2008). *Mathematics curriculum in Pacific Rim countries—China, Japan, Korea, and Singapore: Proceedings of a conference.* Information Age Publishing.

Valverde, G. A., Bianchi, L. J., Wolfe, R. G., Schmidt, W. H., & Houang, R. T. (2002). *According to the book: Using TIMSS to investigate the translation of policy into practice through the world of textbooks.* Kluwer.

ENDNOTE

1. See also the technical reports for the UCSMP secondary curriculum at https://ucsmp.uchicago.edu/secondary/research_reports/downloadable_technical_reports/

CHAPTER 1

INTRODUCTION

Learning About Conducting Curriculum Research

Denisse R. Thompson
University of South Florida

Mary Ann Huntley
Cornell University

Christine Suurtamm
University of Ottawa

This introductory chapter focuses on lessons learned by scholars engaged in mathematics curriculum research. We begin by providing several perspectives on the term curriculum, *and identify different levels of curriculum outlined by researchers and policymakers. We describe how we have organized the volume and briefly discuss the chapters in each section of the volume. We also suggest some connections between chapters and highlight samples of the lessons the authors of the respective chapters learned about conducting research on one or more levels of mathematics curriculum.*

Keywords: conducting curriculum research; curriculum design; enacted curriculum; official curriculum; teacher intended curriculum; textbook analysis

This volume is centered around the process of *conducting* mathematics curriculum research. Generally, publications about curriculum research focus on the *results* of the research, with enough discussion about the research methodology for readers to have faith that the research has integrity. However, little is typically reported about particular frameworks, approaches, or strategies in conducting the research; or issues or challenges that might have arisen in conducting curriculum research, and solutions found to address those issues or challenges. It is these lessons learned from and about conducting mathematics curriculum research that are the focus of this volume. We hope other researchers can learn from and build on these lessons as curriculum research moves forward.

The mathematics curriculum is multifaceted. It incorporates what should be taught from a policy perspective, what is expected to be taught as evidenced in curriculum materials available for students and teachers to use, what teachers intend to teach from those curriculum materials, and what actually is taught as teachers and students interact during classroom instruction. Curriculum can be described as "the complete set of learning experiences and activities that the student undergoes" (Burkhardt et al., 1989, p. 408) or "the plan that expresses these aims and intentions [of educational authorities], which takes them from vision to implementation, and serves as the broad course that runs throughout formal schooling" (Schmidt et al., 1997, p. 4). Later, Remillard and Heck (2014a) described curriculum not only as the end point of educational experiences but as including the path that leads to learning:

> ... we define mathematics curriculum as a *plan for the experiences* that learners will encounter, as well as the *actual experiences* they do encounter, that are designed to help them reach specified mathematics objectives. We use the term "experiences" to signal that curriculum is more than the specification of topics to be covered or objectives to be met; it includes what students are intended to or actually experience to support their learning. (p. 707, italics in the original)

Since the 1980s, there has been considerable interest in mathematics curriculum as educators have recognized the influence of curriculum on student learning. For instance, at the beginning of that decade, the National Council of Teachers of Mathematics (NCTM) developed *An Agenda for Action* with recommendations for the school curriculum in that decade (NCTM, 1980). Looking forward to the 21st century, researchers from varied countries gathered in Australia in the late 1980s to consider issues influencing the future mathematics curriculum, specifically content, technology, teachers, and dynamics (Malone et al., 1989). As part of expanded work related to international assessments, research related to the Third International Mathematics and Science Study (TIMSS) explored the flow of mathematics curricula in participating countries (Schmidt et al., 1997) and how curricular expectations were translated into textbooks used in classrooms (Valverde et al., 2002).

With the development of various curriculum projects in the U.S. to instantiate recommendations outlined in the *Curriculum and Evaluation Standards for School Mathematics* (NCTM, 1989), research was often conducted to ascertain the effectiveness of those curriculum projects (Senk & Thompson, 2003). Yet, in an evaluation of curriculum research, many concerns were raised about the quality of the research and suggestions were provided for future curriculum research (Confrey & Stohl, 2004).

Curriculum research reported over the last few decades has often focused on results from one or more aspects of the curriculum. For instance, Cohen (2003) and Sinclair (2008) explored historical aspects of curriculum, specifically numeracy and geometry, respectively. Some researchers have focused on differences in curriculum standards among various jurisdictions in the U.S. (Reys, 2006; Smith, 2011) or in how textbooks have addressed curriculum content or assessment (Donoghue, 2003; Hunsader et al., 2014; Huntley & Terrell, 2014; Usiskin et al., 2008). Others have focused on the role that teachers play in implementing the curriculum (Huntley & Heck, 2014; Meyer et al., 2008; Remillard et al., 2009; Son & Senk, 2014) or on instruments that have been used to study curriculum enactment (Heck et al., 2012; Ziebarth et al., 2014). Still others have reported on longitudinal results from studying with specific curricula (Romberg & Shafer, 2008; Schoen et al., 2010) or on achievement related to specific content when using curriculum materials (Geiger et al., 2014; Senk et al., 2014; Thompson & Senk, 2014). In addition, design perspectives for developing curricula (Hirsch, 2007), frameworks for studying curriculum (Kaur, 2014; Pepin, 2014; Remillard & Heck, 2014a, 2014b), or potential research agendas for studying curricula (Chval et al., 2014) have been the focus of publications about curriculum.

These research endeavors across different perspectives of curriculum reflect some of the relatively recent interest in curriculum as a focus of mathematics education research (Li & Lappan, 2014). Yet, despite this work and various curriculum reforms worldwide, a recent study by the International Commission on Mathematical Instruction on mathematics curriculum reform efforts internationally reported that scholarly literature on research on mathematics curriculum reform efforts is sparse. In addition, the study directors reported there are "limited writings about appropriate research methodologies for studying this topic [school mathematics curriculum reforms] in all its manifestations, dimensions, levels and components" (Shimizu & Vithal, 2023, p. 4).

This volume addresses this limitation in the research literature noted by Shimizu and Vithal. In particular, the volume considers *how* mathematics curriculum researchers study curriculum in various dimensions and shares what they have learned. For instance, it considers important questions such as: What are the issues and challenges faced when conducting research around mathematics curriculum? What lessons have been learned that might be of use to other researchers? Sharing answers to these questions is the premise for this book.

LEVELS OF CURRICULUM

The mathematics curriculum often undergoes changes between an initial plan for instruction and actual instruction, so various researchers have made distinctions among different levels of curriculum. Burkhardt et al. (1989) identified six levels of curriculum:

- The *ideal curriculum* proposed by experts;
- The *available curriculum* of teaching materials;
- The *adopted curriculum* which schools or districts expect to be taught;
- The *implemented curriculum* that teachers actually teach;
- The *achieved curriculum* that students learn;
- The *tested curriculum* of mandated tests for accountability purposes.

In work related to the Third International Mathematics and Science Study (TIMSS), these levels of curriculum were collapsed into a tripartite model: the *intended curriculum* of intentions, aims, and goals; the *implemented curriculum* of strategies, practices, and activities; and the *attained curriculum* of knowledge, including ideas, constructs, and schemes (Valverde et al., 2002, p. 5). For both sets of categorizations, there are numerous factors that influence curriculum at each level, such as national policy, teacher content knowledge, teacher pedagogical practice, student prerequisite knowledge, or the content of mandated assessments.

Throughout this volume, many chapter authors reference a model of curriculum described by Remillard and Heck (2014a) that considers relationships among the official curriculum, the teacher intended curriculum, instructional materials, the enacted curriculum, and the learned curriculum of student outcomes. Remillard and Heck describe the official curriculum as consisting of the designated curriculum outlined by an educational authority, the curricular goals for learning specified by this authority, and the content of assessments for accountability purposes. Instructional materials are designed to align with the official curriculum and influence teachers' intentions for curriculum implementation, the actual enacted curriculum during instruction, and ultimately student learning.

CURRICULUM LEVELS AND POSSIBLE LESSONS TO LEARN

The word *curriculum* is used in various ways. For instance, curriculum may be an entire plan of educational experiences or reflect a narrower focus on textbooks and other instructional materials. Researchers engaged in mathematics curriculum research often focus on one or more levels of curriculum as described by various researchers in the previous section. Regardless of the level at which they conduct research, numerous factors must be considered to conduct rigorous research able to withstand peer review. The lessons learned in conducting such research provide insights that can be of use to others in the field.

We have grouped the chapters in this volume together in broad ways around levels of curriculum to provide perspectives related to the lessons the authors share. In what follows, we highlight some of the major lessons that arise in each section of the volume.

I. Development and Analysis of the Official Curriculum

The *official curriculum* represents curriculum at a policy level in which the aims and overall expectations of the curriculum are outlined. For countries with a national curriculum, such as many Asian or European countries, the official curriculum might be specified by a Ministry of Education as described in Usiskin and Willmore (2008) or in Thompson et al. (2018). In other countries, such as the U.S. or Canada, the official curriculum is typically specified at the state or provincial level, respectively. However the official curriculum is defined, researching an official curriculum might involve understanding how the official curriculum is developed or how to compare curriculum documents from different countries or different jurisdictions within a country.

Three of the four chapters in this section focus on these issues. Dunphy and Dooley (Chapter 2) discuss four lessons they learned as they helped shape guidelines for the development of an official primary curriculum for Ireland, including lessons about reviewing relevant literature, balancing theoretical and developmental perspectives for primary aged children, addressing emergent themes that influence curriculum design, and then determining how to bridge the official curriculum with the curriculum that is actually operationalized in schools. That is, their work provides lessons about the development of an underlying structure to inform the official curriculum.

Corrêa and Rangel (Chapter 3) as well as Dingman and colleagues (Chapter 5) focus on lessons learned while comparing official curriculum documents, across countries in the case of Corrêa and Rangel and across U.S. states in the case of Dingman and colleagues. To facilitate their work, Corrêa and Rangel developed a technique for studying curriculum documents, *combined reading*, and discuss the lessons they learned as they clarified their process to ensure reliability and useability. Dingman and colleagues discuss numerous lessons in finding ways to compare standards across the 50 U.S. states, including lessons about the grain size of standards, comparing standards of the same type, and ensuring reliability. Both sets of authors learned and share lessons about finding ways to represent results visually and in a way that allows reflection and summarization.

McGarvey and colleagues (Chapter 4) focus on the official curriculum by considering important content that is missing from official documents, what they call the *null curriculum*. They share difficult lessons learned as they sought to design and implement curriculum to address such content, including how to advocate for spending valuable classroom instructional time on topics not explicitly enumerated in official curriculum policy.

II. Textbook Analysis

The official curriculum is instantiated in curriculum materials such as textbooks, often considered to be the *intended curriculum*. In jurisdictions with official curricula, authors of curriculum materials typically want to ensure that their materials address the content and processes envisioned in official policy documents in order to have their materials used. But textbook authors can interpret official documents in different ways, leading to different content and instructional foci of such materials. Curriculum researchers can compare the emphases within different sets of curriculum materials to offer insights to potential users of the materials.

Both Jones (Chapter 6) as well as Huntley and colleagues (Chapter 8) provide lessons for researchers on issues to address when comparing and contrasting textbooks. Some issues they address include selecting which textbooks to study, selecting the features to analyze, and defining tasks to classify, given the different ways that problems are labeled across textbooks and other curriculum materials. Given the enormous quantity of data that can be generated from such analysis, Huntley and colleagues share lessons on issues related to new visual displays of data using the analogy of a mosaic.

Kim and colleagues (Chapter 7) share techniques that can be applied to analyze textbooks in a cross-cultural context to facilitate learning across jurisdictions, an important issue in an increasingly interconnected world. Issues related to determining a unit of analysis or tasks to analyze, as noted by Jones or Huntley and colleagues, are also important in cross-curricular research. Other issues surface when doing cross-cultural textbook analysis, such as the same content being addressed at different grade levels in different countries, or dealing with translations as well as appropriate interpretations of those translations.

III. Curriculum Design and Development

The intended curriculum to be taught in a classroom or with a group of students can take many formats. It might be a curriculum for an entire school year that is expected to be taught to all students. It might be materials designed in close connection with the regular curriculum and to be used in conjunction with it to support enhanced learning. It might be a supplementary curriculum to address content not in the regular curriculum and consist of a set of lessons to be used intermittently throughout the year. The intended curriculum could be a print curriculum or a curriculum with a focus on technology. Issues and lessons in the design and development of curriculum materials, for both full-year and supplementary contexts and with or without technology, are the focus of the six chapters in this section.

Sarama and Clements (Chapter 9) describe lessons learned as they developed a framework to serve as the basis for designing a research-based curriculum for early childhood mathematics. Blanton and colleagues (Chapter 10) referenced this framework as they designed a series of early algebra lessons for use in grades K–5; they share what they learned about the importance of frameworks, having a set of core practices across content domains, and considering the importance of and support for teacher knowledge of content that might not typically be in the curriculum at a given school level.

Fuson (Chapter 11), who developed a curriculum for the entire span of elementary grades, focuses on lessons learned during the design process, considering lessons learned at the levels of the intended curriculum, teacher intended curriculum, enacted curriculum, assessed curriculum, and learned curriculum. The lessons Fuson describes provide opportunities for comparisons with lessons discussed by Sarama and Clements as well as Blanton and colleagues.

The other three chapters in this section focus on lessons learned while designing curriculum in which technology is central to the design. Edson's chapter (Chapter 12) focuses on lessons learned when a technology platform is designed to supplement and complement a full-year print curriculum for middle-grades students. Harel and Yerushalmy (Chapter 13) harness the power of technology to provide curriculum support to Israeli students as they solve mathematics problems; they discuss lessons learned as the technology platform and curriculum interact to provide immediate support to students. Together, the chapters by Edson and by Harel and Yerushalmy discuss lessons when technology is used to support student learning with an entire class.

Movshovitz-Hadar and colleagues (Chapter 14) used technology to give high school students insights into contemporary mathematics content that is not often a part of the official curriculum. They discuss how technology was used to develop a supplementary curriculum and lessons they learned through a four-stage process of implementation of the curriculum in classrooms in Israel. Their work intersects both curriculum design and the enacted curriculum.

IV. Research on the Teacher Intended Curriculum

Implementation of the intended curriculum outlined in a set of curriculum materials is mediated by a teacher. Teachers review the materials and make decisions about what to emphasize during instruction as well as how to engage students with the materials, typically based on their own pedagogical knowledge and views on their students' experiences and needs. This *teacher-intended curriculum* often resides in the minds of instructional leaders, so it can be difficult to research such decisions that are often invisible. Researchers often study questions such as: How do teachers determine what resources to use? To what extent do teachers implement curriculum materials in ways aligned with those of the developers of the materials? That is, does the implementation support the intentions of the developers or thwart those intentions in some way? How does the presence of instructional material online and available for use with a few clicks influence what occurs in the classroom? Lessons about how best to research questions such as these are the focus of the five chapters in this section. The lessons in the chapters within this section of the book suggest there are many techniques that researchers can use to study the intentions of teachers as they plan classroom instruction. The use of technological tools may be of particular interest to many researchers.

Saxton and colleagues (Chapter 15) share some of their strategies as they investigated aspects of the curriculum design and sought to determine the extent to which teachers incorporated these into their teaching. In particular, they focus on core curriculum features and essential curriculum elements and share the development of a suite of instruments to determine whether teachers implement the curriculum in productive ways or in ways that are unproductive in terms of the intentions of the developers.

The teams of Teuscher and colleagues (Chapter 16) and Quaisley and Males (Chapter 17) focus their research on the ways teachers make instructional decisions about the curriculum. Teuscher and colleagues describe their development of a research framework to help describe teachers' curricular reasoning; in particular, they provide a behind-the-scenes look at the messy iterative process of developing a framework and the lessons learned through that process. In contrast, Quaisley and Males use a range of technology tools, such as eye-tracking software, heat maps, and gaze plots, to understand the process by which teachers engage with their curriculum materials in designing instruction.

Sapkota and colleagues (Chapter 18), and the team of Dick and Sawyer (Chapter 19), share lessons learned as they surveyed teachers about their selection of curricular resources. In particular, Sapkota and colleagues discuss lessons learned and benefits of using a data analytics firm for research involving surveys, especially during challenging times such as a pandemic. In contrast, Dick and Sawyer describe issues and lessons learned in researching teachers' use of materials in an online supplementary environment, such as Teachers Pay Teachers.

V. Research on the Enacted Curriculum

Six chapters focus on the lessons learned about researching the *enacted curriculum*, that is, the curriculum actually implemented in classrooms during interactions between teachers and their students. Enactment involves delving into aspects of the curriculum that occur in the classroom, recognizing that researchers cannot be present in the classroom every day of the school year. Enactment is at the intersection of teacher intentions and student learning and can be difficult to investigate. Lessons learned across the six chapters in this section provide insights into ways of conducting research on the enacted curriculum at various grade levels.

Orsini and colleagues (Chapter 20) describe how enactment of a kindergarten numeracy curriculum can influence the design of a curriculum, particularly when teachers are partners with curriculum developers. Their research focuses on the interaction of curriculum design and enactment and illustrates the importance of research to improve the curriculum.

Litster (Chapter 21) discusses technical lessons learned about how to collect and analyze data from curriculum enactment in elementary classrooms. Lessons focus on practical issues, from obtaining consent to using video equipment to analyzing and reporting results.

The work of Stylianou and colleagues (Chapter 22) is at the intersection of curriculum design and enactment, with a focus on whether teachers enact potential openings in the curriculum that provide opportunities for students to engage deeply in mathematics. Specifically, they considered whether teachers enact openings that might be anticipated to occur during instruction and can be planned, as well as whether teachers choose to enact openings that occur spontaneously during instruction. One lesson learned about doing this type of research was the importance of being able to observe implementation of the same lesson taught by multiple teachers; they also learned the importance of understanding teachers' content knowledge, even if they were not able to assess such knowledge in their own work. (The work of Stylianou and colleagues is based on the curriculum described in the chapter by Blanton and colleagues in this volume.)

Spencer and colleagues (Chapter 24) focus their work on the enactment of an algebra support curriculum. They share a variety of lessons related to researching variations in how teachers use a support curriculum for algebra in addition to a regular algebra curriculum. They focus on the need to understand how contextual factors influence the program, as well as how to develop and implement instruments to measure the effectiveness of such a support curriculum.

Both Shafer (Chapter 23) as well as Thompson and Senk (Chapter 25) share lessons from enactments of large-scale curriculum projects, at the middle-school and high-school levels, respectively. Specifically, Shafer focuses on the lessons learned when conducting longitudinal research in urban environments, including developing teacher variables and models to describe enactment. Thompson and Senk describe lessons related to studying enactment in terms of content taught, instructional processes used, and students' achievement. In particular, some lessons they discuss include analyzing content taught from multiple perspectives, the importance of looking at results at individual teacher levels rather than as global summaries, and obtaining information about students' opportunity to learn content appearing on end-of-course assessments as well as using such opportunities to analyze student learning results in different ways.

VI. Personal Reflections on Curriculum Development and Research

The final section includes three chapters that provide personal reflections from researchers about lessons learned throughout their extended work with curriculum. Even though these chapters provide personal reflections, the lessons shared offer important insights that can inform the work of other curriculum researchers.

Usiskin (Chapter 26) shares lessons learned about developing new curriculum or new approaches to curriculum, focusing on potential hindrances to such curriculum development and ways such hindrances might be addressed. These lessons may be of particular interest to those interested in designing innovative curriculum materials and address concerns around development of the intended curriculum.

Ricks (Chapter 27) describes a personal evolution in conducting curriculum research across a series of studies, including those involving cross-cultural perspectives. Complexity theory and the lessons learned relative to that approach may be of particular interest to researchers. Many of the lessons discussed are focused on the enacted curriculum.

Finally, Newton (Chapter 28) focuses on the lessons learned about curriculum research from engagement with the Center for the Study of Mathematics Curriculum. The lessons described in this chapter emphasize the importance of developing a community of scholars at various levels (e.g., undergraduates, doctoral students, classroom teachers) and providing training for future curriculum researchers during graduate studies. Newton's lessons span a range of curriculum levels, including research with the official curriculum, the teacher intended curriculum, and the enacted curriculum.

SUMMARY

As readers read the various chapters in the volume, we encourage them to consider how lessons shared in researching one curriculum level might be adapted for use in researching other levels. Readers might find some similarities in lessons across chapters that seem quite different. For instance, one issue faced by many researchers is how to report a large body of results in visual ways that are easy for others to interpret. Corrêa and Rangel, as well as Dingman and colleagues, developed visual displays for use with the official curriculum, Huntley and colleagues for use with textbook analysis, and Thompson and Senk for use with the enacted curriculum.

McGarvey and colleagues faced issues in developing and implementing topics in the null curriculum at the elementary level, that is, topics that are not addressed in the official curriculum. Movshovitz-Hadar and colleagues faced similar issues when developing and implementing lessons for high school students around contemporary topics in mathematics that are not a typical part of the high school curriculum. Some of the issues these authors faced are comparable to those posed by Usiskin in developing innovative curriculum for use at the high school level. The lessons shared in the three chapters highlight numerous considerations for curriculum designers.

McGarvey and colleagues, Orsini and colleagues, and Movshovitz-Hadar and colleagues all shared lessons as they designed and implemented curriculum in partnership with teachers. In each chapter, the authors share lessons about engaging teachers as part of the design or implementation process. In all three chapters, researchers learned lessons at the intersection of official curriculum, curriculum design, and enactment. The lessons they share highlight the importance of engaging with teachers, which was reflected in numerous other chapters in the volume.

Sample selection cuts across several chapters and levels of curriculum. Depending on the level of curriculum being studied, research might occur in a few schools or districts. Selecting schools and districts for the study sample is often a challenge, especially when working in urban settings. Sample selection is also an issue in textbook analysis.

There is much to be learned about enhancing the mathematics achievement of students by researching the curriculum that guides the path of that learning. This book expands the focus on and knowledge gained from research on the mathematics curriculum by sharing lessons learned by researchers across various curriculum levels and multiple countries.

REFERENCES

Burkhardt, H., Fraser, R., & Ridgway, J. (1989). The dynamics of curriculum change. In J. Malone, H. Burkhardt, & C. Keitel (Eds.), *The mathematics curriculum towards the year 2000: Content, technology, teachers, dynamics* (pp. 403–435). Science and Mathematics Education Centre, Curtin University of Technology.

Chval, K. B., Weiss, I. R., & Taylan, R. D. (2014). Recommendations for generating and implementing a research agenda for studying the enacted mathematics curriculum. In D. R. Thompson & Z. Usiskin (Eds.), *Enacted mathematics curriculum: A conceptual framework and research needs* (pp. 149–176). Information Age Publishing.

Cohen, P. C. (2003). Numeracy in nineteenth-century America. In G. M. A. Stanic & J. Kilpatrick (Eds.), *A history of school mathematics* (pp. 43–76). National Council of Teachers of Mathematics.

Confrey, J., & Stohl, V. (Eds.). (2004). *On evaluating curricular effectiveness: Judging the quality of K–12 mathematics evaluations*. National Research Council.

Donoghue, E. F. (2003). Algebra and geometry textbooks in twentieth-century America. In G. M. A. Stanic & J. Kilpatrick (Eds.), *A history of school mathematics* (pp. 329–398). National Council of Teachers of Mathematics.

Geiger, V., Goos, M., & Dole, S. (2014). Curriculum intent, teacher professional development, and student learning in numeracy. In Y. Li & G. Lappan (Eds.), *Mathematics curriculum in school education* (pp. 473–492). Springer.

Heck, D. J., Chval, K. B., Weiss, I. R., & Ziebarth, S. W. (Eds.). (2012). *Approaches to studying the enacted mathematics curriculum*. Information Age Publishing.

Hirsch, C. R. (Ed.). (2007). *Perspectives on the design and development of school mathematics curricula*. National Council of Teachers of Mathematics.

Hunsader, P. D., Thompson, D. R., Zorin, B., Mohn, A. L., Zakrewski, J., Karadeniz, I., Fisher, E. C., & Macdonald, G. (2014). Assessments accompanying published textbooks: The extent to which mathematical processes are evident. *ZDM Mathematics Education, 46*(5), 797–813. https://doi.org/10.1007/s11858-014-0570-6

Huntley, M. A., & Heck, D. J. (2014). Examining variations in enactment of a grade 7 mathematics lesson by a single teacher: Implications for future research on mathematics curriculum enactment. In D. R. Thompson & Z. Usiskin (Eds.), *Enacted mathematics curriculum: A conceptual framework and research needs* (pp. 21–45). Information Age Publishing.

Huntley, M. A., & Terrell, M. S. (2014). One-step and multi-step linear equations: A content analysis of five textbook series. *ZDM Mathematics Education, 46*(5), 751–766. https://doi.org/10.1007/s11858-014-0627-6

Kaur, B. (2014). Enactment of school mathematics curriculum in Singapore: Whither research! *ZDM Mathematics Education, 46*(5), 829–836. https://doi.org/10.1007/s11858-014-0619-6

Li, Y., & Lappan, G. (2014). Mathematics curriculum in school education: Advancing research and practice from an international perspective. In Y. Li & G. Lappan (Eds.), *Mathematics curriculum in school education* (pp. 3–12). Springer.

Malone, J., Burkhardt, H., & Keitel, C. (Eds.). (1989). *The mathematics curriculum towards the year 2000: Content, technology, teachers, dynamics*. Science and Mathematics Education Centre, Curtin University of Technology.

Meyer, M. R., Langrall, C. W., Arbaugh, F., Webb, D. C., & Hoover, M. B. (Eds.). (2008). *A decade of middle school mathematics curriculum implementation: Lessons learned from the Show-Me Project*. Information Age Publishing.

National Council of Teachers of Mathematics [NCTM]. (1980). *An agenda for action: Recommendations for school mathematics of the 1980s*.

National Council of School Mathematics [NCTM]. (1989). *Curriculum and evaluation standards for school mathematics*.

Pepin, B. (2014). Re-sourcing curriculum materials: In search of appropriate frameworks for researching the enacted mathematics curriculum. *ZDM Mathematics Education, 46*(5), 837–842. https://doi.org/10.1007/s11858-014-0628-5

Remillard, J. T., & Heck, D. J. (2014a). Conceptualizing the curriculum enactment process in mathematics education. *ZDM Mathematics Education, 46*(5), 705–718. https://doi.org/10.1007/s11858-014-0600-4

Remillard, J. T., & Heck, D. J. (2014b). Conceptualizing the enacted curriculum in mathematics education. In D. R. Thompson & Z. Usiskin (Eds.), *Enacted mathematics curriculum: A conceptual framework and research needs* (pp. 121–148). Information Age Publishing.

Remillard, J. T., Herbel-Eisenmann, B. A., & Lloyd, G. M. (Eds.). (2009). *Mathematics teachers at work: Connecting curriculum materials and classroom instruction*. Routledge.

Reys, B. J. (Ed.). (2006). *The intended mathematics curriculum as represented in state-level curriculum standards: Consensus or confusion?* Information Age Publishing.

Romberg, T. A., & Shafer, M. C. (2008). *The impact of reform instruction on student mathematics achievement: An example of a summative evaluation of a Standards-based curriculum*. Routledge.

Schmidt, W. H., McKnight, C. C., Valverde, G. A., Houang, R. T., & Wiley, D. E. (1997). *Many visions, many aims: A cross-national investigation of curricular intentions in school mathematics* (Volume 1). Kluwer.

Schoen, H. L., Ziebarth, S. W., Hirsch, C. R., & BrckaLorenz, A. (2010). *A five-year study of the first edition of the Core-Plus mathematics curriculum*. Information Age Publishing.

Senk, S. L., & Thompson, D. R. (2003). *Standards-based school mathematics curricula: What are they? What do students learn?* Lawrence Erlbaum.

Senk, S. L., Thompson, D. R., & Wernet, J. L. W. (2014). Curriculum and achievement in algebra 2: Influences of textbooks and teachers on students' learning of functions. In Y. Li & G. Lappan (Eds.), *Mathematics curriculum in school education* (pp. 515–540). Springer.

Shimizu, Y., & Vithal, R. (2023). School mathematics curriculum reforms: Widespread practice but under-researched in mathematics education. In Y. Shimizu & R. Vithal (Eds.), *Mathematics curriculum reforms around the world: The 24th ICMI Study* (pp. 3–21). Springer.

Sinclair, N. (2008). *The history of the geometry curriculum in the United States*. Information Age Publishing.

Smith, J. P., III. (2011). *Variability is the rule: A companion analysis of K–8 mathematics standards*. Information Age Publishing.

Son, J.-W., & Senk, S. L. (2014). Teachers' knowledge and the enacted mathematics curriculum. In D. R. Thompson & Z. Usiskin (Eds.), *Enacted mathematics curriculum: A conceptual framework and research needs* (pp. 75–95). Information Age Publishing.

Thompson, D. R., Huntley, M. A., & Suurtamm, C. (Eds.). (2018). *International perspectives on mathematics curriculum*. Information Age Publishing.

Thompson, D. R., & Senk, S. L. (2014). The same geometry textbook does not mean the same classroom enactment. *ZDM Mathematics Education, 46*(5), 781–795. https://doi.org/10.1007/s11858-014-0622-y

Usiskin, Z., Griffin, J., Witonsky, D., & Willmore, E. (Eds.) (2008). *The classification of quadrilaterals: A study of definition*. Information Age Publishing.

Usiskin, Z., & Willmore, E. (Eds.) (2008). *Mathematics curriculum in Pacific Rim countries—China, Japan, Korea, and Singapore: Proceedings of a conference*. Information Age Publishing.

Valverde, G. A., Bianchi, L. J., Wolfe, R. G., Schmidt, W. H., & Houang, R. T. (2002). *According to the book: Using TIMSS to investigate the translation of policy into practice through the world of textbooks*. Kluwer.

Ziebarth, S. W., Fonger, N. L., & Kratky, J. L. (2014). Instruments for studying the enacted mathematics curriculum. In D. R. Thompson & Z. Usiskin (Eds.), *Enacted mathematics curriculum: A conceptual framework and research needs* (pp. 97–120). Information Age Publishing.

SECTION I

DEVELOPMENT AND ANALYSIS OF THE OFFICIAL CURRICULUM

CHAPTER 2

TOWARDS THE DEVELOPMENT OF A PRIMARY MATHEMATICS CURRICULUM IN IRELAND

Elizabeth Dunphy and Thérèse Dooley
Dublin City University, Ireland

This chapter is based on research commissioned by the National Council for Curriculum and Assessment in the Republic of Ireland. The purpose of our research was to help shape a redevelopment of the official primary mathematics curriculum for children aged 3–8 years. We discuss four challenges that arose as we conducted the research: deciding on the nature of the literature review; balancing theoretical perspectives on mathematical learning and development; aligning emerging themes to support curriculum design; and bridging the official and operational curricula. In each case, we identify the issue of concern, we describe how we resolved it, we highlight what we learned in the process, and we indicate how our research appears to be influencing curriculum documentation. Finally, we bring together lessons learned with the aim of assisting others engaged in mathematics curriculum research.

Keywords: aligning theories; curriculum model; learning paths; metapractices; narrative review

In the Republic of Ireland, the *Primary school curriculum: Mathematics* (Government of Ireland, 1999) for children ages 3–8[1] has now been in place for over two decades. During these two decades, a number of factors have emerged to influence curriculum development and renewal. First, as elsewhere, there is a growing recognition that a sophisticated understanding of mathematics is fundamental to participation in everyday life, the workplace, and modern democracies (e.g., Stemhagen & Henney,

2021). Second, Ireland continues to experience significant demographic changes resulting in increased participation of children in the education system who do not speak the language of instruction (English or Gaeilge[2]) at home. Third, research has suggested there have been problems related to the enactment of the mathematics curriculum. For example, a decade after its introduction, Dunphy (2009) reported on a large-scale study that asked teachers of four- and five-year-old children about the greatest challenges for mathematics learning and teaching. The findings revealed the extent to which teachers struggled to engage children in problem solving and in the general processes of mathematics (e.g., reasoning, proof, communication, justifying mathematical ideas, making mathematical arguments). Teachers also reported challenges with developing the mathematical understanding of children from diverse language backgrounds, supporting children experiencing difficulties with mathematics, and documenting children's learning. Fourth, a consistent outcome of national assessments in mathematics—involving those aged 8, 10, and 12 years—was that many children experienced difficulty with mathematical processes, but did well on items that assessed their understanding and recall of, for example, numerical procedures (e.g., Eivers et al., 2010; Shiel et al., 2014). In other words, process goals had, to a large extent, been overshadowed by content goals.

Due to unanticipated matters, there was a delay in the process of redeveloping the primary mathematics curriculum (PMC), but it was finally published in September 2023. The specification for the official PMC is guided by a primary curriculum framework that forms the basis for high-quality teaching, learning, and assessment for all children attending primary and special schools (Department of Education, 2023). Curriculum development in the Republic of Ireland is managed by the National Council for Curriculum and Assessment (NCCA). This body advises the Minister for Education[3] on all matters related to curriculum and assessment from early childhood to the end of second-level education (age 18). The process of curriculum development by NCCA is research-informed, and a consultative and deliberative approach shapes all decisions related to policy, research, and practice. Consequently, proposals and advice on curriculum and assessment draw on network activity, consultation, and discussion, as well as research.

In writing this chapter, we consulted key curriculum documentation to examine how the commissioned research supported the NCCA in their design of the redeveloped PMC. The rationale, aims, strands and elements, and learning outcomes are presented in the *Primary Mathematics Curriculum: For Primary and Special Schools* (NCCA, 2023a), (referred to as the *Primary Mathematics Curriculum* hereon). Practical support for teachers in building rich mathematical learning experiences for children is provided in the

Primary Mathematics Toolkit (NCCA, 2023b). There are four components in the latter document: (i) mathematical concepts; (ii) progression continua; (iii) examples of children's learning; and (iv) support materials for teachers.

Our involvement in the redevelopment of the PMC began in 2013 when we were commissioned to present a review of research on mathematics in early childhood and primary education (3–8 years). This was a five-month desk-based research project that culminated in two reports (Dooley et al., 2014; Dunphy et al., 2014). The project was guided by a "brief," that is, key areas and questions presented by the NCCA. The first report, the focus of which is summarized in Table 2.1, concerned definitions, theories, stages of development, and progression; the second report, as seen in Table 2.2, focused on teaching and learning. The reports, alongside several other relevant reports, were used to inform the redevelopment of the curriculum.

Table 2.1

Key Areas and Questions as Presented by the National Council for Curriculum and Assessment (Report 1)

Area	Questions
Definitions	1. How does the research define **mathematics education** for children aged 3-8 years? 2. How does it define **numeracy**? 3. How does the research conceptualize the **relationship between the two**?
Theoretical Perspectives	4. What are the **main theoretical perspectives** underpinning recent and current research on children's mathematical learning and development?
Stages of Development	5. Does research propose **stages of development** in children's mathematical learning? If so, how are these stages defined and what are the essential **indicators** at each stage?
Assessing and Planning for Progression	6. What practical guidance does the research offer on **assessing** and **planning for progression** in children's mathematical development: • at teacher/classroom level (and in the case of a diverse range of mathematical abilities)? • in an immersion setting? • at school level?
Curriculum Development	7. What are the key implications of the research for the development of the Primary School Mathematics Curriculum in Ireland? 8. What might be the implications for: • initial teacher education? • teacher continuing professional development?

Note: Table 2.1 is taken from the request for tender document related to the commissioned research, published on etenders.gov.ie with a closing date of 18.01.2013. (Reprinted with permission from NCCA, July 2023.)

Table 2.2

Key Areas and Questions as Presented by the National Council for Curriculum and Assessment (Report 2)

Area	Questions
Pedagogy	1. According to research, what are the **features** of **good mathematical pedagogy** for children aged 3–8 years: • at teacher/classroom level? • in an immersion setting? • in differentiating and supporting a wide range of children's mathematical abilities? • at school level? 2. What **strategies** does the research highlight as being particularly effective in supporting **all** children when learning in their mother tongue and when learning in a second language, to: • develop **key mathematical skills**? What are these skills? • understand and use mathematical **concepts**? What are these concepts? To what extent, if any, are **number limits** for computational purposes helpful? • develop **positive dispositions** in mathematics? 3. What strategies, if any, does the research identify as being particularly effective in helping children's development of **mathematical language** in **immersion settings**? 4. What does the research say about teaching methods that are particularly effective in **engaging and motivating children** in mathematics?
Curriculum Integration	5. What **strategies** and **examples** does research show as being effective in bringing about **meaningful integration** of mathematics across the curriculum? 6. What advice, if any, does research offer on curriculum **time** allocations for mathematics?
Partnerships with Parents	7. What does research say about good practice in **working with parents** and the wider community to support children's mathematical learning? Give some **illustrations of how** schools can work with parents and the wider community.
Curriculum Development	8. What are the key **implications** of the research for the development of the Primary School **Mathematics Curriculum** in Ireland? 9. What might be the implications for: • initial teacher education? • teacher continuing professional development?

Note: Table 2.2 is taken from the request for tender document related to the commissioned research, published on etenders.gov.ie with a closing date of 18.01.2013. (Reprinted with permission from NCCA, July 2023.)

During our research, we encountered a number of issues. This chapter addresses four of these, each of which created a conundrum in terms of how best to present the research for maximum impact on the redeveloped curriculum. We outline our thinking as we worked through the questions posed, and we describe the various steps we took to ensure that our work was as clear and comprehensive as possible. We begin with how we approached the literature review.

DECIDING ON THE NATURE OF THE LITERATURE REVIEW

As we embarked on our research project, a challenge arose for us regarding the type of literature review that would be appropriate given our particular brief. Although there are varied approaches to conducting research literature reviews (Suri & Clarke, 2009), most debate centers around the relative merits of systematic and narrative methods (e.g., Biesta, 2007; Chalmers, 2003; Davies, 2000; Hammersley, 2008, 2020; MacLure, 2005). A systematic review is defined as "a review of existing research using explicit, accountable rigorous research methods" (Gough et al., 2017, p. 4). Such reviews have gained traction in the field of education in many countries because of the drive to improve national economic performance through raising educational standards. In this context, they are seen as effective in giving guidance to translate research into practice. Narrative literature reviews, in contrast, are more traditional—they involve analyzing findings that might be complementary or contradictory. Their aim is not so much to find a solution to a particular research question but to identify key landmarks of an issue in question or to identify significant gaps in the literature (Hammersley, 2020). Proponents of systematic reviews argue for their exhaustiveness, transparency, and lack of bias and opacity; others maintain that such reviews, by their nature, focus on learning that is easily assessed, and thus promote an instrumentalist view of teaching and learning. Conversely, those committed to narrative literature reviews claim that they allow for creativity and author judgement. However, it is generally conceded that the kind of review conducted depends on the nature of the research questions being addressed. Systematic reviews usually examine questions about impacts and effects; the more traditional narrative review looks at research about meaning and interpretation to explore and develop theory (Newman & Gough, 2020). In practice, researchers adopt a balanced approach; even though they might choose a particular genre, they adopt aspects of the other. For example, Nind (2006) describes how, in conducting a systematic review, she did not cast aside her interpretive judgement on the findings that emerged. In a broad overview of literature reviews, Suri and Clarke (2009) argue that all research synthesists, including those engaged in a traditional narrative review, must take account of any biases brought to bear on their work and be appropriately transparent for the intended audience.

Our brief from the NCCA was quite challenging. It was broad in scope and addressed some theoretical and philosophical issues as well as the implications of these for practice. It is possible that some form of systematic review would have been useful for this purpose. However, there were practical reasons that caused us to decide against undertaking this form of review. First, our timescale was short and considerable time and effort are required to undertake a quality systematic review, time and effort that might usefully be deployed on other aspects of the review process (Hammersley, 2020). Even if time had been at our disposal, it is unlikely that our decision would have been different because most systematic reviews are based on large-scale random control trials, of which there are historically few in mathematics education. We were also dubious that a systematic review would capture some important aspects of mathematics teaching and learning. As pointed out by Bakker et al. (2015), efforts to undertake empirical research in multifaceted aspects of mathematics teaching and learning can lead to a diminution of the concept in question. It seemed then that a narrative review was a better option for us. However, such reviews, if they are to be rigorous and comprehensive, also take considerable effort.

To address the possible limitations of a narrative review, and at the same time conform to our brief in the time allowed, several steps were warranted. First, we gathered colleagues with expertise in a range of different areas—early childhood and primary mathematics education, mathematics learning and teaching in different language contexts, inclusive mathematics education, digital learning, educational theories, large-scale assessments of mathematics, community and parental involvement in education, and professional development in mathematics education. We invited them to contribute to relevant sections of the research reports. These colleagues were established in their various fields and could make trustworthy judgements about lead studies and thinking in their areas of expertise.

As evident in our final reports, we also consulted a number of best evidence syntheses and seminal publications that had been developed in other jurisdictions, as well as widely referenced handbooks related to mathematics education. For example, to examine the principles underpinning good mathematics pedagogy, we considered two important bodies of work. The first was a synthesis of international research on effective pedagogy in mathematics conducted by New Zealand researchers, Anthony and Walshaw (2007, 2009). The second was the United States National Research Council (NRC) Report (2005), a landmark report that focused on how learning theory can be applied to mathematics education. From these, we identified a comprehensive account of the features of good mathematics pedagogy embracing the elements of people and relationships, the learner, and the learning environment.

To avoid subjectivity, the lead authors and coauthors met regularly and engaged in ongoing critical review of drafts of the reports. However, we cannot claim that we were working from a theoretically neutral stance. One reason was that the redeveloped mathematics curriculum needed to be aligned with the general direction of the primary curriculum framework, as well as curricula being devised in other subject areas. For example, "the social and cultural dimension of learning and learners" (Conway & Sloane, 2005, p. 157) was central to earlier reform in the post-primary mathematics curriculum in Ireland. Furthermore, the international advisor chosen by us represented a strongly sociocultural perspective on the teaching and learning of mathematics for children aged 3–8 years. This was an important factor given what was known about the enactment of the 1999 mathematics curriculum.

As we developed the reports, we were particularly cognisant that content goals had received greater emphasis than process goals in the enactment of the 1999 PMC. Consistent with the sociocultural perspective on learning promoted in the research reports, one of the recommendations we made was that goals should relate to both processes and content. We see this recommendation operationalized in the progression continua in the *Primary Mathematics Toolkit* (NCCA, 2023b). Within each progression continuum, there are eleven proposed milestones where learning is described in terms of both processes (named as "elements") and mathematical content. Critically, processes are intertwined with content.

In responding to the research questions posed, we took a pragmatic approach to reviewing the literature. The lesson for others is this: *In developing research to underpin a mathematics curriculum, account should be taken of the nature of the questions to be addressed, the parameters and purpose of research, and the local mathematics teaching and learning context.* The complexity of developing research to underpin a mathematics curriculum needs to be acknowledged, as does the recognition that there is no *best way* to conduct this research.

BALANCING THEORETICAL PERSPECTIVES ON MATHEMATICAL LEARNING AND DEVELOPMENT

A question posed to us in the brief related to the identification of the main theoretical perspectives underpinning recent and current research on children's mathematical learning and development. From our readings, it appeared that, in general terms, a cognitive science approach to curriculum design leads to an emphasis on content, whereas a focus on mathematical processes is consistent with a sociocultural approach. We deduced that a powerful theoretical framework for mathematics education for young children was one that recognized the contributions of both perspectives. As a result, we positioned ourselves alongside those who saw cognitive perspectives as helpful when focusing on individual learners and sociocultural/situated perspectives as appropriate when focusing on issues such as pedagogy (Cobb, 2007; Cobb & Yackel, 1996). A dilemma arose then as to how we could signal a balanced, theoretical stance on content and pedagogy within the reports.

One opportunity to illustrate relative balance arose as we considered how we might respond to the question posed regarding stages of development in children's mathematical learning. Within cognitive science literature, the idea of stages of development had been replaced by the idea of mathematical learning trajectories (LTs). We saw that LTs focus on how individual children's thinking becomes increasingly more sophisticated as they develop a particular concept. These thinking paths could potentially be based on developmental progressions in several central aspects of mathematics (number, shape and space, measure, data, and geometry). As we reviewed the literature, it became clear that LTs were becoming increasingly important as a supportive framework for mathematics learning, teaching, and assessment in countries as diverse as Japan, Korea, and Australia, as well as in Europe and the United States (Bobis et al., 2005; Daro et al., 2011; Griffin, 2004; Lewis & Tsuchida, 1998; Stigler & Thompson, 2012; van den Heuvel-Panhuizen, 2008). Confusingly at first, the terms *learning trajectory*, *learning progression*, *learning pathway*, and *general learning path* were often conflated in the literature. We struggled to put some order on our presentation of the learning trajectories in a way that would help those designing the official curriculum. Our eventual resolution focused on comparing three indicative perspectives on LTs, as described below. Critically, in terms of addressing theoretical balance, we reviewed how these were being conceptualized theoretically. In presenting different perspectives on LTs, we could exemplify how particular presentations reflect a position on the continuum of theoretical perspectives.

These distinct approaches to the presentation of LTs served to make our point that theoretical stance was the key factor in how they were presented. First, we noted how Simon (1995) applied constructivist theories to teaching approaches wherein the teacher, in interaction with the children and with a specific goal in mind, develops a working hypothesis of how children's learning might be assisted through interactions in a given task. As a child's trajectory emerges in response to the task, the teacher responds and tailors the initial hypothesis of the learning process on a contingency basis. In a development of the trajectory idea, Simon and Tzur's (2004) theoretical notion of hypothetical learning trajectory (HLT) moved away from any suggestion that learning progresses in a linear way.

Second, we interpreted Sarama and Clements's (2009) presentation of LTs as derived from a cognitive science perspective and linear in nature. We reported how they used the available research; but when unavailable, they used judgement and best guesses to suggest a hypothetical path for learning in specific mathematical domains. The resultant LTs can be regarded as "invented cultural artefacts" that have been constructed to "help students get from point A to point B" (Stigler & Thompson, 2012, p. 192).

Third, we saw that the learning-teaching trajectories from the Realistic Mathematics Education (RME) perspective (van den Heuvel-Panhuizen, 2008) had much in common with sociocultural perspectives where practices such as establishing an appropriate classroom culture for successful learning in mathematics are emphasized. The RME trajectories are less detailed than Sarama and Clements's developmental progressions and recognize that development can follow different paths. In common with Simon and Tzur, the RME perspectives emphasized working in a contingent way with children's ideas.

In the *Primary Mathematics Curriculum*, progression continua are described as suggesting "a sample learning journey in mathematics at primary level, [but] they are not intended to be prescriptive or exhaustive" (NCCA, 2023a, p. 39). Their role is seen as one that supports teachers in planning learning activities and possible progression steps for children in their classroom. Thus, our presentation, review, and discussion of different approaches to working with learning trajectories helped inform the approach taken by NCCA in developing the curriculum documentation.

In responding to the question on theoretical perspectives, we deduced that a balanced view of theories provides underpinnings that can best serve to shape a mathematics curriculum. Different theories give rise to varying and sometimes divergent perspectives on common themes. The researcher's task is to sift through these perspectives and to judge how they might be incorporated in a mathematics curriculum. The lesson for others is: *An openness to the potential contribution of different theoretical perspectives can ensure that a balanced and nuanced view of mathematics learning and development is reflected in the curriculum.*

ALIGNING EMERGING THEMES
TO SUPPORT CURRICULUM DESIGN

As we worked through the questions for the first report, we noted numerous themes that were receiving considerable attention and support in the literature. Some were novel in the Irish context, and critically they were consistent with the direction of our reports, and with the balanced, nuanced theoretical position we were promoting. Our overriding concern at this point was how best to respond to the question posed on key implications of our research for the development of the PMC. We wanted to foreground the main themes we had found in the literature and which we were now viewing as essential components of the PMC. Furthermore, we also needed to convey the interrelationships between these different components. After considerable thought, we devised a visual model to encompass the crucial components we had identified (as presented in Figure 2.1).

Figure 2.1

Visual Model of Interrelated Official Curriculum Components

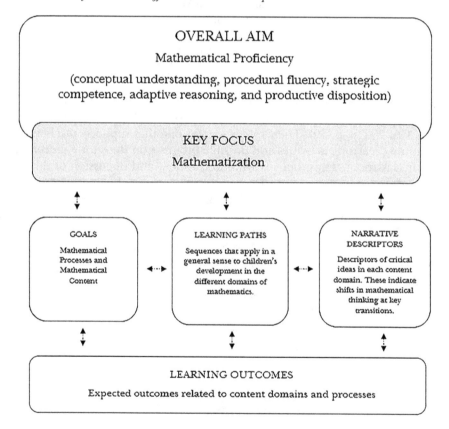

Mathematical proficiency with its intertwining strands of adaptive reasoning, strategic competence, conceptual understanding, productive disposition, and procedural fluency (National Research Council, 2001) had emerged as an important theme in our research. Given the breadth and depth of mathematical proficiency, and the general consensus around its usefulness, we proposed that it should be adopted as a key aim of the redeveloped PMC. We reported how many countries had already adopted or were promoting mathematical proficiency as a key aim of mathematics education. We also reported that children's engagement with processes such as representing, communicating, reasoning, justifying, and generalizing was highlighted in the literature as the principal means by which mathematical proficiency is developed. We noted the arguments that engagement in these processes enables children to go back and forth between abstract mathematics and real situations around them. We identified the term *mathematization* as being widely used to describe children's sense making of abstract mathematics and their formulation of real situations in abstract terms, and we indicated the role of a process-rich classroom culture to achieve this. Consequently, we made the case for making mathematization a key focus in the PMC, with a recommendation for specific and sustained attention by teachers to developing children's engagement in mathematical processes.

Having presented the case that LTs can provide a solid basis for deciphering curriculum goals (key mathematical ideas) at the content level, we wanted to see how approaches to breaking down content goals could be aligned with a process-rich classroom culture. We reported how goals across the range from kindergarten through Grade 8 had been specified and detailed in the curriculum focal points in the United States (National Council of Teachers of Mathematics, 2006). The focal point approach was to identify the most significant mathematical concepts and skills critical for each grade level. These were then presented using narrative descriptors to show how the critical ideas can be broken down into important transitions that indicate shifts in reasoning and framed as learning outcomes. The bidirectional arrows in the visual model indicate the interconnectivity of its components.

It was essential that our final research reports should be characterized by internal consistency, and that the various components fit together theoretically and conceptually, as well as reflecting the emerging consensus evident in the literature on mathematics education. As we drew our reports together, uppermost in our minds was their purpose: that they would help shape a redevelopment of the mathematics curriculum for primary schools. As we sought to identify the key implications of the research for the PMC, the unpacking of relevant issues did not occur concurrently. Given the tight deadline, we concentrated first on working systematically through the brief for the first report. Consequently, a draft of that report was with the NCCA while we worked on the questions for the second report. This gave rise to some challenges. For example, having identified themes such as mathematical proficiency and mathematization while working on the first report, we were unsure how these would be positioned as we addressed the issues of pedagogy, curriculum integration, and partnerships in the second report. We resolved this confusion by devising, as part of the work for that report, the curriculum model described above. It allowed us to synthesize the key implications of the research for the PMC.

Although the visual model does not appear in the curriculum documentation, its components feature prominently. For instance, mathematical proficiency is clearly identified as the overarching aim of the curriculum. There is no explicit mention of the overarching process of mathematization; however, reference is made in the *Primary Mathematics Curriculum* (NCCA, 2023a) to the role of five key pedagogical practices that teachers can use to promote children's mathematization. The progression continua are informed by learning paths and narrative descriptors. Reflecting the thrust of curriculum goals as presented in the visual model, there is a focus on the integration of processes and content in the redeveloped PMC. Learning outcomes, describing the expected mathematical learning and development for children at the end of a two-year period, are also detailed.

In summary, in responding to the question on key implications of our research for the development of the PMC, we identified the main components of an emerging curriculum model. We devised a visual model to communicate the interrelationships between the components and to help guide the design of the mathematics curriculum. The lesson for others is: *Synthesis, by way of a model, can bring together findings from various stages of the research process and clarify how these are situated in relation to each other.*

BRIDGING THE OFFICIAL AND OPERATIONAL CURRICULUM

The brief for the second report called for direction and recommendations on pedagogy. Thus, in writing that report, our attention shifted from the official curriculum to the operational curriculum, described by Remillard and Heck (2014) as what happens in classrooms. We took the stance that teachers are active interpreters of curriculum guides and materials and that they modify these as they work with them (Remillard, 2005). Consequently, we sought to ensure that any advice that might emanate from the report should not be prescriptive. We wanted the advice to be appropriate in terms of context, both local and national. Our challenge was to provide guidance on pedagogy that reflected current thinking about mathematics teaching and learning for children ages 3–8 years. At the same time, the guidance needed to leave room for teacher judgement and agency in how this was enacted in classrooms.

Arising from our review of the literature for the report, we identified five metapractices as pivotal in promoting mathematical thinking and understanding: *promotion of math talk*; *development of productive disposition*; *emphasis on mathematical modeling*; *use of cognitively challenging tasks*; and *integration of formative assessment*. We were influenced in our thinking about these practices by the work of Kemmis (2012), who described practices as living things that are interrelated and diffusive in nature. In particular, he spoke about metapractices as practices that shape and are shaped by other practices in a network of practices.

We conceptualized each of our metapractices as expansive in scope, and this conceptualization enabled us to harness linkages among them. For example, in considering the role of language in mathematics learning, we paid attention not only to the learning of mathematical vocabulary, but also to Sfard's (2007) commognitive approach, where the learning of mathematics is seen as modifying and extending mathematical discourse. In exploring the emphasis on mathematical modelling, we focused not so much on teacher modelling where children are presented with models of mathematical situations but on children's opportunity to model "real" mathematical situations in ways that make sense to them (e.g., English, 2006; van den Heuvel-Panhuizen, 2003). As a consequence, we were able to communicate how, in practice, the boundaries among the pedagogical practices are blurred and indistinct. It seemed to us, for example, that if children model a mathematical situation in a way that represents their thinking, they may be more productively disposed to mathematics, and to see it as sensible and useful.

We recognized teachers' abilities to identify mathematics that engage young children in playful contexts, for example, play, picture-book reading, and project work. We saw the metapractices as supporting teachers in developing children's mathematical proficiency within such contexts. We presented the idea that, at times, the opportunities for mathematics learning are obvious and, in such instances, the metapractices can deepen and extend these opportunities. We also promoted the idea that sometimes the playful context is not mathematically productive; in these instances, the metapractices can enable the teacher to support and encourage children's mathematical engagement. However, we also emphasized that this is contingent on teachers recognizing opportunities as they arise and responding appropriately to children. Here we wanted to convey the message that the metapractices are intrinsic to *intentional teaching*, that is, the skill of "adapting teaching to the content, type of learning experience, and individual child with a clear learning target as a goal" (National Research Council, 2009, p. 226).

Although other metapractices might also be identified, we concluded that the five named practices were particularly suitable for the Irish context. This is not to say that they were entirely absent in Irish classrooms; but in our reports, we brought the metapractices together in a network. We hoped this would encourage teachers to use them in a related and fluid manner towards achieving the aim of mathematics education as communicated in our visual model. We envisaged that teachers' understanding of the metapractices would be deepened by utilizing them (as appropriate) for their classrooms and seeing their effect on children's mathematical development. In this way, it seemed that the metapractices would support teachers to revise and reinterpret their understanding of PMC on an ongoing basis; that is, they would support the role of teachers as active interpreters of the curriculum.

The metapractices are described in the *Primary Mathematics Curriculum* (NCCA, 2023a), and there are also suggestions on how they can be developed in classrooms. Moreover, many of the metapractices feature prominently in the support materials contained in the *Primary Mathematics Toolkit* (NCCA, 2023b). These support materials include "important resources and reference material to support teachers to enact the curriculum in a meaningful way" (NCCA, 2023a, p. 39). Moreover, it is envisaged that these materials will be modified as the needs of children, teachers, and school communities change. In the curriculum documentation, account is clearly taken of the participatory role of teachers in their relationship with the curriculum and the key part that metapractices play in supporting this relationship.

In summary, the pedagogical strategies that we highlighted can enable teachers to work in a fluid, flexible, and contextualized way towards operationalizing the official curriculum. The lesson for others is: *In considering pedagogy, a recognition of the active nature of the teacher-curriculum relationship is the first step*. Conclusions and recommendations about the teaching and learning of mathematics should be consistent with that relationship.

DISCUSSION

In this chapter, we highlighted four issues that arose as we worked on two research reports to help shape a redeveloped mathematics curriculum for ages 3–8 in Ireland: the nature of the literature review; the balance of theoretical perspectives; the alignment of themes; and the bridging of the official and operational curricula. We discussed the ways in which we resolved these issues in line with our research questions and the wider context of mathematics curriculum enactment in Ireland. Understandably, others will resolve similar issues in ways appropriate to their specific situation, for example, their research questions, their national and local mathematics education contexts, and the nature of the official mathematics curriculum in their jurisdiction. However, we argue that the processes we followed in resolving issues we encountered can assist others in negotiating the intricacies of mathematics curriculum research.

Most would agree that doing research on mathematics curriculum is a complex endeavour and that there are no easy solutions to the conundrums presented in the course of the research. Open-mindedness to research findings and their potential contribution is paramount. The ways in which research might speak to the design of the official mathematics curriculum is a key concern. Moreover, the ongoing relationship between the teacher and the curriculum is central to how the research is conceived and presented. Importantly, the development of research to underpin an official mathematics curriculum is relatively novel. Thus, the sharing of lessons learned in conducting such research is a critical aspect contributing to the quality and diversity of this type of work.

REFERENCES

Anthony, G., & Walshaw, M. (2007). *Effective pedagogy in mathematics/Pangarau/Mathematics: Best evidence synthesis iteration (BES)*. New Zealand Ministry of Education. https://www.educationcounts.govt.nz/__data/assets/pdf_file/0007/7693/BES_Maths07_Complete.pdf

Anthony, G., & Walshaw, M. (2009). Mathematics education in the early years: Building bridges. *Contemporary Issues in Early Childhood, 10*(2), 107–121. https://doi.org/10.2304/ciec.2009.10.2.107

Bakker, A., Smit, J., & Wegerif, R. (2015). Scaffolding and dialogic teaching in mathematics education: Introduction and review. *ZDM Mathematics Education, 47*(7), 1047–1065. https://doi.org/10.1007/s11858-015-0738-8

Biesta, G. (2007). Why "what works" won't work: Evidence-based practice and the democratic deficit in educational research. *Educational Theory, 57*(1), 1–22. https://doi.org/10.1111/j.1741-5446.2006.00241.x

Bobis, J. B., Clarke, D., Clarke, G., Thomas, B., Wright, B., Young-Loveridge, J., & Gould, P. (2005). Supporting teachers in the development of young children's mathematical thinking: Three large scale cases. *Mathematics Education Research Journal, 16*(3), 27–57. https://doi.org/10.1007/BF03217400

Chalmers, I. (2003). Trying to do more good than harm in policy and practice: The role of rigorous, transparent, up-to-date evaluations. *Annals of the American Academy of Political and Social Science, 589*(1), 22–40. https://doi.org/10.1177/0002716203254762

Cobb, P. (2007). Putting philosophy to work: Coping with multiple theoretical perspectives. In F. K. Lester, Jr. (Ed.), *Second handbook of research on mathematics teaching and learning* (pp. 3–38). Information Age Publishing.

Cobb, P., & Yackel, E. (1996). Constructivist, emergent and sociocultural perspectives in the context of developmental research. *Educational Psychologist, 31*(3/4), 175–190. https://doi.org/10.1080/00461520.1996.9653265

Conway, P., & Sloane, F. (2005). *Research Report No.5: International trends in post primary mathematics education: Perspectives on learning, teaching and assessment.* National Council for Curriculum and Assessment. ncca.ie/media/1491/international_trends_in_postprimary_mathematics_education_rr5.pdf

Daro, P., Mosher, F., & Corcoran, T. (2011). *Learning trajectories in mathematics: A foundation for standards, curriculum, assessment, and instruction:* Research Report (No. 68). Consortium for Policy Research in Education. https://eric.ed.gov/?id=ED519792

Davies, P. (2000). The relevance of systematic reviews to educational policy and practice. *Oxford Review of Education, 26*(3/4), 365–378. https://doi.org/10.1080/713688543

Department of Education. (2023). *Primary curriculum framework: For primary and special schools.* https://www.curriculumonline.ie/getmedia/84747851-0581-431b-b4d7-dc6ee850883e/2023-Primary-Framework-ENG-screen.pdf

Dooley, T., Dunphy, E., Shiel, G., Butler, D., Corcoran, D., Farrell, T., Nic Mhuirí, S., O'Connor, M., & Travers, J. (2014). *Research report No.18: Mathematics in early childhood and primary education (children aged 3–8 years): Teaching and learning.* National Council for Curriculum and Assessment. https://ncca.ie/media/2147/ncca_research_report_18.pdf

Dunphy, E. (2009). Early childhood mathematics teaching: Challenges, difficulties and priorities of teachers of young children in primary schools in Ireland. *International Journal of Early Years Education, 17*(1), 3–16. https://doi.org/10.1080/09669760802699829

Dunphy, E., Dooley, T., Shiel, G., Butler, D., Corcoran, D., Ryan, M., & Travers, J. (2014). *Research report No.17: Mathematics in early childhood and primary education (children aged 3–8 years): Definitions, theories, stages of development and progression.* National Council for Curriculum and Assessment. https://ncca.ie/media/1494/maths_in_ecp_education_theories_progression_researchreport_17.pdf

Eivers, E., Close, S., Shiel, G., Clerkin, A., Gilleece, L., & Kiniry, J. (2010). *The 2009 national assessments of mathematics and English.* Educational Research Centre, Ireland. https://www.erc.ie/documents/na2009_report.pdf

English, L. D. (2006). Mathematical modeling in the primary school: Children's construction of a consumer guide. *Educational Studies in Mathematics, 63*(3), 303–323. https://doi.org/10.1007/s10649-005-9013-1

Gough, D., Oliver, S., & Thomas, J. (2017). Introducing systematic reviews. In D. Gough, S. Oliver, & J. Thomas (Eds.), *An introduction to systematic reviews* (2nd ed., pp. 1–17). SAGE.

Government of Ireland. (1999). *Primary school curriculum: Mathematics.* The Stationery Office. https://www.curriculumonline.ie/getmedia/c4a88a62-7818-4bb2-bb18-4c4ad37bc255/PSEC_Introduction-to-Primary-Curriculum_Eng.pdf

Griffin, S. (2004). Building number sense with Number Worlds: A mathematics programme for young children. *Early Childhood Research Quarterly, 9*, 173–180. https://doi.org/10.1016/j.ecresq.2004.01.012

Hammersley, M. (2008). Paradigm war revived? On the diagnosis of resistance to randomized controlled trials and systematic review in education. *International Journal of Research & Method in Education*, *31*(1), 3–10. https://doi.org/10.1080/17437270801919826

Hammersley, M. (2020). Reflections on the methodological approach of systematic reviews. In O. Zawacki-Richter, M. Kerres, S. Bedenlier, M. Bond, & K. Buntins (Eds.), *Systematic reviews in educational research: Methodology, perspectives and application* (pp. 23–39). Springer. https://link.springer.com/chapter/10.1007/978-3-658-27602-7_2

Kemmis, S. (2012). Researching educational praxis: Spectator and participant perspectives. *British Educational Research Journal*, *38*(6), 885–905. https://doi.org/10.1080/01411926.2011.588316

Lewis, C., & Tsuchida, I. (1998). A lesson is like a swiftly flowing river: How research lessons improve Japanese education. *American Educator*, *12*(14–17), 50–52.

MacLure, M. (2005). 'Clarity bordering on stupidity': Where's the quality in systematic review? *Journal of Education Policy*, *20*(4), 393–416. https://doi.org/10.1080/02680930500131801

National Council for Curriculum and Assessment. (2023a). *Primary mathematics curriculum: For primary and special schools.* https://www.curriculumonline.ie/getmedia/484d888b-21d4-424d-9a5c-3d849b0159a1/PrimaryMathematicsCurriculum_EN.pdf

National Council for Curriculum and Assessment. (2023b). *Primary mathematics toolkit.* https://www.curriculumonline.ie/Primary/Curriculum-Areas/Mathematics/

National Council of Teachers of Mathematics. (2006). *Curriculum focal points for Prekindergarten through Grade 8 mathematics: A quest for coherence*. https://www.nctm.org/curriculumfocalpoints/

National Research Council. (2001). *Adding it up: Helping children learn mathematics*. J. Kilpatrick, J. Swafford, & B. Findell (Eds.). Mathematics Learning Study Committee, Center for Education, Division of Behavioral and Social Sciences and Education. National Academies Press. https://doi.org/10.17226/9822

National Research Council. (2005). *How students learn: History, science and mathematics in the classroom*. M. S. Donovan & J. Bransford. (Eds.). Division of Behavioural and Social Sciences and Education, Committee on How People Learn. National Academies Press. https://doi.org/10.17226/10126

National Research Council. (2009). *Mathematics learning in early childhood: Paths towards excellence and equity*. C. Cross, T. Woods, & H. Schweingruber (Eds.). Committee on Early Childhood Mathematics Center for Education. Division of Behavioural and Social Sciences and Education. National Academies Press.

Newman, M., & Gough, D. (2020). Systematic reviews in educational research: Methodology, perspectives and application. In O. Zawacki-Richter, M. Kerres, S. Bedenlier, M. Bond, & K. Buntins (Eds.), *Systematic reviews in educational research: Methodology, perspectives and application* (pp. 3–22). Springer. https://doi.org/10.1007/978-3-658-27602-7_1

Nind, M. (2006). Conducting systematic review in education: A reflexive narrative. *London Review of Education*, *4*(2), 183–195. https://doi.org/10.1080/14748460600855500

Remillard, J. T. (2005). Examining key concepts in research on teachers' use of mathematics curricula. *Review of Educational Research, 75*(2), 211–246. https://doi.org/10.3102/00346543075002211

Remillard, J. T., & Heck, D. (2014). Conceptualising the curriculum enactment process in mathematics education. *ZDM Mathematics Education, 46*(5), 705–718. https://doi.org/10.1007/s11858-014-0600-4

Sarama, J., & Clements D. (2009). *Early childhood mathematics education research: Learning trajectories for young children*. Routledge.

Sfard, A. (2007). When the rules of discourse change, but nobody tells you: Making sense of mathematics learning from a commognitive standpoint. *The Journal of the Learning Sciences, 16*(4), 565–613. https://doi.org/10.1080/10508400701525253

Shiel, G., Kavanagh, L., & Millar, D. (2014). *The 2014 national assessments of English reading and mathematics. Volume 1: Performance report*. Educational Research Centre, Ireland.

Simon, M. (1995). Reconstructing mathematics pedagogy from a constructivist perspective. *Journal for Research in Mathematics Education, 26*(2), 114–145. https://doi.org/10.2307/749205

Simon, M., & Tzur, R. (2004). Explicating the role of mathematical tasks in conceptual learning: An elaboration of the hypothetical learning trajectory. *Mathematical Thinking and Learning, 6*(2), 91–104. https://doi.org/10.1207/s15327833mtl0602_2

Stemhagen, K., & Henney, C. (2021). *Democracy and mathematics education: Rethinking school math for our troubled times*. Routledge.

Stigler, J., & Thompson, B. (2012). You can't play twenty questions with mathematics teaching and learning, and win. In J. Carlson & J. Levin (Eds.), *Instructional strategies for improving students' learning: Focus on early reading and mathematics* (pp. 187–195). Information Age Publishing.

Suri, H., & Clarke, D. (2009). Advancements in research synthesis methods: From a methodologically inclusive perspective. *Review of Educational Research, 79*(1), 395–430. https://doi.org/10.3102/0034654308326349

van den Heuvel-Panhuizen, M. (2003). The didactical use of models in realistic mathematics education: An example from a longitudinal trajectory on percentage. *Educational Studies in Mathematics, 54*(1), 9–35. https://doi.org/10.1023/B:EDUC.0000005212.03219.dc

van den Heuvel-Panhuizen, M. (2008). *Children learn mathematics: A learning teaching trajectory with intermediate attainment targets for the lower grades in primary school*. Sense.

ENDNOTES

1. Early childhood education in the Republic of Ireland is provided in preschools (children aged 3–6 years) and/or in Junior or Senior Infant classes in primary schools (children aged 4–6 years). Statutory school starting age is six years. Generally, the duration of primary education is eight years.

2. Gaeilge (Irish) is the national and first official language of Ireland. Although English is the primary language of use, there are regions known as the Gaeltacht where Irish is the main language, both in businesses and in families.
3. The Minister for Education is the senior government minister at the Department of Education in the Government of Ireland.

CHAPTER 3

COMBINED READING

A Methodological Approach for Documental Curricular Analysis

Priscila Dias Corrêa
University of Windsor, Canada

Leticia Rangel
Federal University of Rio de Janeiro (UFRJ), Brazil

The present research focuses on qualitative analysis of official curriculum documents, seeking to promote reflections and learnings about mathematics teaching in elementary education. The study presents an original research methodology developed for this purpose, combined reading. *We understand combined reading as a methodology that supports documental analysis processes, establishes a relationship between different curricular guidelines, and reveals aspects and considerations that are not specific to analyzed documents but to mathematics teaching in general. These characteristics of combined reading contribute to the development of mathematics knowledge for teaching. We highlight learnings from two investigations of mathematics curricula from distinct countries. One investigation explores mathematics teaching and learning processes based on Brazilian and Canadian curriculum documents. The other investigates curricular terms and expressions used in curriculum documents from seven countries. The chapter presents the lessons learned from these studies to contribute to curriculum research in the field of mathematics education.*

Keywords: combined reading; curriculum analysis; mathematics curriculum; mathematics elementary education; mathematics knowledge for teaching

INTRODUCTION

Discussions about curriculum have been frequent in mathematics education, with many countries, cities, and regions establishing reforms in their education systems. However, the curriculum has only recently gained prominence in mathematics education research literature (Li & Lappan, 2014; Shimizu & Vithal, 2018; Thompson et al., 2018). Curriculum studies comprise curricular documents and practices from different countries and present distinct approaches, such as identifying curricular particularities (e.g., Acar & Serçe, 2021; Cerqueria & Silva, 2020; Pires, 2013; Wang & McDougall, 2019), investigating how curricula are implemented in classrooms or teaching materials (e.g., Lui & Leung, 2013; Wang & Fan, 2021), exploring the acceptance or resistance of teachers to curricular proposals (e.g., Charalambous & Philippou, 2010), and comparing education across systems (e.g., Son et al., 2017). Our research does not investigate curricula along the lines of comparative education (Dias & Gonçalves, 2017); it does not aim at best practices (e.g., Villalobos Torres & Trejo Sánchez, 2015); it is not focused on social, cultural, economic, and political aspects (e.g., Bickmore et al., 2017); nor does it suit a comparative historical analysis of curricula (e.g., Bessot & Comiti, 2006). Rather, our investigative work is developed based on qualitative analysis of curricula, aiming to improve mathematics teachers' knowledge.

To contribute to the reflective process inherent to teaching practice and the development of mathematics knowledge for teaching (Ball et al., 2008), our curriculum research focuses on analyzing official documents and curricular guidelines. These curricular documents offer conceptions of school mathematics, reveal how school mathematics is organized, and indicate what should be taught and learned during the schooling process. We are not limited to conclusions about the examined curricular documents. We expand our analysis to aspects that transcend these documents, which emerge from a reflective methodological process based on qualitative comparisons. We call this process *combined reading*. This chapter presents this original methodology and the lessons learned while developing curricular research in elementary mathematics education. In particular, we highlight lessons learned from two studies: (1) a study about the teaching of numbers from reviewing official curriculum guidelines in Brazil and Canada (Corrêa & Rangel, 2021); and (2) an ongoing study that investigates how school mathematics and its teaching and learning are understood through the identification of terms, expressions, and their meanings in curriculum documents from seven different countries.

COMBINED READING: METHODOLOGICAL LEARNINGS

We named the methodology we designed *combined reading*. In this approach, we intend to establish and sustain qualitative relationships between curricula anchored in correlated elements from the documents under analysis. Combined reading seeks to identify parities, contrasts, and singularities to bring out reflections and aspects of mathematics teaching in elementary education. The focus is not on particular implications of the curricula for mathematics teaching in each of the involved countries but on general aspects that emerge from the analysis—for example, identifying potential biases in the teaching of specific topics, such as in the teaching of numbers or, more specifically, of fractions (Corrêa & Rangel, 2021). We seek to understand how school mathematics is organized into strands, or what expressions and terms permeate and characterize the understanding of school mathematics and its teaching and learning. Combined reading supports the documental analysis and highlights essential aspects in the study of the documents. Both the development of the combined reading methodology and its use have provided much learning.

The combined reading methodology encompasses sequential development emphases (Figure 3.1). Based on a research question, we begin the process with a parallel reading of the documents, aiming to get familiar with the curricular references and to identify correlated elements that will support the analysis. At this stage, we take a panoramic view of the documents. Then, from the analysis of correlated elements, parities, contrasts, and singularities between the documents are identified. Next, a schematic representation is created to illustrate and summarize the outcomes of the combined reading. Such representation is characterized both as a product of the initial analysis and an instrument for further analysis, bringing forward emerging considerations about mathematics teaching. In the following sections, we describe each of these emphases and its respective constructs.

Parallel Reading and Identification of Correlated Elements

The parallel reading of documents refers to reading one and another document concurrently. This concomitant reading allows familiarization with the curriculum documents and the search for possible correlated elements that will serve as the basis for the qualitative analysis. Correlated elements correspond to elements of the curriculum documents that, in the context of the investigation, have similar functions. Identifying correlated elements is an essential part of the combined reading methodology and is

Figure 3.1

Combined Reading Methodological Process

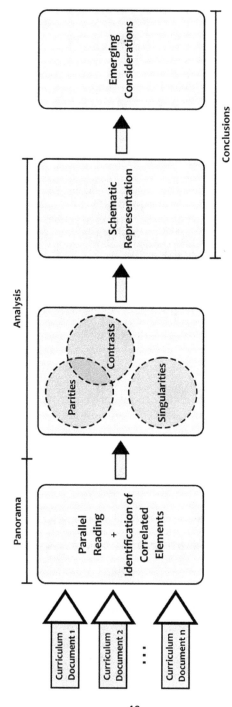

related to the research question. Examples of correlated elements are the distinction between school years and the organization based on thematic units (e.g., numbers, geometry, statistics and probability). The existence of correlations between specific elements of the curriculum documents is necessary to enable the analysis and support the qualitative investigation inherent to the combined reading. Thus, we learned the following lesson: *Identifying correlated elements is necessary to establish a relationship between curriculum documents, enabling the combined reading.*

The correlation between such elements may be evident or established from the first parallel reading. For example, research may focus on the development of numbers throughout the first years of schooling (e.g., Corrêa & Rangel, 2021). In this case, identifying correlated elements can be immediate because number is a common theme in curricular guidelines and pertinent to the first years of schooling; so, the correlated elements would be learning goals per school year. Whereas for topics that do not follow a certain standard in the composition of curricula, such as financial mathematics, the identification of correlated elements can attend to the lists of progressive learning goals, without relating them to school years.

There are cases in which the identification of correlated elements is not possible, which, in itself, can be considered a result that emerges from the parallel reading. For instance, suppose the goal is to investigate the teaching of computational thinking. In this case, there is no way to establish a parallel reading between a curriculum that includes the topic in its framework and another that does not mention it. Having identified the correlated elements that support the investigation, we proceed to the next emphasis of the combined reading: to determine parities, contrasts, and singularities between the documents under analysis.

Determination of Parities, Contrasts, and Singularities

This process, based on the analysis of correlated elements, is eminently qualitative and reflexive. *Parities* are identified from correlated elements that are equivalent, have significant similarities, or deal with the same aspect or topic, depending on the focus of the investigation. For example, in a study that identifies learning goals per school year as correlated elements, the introduction of the representation of decimal numbers in corresponding school years can be considered a parity. Or, if the correlated elements refer to methodological guidelines, the suggestion to focus the teaching approach on problem solving can be viewed as a parity. A parity can also be identified from absences when the documents under analysis do not address a topic, such as the development of computational thinking.

Contrasts are identified from correlated elements that reveal structural or fundamental differences between the analyzed documents. For example, the difference between school years in which a given topic is taught for the first time or the inclusion (or not) of statistics and probability as a content strand can be contrasted. Both parities and contrasts can be identified from the same correlated element. For example, Corrêa and Rangel's (2021) research, which considers learning goals per school year as correlated elements, recognizes that the documents under analysis propose the initial approach of multiplication of whole numbers in different school years, a contrast. However, both recommend limiting multiplication factors to numbers less than or equal to 5, a parity.

Finally, a *singularity* reveals a specific particularity of one or more curricular documents. For example, observing as correlated elements a strand-based structure in the documents, a potential singularity is the indication of Financial Literacy and Social-Emotional Learning Skills as a content strand in a single curriculum document from a set of documents analyzed. This is the case of the Ontario curriculum (Ontario Ministry of Education, 2020).

Identifying parities, contrasts, and singularities is a substantive part of our analysis, anchored by a critical-reflexive process inherent to the teaching practice. Refining or broadening the discussion about parities, contrasts, and singularities is a legitimate and natural process resulting from a combined reading and can reveal teaching aspects not evident in an isolated reading of a curricular document. The value of this analytic process lies in the reflective process, not in the identification or categorization itself. From this process, the schematic representation emerges, and conclusions and considerations from the study are established. This leads to another lesson: *In a combined reading, the identification and analysis of parities, contrasts, and singularities may reveal demands of teachers' mathematics knowledge not evident in an isolated reading of a curricular document.*

Schematic Representation of the Analysis and Emerging Considerations

After the qualitative analysis of the correlated elements, a new stage of the combined reading is reached: the schematic representation of the analysis. The goal is to offer a visual model that allows a concomitant overview of the investigated documents, synthesizes the analysis performed, and makes parities, contrasts, and singularities prominent. Such a schematic representation (as discussed in a later section and partially illustrated in Figure 3.2) will sustain and determine emerging reflections and support the study's conclusions, a closing stage of the combined reading. For example, in Corrêa and Rangel (2021), the schematic representation was established as a unique grid that portrays the progression of content over the school years in Brazil and Canada. This representation was named *curricular trajectory*. From its reading, and considering parities, contrasts, and singularities, conclusions of the study were established. This leads to another lesson: *A schematic representation, which summarizes and highlights the analysis stage of a combined reading, triggers emerging reflections.*

We believe that the proposed composition of a schematic representation based on a visual model is itself a contribution to curriculum studies. As conceived, a schematic representation, in addition to supporting the study reflections, can (1) prompt new investigations, (2) be a reference for specific comparisons between the analyzed documents and other curricular documents, (3) be expanded, within the same scope, by including new curricular guidelines, or (4) be adapted to contemplate other themes or questions.

Combined reading is a methodology that underlines the relationship between diverse curricular orientations and prompts thoughts and reflections that are not specific from the analyzed documents but relate to mathematics teaching in general. This feature of the combined reading contributes, in particular, to the development of teachers' mathematics knowledge for teaching (Ball et al., 2008). Teachers can question, confirm, or refute their teaching practice based on the distinction of parities, contrasts, and singularities. Our lesson is: *The analysis that emerges from the combined reading methodology can transcend curriculum documents revealing general aspects of mathematics teaching.*

INVESTIGATIONS: EMERGING LEARNINGS

In this section, we describe curricular investigations grounded in the combined reading methodology and the learning stemming from these studies, configuring a meta-study. We highlight two investigations with distinct characteristics. The first research study sought to explore the mathematics teaching and learning processes from the combined reading of Brazilian and Canadian curriculum documents. The second research was motivated by the first and is an ongoing study based on the combined reading of curricular documents from seven countries; it explores how school mathematics and its teaching and learning are understood by identifying terms, expressions, and their related meanings.

The Teaching of Numbers in Curriculum Documents: Brazil and Canada

The first investigation refers to a documental analysis based on official curriculum documents from Brazil (Brasil, 2018) and Canada (Alberta

Education, 2006) that reflects on the teaching of numbers and, in particular, on the teaching of fractions (Corrêa & Rangel, 2021). This study is important because it resulted in the development of the combined reading methodology based on qualitative relational analysis among curriculum documents.

The correlated elements that supported this study have a structuring role in the curricular documents; therefore, they were called *structuring correlated elements*, or *structuring elements*, for short. These elements list what to teach by means of learning goals per school year. The focus is on content, without specific intention in pedagogical guidance, that is, on how to teach. In the Brazilian document, the structuring elements were knowledge topics and skills; in the Canadian document, they were specific outcomes and achievement indicators (Table 3.1).

Table 3.1

Example of Structuring Correlated Elements – Whole Numbers

Brazilian Curriculum: Grade 2 (Brasil, 2018, p. 283)	Canadian Curriculum: Grade 2 (Alberta Education, 2006, p. 64)
Knowledge Topics: Reading, writing, comparing, and ordering numbers up to hundreds by understanding the features of the decimal number system (place value and the role of the number zero). *Skills:* (EF02MA01) Compare and order natural numbers (up to hundreds) by understanding the characteristics of the decimal number system (place value and the function of the zero).	*Specific Outcomes:* 5) Compare and order numbers up to 100. *Achievement Indicators:* Order a given set of numbers in ascending or descending order and verify the result using a hundred chart, number line, ten frames or by making references to place value. • Identify errors in a given ordered sequence. • Identify missing numbers in a given hundred chart. • Identify errors in a given hundred chart.

Based on the structuring elements, the qualitative analysis of the combined reading identified parities and contrasts, revealing equivalences, similarities, and differences between the two curriculum documents. As shown in Table 3.1, which refers to the teaching of whole numbers in Grade 2, the comparison and ordering of whole numbers are introduced in Grade 2 in both documents, therefore a parity. However, in the Brazilian document, there is a limitation of numbers up to 999, whereas in the Canadian document, the limitation is up to 100, which is a contrast. Following the analysis process that led to the identification of parities and contrasts, and aiming at designing the schematic representation, it was necessary to determine elementary aspects of mathematics teaching present in the structuring elements of the documents, as the full descriptions of the structuring elements are too extensive for a schematic representation. Elementary aspects are concepts and understandings that support the learning of a topic. In Table 3.1, we observe elementary aspects that emerged from the analysis of the structuring elements, for instance, representation of the decimal system and comparison and ordering. Once the elementary aspects were identified, they were used to constitute the body of the schematic representation, named in this particular study as curricular trajectory (Figure 3.2).

Figure 3.2

Excerpt of the Curricular Trajectory—Whole Numbers

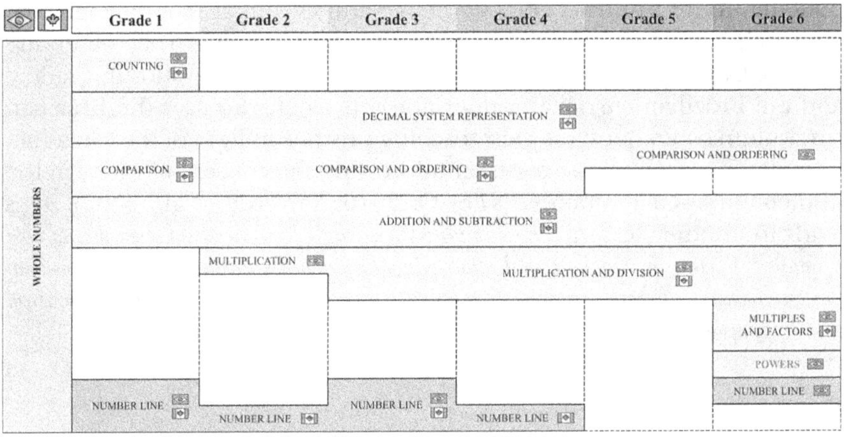

WHOLE NUMBERS	Grade 1	Grade 2	Grade 3	Grade 4	Grade 5	Grade 6
	COUNTING					
				DECIMAL SYSTEM REPRESENTATION		
	COMPARISON		COMPARISON AND ORDERING		COMPARISON AND ORDERING	
				ADDITION AND SUBTRACTION		
		MULTIPLICATION		MULTIPLICATION AND DIVISION		
					MULTIPLES AND FACTORS	
					POWERS	
					NUMBER LINE	
	NUMBER LINE	NUMBER LINE	NUMBER LINE	NUMBER LINE		

Source: Adapted from Corrêa and Rangel (2021, p. 4478).

This representation relates the two documents, highlighting: (1) the temporal variation from grades 1 to 6, (2) the main topics about numbers (whole numbers, rational numbers, and integers), (3) the elementary aspects of the teaching of numbers that emerged from the analysis, and (4) the country to whose curriculum the elementary aspects correspond. The curricular trajectory was a product of the initial research analysis that illustrated the curriculum data from both documents, enabling the subsequent analysis of the data. This leads to the following lesson: *In a combined reading, it is not trivial to associate correlated elements from their full descriptions; identifying elementary aspects sustains the schematic representation and highlights parities, contrasts, and singularities.*

From the curricular trajectory, we started a cross-sectional analysis of Brazilian and Canadian curriculum documents. This analysis confirms that intended mathematics curricula (United Nations Educational, Scientific and Cultural Organization [UNESCO], 2016; Valverde et al., 2002) do not necessarily present the same set or order of elementary aspects; also, intended curricula may reveal characteristics that are equivalent or that complement each other. Observing these variations in curricular documents opens space for deepening, questioning, and improving one's mathematics knowledge of teaching through the lens of teaching and learning processes. For example, Figure 3.3 presents an excerpt of the curricular trajectory concerning the teaching of fractions. The Canadian curriculum addresses the construct part/whole in grades 3 and 4, and the Brazilian curriculum addresses this same construct in grades 5 and 6. Both curricula present this same elementary aspect (construct part/whole), despite teaching it in different grades. However, the Brazilian curriculum also addresses the quotient construct, which is not mentioned in the Canadian curriculum. This feature of the Brazilian curriculum has the potential to inform teachers' knowledge and eventually complement the approach proposed by the Canadian curriculum. Similarly, by observing that the Canadian curriculum introduces the concept of ratios in grade 6 and the Brazilian curriculum does not, a teacher who uses the Brazilian curriculum could question their teaching practice in light of the Canadian intended curriculum and consider addressing the concept of ratios in the implemented curriculum (UNESCO, 2016; Valverde et al., 2002). This leads to another lesson: *The combined reading of curriculum documents can promote reflection by teachers and other educators about specific teaching aspects of the curriculum and improve the process of transforming the intended curriculum into the implemented curriculum.*

Figure 3.3

Excerpt of the Curricular Trajectory—Fractional Representation of Rational Numbers

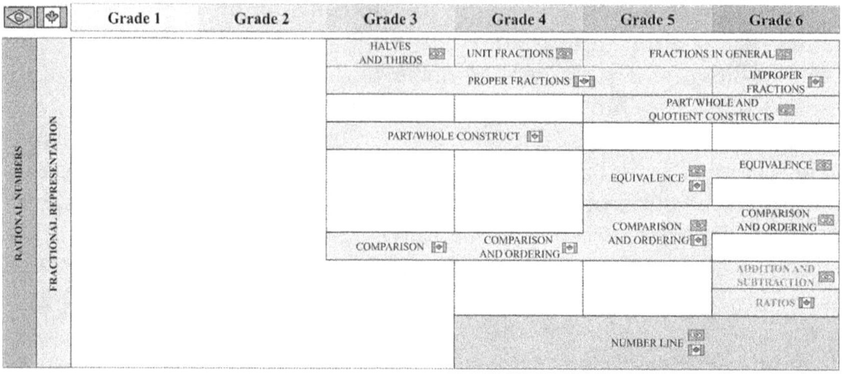

	Grade 1	Grade 2	Grade 3	Grade 4	Grade 5	Grade 6
RATIONAL NUMBERS — FRACTIONAL REPRESENTATION			HALVES AND THIRDS	UNIT FRACTIONS	FRACTIONS IN GENERAL	
				PROPER FRACTIONS		IMPROPER FRACTIONS
					PART/WHOLE AND QUOTIENT CONSTRUCTS	
			PART/WHOLE CONSTRUCT			
					EQUIVALENCE	EQUIVALENCE
					COMPARISON AND ORDERING	COMPARISON AND ORDERING
			COMPARISON	COMPARISON AND ORDERING		
					ADDITION AND SUBTRACTION	
						RATIOS
					NUMBER LINE	

Source: Adapted from Corrêa and Rangel (2021, p. 4478).

We understand that this analysis of one's practice based on curricular documents promotes the development of mathematics knowledge for teaching (Ball et al., 2008). Intended curricula offer a form of reflection that requires evaluating the content from students' points of view, focused on the construction of knowledge. Thus, teachers can develop and improve their own knowledge of teaching to promote means for students' mathematics learning.

Terms and Expressions in Curriculum Documents: Australia, Brazil, Canada, Chile, United States, Singapore, and Portugal

During the investigation described in the previous section, we observed the diversity of expressions in the mathematics curriculum documents of Brazil (Brasil, 2018) and Canada (Alberta Education, 2006). We noted that different curricula use different terms to refer to the same understandings, similar terms to refer to different understandings, or even different fundamental terms of greater scope. We observed expressions such as *fundamental ideas*, *nature of mathematics*, *mathematical processes*, *mathematical literacy*, *skills*, *specific outcomes*, and *achievement indicators*. These terms relate to how mathematics and its learning are understood. For example, in the Brazilian curriculum, the expression *fundamental ideas* refers to ideas of equivalence, order, proportionality, interdependence, representation, variation and approximation, which, according to the document, promote the articulation between the different areas of mathematics. The Canadian curriculum expression *nature of mathematics* identifies components intrinsic to mathematics: variation, conservation, numerical sense, patterns, relationships, spatial sense, and uncertainty. Drawing an equivalence between such expressions is not always possible or even desirable. However, the diversity of terms and expressions led us to reflect on what can be revealed from using these terms and expressions in different official curricula. It was natural to establish questions such as: What terms are used to identify areas of mathematics in elementary education? How are learning outcomes organized and identified? Is there a pattern? Is the expression *mathematical processes* used in other curricula? What about the expression *mathematical literacy*?

Based on this concern, we started a second investigation to explore and analyze the different terminologies used in curriculum documents from diverse countries. To develop the research, the main challenges faced were choosing countries and identifying the most current curricula. It was important to consider countries that have recently reviewed or established their curriculum guidelines, such as Singapore. Among the curricular documents analyzed to date, the U.S. document (National Governors Association Center for Best Practices & Council of Chief State School Officers [NGA & CCSSO], 2010) is the oldest one, dating to 2010, about a decade old. In addition, we decided to focus our research on guidelines corresponding to elementary school. We started exploring the information available on the TIMSS & PIRLS International Study Center portal (https://timssandpirls.bc.edu/index.html). TIMSS (Trends in International Mathematics and Science Study) and PIRLS (Progress in International Reading Literacy Study) are international assessments that monitor trends in students' academic results in mathematics, science, and reading. The TIMSS & PIRLS website (2020) provides a summary table on the countries' curricula that are part of the evaluation processes. From the information offered by this table, we began to investigate official data on the websites of the Ministries of Education of various countries. In some countries, access to curricula was immediate, such as in Australia. However, in other countries, we have yet to identify the current curriculum or access the documents, as for France. As a result of this process, we selected official curriculum documents from seven countries with varied geographical locations: Australia (Australian Curriculum, Assessment and Reporting Authority [ACARA], 2018), Brazil (Brasil, 2018), Canada (Ontario Ministry of Education, 2020), Chile (Gobierno de Chile, 2018), United States (NGA & CCSSO, 2010), Singapore (Singapore Ministry of Education, 2012), and Portugal (República Portuguesa, 2013). In the case of Canada, which does not have a national curriculum, we selected the Ontario Curriculum due to its recent implementation in 2020. In the case of the United States, we selected the *Common Core State Standards for Mathematics* (*CCSSM*), a document that informed many state standards in the United States of America (Council of Chief State School Officers, 2022).

In this study, the focus is not on investigating a specific topic, such as numbers, geometry, or the introduction of statistics and probability; instead, it focuses on general aspects of curriculum organization. Based on the combined reading methodology, we investigate the terms used in curricula supported by the research question: What aspects of school mathematics and its teaching and learning can be revealed by the diversity of terms and expressions that underpin curriculum guidelines? The investigation required the establishment of correlated elements, which, in this case, did not have a structuring role as in the research described in the previous section, but instead indicated essential aspects of a more subjective nature, such as values, principles, beliefs, and conceptions. Therefore, the correlated elements that support this study have a foundational role and are called *foundational correlated elements*, or *foundational elements*, for short.

The foundational characteristic of these elements imposed a significant challenge in establishing correlations. The parallel reading of the documents allowed us to identify terms and expressions from the curricula and then correlate them according to their nature, function in the document, or interpretation. For example, in all analyzed documents, processes or practices that intrinsically characterize doing mathematics are somehow identified explicitly. Reasoning and proving, for instance, is regarded as a mathematical process. However, some documents refer to and describe this same process in a way not directly related to doing mathematics, but from the student's learning perspective, as a skill or goal to be achieved in the learning process, as in the U.S. curriculum, *construct viable arguments and critique the reasoning of others*. Despite the differences in how such processes are presented in the different curricular documents, they were considered foundational elements due to their nature and function in the documents.

A schematic representation of this investigation reflects the parallelism between the foundational elements of the different countries. In particular, the use of a schematic representation served both as a product of the analysis and as an instrument of further investigation. Based on this schematic representation, we observed that school mathematics is organized and grounded in thematic areas denominated by similar expressions: *content strands, thematic units, strands, content standards by domain, content domains* (Figure 3.4). Two documents identify three thematic areas (Australia and Singapore), three documents identify five thematic areas (Brazil, Chile, and Portugal), one document identifies six thematic areas (Canada), and one document identifies 10 thematic areas (United States). We observed that as much as the number of thematic areas might differ, all documents, except the curriculum of Ontario, Canada, present some variation of the thematic areas of numbers, algebra, geometry, and statistics and probability, that is, a parity.

Numbers and algebra may have slightly different terminologies and be considered as two distinct thematic areas (Brazil, Canada, Chile, and Portugal) or as a single thematic area (Australia and Singapore). The same is observed for geometry and measurement, which can vary in terminology and appear together or separately. In the thematic area of statistics and probability, we observed a single thematic area that encompasses the theme in all curriculum documents. However, it is the thematic area that presents the most variations in terminology: statistics and probability; data; data and probabilities; statistics; and data organization and processing. We understand that the consistency observed in formulating the thematic areas of the different curricula indicates a convergent understanding of mathematics teaching in elementary education. The lesson we take away from this analysis is: *The slight variation of curricular terms and expressions used in formulating thematic areas indicates a convergence in mathematics teaching*.

Figure 3.4

Schematic Representation—Mathematics Content by Thematic Areas for Seven Countries

Australia 2018	Brazil 2018	Canada (ON) 2020	Chile 2018	USA (CCSSM) 2010		Singapore 2012	Portugal 2013
Content Strands	Thematic Units	Content Strands	Strands	Content Standards by Domain		Content Strands	Content Domains
Number & Algebra	Numbers	Number	Numbers & Operations	Counting & Cardinality	K	Number & Algebra	Numbers & Operations
				Numbers & Operations	K - 5		
				Numbers System	6 - 8		
	Algebra	Algebra	Patterns & Algebra	Operations & Algebraic Thinking	K - 5		Algebra
				Ratios & Proportional Relationships	6 - 7		
				Expressions & Equations	6 - 8		
				Functions	8		Functions, Sequencies & Series
Measurement & Geometry	Geometry	Spatial Sense	Geometry	Measurement & Data	K - 5	Measurement & Geometry	Geometry & Measurement
	Magnitudes & Measurement		Measurement	Geometry	K - 8		
Statistics & Probability	Statistics & Probability	Data	Data & Probabilities	Statistics & Probability	6 - 8	Statistics	Data Organization & Processing
		Financial Literacy					
		Social-Emotional Learning					

Nonetheless, the division into thematic areas in Ontario (Ontario Ministry of Education, 2020) can be regarded as a singularity identified in the combined reading process. Ontario's curriculum is divided into six thematic areas: number, algebra, spatial sense, data, financial literacy, and social-emotional learning. The first four areas are equivalent to other curricula and use similar expressions. However, the financial literacy and social-emotional learning thematic areas are unique compared to the curricula in the other six countries. This singularity is verified by the expressions used to characterize the thematic areas and can indicate progress in comprehending mathematics education.

The same can be seen in the U.S. curriculum document (NGO & CCSSO, 2010), which presents 10 thematic areas: counting and cardinality, operations and algebraic thinking, numbers and operations, geometry, measurement and data, ratios and proportional relationships, number systems, expressions and equations, functions, and statistics and probability. These thematic areas permeate school years in different ways. For example, counting and cardinality appear only in kindergarten, geometry appears from kindergarten to Grade 8, and statistics and probability appear from grades 6 to 8. We understand that the variety of nomenclatures in the U.S. document reflects not only a recommended approach but also a proper refinement of each stage. For example, numbers are addressed progressively considering three thematic areas: counting and cardinality, numbers and operations, and number systems. Such organization of the thematic areas is unique among the documents analyzed, therefore, a singularity. This lesson learned is: *Unique curricular terms and expressions indicate singularities in curriculum documents that may speak to the advancement and refinement of mathematics curricula.*

This second investigation also directed our attention to terms and expressions used to generally name the set of mathematics learning processes, such as *key ideas, mathematical processes, skills, mathematical practices, process standards,* and *goals*. These expressions did not necessarily seem to refer to the same things. However, when analyzing the specificities of each of them, we could find convergences. For example, although the expression *key ideas* and the term *goals* admit different interpretations, they are used in two different curricula with the same objective: to guide the processes of teaching and learning mathematics. *Key ideas* refers to problem-solving, reasoning, understanding, and fluency, whereas *goals* refers to problem-solving, mathematical reasoning, knowledge of facts and procedures, the vision of mathematics as a whole, and communication. The two terminologies involve the processes of problem-solving, reasoning, and fluency, therefore, a parity. Nevertheless, we also notice contrasts; for example, *goals* involve communication processes not specified in *key ideas*. The lesson here was: *Curricular terms and expressions that admit different interpretations may communicate the same intentions.*

Another schematic representation of this investigation is illustrated in Figure 3.5. It shows that, in addition to using different terms and expressions to name the set of mathematics learning processes (second row from the top), supplementary terms are used to describe each of the mathematics learning processes individually (left column). Some terms are widely used, but others may be specific and used uniquely in certain documents. For example, the expression *problem solving* appears unanimously in all seven analyzed documents, indicating a parity—such parity suggests the advancement of mathematics teaching in line with research developments in mathematics education. Other terms are also used relatively consistently in curriculum documents: *reasoning* used in six of the seven analyzed documents; and *communicating* used in five of the seven documents. Sometimes, the term used may not be exactly *reasoning* or *communicating*; however, the description of the associated term or expression indicates the same purpose. For example, the Portuguese curriculum describes one of its goals as *construction and development of mathematical reasoning* rather than indicating *reasoning* as one of its processes. This leads to another lesson: *Unanimity in the use of a specific expression, for example, "problem solving," may indicate a potential advance in mathematics teaching in line with mathematics education research.*

Figure 3.5

Schematic Representation—Terms and Expressions Used to Describe the Set of Mathematics Learning Processes and Each Learning Process Individually

	Australia	Brazil	Canada (ON)	Chile	USA (CCSSM)	Singapore	Portugal
	Key Ideas	Mathematical Processes	Mathematical Processes	Skills	Mathematical Practices	Mathematical Processes	Goals
Problem Solving	x	x	x	x	x	x	x
Reasoning	x		x	x	x	x	x
Communicating			x	x	x	x	x
Modeling		x		x	x	x	
Connecting		x				x	x
Representing		x		x			
Fluency	x						x
Selecting Tools and Strategies			x			x	
Understanding	x						
Reflecting			x				
Thinking Skills						x	
Investigation		x					
Structuring					x		
Patterning					x		
Precision					x		
Heuristics						x	
Project Development		x					

Other terms and expressions are used uniquely in the curriculum documents analyzed. This is the case of the terms *understanding*, *reflecting*, *thinking skills*, *investigation*, *structuring*, *patterning*, *precision*, *heuristics*, and the expression *project development*. Such terms and expressions might present commonalities with other terms and expressions more widely used in the documents. For example, *patterning* could be considered in *modeling*, but *modeling* comprises a broader range of processes than *patterning*. Similarly, *investigation* could be regarded as *reasoning*, but differences can be pointed out between such terms. The diversity of terms and expressions may indicate a detachment between the curricular documents; alternatively, this diversity may also indicate a refinement of mathematics teaching and learning processes. The lesson learned was: *Diversity in terminology may indicate a refinement of mathematics teaching and learning processes*.

FINAL CONSIDERATIONS

Our curriculum research originated from and is focused on mathematics teaching and learning processes, in particular, on mathematics knowledge for teaching (Ball et al., 2008). Motivated by curricular changes in Brazil and Canada, where we work, we decided to go beyond observing the documents in isolation. We chose to work collaboratively through a reflective process based on identifying convergences and divergences in the documents. As a result, the combined reading methodology was developed, a learning process that we present in this chapter highlighting lessons learned and challenges faced. The combined reading has become our standard way of interacting with curriculum documents. When we come across curriculum documents or guidelines, it becomes natural to reflect on what a combined reading would reveal about mathematics teaching. A combined reading can serve as a methodological guideline for other studies (e.g., Rangel et al., 2022), positively contributing to qualitative curricular research. We conclude with another lesson: *The importance of venturing into the qualitative research process*. We allowed ourselves to create a methodological approach that helped in the documental research, went beyond the documents, and provided reflections that reached the mathematics knowledge needed for teaching.

REFERENCES

Acar, F., & Serçe, F. (2021). A comparative study of secondary mathematics curricula of Turkey, Estonia, Canada, and Singapore. *Journal of Pedagogical Research*, 5(1), 216–242. https://doi.org/10.33902/JPR.2021167798

Australian Curriculum, Assessment and Reporting Authority [ACARA]. (2018). *The Australian curriculum: Mathematics* (Version 8.4).

Alberta Education. (2006). *The common curricular framework for K-9 mathematics*.

Ball, D., Thames, M. H., & Phelps, G. (2008). Content knowledge for teaching: What makes it special? *Journal of Teacher Education*, 59(5), 389–407.

Bessot, A., & Comiti C. (2006). Some comparative studies between French and Vietnamese curricula. In F. K. S. Leung, K. D. Graf, & F. J. Lopez-Real (Eds.), *Mathematics education in different cultural traditions—A comparative study of East Asia and the West*. New ICMI Study Series (Vol. 9, pp. 159–179). Springer. https://doi.org/10.1007/0-387-29723-5_10

Bickmore, K., Hayhoe, R., Manion, C., Mundy, K., & Read, R. (2017). *Comparative and international education issues for teachers* (Second edition). Canadian Scholars.

Brasil. (2018). *Base nacional comum curricular [National common curricular framework]*. Ministério da Educação.

Council of Chief State School Officers [CCSS]. (2022). *Standards in your state*. Retrieved from http://www.thecorestandards.org/standards-in-your-state/

Cerqueria, D. S., & Silva, M. N. da. (2020). Sistemas educacionais do Brasil, Chile e México: Análise dos currículos prescristos de Matemática [Educational systems in Brazil, Chile and Mexico: Analysis of prescribed mathematics curriculum]. *Ensino em Re-Vista*, 27(3), 1005–1028. https://doi.org/10.14393/er-v27n3a2020-10

Charalambous, C. Y., & Philippou, G. N. (2010). Teachers' concerns and efficacy beliefs about implementing a mathematics curriculum reform: Integrating two lines of inquiry. *Educational Studies in Mathematics*, *75*(1), 1–21. https://doi.org/10.1007/s10649-010-9238-5

Corrêa, P. D., & Rangel, L. (2021). The teaching of fractions – Emerging questions from the combined reading of Brazilian and Canadian curricular documents. *International Journal for Cross-Disciplinary Subjects in Education*, *12*(2), 4473–4483. https://doi.org/10.20533/ijcdse.2042.6364.2021.0547

Dias, A. L. B., & Gonçalves, H. J. L. (2017). Contribuições da educação comparada para investigações em currículos de matemática [Contributions of comparative education to research on mathematics curricula]. *Educação Matemática Pesquisa: Revista do Programa de Estudos Pós-Graduados em Educação Matemática*, *19*(3), 230–254. https://doi.org/10.23925/1983-3156.2017v19i3p230-254

Gobierno de Chile, Ministério de Educación. (2018). *Bases curriculares: Primero a sexto básico* [Curricular framework: First to sixth grade]. Unidad de Curriculum y Evaluación.

Li, Y., & Lappan, G. (Eds.). (2014). *Mathematics curriculum in school education*. Springer.

Lui, K. W., & Leung, F. K. S. (2013). Curriculum traditions in Berlin and Hong Kong: A comparative case study of the implemented mathematics curriculum. *ZDM Mathematics Education*, *45*(1), 35–46. https://doi.org/10.1007/s11858-012-0387-0

National Governors Association Center for Best Practices & Council of Chief State School Officers. [NGA & CCSSO]. (2010). *Common core state standards for school mathematics*. http://www.thecorestandards.org/Math

Ontario Ministry of Education. (2020). *The Ontario curriculum, grades 1–8: Mathematics*. Queen's Printer for Ontario.

Pires, C. M. C. (2013). Pesquisas comparativas sobre organização e desenvolvimento curricular na área de educação matemática, em países da América Latina [Comparative research on the organization and curriculum development in the field of mathematics education, in Latin American countries]. *Educação Matemática Pesquisa*, *15*(2), 513–542.

Rangel, L., Landim, F. M. P. F., Novaes, A. M., Baccar, M. H., & Leal, V. M. (2022). Orientações curriculares e letramento estatístico: Uma leitura combinada da BNCC e do GAISE [Curriculum guidelines and statistical literacy: A combined reading of BNCC and GAISE]. *Proceedings of the XIV Encontro Nacional de Educação Matemática*, Brazil.

República Portuguesa. (2013). *Programa e metas curriculares: Matemática, ensino básico* [Curricular program and goals: Mathematics, basic education]. Direção-Geral da Educação.

Singapore Ministry of Education. (2012). *Mathematics syllabus: Primary one to six*. Curriculum Planning and Development Division.

Shimizu, Y., & Vithal, R. (Eds.). (2018). *Proceedings of The Twenty-fourth ICMI Study School mathematics curriculum reforms: Challenges, changes and opportunities*. University of Tsukuba.

Son, J.-W., Watanabe, T., & Lo, J.-J. (2017). *What matters? Research trends in international comparative studies in mathematics education.* Springer. https://doi.org/10.1007/978-3-319-51187-0

Thompson, D. R., Huntley, M. A., & Suurtamm, C. (2018). *International perspectives on mathematics curriculum.* Information Age Publishing.

TIMSS & PIRLS. (2020). *TIMSS 2019 Encyclopedia: Countries' Chapters.* https://timssandpirls.bc.edu/timss2019/encyclopedia/index.html

United Nations Educational, Scientific and Cultural Organization [UNESCO]. (2016). *Glossário de terminologia curricular* [Glossary of curriculum terminology]. International Bureau of Education.

Valverde, G., Bianchi, L. J., Wolfe, R. G., Schmidt, W. H., & Houang, R. T. (2002). Textbooks and educational opportunity. In G. Valverde, L. J. Bianchi, R. G. Wolfe, W. H. Schmidt, & R. T. Houang (Eds.), *According to the book: Using TIMSS to investigate the translation of policy into practice through the world of textbooks* (pp. 1–20). Springer Science + Business Media.

Villalobos Torres, E. M., & Trejo Sánchez, C. M. (2015). Fundamentos teórico-metodológicos para la educación comparada [Theoretical-methodological foundations for comparative education]. In M. A. Navarro Leal & Z. Navarrete Cazales (Eds.), *Educación comparada internacional y nacional* [National and international comparative education] (pp. 19–27). Plaza y Valdes Editores.

Wang, Y., & Fan, L. (2021). Investigating students' perceptions concerning textbook use in mathematics: A comparative study of secondary schools between Shanghai and England. *Journal of Curriculum Studies, 53*(5), 675–691. https://doi.org/10.1080/00220272.2021.1941265

Wang, Z., & McDougall, D. (2019). Curriculum matters: What we teach and what students gain. *International Journal of Science and Mathematics Education, 17*(6), 1129–1149. https://doi.org/10.1007/s10763-018-9915-x

CHAPTER 4

CREATING SPACE FOR A TOPIC IN THE NULL CURRICULUM

Lynn M. McGarvey
University of Alberta, Canada

Jennifer S. Thom
University of Victoria, Canada

Josh Markle
University of Alberta, Canada

Mathematics curricula continue to become more standardized worldwide, which makes researching topics not appearing in the explicit curriculum challenging. In this chapter, we describe four lessons learned while conducting classroom-based research in Canada on a topic in the null curriculum. Despite its potential to support geometric and spatial reasoning, research in projective geometry with children was abandoned decades ago. Through our project, we learned multiple lessons while resurrecting, updating, inventing, implementing, and conducting classroom-based research within the null space of projective geometry. These lessons include: (1) locating relevant curriculum resources for a non-existent topic in the elementary-grades curriculum; (2) inventing ways to create, pilot, and use unfamiliar tasks in a classroom-based setting; (3) rationalizing the value of the topic to school district personnel, teachers, parents, and students; and (4) generating reciprocal growth in understanding amongst teachers and researchers. The insights gained offer ways to explore and justify the research, teaching, and learning of topics outside of the explicit curriculum.

Keywords: elementary grades; null curriculum; projective geometry; spatial reasoning

INTRODUCTION

With increased attention to international tests, the learning objectives in official mathematics curriculum documents continue to become globally standardized. In Canada, efforts to change or update mathematics curricula and pedagogical practices based on current research are often met by the media and public with skepticism and even hostility (Chorney et al., 2016; McFeetors & McGarvey, 2019). This is particularly true in the elementary grades, where the pressure to return to the "basics" arises frequently. This push to a traditional curriculum, with its emphasis on arithmetic skills and procedures, makes it challenging to explore or research areas of mathematics learning that have the potential to support children's learning, understanding, achievement, and success in mathematics, but are not explicitly covered by curricular objectives.

One such area that has received attention in recent years is spatial reasoning. Ongoing evidence suggests there is a strong link between spatial reasoning and success in mathematics (Frick, 2019; Mix et al., 2016; Xie et al., 2020). More importantly, intervention studies indicate that training spatial skills improves mathematics performance (Hawes et al., 2017; Lowrie et al., 2019). Research is still needed to understand how spatial reasoning mediates mathematical understanding and what spatial actions are most helpful. Yet, many aspects of spatial thinking, such as interpreting diagrams, constructing models, and object and mental manipulation and deconstruction, are often absent in the elementary mathematics curriculum, even though they are needed for nearly every mathematical action (Davis & Spatial Reasoning Study Group [SRSG], 2015). Instead, spatial concepts in the elementary grades tend to be addressed primarily within topics of geometry, which are traditionally oriented toward categorizing shapes and objects by their attributes, such as side lengths or measures of interior angles. Although visualization may be included as a process or competency in some curricula in Canada, the lack of associated learning outcomes often means that it does not receive explicit attention in the classroom.

Our research team sought to explore an alternative approach to studying spatial reasoning through a topic with defined concepts and the potential to elicit spatial actions. One such topic is *projective geometry*. Although projective geometry is an active area of research in contemporary mathematics, it does not appear as a named topic, content, or outcome in any elementary curricula in Canada. Based on a previous project (McGarvey et al., 2018), we noted commonalities between spatial skills that predict success in engineering education, and the dynamic spatial reasoning needed in projective geometry, including translation between 2D representations of 3D objects, rigid transformations through physical and mental rotation, and dynamic transformations of paperfolding and cross-cutting. Thus, our research team began exploring projective geometry with elementary school students. In this chapter, we speak to the challenges of engaging in classroom-based research for a topic that does not formally exist in the official or intended curriculum; that is, it exists in the null curriculum.

NULL CURRICULUM

In 1979, Eisner proposed three types of curricula. The *explicit curriculum* is that which is purposefully taught and appears in official documents and teaching resources, such as addition and subtraction facts. The *implicit* or *hidden curriculum* is unwritten and unofficial, but nonetheless learned by students, including implicit rules of turn-taking or the common belief that "math is hard." The third type is the *null curriculum*, which is neither taught nor learned. There is insufficient time to study every topic or subject, so what gets excluded? There is no definitive list of possible curriculum topics from which to choose or exclude. In mathematics education literature, the null curriculum is often associated with topics that typically do not receive any consideration, such as applying mathematics and statistics to sociocultural problems, inequities, and policies (e.g., Applebaum & Stathopoulou, 2016). Some exclusions are intentional decisions based on topics that are deemed immaterial, unnecessary, or obsolete, such as division with three-digit divisors and manually calculating square roots. In the case of projective geometry, its introduction as a coherent course of study often occurs at the postsecondary level with topics such as collineation, homogeneous coordinates, duality, cross-ratio, Desargues' theorem, and so on. Entering projective geometry at this level certainly appears developmentally inappropriate for elementary school children.

Yet, projective geometry can also be thought of as a simplified form of geometry. It emphasizes points and lines and ignores distances and angles. As a brief example, the shadow (i.e., projection) of a 2D triangle onto a flat surface will still look like a triangle with three straight sides regardless of how it is held (unless perpendicular to the surface). However, as the triangle is manipulated, the side lengths and vertex measures of the projected shadow will change. This activity is a dynamic way to investigate the attributes of geometric shapes from a projective perspective. Yet, elementary mathematics provincial curricula across Canada do not reference projective geometry or even the term *projective*. Our assumption, then, is that projective geometry for the elementary grades exists in the null curriculum as a topic that is unknown, not valued, or believed to be developmentally inappropriate by policymakers, educators, and mathematics experts.

ORIGINS OF PROJECTIVE GEOMETRY AND ITS PLACE IN ELEMENTARY CURRICULA

It was not until 600 years ago during the Renaissance period that artists and painters attempted to create more realistic images in their art using perspective. However, doing so required a different type of geometry, one in which some parallel lines meet at a point rather than remain equidistant apart as they are in Euclidean geometry. This was a radical shift in thinking. As a simple example, Figure 4.1 is a photograph (i.e., projection) of a 3D cube. We know in real life the cube has equal edge lengths, 90-degree angles at each vertex, and equivalent surface areas for its six faces. As photographed, these measurement attributes are not maintained in the 2D representation. In projective geometry, straight lines remain straight lines and points remain points, but lengths, angles, and areas are not preserved under projective transformation.

Figure 4.1

90-Degree Angles Not Preserved on the Projection of a Cube

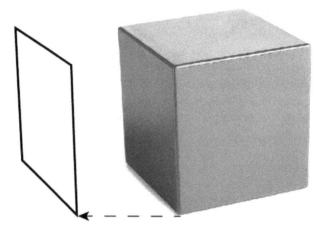

Perhaps one reason projective geometry exists within the null space has to do with the role of the body in knowing and doing mathematics. Ideas from embodied cognition, an interdisciplinary field that studies the complex interactions between brain and body, are prominent in contemporary mathematics education discourse. Moreover, embodied actions, such as gestures (Alibali, 2005; Kim et al., 2011), are acknowledged as critical aspects of classroom mathematics; yet historically, the body has been a contentious site of knowing and doing mathematics (Thom et al., 2015).

Projective geometry in the Renaissance was as much a practical tool, used to affect verisimilitude in painting, for example, as it was a philosophical approach. The art historian Mari Yoko Hara elaborated on this duality in her discussion of Brunelleschi's *Basilica of San Lorenzo*, in Florence, Italy (Hara, 2020). Brunelleschi physically inscribed lines on the floor, ceiling, and walls of the structure, all of which converge to a vanishing point behind a high altar at one end of the space. Similar to the perspective painting of Da Vinci and others, Hara noted that the effect "blurs the boundaries between mental space, the space of everyday experience, and the spiritual realm, representing the harmony of the cosmos as well as our place within it" (Hara, 2020, p. 180). Key to this view is the notion that perspective within projective geometry is somehow akin, as Hara (2020) noted, to our own everyday embodied experience.

This description of geometry will likely resonate with teachers of art, not mathematics. In fact, as we noted above, the explicit elementary school mathematics curricula across Canada emphasize Euclidean geometry through topics such as the categorization of shapes and objects by their measurable attributes. Instead, projective geometry eschews a focus on attributes, such as the careful measurement of angles and their comparison, to focus instead on the variant and invariant relationships between the most fundamental geometric objects: points, lines, and planes. Of course, this is not to say one of these fields of inquiry is superior to the other, nor are they mutually exclusive. In fact, as mathematics educators, we highly value the capacity to discern shapes and objects by their measurable attributes, and we emphasize the need to develop this capacity in mathematics classrooms. But we argue that projective geometry, with its focus on the dynamic relationships between objects in space, enlists a distinct set of spatial understandings and transformations.

Although projective geometry arose from the beginnings of perspective drawing, it continues to exist as an important field of study in mathematics with applications to other fields such as architecture, computer graphics, and robotics. Moreover, it offers an opportunity to make rich connections between mathematics and students' everyday embodied experiences. For these reasons, we were compelled to engage in the study of projective geometry with children, despite its absence from the traditional elementary school curriculum.

LESSONS LEARNED WORKING IN THE SPACE OF THE NULL CURRICULUM

The essence of our research was to identify a mathematical area of study that might explicitly promote spatial reasoning through embodied experiences in the elementary classroom—not only as a process of visualization, but as a topic of learnable concepts. Projective geometry offered the potential to reach that goal. Nearly 50 years ago, Lesh (1976) wrote:

> Few mathematical topics can compare with the simplicity, power, and elegant beauty of projective geometry, and few mathematical topics are so firmly rooted in concrete experience; yet, few laboratory activities have been developed to exploit the intuitive origins of projective geometry. (p. 202)

While reviewing decades old literature on projective geometry, we felt we had stumbled upon a forgotten site of spatial inquiry, a topic relegated to the null curriculum. The absence of research and resources led to lessons learned throughout the research process—from an initial search for meaning to its implementation in the classroom. The following four lessons outline our journey into the null curriculum.

Lesson 1: Locating Curriculum Resources for a Non-Existent Topic

Because the vast majority of resources in projective geometry are written for the postsecondary level, our initial challenge was to become familiar with the introductory concepts to see how they might make sense at an elementary school level. We began with our own learning of basic concepts, including projectivities of points and lines, the theorems of Desargues and Pappas, and duality and conics in projective spaces, by pulling out postsecondary textbooks to read, watching YouTube videos, and working through the mathematical concepts through self-study (see Figures 4.2a and b). Learning (and re-learning) the mathematics was both difficult and rewarding. Imagining how these ideas could be explored with children was not immediately apparent.

To explore the null space further, we conducted a systematic search of projective geometry literature relevant to elementary and secondary schooling. This resulted in a small body of professional and research literature stemming initially from Piaget and Inhelder's (1967) conjecture that children's understanding of geometry begins with topology, then projective geometry, and finally the Euclidean system. This theory spawned a smattering of research in the decades that followed, but little was written once the van Hiele model became the prominent driver for curriculum development of geometry in the 1980s and beyond. A direct search for projective geometry research and resources with elementary grades resulted in a limited number of hits.

Creating Space for a Topic in the Null Curriculum 63

Figure 4.2a

Self-Study of Projective Geometry

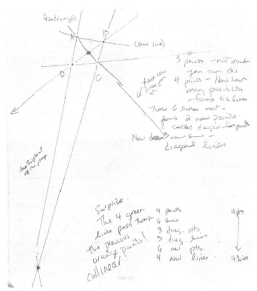

Figure 4.2b

Self-Study of Projective Geometry

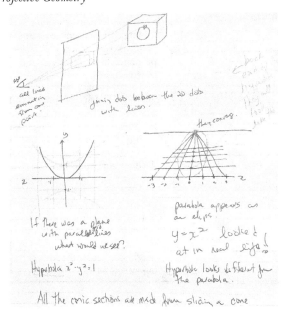

To expand our search, we generated keywords of ideas and activities associated with projective geometry to locate research and resources. Using an iterative approach, we continued to expand our search based on new topics and activities that appeared in the literature, and we attempted to map projective geometry topics in both research and professional literature that provided a basis for classroom activities. Table 4.1 provides examples of several projective geometry topics we thought were appropriate for the elementary grades.

Table 4.1

Projective Geometry Topics for the Elementary Grades

Activities	Visual	Description
Matching 2D Representations of 3D Objects		A taken-for-granted skill is the recognition and matching of 2D representations of 3D objects that may appear on worksheets, textbooks, posters, etc.
Shadow Images		A shadow is created on a plane when an object blocks a light source. The shape of the shadow depends on the distance and angle between the light source, the object, and the plane on which the shadow is projected.
Straight-Edge Constructions / Paper Folding		Because projective geometry is not concerned with measurement, it can be studied through straight-edge constructions, such as paper folding. For example, origami crease patterns (such as a crane, as shown) create perpendiculars, bisectors, and other geometric constructions.
Aerial Views / Mapping		Aerial views can be captured using photography, remote sensing, satellite imagery, or simply drawings. They are used to create maps and examine land use patterns, transportation networks, and urban development. This bird's-eye view provides a flat, 2D image of our 3D world.
Coded Plans		Coded plans are often used in engineering to represent features and dimensions of a 3D object on a 2D surface. In the example shown, the coded plan on the right records the various heights as well as the overall outline of the object on the left.

(Table continued on next page)

Table 4.1 (Continued)

Projective Geometry Topics for the Elementary Grades

Activities	Visual	Description
Computer Modeling		Computer modeling can be used to create virtual 2D representations prior to building an actual model. The computer model allows us to explore and test the model from multiple perspectives, and allows for changes more easily than concrete structures.
Orthogonal Drawings / Projection		Orthogonal drawings are used in engineering and architecture. An object is projected onto perpendicular planes to show what the object looks like from different angles, usually front, side, and top.
Perspective Drawings		Perspective drawings are used in art, architecture, and modeling to create realistic images of 3D objects onto a 2D surface. Unlike orthogonal drawings, a one-point perspective drawing (as shown) illustrates how parallel lines meet at a point at infinity along a horizon.
Cross Sectioning as Projection		Cross-sectioning as projection is to imagine a plane passing through an object and having the 2D image projected onto a plane

Many of these ideas were certainly familiar to us. Some of the projective geometry topics appear in curricula at the secondary level (e.g., cross-sectioning) or in other subjects (e.g., light and shadows in science; perspective drawing in art). Some of the topics appear in elementary curricula, but not necessarily with a projective geometry focus (e.g., mapping). We learned that *we needed to look at these familiar topics through a new lens. What was prominent across the topics was the need for dimension shifting between 3D objects and 2D representations*. Yet this overarching idea seems to be taken for granted in the early grades. Our overall process of locating resources through the iterative keyword search and topic mapping proved valuable in outlining possibilities for the curriculum development portion of our research.

Lesson 2: Creating Classroom Activities

Now that we had identified a set of possible topics to explore, our next challenge was to determine how to reinvent old and create new classroom activities that emphasized projective geometry concepts while allowing us to understand children's spatial thinking. We adapted a few activities from the literature, such as perspective drawings (Salisbury, 1983) and shadow activities (Mansfield, 1985), but some activities were quite dated. For example, completely missing from these older resources was any use of computer technology. Today, computer graphics and modeling are valuable tools for, and examples of, projective geometry; these tools simply were not readily available when projective geometry in the elementary years was explored in the 1960s to 1990s.

We also needed to take topics that were familiar to us from Table 4.1, but see them from a projective perspective. For example, two of the authors developed tasks in cross-sectioning. One of these was called Building Skyscrapers. In this task, we anticipated that students would build a structure on a 2 × 2 base using a limited number of cubes. Once built, they would imagine a cross section of their skyscraper (see Figure 4.3). This led to a debate amongst our research team over the different meanings of cross-sectioning, that is, the 2D shape resulting from the intersection of a plane with a solid. Our most familiar meaning was thinking about that plane as slicing through a solid, and the cross section was the 2D shape revealed by slicing. However, cross-sectioning from a projective perspective may also be thought of as the shape that is projected onto that intersecting plane (see cross-sectioning as projection in Table 4.1). Although the resulting shape is the same, the process is somewhat different.

Figure 4.3

Building and Cross-Sectioning a Skyscraper

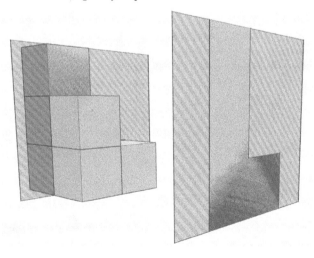

Throughout our two-year project, we created nine lessons for each of the four grades in the project—2, 3, 4, and 5.[1] The lessons included multiple activities that involved, for example, perspective drawing, orthogonal drawing, shadow tasks, 3D computer modeling, cross-sectioning, and points at infinity. *In the creation of classroom-based activities for the project, we needed to return to advanced mathematics to aid our understanding; be open to different interpretations of familiar mathematical topics; and engage in genuine explorations along with the willingness to be surprised.* Exploring within the null space requires the mathematical dispositions we ask of our own students—to engage in inquiry, take risks, be open to learning despite uncertainty, and share our learning with others.

Lesson 3: Convincing School Districts, Teachers, Students, and Parents of Our Research

Lesson 3 involved locating school districts and convincing principals to support the two-year classroom study. During the first year of sessions, we planned to work with Grade 2 and Grade 4 teachers and students who were interested in participating. The following year, we wanted to continue with a Grade 3 and a Grade 5 class, preferably in the same school(s).

Conducting classroom-based research on spatial reasoning from an embodied perspective necessitates the use of video to observe object manipulation, gestures, and body movement. However, many school districts in Canada now restrict the use of videos and photographs for research. This restriction was compounded by the timing of the data collection, which began in the fall of 2021 when many schools no longer allowed visitors due to COVID restrictions. Fortunately, we located school districts still allowing in-school research near the second author's location in British Columbia. Research applications to three school districts were submitted. Two districts granted permission and we decided to proceed with one of the districts. This decision was based on student demographics as well as having prior experience working with some of the school administrators and teachers.

Although we felt fortunate to receive school district support for the project, we still needed consent from principals, teachers, and parents or guardians. With concerns already raised due to the impact of COVID on children's learning, we had to be prepared to answer the most difficult question: Why teach mathematics concepts that are not in the provincially prescribed curriculum? The question was anything but trivial. It could not be answered simply or once and for all. Inherent aspects of the question included the importance of spatial reasoning, its relationship with projective geometry, and why, other than for the sake of novelty or enrichment, these topics should be emphasized in grades 2 through 5 mathematics curricula for elementary students.

This situation reminded us of a similar dilemma in which Davis (2001) explored the question, *Why teach mathematics to all students?* By posing three different yet related queries, Davis offered more critical and focused responses. Following Davis's lead, we chose to anticipate three variations of the question: one as a teacher-relevant question, one that was more rhetorical, and one taken from a hermeneutic stance (Davis, 2001; Gadamer, 1990), respectively:

- Why teach spatial reasoning and projective geometry to elementary students?
- Why would we teach spatial reasoning and projective geometry to elementary students?
- Why should we teach spatial reasoning and projective geometry to elementary students?

Many insights were gained as a result of exploring our question from the three perspectives. What we learned not only provided compelling responses for us, principals, teachers, students, and their parents, but also, important considerations for how we could proceed with our mathematics curriculum research.

Why Teach Spatial Reasoning and Projective Geometry to Elementary Students?

From a teacher's perspective, this question anticipates an explanation, justification, or rationale. Such *types* of responses are commonly found "in policy statements, textbooks, curriculum documents, and so on" (Davis, 2001, p. 18). However, unlike Davis's question, which concerned mathematics more generally, our more specific focus involved the absence of content from current mathematics curricula, meaning there would be no arguments validating or advocating for spatial reasoning and projective geometry in the elementary years. Although this was the case, the age-old argument that "knowledge of mathematics is necessary for every citizen of today's world. It is useful" (Davis, 2001, p. 18) certainly applied to our research and its importance in elementary grades.

Today, more than ever, spatial skills are essential for entry into and success in STEM (Science, Technology, Engineering, and Mathematics) jobs. Research findings show that developing these skills, especially in early elementary grades, not only leads to improved performance in mathematics, but also predicts future school and career success in STEM fields as well as in the arts (Verdine et al., 2014; Wai et al., 2009). What is more, although spatial skills underlie all STEM-related areas, it is mathematics and geometry, including projective geometry, that have the greatest potential for providing educational experiences in spatial reasoning (Sarama & Clements, 2009). Given the continued global demand for STEM skills, the identified need for curriculum research concerning spatial-geometric skills in the elementary grades becomes even more important.

Why Would We Teach Spatial Reasoning and Projective Geometry to Elementary Students?

From a rhetorical perspective, this question calls for a response that speaks to mainstream culture. In Western society (rightly, wrongly, or otherwise), mathematics is used to organize and structure everyday life, and thus, our experiences (Davis, 2001).

Consider the fundamental ways that projective geometry is used and is necessary for computer modeling, 3D printing, digital photography, perspective drawing, engineering designs, and other imaging applications. Connect this to earlier research by Piaget and Inhelder (1967) that shows children ages 7–12 years old engaging in projective geometry by representing objects from different viewpoints, including concepts involving perspective in their drawings. Now take these two points and add the fact that current studies of projective geometry with young children are virtually non-existent in the literature today. This makes our research especially relevant. The current and future economic demands for STEM workers (Government of Canada, 2018a, 2018b), as well as our previous research demonstrating that STEM subjects rely on dimension shifting between 3D models and 2D representations (McGarvey et al., 2018), suggest that projective geometry would be a useful topic for students to study.

Why Should We Teach Spatial Reasoning and Projective Geometry to Elementary Students?

From a hermeneutic perspective, asking this question calls for a more interpretative stance to consider how the two topics might "expand the sphere of the known" (Davis, 2001, p. 18). In other words, what does spatial reasoning and projective geometry offer elementary students that is more than or different from what is already prescribed in current mathematics curricula?

We find projective geometry to be particularly intriguing as a means to support spatial reasoning. Projective geometry arises from our interactive and visual experiences. This quality alone opens tremendous opportunities for elementary teachers and students to "exploit [its] intuitive origins" (Lesh, 1976, p. 202)—how it is that we perceive, make sense of, and represent the world, including ourselves in relationship with it (Thom et al., 2021). This action-oriented component is fundamental to projective geometry. Rather than examining geometric attributes from a static state, projective geometry requires dynamic transformations, such as rotating, sliding, enlarging, shrinking, shearing, or slanting (Birchfield, 1998) as part of examining what changes and what stays the same.

Lesson 4: Reciprocal Growth in Understanding of Null-Curriculum Content

There are many settings in which elementary children can and do engage in projective concepts and spatial actions. However, projective geometry as null curriculum content precludes teachers from exploring and developing their understanding of such concepts, which in turn means lost opportunities for student learning.

Our project involved four classroom teachers working in grades 2, 3, 4, and 5 over a two-year period. An essential part of the research was to provide teachers with experiences allowing them to learn projective ideas in the tasks we developed, as well as to understand their relationship to outcomes in the curriculum and to other relevant concepts and applications. The results of this process were not unidirectional. As we supported the teachers' learning, they too expanded our own understanding of ideas, not only within mathematics but across curricular areas.

To introduce the teachers to projective geometry ideas, we created a virtual tutorial in *Tinkercad* (see tinkercad.com). *Tinkercad* is an online design program for 3D modeling, basic coding, and electronic circuitry. We chose *Tinkercad* for its accessibility; it is a free web application that is user-friendly and well suited for use in schools. We were also impressed by how well the software animates projective concepts and spatial actions for a person as they play (i.e., tinker). Using *Tinkercad*, the teachers created 3D models on the 2D screen and used features of *Tinkercad* to examine the structure from multiple perspectives. For example, by building a simple rectangular prism, teachers could examine it from the front-side-top view as orthogonal projections (see Figure 4.4). They could also see how shifting from an orthogonal side view to a front corner view dynamically changes the edge lengths and vertex angles of the side face. By including lines that extend the parallel edges at the top and bottom, we can see that the parallel lines would meet at what is called a point at infinity in projective geometry (see Figure 4.5). Through the process of 3D modeling in a 2D environment, the teachers gained firsthand experience of projective principles.

Figure 4.4

Front-Side-Top View of a Rectangular Prism in Tinkercad

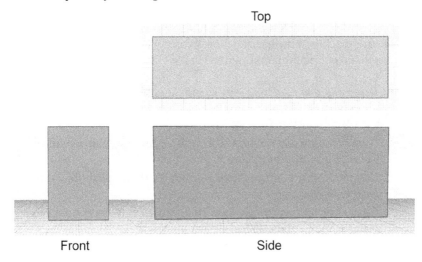

Figure 4.5

Examining Parallel Lines in Projective Geometry

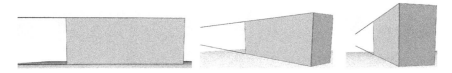

Concepts and spatial actions essential to making sense of the world and our relationships within it are often minimized in school mathematics curricula. Understanding how we come to perceive, imagine, and represent the world; move and live in it; as well as the ways we interact with it *is* important—not only in terms of mathematics but in a most basic everyday sense. Therefore, from a curricular perspective, projective geometry has tremendous potential to contribute to mathematics learning.

In the provincial curriculum where this project took place, the first-time students are expected to examine the side-front-top views of a solid is Grade 8, where it appears within the content related to surface area, nets, and volume (British Columbia Ministry of Education, 2016). Yet, by exploring orthogonal projections, students in elementary school can begin to generalize the attributes of 3D objects through 2D representations without having to focus on measurements. Our research enabled teachers to recognize how projective concepts and spatial actions directly relate to and can support elementary students' understanding of explicit curriculum content such as the "comparison of 2D shapes and 3D objects" (Grade 1), "multiple attributes of 2D shapes and 3D objects" (Grade 2), "construction of 3D objects" (Grade 3), and "classification of prisms and pyramids" (Grade 5). Further, the teachers emphasized how projective activities supported the four provincial curriculum competencies for Grades K–5:

- *Reasoning and analyzing*—by exploring and modeling mathematics experientially and using technology (e.g., building structures with cubes and in *Tinkercad*);
- *Understanding and solving*—through visualizing mathematical ideas (e.g., point at infinity);
- *Communicating and representing*—with mathematical objects in a variety of ways; and
- *Connecting and reflecting*—how mathematical ideas connect to other topics, subjects, and personal activities.

In our study, we gained new insights into how projective geometry activities fostered the fourth competency as students connected to other experiences and interests. For example, many students were motivated by perspective drawings and spontaneously engaged in drawing at other times during the school day (see Figures 4.6a and b).

Figure 4.6a

Students Spontaneously Drawing 2D Representations of 3D Objects

Creating Space for a Topic in the Null Curriculum 73

Figure 4.6b

Students Spontaneously Drawing 2D Representations of 3D Objects

Some weeks later, projective ideas were brought forth by students in one of their social studies lessons. The class was learning about the Maritime provinces. Looking at a photograph of downtown Newfoundland (Figure 4.7), they were asked to respond to the question, "What do you notice, what do you wonder?"

Figure 4.7

Photograph of Newfoundland (iStock.com/gnagel)

The children were especially curious as to why the homes in the photograph appeared 3D, but part of the large building in the top left looked flat. As the class discussion grew, Flynn[2] (a student) explained: "We can only see the front of the [building]. That's why it looks 2D. We can see the other sides of the smaller homes. That's why they look 3D." Their teacher was taken completely by surprise. She had no idea that the class would notice or choose to focus on projective concepts of perspective and dimension shifting. She also remarked, "I love how math and spatial reasoning found its way into our social studies lesson today."

Three months later, the children occasioned another instance of projective geometry. This time it was during a guided drawing lesson in art. The teacher provided step-by-step instructions towards creating an image that resembled Vincent van Gogh's *Sunflowers* paintings (1888–89). Their teacher described to us that she had done the activity for a number of years, but this time, the results were different. The children were instructed to draw a line roughly halfway down and across the outlined vase (horizontally) so that it "swoops down and up." In previous years, the teacher explained that most students had difficulty with this instruction. Students often drew a straight line across the vase, rather than a curved line. Although both straight and curved lines divide the vase, in the past, students did not recognize it is the curved line that creates a 3D effect. This year, however, every student, with the exception of one, drew a curved line on their vase (see Figures 4.8a and b). The students appeared to enjoy creating the 3D effect and drew many more curved lines than in previous years.

Figure 4.8a

Children's Drawings of Sunflowers in a Vase

Figure 4.8b

Children's Drawings of Sunflowers in a Vase

Together, the series of lessons and the ongoing teacher explorations before, during, and after them proved useful and enriching to the teachers as well as researchers. As one of the teachers expressed, the projective geometry "lessons this year ... [and] learning from our conversations in connection to spatial reasoning [apply] to the details in student artwork, and our 'notic[ings] and wonders' in other subject areas such as social studies." These connections to other areas were equally rewarding to us as researchers.

CLOSING REMARKS

As mathematics educators working with children in classrooms, we are often expected to adhere to topics in the official curriculum; yet, staying in that realm perpetuates a set of learning outcomes that does not necessarily serve our children well. Working in the void space of the null curriculum proved to be both challenging and rewarding. With few resources to draw on and without the confines of curriculum learning outcomes, we were free to be creative and exploratory, thus expanding our own learning of an unfamiliar mathematical topic. Our school partners agree that projective geometry as a topic was valid and worthwhile. As a whole, the concepts were outside of the elementary curriculum, and making space for them was initially challenging. However, this work proved highly valuable to the children's spatial thinking and contributed to topics outside of the explicit mathematics curriculum.

ACKNOWLEDGMENTS

This work was funded by a grant from the Social Sciences and Humanities Research Council of Canada.

REFERENCES

Alibali, M.W. (2005). Gesture in spatial cognition: Expressing, communicating, and thinking about spatial information. *Spatial Cognition and Computation, 5*(4), 307–331. https://doi.org/10.1207/s15427633scc0504_2

Applebaum, P., & Stathopoulou, C. (2016). Mathematics education as a matter of curriculum. In M. A. Peters (Ed.), *Encyclopedia of educational philosophy and theory* (pp. 1–6). Springer. https://doi.org/10.1007/978-981-287-532-7_515-1

Birchfield, S. (1998). *An introduction to projective geometry* (for computer vision). http://robotics.stanford.edu/~birch/projective/

British Columbia Ministry of Education. (2016). *BC's new curriculum: Mathematics K–12* [Program of Studies]. https://curriculum.gov.bc.ca/curriculum/mathematics

Chorney, S., Ng, O.-L., & Pimm, D. (2016). A tale of two more metaphors: Storylines about mathematics education in Canadian national media. *Canadian Journal of Science, Mathematics and Technology Education, 16*(4), 402–418. http://dx.doi.org/10.1080/14926156.2016.1235747

Davis, B. (2001). Why teach mathematics to all students? *For the Learning of Mathematics, 21*(1), 17–24.

Davis, B., & Spatial Reasoning Study Group [SRSG]. (2015). *Spatial reasoning in the early years: Principles, assertions, and speculations.* Routledge.

Eisner, E. (1979). *The educational imagination: On the design and evaluation of school programs.* MacMillan.

Frick, A. (2019). Spatial transformation abilities and their relation to later mathematics performance. *Psychological Research, 83*, 1465–1484. https://doi.org/10.1007/s00426-018-1008-5

Gadamer, H.-G. (1990). *Truth and method.* Continuum.

gnagel. (2018). St. John's, Newfoundland, Canada—July 25, 2018: Colorful buildings of historic downtown cityscape of St. John's from Signal Hill [Photograph]. iStock ID 1028996194.

Government of Canada. (2018a). *The Government of Canada and STEM.* Retrieved from the Government of Canada website: https://www.ic.gc.ca/eic/site/013.nsf/eng/00014.html

Government of Canada. (2018b). *What new ways of learning, particularly in higher education, will Canadians need to thrive in an evolving society and labour market?* Retrieved from the Social Sciences and Humanities Research Council website: https://www.sshrc-crsh.gc.ca/society-societe/community-communite/ifca-iac/01-new_ways_of_learning-nouvelles_methodes_d_apprentissage-eng.aspx

Hara, M. Y. (2020). Space, vision, and faith. In A. Janiak (Ed.), *Space: A history* (pp. 176–183). Oxford University Press.

Hawes, Z., Moss, J., Caswell, B., Naqvi, S., & MacKinnon, S. (2017). Enhancing children's spatial and numerical skills through a dynamic spatial approach to early geometry instruction: Effects of a 32-week intervention. *Cognition and Instruction, 35*(3), 236–264. https://doi.org/10.1080/07370008.2017.1323902

Kim, M., Roth, W.-M., & Thom, J. (2011). Children's gestures and the embodied knowledge of geometry. *International Journal of Science and Mathematics Education, 9*(1), 207–238.

Lesh, R. (1976). Transformational geometry in elementary school: Some research issues. In J. L. Martin & A. David (Eds.), *Space and geometry. Papers from a research workshop* (pp. 185–243). Ohio State University.

Lowrie, T., Logan, T., & Hegarty, M. (2019). The influence of spatial visualization training on students' spatial reasoning and mathematics performance. *Journal of Cognition and Development, 20*, 729–751. https://doi.org/10.1080/15248372.2019.1653298

Mansfield, H. (1985). Projective geometry in the elementary school. *The Arithmetic Teacher, 32*(7), 15–19.

McFeetors, P. J., & McGarvey, L. M. (2019). Public perceptions of the basic skills crisis. *Canadian Journal of Science, Mathematics and Technology Education, 19*, 21–34. https://doi.org/10.1007/s42330-018-0016-1

McGarvey, L., Luo, L., Hawes, Z., & Spatial Reasoning Study Group. (2018). Spatial skills framework for young engineers. In L. English & T. Moore (Eds.), *Early engineering learning* (pp. 53–81). Springer.

Mix, K. S., Levine, S. C., Cheng, Y. L., Young, C., Hambrick, D. Z., Ping, R., & Konstantopoulos, S. (2016). Separate but correlated: the latent structure of space and mathematics across development. *Journal of Experimental Psychology General, 145*(9), 1206–1227. https://doi.org/10.1037/xge0000182

Piaget, J., & Inhelder, B. (1967). *The child's conception of space.* Routledge & Kegan Paul.

Salisbury, A. J. (1983). *Projective geometry in the primary school curriculum: Children's spatial-perceptual abilities* [Unpublished Doctoral dissertation, University of London].

Sarama, J., & Clements, D. H. (2009). *Early childhood mathematics education research: Learning trajectories for young children.* Routledge.

Thom, J. S., D'Amour, L., Preciado, P., & Davis, B. (2015). Spatial knowing, doing, and being. In B. Davis & Spatial Reasoning Study Group (Eds.), *Spatial reasoning in the early years* (pp. 63–82). Routledge.

Thom, J. S., McGarvey, L. M., & Lineham, N. D. (2021). Perspective taking: Spatial reasoning and projective geometry in the early years. In Y. H. Leong, B. Kaur, B. H. Choy, J. B. W. Yeo, & S. L. Chin (Eds.), *Excellence in mathematics education: Foundations and pathways (Proceedings of the 43rd Annual Conference of the Mathematics Education Research Group of Australasia),* (pp. 385–392). Mathematics Education Research Group of Australasia.

Verdine, B. N., Golinkoff, R. M., Hirsh-Pasek, K., & Newcombe, N. S. (2014). Finding the missing piece: Blocks, puzzles, and shapes fuel school readiness. *Trends in Neuroscience and Education, 3*(1), 7–13. https://doi.org/10.1016/j.tine.2014.02.005

Wai, J., Lubinski, D., & Benbow, C. P. (2009). Spatial ability for STEM domains: Aligning over 50 years of cumulative psychological knowledge solidifies its importance. *Journal of Educational Psychology, 101*(4), 817–835. https://doi.org/10.1037/a0016127

Xie, F., Zhang, L., & Chen, X. (2020). Is spatial ability related to mathematical ability: A meta-analysis. *Educational Psychology Review, 32*(1), 113–155. https://doi.org/10.1007/s10648-019-09496-y

ENDNOTES

1. Students in these grades would range from ages 6–11.
2. Flynn is a pseudonym.

CHAPTER 5

ANALYZING THE OFFICIAL AND INTENDED CURRICULA

Methods and Techniques to Study Mathematics Standards and Textbooks

Shannon Dingman
University of Arkansas

Dawn Teuscher
Brigham Young University

Lisa Kasmer
Grand Valley State University

Travis Olson
University of Nevada-Las Vegas

In this chapter, we reflect on our collective experiences researching mathematics standards and textbooks to provide advice stemming from the lessons we learned. We begin with lessons learned from researching mathematics standards, including how methods and data analysis techniques can be selected to facilitate analyses of standards. These comparative methodologies and data analysis techniques provide a number of approaches curriculum researchers can use depending upon their desired focus. We then detail lessons from our study of mathematics textbooks through the lens of standards documents, including how textbooks as well as specific content areas can be selected, and how data from the standards and textbooks were analyzed. Our goal is to provide insight to future generations of curriculum researchers, who will continue to advance research in these important areas.

Keywords: curriculum; standards; textbooks

Historically, mathematics textbooks have held a prominent role in the U.S. in aiding teachers' curricular decisions and therefore influencing the content that students have an opportunity to learn. Research has demonstrated how textbooks in the U.S. have traditionally served as a "de facto curriculum" (Finn et al., 2004), providing the basis for a majority of homework and classroom activities in many schools. Although mathematics textbooks hold a ubiquitous role in mathematics teaching and learning (Tyson-Bernstein & Woodward, 1991), researchers have also documented deficiencies in textbooks, including repetitive material across grade levels, the lack of coherency across topics, and the unclear impact textbooks have on increased student learning (Flanders, 1987; McKnight et al., 1987; Schmidt et al., 1997).

Over the past 40 years, the role of curriculum standards in the U.S. in determining what is taught in the mathematics classroom has greatly increased due to accountability measures associated with learning standards and related mandated assessments. Throughout the 1980s and 1990s, curriculum standards were used in many states to specify what students should know and be able to do. This movement culminated in the 2001 passage of the No Child Left Behind (NCLB) Act, which mandated content standards and annual assessments be developed in all states for grades 3–8 in reading and mathematics (Linn et al., 2002). This legislation propelled the writing or revising of mathematics standards for all grade levels across the U.S., with varying degrees of quality and agreement (Reys, 2006; Smith, 2011). To provide greater coherence and clarity to the mathematics curriculum across the U.S., the National Governors Association and the Council of Chief State School Officers (NGA & CCSSO, 2010) established the *Common Core State Standards Initiative* (*CCSSI*), which brought together stakeholders from across the U.S. to create standards that offered a shared vision of what students should know and be able to do at each grade level. The result was the *Common Core State Standards for Mathematics* (*CCSSM*), which was subsequently adopted within a few years of its 2010 release by 45 states as well as the District of Columbia, the U.S. Virgin Islands, Guam, and the Northern Mariana Islands. *CCSSM* has further cemented the place of mathematics standards in guiding and informing mathematics curriculum in the U.S.

The significance of mathematics standards and mathematics textbooks in determining what students have an opportunity to learn necessitates the importance of creating and using methodologies to study these types of curricula. Subsequently, the close relationship between standards and textbooks provided our research team a number of avenues to study the contents of standards and textbooks, ranging from examinations of the level of agreement of standards across the U.S. (Reys, 2006; Reys et al., 2007; Smith, 2011), as well as the alignment between mathematics textbooks and state standards prior to the *CCSSM* (Dingman, 2010), to studies analyzing standards and textbooks through the lens of *CCSSM* (Dingman et al., 2022; Teuscher et al., 2016). Along the trajectory of this research agenda, our experiences have informed and spurred our subsequent areas of study, including the guiding of methodologies and data analysis techniques we employed and the questions that have initiated broader study of how mathematics standards and textbooks guide teachers' curricular decision making. In this chapter, we describe some of the lessons learned in studying mathematics standards and textbooks over the past several decades, offering insight into the challenges faced and suggestions for other researchers pursuing curriculum research.

LESSONS LEARNED IN RESEARCHING MATHEMATICS STANDARDS[1]

Our research team has conducted numerous studies examining curriculum standards since the 2001 passage of NCLB. These studies have focused on many issues concerning standards, including:

- The variance of grade-level learning expectations (GLEs) with respect to grades at which certain mathematics content is specified in different state standards prior to *CCSSM* (Reys, 2006; Smith, 2011);
- The focus and emphasis on calculators in grades K–8 mathematics across state standards (Chval et al., 2006);
- The lack of consensus in fourth-grade mathematics standards across the U.S. (Reys et al., 2007);
- The comparison of *CCSSM* to state standards used prior to the release of *CCSSM* (Dingman et al., 2013; Teuscher et al., 2014; Teuscher et al., 2015; Tran et al., 2016); and
- The comparison of *CCSSM* to standards from states that did not adopt *CCSSM* (Dingman et al., 2022).

Throughout our study of mathematics standards, we have faced methodological issues involving research design and implementation, which impacted our ability to describe the phenomena under investigation. We now discuss some of the lessons we learned along our research journey that may assist other researchers in their study of mathematics standards and textbooks.

Establish a Common Grain Size for Standards

Mathematics standards are written in varying levels of detail, ranging from simplistic descriptions of one idea to more complex specifications of multiple mathematics concepts. For example, in our analysis of probability standards in pre-*CCSSM* state standards (Smith, 2011), the following Grade 7 standards are examples of the varying levels of detail provided in standards that address the same content:

- Conducts an experiment or simulation with a compound event composed of two independent events including the use of concrete objects; records the results in a chart, table, or graph; and uses the results to draw conclusions and make predictions about future events. (Kansas State Department of Education, 2003)
- Design and conduct an experiment to test predictions. (New York State Education Department, 2005)

To allow effective comparisons across these different formats, we identified the main focus of a standard and developed a "generalized learning expectation" that captured a succinct description of what the standard depicted (Tran et al., 2016). For example, the *CCSSM* standard 6.EE.8 specifies that students

Write an inequality of the form $x > c$ or $x < c$ to represent a constraint or condition in a real-world or mathematical problem. Recognize that inequalities of the form $x > c$ or $x < c$ have infinitely many solutions; represent solutions of such inequalities on number line diagrams. (NGA & CCSSO, 2010)

To target the multiple concepts articulated in this standard, we broke it into smaller, generalized learning expectations that captured individual concepts:

- Write an inequality of the form $x > c$ or $x < c$ to represent a constraint or condition in a real-world or mathematical problem;
- Recognize that inequalities of the form $x > c$ or $x < c$ have infinitely many solutions;
- Represent solutions of inequalities ($x > c$ or $x < c$) on number line diagrams. (Dingman et al., 2022, p. 23)

Utilizing a smaller grain size in these generalized learning expectations allowed for easier comparison across multiple state standards to compare content and examine similarities and differences across standards documents.

Target Research to Specific Mathematical Topics in Small Grade Bands

Our initial work analyzing state standards focused on the mathematical content and processes outlined in grades K–8. We considered extending our work into the secondary grades to capture the similarities and differences in mathematical expectations at the grades 9–12 level. However, the variability with respect to the organization of secondary grades standards made this effort challenging, as reported in Teuscher et al. (2008). Secondary grades state standards may have grade-band or grade-level standards, which stipulate what students should know and be able to do across a set of grades (e.g., grades 9–10 and 11–12, grades 9–11, grades 9–12) or at specific grade levels. Or, secondary standards may be course-based, which are learning expectations organized into specific courses (Algebra I, Geometry, Integrated 1). Adding to this complexity is the array of course expectations across states, ranging from the traditional high-school course sequence of Algebra I, Geometry, Algebra II to standards for Integrated Mathematics I, II, and III courses. Overall, Teuscher et al. (2008) reported that 25 states articulated grade-band standards for grades 9–12 and 19 states offered course-based standards for secondary mathematics. Additionally, at that time, one state (Louisiana) provided both grade-level standards as well as course-based standards.

In contrast to the varying methods of organizing grades 9–12 standards, across states there was not much variation in the organization of grades K–8 standards, with 43 states specifying grade-level specific standards (Reys, 2006). Additionally, given the wide array of mathematical content taught in grades 9–12, we chose instead to target our efforts on documenting how and when specific mathematical concepts were presented in state standards. For example, arithmetic operations with fractions is a common topic taught in elementary and middle-grades mathematics. We therefore chose to examine the variability in grade level at which students were expected to be proficient with a topic, such as adding and subtracting fractions with unlike denominators. We found in state standards across the U.S. that this concept was expected to be taught to students anywhere from Grade 4 to Grade 7 (Reys, 2006).

Compare Standards of the Same Type

After the 2001 passage of the NCLB states set off on an ambitious effort to create or revise mathematics standards documents. However, these efforts often were led by policymakers within individual states with little collaborative effort across states (Reys et al., 2005). As a result, state mathematics standards varied with respect to the organization and level of specificity.

For example, some states set *content standards* that describe the general mathematical content that was to be taught. These types of standards were generally not specific to a grade level, but rather, were used to organize more detailed learning expectations. For example, prior to the adoption of *CCSSM*, Delaware was one of eight states that specified content standards by grade bands, organizing their standards into grades K–3, 4–5, 6–8, and 9–10 (Reys, 2006). This provided flexibility for local districts to organize standards based on local input. Many of the states that used content standards also stipulated *grade-level learning expectations* (GLEs) that provided further detail on what a student should learn at a specific grade level. Some states outlined *performance/assessment standards* that articulated the mathematical content and processes that would be the focus of state-mandated assessments, which were an accountability component of NCLB. Typically, assessment standards were a subset of content standards or GLEs, yet in several cases, assessment standards only served to guide teachers' instruction.

With the variety of ways to organize standards for mathematical learning, it is important to compare similar sets of standards when analyzing standards. In our work, we most often compared GLEs across states as the most specific forms of expectations for what is to be taught at each grade level. By analyzing standards with similar goals and purposes, we were able to provide a descriptive analysis of the similarities and differences across standards documents.

Develop Methods for Ensuring Reliability

Standards are statements of goals and objectives, often written by policymakers and implemented by teachers. At their core, these statements can be interpreted differently by teachers, resulting in variability in their implementation. Additionally, researchers can interpret these statements differently, thus requiring effective methods to ensure consensus as to the meaning of standards and reliability in a study's outcomes.

As we participated in various standards analyses, the issue of reliability, or the degree to which our coding was consistent across different researchers, became a critical piece of our work. To ensure reliability, our team would collaboratively develop a series of codes to highlight the critical pieces we wanted to study. For example, for our analysis of fraction concepts and computation (Reys, 2006), we developed codes to capture the focus of the standards as conceptual development and/or computational fluency (see Table 5.1). Once this list of codes was complete, together we then coded a subset of standards to test whether our list of codes accurately captured what we desired. If codes were missing, we would revise our list and then recode those standards. After we were comfortable with the coding scheme, we each individually coded the standards included in our study and compared our results. Where differences existed, we then discussed the differing codes and came to an agreement with respect to the consensus codes for that standard. By utilizing this team approach to coding, we were able to reach reliable and accurate findings with respect to the standards within our coding schemes.

Table 5.1

Coding Categories and Example Grade-Level Learning Expectations (GLEs) for Fraction Computation and Concepts

	Coding Categories	Example Learning Expectation
Fraction Computation	*Add Fractions*	Add or subtract fractions with like denominators (halves, thirds, fourths, eighths, and tenths) appropriate to grade level. (AZ, gr.3)
	Subtract Fractions	Using concrete materials or pictures, add and subtract halves, thirds, and fourths. (CO, gr.2)
	Multiply Fractions	Multiply and divide fractions and mixed numbers. (GA, gr. 6)
	Divide Fractions	Divide fractions and mixed numbers. (MD, gr. 7)
	Effect of Operations on Fractions	Describe the effects of addition and subtraction on fractions and decimals. (MO, gr. 6)
	Estimate with Fractions	Use strategies to estimate computations involving fractions and decimals (NM, gr. 4)
Fraction Concepts	*Meaning of Fraction*	Use words, numerals, and physical models to show an understanding of fractions and their relationship to a whole. (DoDEA, gr. 2)
	Model/Represent Fractions	Identify alternative representations of fractions and decimals involving tenths, fourths, halves, and hundredths. (SD, gr. 5)
	Judge Size of Fractions	Compare and order whole numbers, fractions (rational numbers), and decimals and find their approximate locations on a number line. (UT, gr. 7)
	Equivalence	Apply factors and multiples to express fractions in lowest terms and identify fraction equivalents. (OR, gr. 6)

Source: Reys (2006, p. 43). Reprinted with permission of Information Age Publishing

Find Novel Methods to Illustrate Findings

After the coding and analysis of standards was complete, it was time to convey the findings of the analysis. Although some findings could be disseminated with descriptions of the standards, the large number of standards documents necessitated pictorial displays of our results. This precipitated the creation of novel graphs that displayed how various state standards compared with respect to a particular idea.

For example, in studying the grade placement of standards related to the addition and subtraction of fractions across 42 state standards documents (Reys, 2006), we discovered large variation across states regarding initial coverage of this concept, when proficiency was expected, and the developmental trajectory of this topic in grades K–8. The challenge was how to portray the large amount of data in a display that could allow a reader to digest conclusions from our analysis. In the end, we used Microsoft Excel to develop a graphical display, as shown in Figure 5.1, that highlights initial learning expectations for each state for this topic (depicted as a triangle), additional standards that further develop this idea (depicted as a circle), and the grade level at which students were expected to add and subtract fractions with unlike denominators fluently (depicted with a dark square). Additionally, in some states, teachers were to continue revisiting the topic after fluency was expected, which we depicted with a lighter square. These displays were used extensively across the study of various topics (Reys, 2006; Smith, 2011) and allowed us to display our data in powerful visual representations.

RESEARCH ON MATHEMATICS TEXTBOOKS

In addition to our work analyzing mathematics standards, our research team has also studied mathematics textbooks in relation to standards. These studies have centered on several issues pertaining to textbooks, including:

- The extent to which mathematics textbooks are aligned with state standards for the concept of fractions (Dingman, 2010);
- How middle-grades mathematics textbooks differ with respect to the learning trajectories offered for mathematics content (Olson, 2010); and
- How middle-grades mathematics textbooks provide coverage to new topics in *CCSSM* geometry that were moved from the secondary level in past standards (Teuscher et al., 2016).

Figure 5.1

Grade-Level Progression of Initial to Culminating Learning Expectations for Addition and Subtraction of Fractions Across 42 States

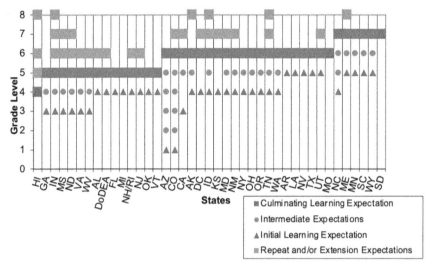

Source: Reys (2006, p. 43). Reprinted with permission of Information Age Publishing.

Similar to our work with mathematics standards, our research with mathematics textbooks has forced us to confront a number of issues related to sample selection, design, and data analysis. Below are some of the critical lessons learned throughout our experiences.

Allow Standards to Drive the Study of Textbooks

Mathematics standards and textbooks are closely intertwined; both provide guidance to teachers regarding the mathematical content students are expected to learn. As standards provide the grade-to-grade explanations of what students should know and be able to do as a result of their learning experiences, mathematics textbooks generally drive the day-to-day development of that knowledge. Thus, because standards greatly influence the topics and learning experiences found in textbooks, so too should research on standards provide a lens through which textbooks can be investigated.

For example, as previously discussed, Reys (2006) documented the considerable variation across state standards with respect to the grade placement of topics, including concepts related to fractions (e.g., computation with fractions, equivalence of fractions, comparing and ordering fractions). Using this standards analysis as a guide, Dingman (2010) examined the alignment of state standards related to fractions from ten populous states with two elementary (K–5) and two middle-grades (6–8) textbook series. These states differed with respect to the grade placement of topics related to fractions. The study's finding suggested that, although a large percentage of standards regarding fractions were aligned to content found in mathematics textbooks, each textbook series had considerable "extra" material not aligned to each state's standards. This study provided evidence to support the hypothesis that mathematics textbooks are often large and unfocused, overloaded with a wide array of mathematical content to provide coverage of standards across multiple states.

Establish a Common Unit of Analysis for Textbooks

As with our work with standards, determining a common unit of analysis when studying textbooks is an essential element of research on textbooks. The differences in how textbooks are organized necessitates using a method that compares the content of textbooks and documents similarities and differences in approaches for specific mathematical topics.

Dingman (2010) used *instructional segments* to provide a standard unit of analysis when examining the contents of elementary and middle-grades mathematics textbooks. This unit of analysis designated a short portion of material in a textbook that offered coverage for a specific mathematical topic. Each instructional segment was classified as one of five distinct forms: a lesson, a pre-lesson, an end-of-lesson extra feature, an end-of-chapter feature, or a game. The portions of each textbook in the sample pertaining to fraction concepts and computation were separated into instructional segments and then coded for both the type of instructional segment and for the generalized learning expectation(s) aligned to that instructional segment. This allowed for close comparisons across textbooks that varied in their organization and in their approach to mathematics teaching and learning.

Teuscher et al. (2016) used a *textbook instance* as the unit of analysis in a study of isometry and congruence in eighth-grade mathematics textbooks. A textbook instance was defined as "the way in which textbook content is delimited or sectioned by authors for the purpose of communicating ideas or providing students opportunities to engage with the mathematics content" (p. 4). The textbook instances were determined by the type of text seen in the textbook: author text, author example, student problem, and student super problem (i.e., a student problem with more than one part or question). This unit of analysis provided a basis for coding six textbooks to compare the isometry and congruence content across the different books.

Target Certain Topics

Our findings from research on state mathematics standards suggested that grade placement and approach to content and mathematical strands had changed over various iterations of state standards. The changes to approach and emphasis of these topics spurred natural questions as to how textbook series had captured these changes and subsequently changed learning opportunities for students.

For example, our comparison of *CCSSM* with pre-*CCSSM* standards in the middle grades (6–8) (Teuscher et al., 2015) illustrated that topics in strands of geometry and probability and statistics were moved from the secondary curriculum in pre-*CCSSM* standards to the middle-grades level in *CCSSM*. Additionally, authors of *CCSSM* introduced the idea of congruence and similarity through the concept of geometric transformations in the middle grades. The use of geometric transformations as a unifying topic in the middle-grades mathematics curriculum was a dramatic change from past approaches to introducing congruence and similarity through static approaches with triangle congruence and similarity properties (i.e., Side-Angle-Side congruence) (Seago et al., 2013). Given these shifts in the mathematics curriculum, we chose to study approaches to geometric transformations in popular middle-grades mathematics textbook series (Teuscher et al., 2016), documenting the differences and deficiencies in these materials related to properties of isometries (i.e., reflections, rotations, translations), congruence in terms of isometries, and the orientation of geometric figures. By targeting the approach to geometric transformations, we were able to highlight important issues that would impact teachers' implementation of *CCSSM*.

Create Samples That Reflect the Diversity of Textbooks

In analyzing mathematics textbooks, it is important that the textbook samples reflect the diversity of approaches used to produce textbook series. Dingman (2010) analyzed the topics related to fraction concepts and computation found in elementary (K–6) and middle-grades (6–8) mathematics textbooks. Using market share data (Education Market Research, 2005), two elementary series were selected—*Scott Foresman-Addison Wesley Mathematics* (Charles et al., 2004) and *Everyday Mathematics* (Bell et al., 2004)—that comprised approximately 30% of the U.S. market share, and two middle-grades series—Glencoe's *Mathematics: Applications and Concepts* (Collins et al., 2001) and the *Connected Mathematics Project* (*CMP*) (Lappan et al., 2002)—that held approximately 20% of the U.S. middle-grades market share. The same selection in each grade band was purposeful. It included one publisher-generated series, that is, textbooks that are often produced by teams of writers for publishing companies to meet state deadlines and purchasing guidelines. It also included one National Science Foundation (NSF)-supported series; these are textbooks that reflected the vision of the *Curriculum and Evaluation Standards for School Mathematics* (National Council of Teachers of Mathematics [NCTM], 1989) and typically created at universities with research on the textbooks' effectiveness. Publisher-generated series have traditionally held large shares of the textbook market yet have been criticized by researchers as large and unfocused, with imperfect alignment to any state's standards (Tyson-Bernstein & Woodward, 1991). NSF-supported textbook series were authored by teams of researchers to build from existing research on mathematics teaching and learning and were revised after field-testing before their eventual publication. The selection of this textbook sample allowed Dingman (2010) to look for differences across textbook series (publisher-generated versus NSF-supported) with respect to the alignment between the textbooks and state standards.

Teuscher et al. (2016) included six textbooks in their study of geometric transformations in popular middle-grades textbooks. In addition to including both publisher-generated series and NSF-supported series, the research team also included open-source textbooks (textbooks provided freely to schools and accessible from online sites); such selection enabled them to examine the similarities and differences across different textbooks with respect to the approach and connections between geometric transformations and congruence in Grade 8. The textbook sample included *Big Ideas Math* (Larson & Boswell, 2014), *Glencoe Math* (Carter et al., 2013), *Go Math* (Burger et al., 2014), *Math in Focus* (Cavendish, 2011), *Connected Mathematics Project* (Lappan et al., 2013), and the open-source *Utah Middle School Mathematics* (Rossi et al., 2013). Teuscher et al. (2016) found that these series varied on basic definitions of congruence and orientation and were lacking in the discussion of properties of isometries, congruence in terms of isometries, and in the inclusion of orientation of figures.

CONCLUSION

Research on mathematics standards and textbooks has assumed greater prominence as their role in mathematics education has grown (Ball & Cohen, 1996; National Research Council, 2002; Reys, 2006). Research on mathematics standards has served to highlight the incoherence and variation of pre-*CCSSM* state standards (Reys, 2006; Smith, 2011) that laid the foundation for the widespread adoption of *CCSSM*. Once *CCSSM* was widely adopted throughout the U.S., research on mathematics standards helped to underscore areas where mathematics teachers may need additional support due to the shift in standards from previous state standards, for example, in situations where content is now introduced at earlier grade levels (Teuscher et al., 2014). Additionally, research findings on mathematics textbooks have illustrated areas in which prominent textbooks held potential deficiencies that challenge teachers in teaching the new standards (Teuscher et al., 2016).

The important position of standards and textbooks in assisting teachers to enact curriculum is well-established (Reys, 2006; Smith, 2011; Stein et al., 2007). Despite the controversies that surrounded the implementation of *CCSSM* (Kanbir, 2016), as well as the multitude of open-source materials of questionable quality that are currently available to teachers, standards and textbooks hold considerable influence on what is taught by teachers and what students have an opportunity to learn. It is these two critical forms of curricula that assist teachers in creating trajectories for the enacted curriculum that plays out in mathematics classrooms across the United States.

Given their central place in mathematics education, it is important that mathematics standards and textbooks are constructed on solid, research-based foundations. Research on mathematics standards can serve to inform future iterations and revisions of current standards, leading to continued strengthening and enhancement of the official curriculum. Likewise, research on mathematics textbooks can act to support the creation of high-quality materials that engage students through improved forms of the intended curriculum. Taken together, research on mathematics standards and textbooks can support mathematics teachers in their quest to provide meaningful learning experiences and thus support stronger mathematical knowledge for their students.

REFERENCES

Ball, D. L., & Cohen, D. K. (1996). Reform by the book: What is—or might be—the role of curriculum materials in teacher learning and instructional reform? *Educational Researcher, 25*(9), 6–8, 14. https://doi.org/10.3102/0013189X025009006

Bell, M., Bell, J., Freedman, D., Goddsell, N. G., Hanvey, N., & Morrison, K. (2004). *Everyday mathematics* (2nd ed.). McGraw Hill-Wright Group.

Burger, E. B., Dixon, J. K., Kanold, T. D., Larson, M. R., Leinwand, S. J., & Sandoval-Martinez, M. E. (2014). *Go math: Middle school accelerated grade 7*. Houghton Mifflin Harcourt.

Carter, J. A., Cuevas, G. J., Malloy, C., & Day, R. (2013). *Glencoe math: Your common core edition*. McGraw-Hill.

Cavendish, M. (2011). *Math in focus: Singapore math*. Houghton Mifflin Harcourt.

Charles, R. I., Crown, W., Fennell, F., Caldwell, J. H., Cavanagh, M., & Chancellor, D. (2004). *Scott Foresman-Addison Wesley mathematics*. Pearson-Scott Foresman.

Chval, K., Reys, B., & Teuscher, D. (2006). What is the focus and emphasis on calculators in state-level K-8 mathematics curriculum standards documents? *NCSM Journal of Mathematics Education Leadership, 9*(1), 3–13.

Collins, W., Dritsas, L., Frey, P., Howard, A. C., McClain, K., Molina, D., Moore-Harris, B., Ott, J., Pelfrey, R. S., Price, J., Smith, B., & Wilson, P. S. (2001). *Mathematics: Applications and connections*. Glencoe McGraw-Hill.

Dingman, S. (2010). Curriculum alignment in an era of standards and high-stakes testing. In B. J. Reys, R. E. Reys, & R. Rubenstein (Eds.), *The K-12 mathematics curriculum: Issues, trends, and future directions (NCTM's 72nd yearbook)* (pp. 103–114). National Council of Teachers of Mathematics.

Dingman, S. W., Namakshi, N., Homem, S., & Pullin, H. (2022). What is common in middle grade mathematics? An examination of common core and non-common core mathematics standards. *Middle Grades Research Journal, 13*(1), 21–33.

Dingman, S. W., Teuscher, D., Newton, J. A., & Kasmer, L. (2013). Common mathematics standards in the United States: A comparison of state and common core standards. *Elementary School Journal, 113*(4), 541–564. https://doi.org/10.1086/669939

Education Market Research. (2005). *Mathematics market, grades K–12: Teaching methods, textbooks/materials used and needed, and market size*.

Finn, C. E., Ravitch, D., & Whitman, D. (2004). *The mad, mad world of textbook adoption*. Thomas B. Fordham Institute.

Flanders, J. R. (1987). How much of the content in mathematics is new? *Arithmetic Teacher, 35*(1), 18–23.

Kanbir, S. (2016). Major curriculum reforms and controversies in the United States. *International Journal of Educational Studies in Mathematics, 3*(1), 1–8.

Kansas State Department of Education. (2003). *Kansas curricular standards for mathematics*.

Larson, R., & Boswell, L. (2014). *Big ideas math: A common core curriculum*. Big Ideas Learning.

Lappan, G., Fey, J. T., Fitzgerald, W. M., Friel, S. N., & Phillips, E. D. (2002). *Connected mathematics project*. Dale Seymour.

Lappan, G., Phillips, E. D., Fey, J. T., & Friel, S. N. (2013). *Connected mathematics 3: Butterflies, pinwheels, and wallpaper: Symmetry and transformations*. Pearson Prentice Hall.

Linn, R. L., Baker, E. L., & Betebenner, D. W. (2002). Accountability systems: Ramifications of the No Child Left Behind Act of 2001. *Educational Researcher, 31*(6), 3–6.

McKnight, C. C., Crosswhite, F. J., Dossey, J. A., Kefir, E., Swafford, J. O., Travers, K. J., & Cooney, T. J. (1987). *The underachieving curriculum: Assessing U.S. school mathematics from an international perspective*. Stipes.

National Council of Teachers of Mathematics [NCTM]. (1989). *Curriculum and evaluation standards for school mathematics.*

National Governors Association Center for Best Practices & Council of Chief State School Officers [NGA & CCSSO]. (2010). *Common core state standards for school mathematics.* https://www.thecorestandards.org/Math/

National Research Council. (2002). *Investigating the influence of standards: A framework for research in mathematics, science, and technology education.* National Academies Press. https://doi.org/10.17226/10023

New York State Education Department. (2005). *New York learning standards for mathematics.*

Olson, T. A. (2010). *Articulated learning trajectories related to the development of algebraic thinking that follow from patterning concepts in middle grades mathematics* [Unpublished Doctoral dissertation, University of Missouri-Columbia]. https://mospace.umsystem.edu/xmlui/handle/10355/8318?show=full

Reys, B. J. (Ed.). (2006). *The intended mathematics curriculum as represented in state-level curriculum standards: Consensus or confusion?* Information Age Publishing.

Reys, B. J., Chval, K., Dingman, S., McNaught, M., Regis, T., & Togashi, J. (2007). Grade-level learning expectations: A new challenge for elementary mathematics teachers. *Teaching Children Mathematics, 14*(1), 6–11.

Reys, B. J., Dingman, S., Sutter, A., & Teuscher, D. (2005). *Development of state-level mathematics curriculum documents: Report of a survey.* https://mospace.umsystem.edu/xmlui/handle/10355/2242

Rossi, H., Bodrero, J., Walker, C., Cummings, M., Eischeid, C., & Serr, E. (2013). *The Utah Middle School Math Project.* http://utahmiddleschoolmath.org

Schmidt, W. H., McKnight, C. C., Valverde, G. A., Houang, R. T., & Wiley, D. E. (1997). *Many visions, many aims: A cross-national investigation of curricular intentions in school mathematics.* Springer.

Seago, N., Jacobs, J., Driscoll, M., Nikula, J., Matassa, M., & Callahan, P. (2013). Developing teachers' knowledge of a transformations-based approach to geometric similarity. *Mathematics Teacher Educator, 2*(1), 74–85.

Smith, J. P., III.(Ed.). (2011). *Variability is the rule: A companion analysis of K–8 mathematics standards.* Information Age Publishing.

Stein, M. K., Remillard, J., & Smith, M. S. (2007). How curriculum influences student learning. In F. K. Lester, Jr. (Ed.), *Second handbook of research on mathematics teaching and learning* (pp. 319–370). Information Age Publishing.

Teuscher, D., Dingman, S. W., Nevels, N. N., & Reys, B. J. (2008). Curriculum standards, course requirements, and mandated assessments for high school mathematics: A status report of state policies. *NCSM Journal of Mathematics Education, 11*(2), 50–55.

Teuscher, D., Reys, B., Dingman, S., & Thomas, A. (2014). Transitioning to common mathematics standards: Tracing the movement of computational fluency in the K–5 curriculum. In K. Karp & A. Roth McDuffie (Eds.), *Using research to improve instruction* (*NCTM Annual Perspectives in Mathematics Education*) (pp. 3–12). National Council of Teachers of Mathematics.

Teuscher, D., Tran, D., & Reys, B. J. (2015). Common core state standards in the middle grades: What's new in the geometry domain and how can teachers support student learning? *School Science and Mathematics, 115*(1), 4–13.

Teuscher, D., Kasmer, L., Olson, T., & Dingman, S. (2016). *Isometries in new U.S. middle grades textbooks: How are isometries and congruence related?* https://www.researchgate.net/publication/308136979

Tran, D., Reys, B. J., Teuscher, D., Dingman, S., & Kasmer, L. (2016). Analysis of curriculum standards: An important research area. *Journal for Research in Mathematics Education, 47*(2), 118–133.

Tyson-Bernstein, H., & Woodward, A. (1991). Nineteenth century policies for twenty-first century practice: The textbook reform dilemma. In P. G. Altbach, G. P. Kelly, H. G. Petrie, & L. Weis (Eds.), *Textbooks in American society: Politics, policy, and pedagogy* (pp. 91–104). State University of New York.

ENDNOTE

1. In the U.S., elementary schools are typically for grades K–5 students (ages 5–11), middle grades for grades 6–8 students (ages 11–14), and high school for grades 9–12 students (ages 14–18).

SECTION II
TEXTBOOK ANALYSIS

CHAPTER 6

CHOOSING SAMPLES IN MATHEMATICS TEXTBOOK RESEARCH

Textbooks, Chapters, Pages, and Tasks

Dustin L. Jones
Sam Houston State University

The focus of this chapter is on lessons learned about research on mathematics textbooks, which mediate the intended curriculum and the implemented curriculum (Valverde et al., 2002). Based on the author's experiences in the U.S., several suggestions are provided regarding the process of choosing the sample for a study. This process includes determining which textbooks to analyze, as well as selecting the specific chapters, pages, tasks, or other component structures of a textbook. Examples are provided to illustrate various choices that can be made when conducting research on mathematics textbooks. In general, the sampling choices for a study should be based on the research questions. Although for his research, the author tends to include all chapters, pages, and tasks in a sample, there are valid reasons for not doing so.

Keywords: mathematics curriculum; mathematics textbooks; research methods; samples; sampling methods

I have engaged in research on mathematics textbooks over the past two decades, starting with my dissertation study (Jones, 2004). Fortunately, I was working at the University of Missouri (U.S.) when it housed the National Science Foundation (NSF)-funded Show-Me Center Project (Meyer & Langrall, 2008; NSF, 2023a) and later the Center for the Study of Mathematics Curriculum (NSF, 2023b).[1] Although the focus of these Centers and their associated faculty members played a major role in shaping my research interests, I have been interested in mathematics textbooks for nearly my entire life. In second grade, I would try (sometimes unsuccessfully) to work ahead in the arithmetic practice workbook. In fourth grade, my teacher let me work independently for a few weeks to learn statistics using an eighth-grade mathematics textbook. In my ninth-grade geometry course, my teacher paired me with another student, and we learned the content by working through the textbook together. These formative experiences helped me view the mathematics textbook as a helpful guide for the learner.

Later, when I became a high-school mathematics teacher, the textbooks for my classes were helpful resources as I planned and prepared for classes. As a doctoral student, I was in a study group with two high-school teachers, and we worked through both volumes of Course 4 of *Contemporary Mathematics in Context* (Coxford et al., 2001a, 2001b) from the Core-Plus Mathematics Project as if we were high-school students. Taken as a whole, these formative experiences helped introduce me to the diverse ways that students and teachers may interact with mathematics textbooks. I also began to understand that not all mathematics textbooks are the same.

Since my dissertation study, I have conducted and been a collaborator on numerous mathematics textbook analyses. There are many mathematics textbooks available for research studies, and I quickly learned that it was not feasible to examine every relevant one for a study. In this chapter, I provide details from several studies, emphasizing how one goes about choosing the sample for a study. In doing so, I address the following questions:

- Once a textbook has been included in a sample for a study, what part(s) or chapter(s) will be examined?
- How do researchers choose which pages or tasks to examine within a sample?

In this chapter, I describe some lessons I learned while addressing these questions in the context of my research. I will not recommend one particular sampling method over another, as the selection of the sample (at various levels) should be a purposeful choice to address the research questions in the study.

CHOOSING TEXTBOOKS FOR THE SAMPLE

Textbooks have multiple components. From a structural perspective, textbooks contain pages. From an organizational perspective, textbooks often contain lessons within chapters, which may include explanatory text, activities, problems, exercises, and other features. Textbooks may have additional sections not directly related to instruction, such as a table of contents, an index, a glossary, or appendices.

Valverde et al. (2002) describe how mathematics textbooks serve as mediators between the intentions, aims, and goals of the *intended curriculum* and the strategies, practices, and activities of the *implemented curriculum*. In their report on textbooks used in countries that participated in the Third International Mathematics and Science Study, they stated:

> Since textbooks are designed as templates for action, they are more closely linked with enacted teaching and learning activities than documents such as content standards, frameworks or programs of study. In view of this, characterizations of the pedagogical features of textbooks are especially important in understanding the nature of the provided educational opportunities in different countries. (Valverde et al., 2002, p. 12)

To describe and understand students' opportunities to learn mathematics, I have engaged in several content analyses of mathematics textbooks. The following section contains examples of my own work, and lessons learned related to choosing the sample for a study. To provide a broader view of sampling choices, examples from other researchers of mathematics textbooks are also included.

Lessons Learned for Choosing Textbooks for the Sample

In my own research on mathematics textbooks intended for students in grades 1–12 or for college undergraduates, I have sought to describe students' opportunities to learn mathematics as afforded by textbooks. Some of these studies, including my dissertation study, were historical in nature; in other studies, I examined textbooks that were currently in use. Across these studies, I attempted to select textbooks for the sample that were *popular* (i.e., commonly used) within a population of students in a given location and period of time. Identifying these popular textbooks to include in the sample has been a key challenge. Below are five strategies to overcome this challenge.

The first strategy is to *examine market research data*. When this information is available, it provides figures for textbook sales during a given year.

Textbook publishers may also have this data, but companies may not be willing to share it. Market intelligence companies also publish reports of market share data, but they can be costly. For example, the report titled *Mathematics/STEM Market K–12: Teaching Methods, Traditional and Digital Materials Used and Needed, and Market Size* can be purchased for $1,600 USD from Simba Information (2022). I have not purchased this information for my research.

A second recommended strategy is to *examine nationwide surveys and reports*. Some nationwide surveys of mathematics education in the U.S. (e.g., Banilower et al., 2018; Weiss, 1978; Weiss et al., 2001), or national reports such as the one from the National Council on Teacher Quality (Greenberg & Walsh, 2008), include questions about which textbooks are used in classrooms. These reports were helpful in identifying popular textbooks from the 1970s, 1980s, and 1990s for my dissertation study (Jones, 2004), and also for identifying textbooks that were commonly used in mathematics content courses for prospective elementary teachers (Jones & Jacobbe, 2014; Jones et al., 2017). These reports are usually available at no cost.

A third strategy for identifying popular textbooks is to *consider lists of textbooks approved for adoption*. Some states in the U.S. provide school districts with a set amount of funding to purchase textbooks, along with a list of textbooks that have been approved for adoption. District administrators in those states may only spend state funds on textbooks that appear on the approved list. This practice has had interesting effects on the mathematics textbook publishing industry (Seeley, 2003). In practice, school district administrators in states with textbook adoption policies overwhelmingly tend to purchase books from their list. In an NSF-funded Research Experiences for Undergraduates (REU) project, a group of prospective teachers and their mentor analyzed the statistics content of textbooks for students in grades 1–5 (Jones et al., 2015) using all five mathematics textbook series that were approved for adoption in Texas, the geographic location of the project.

A fourth recommended strategy for selecting a sample is to *talk to experts in the field*. This strategy can be useful when documented sources of popular textbooks are not available, or if the study will include textbooks that are not necessarily popular. For my dissertation study, I had difficulty finding market research or other reports related to mathematics textbook use for certain decades. Fortunately, several mathematics education curriculum researchers who were aware of national trends in mathematics education during that decade agreed to share their expertise and experiences. Interviewing this panel of experts allowed for identification of a textbook series that was likely the most popular during that time. This strategy of finding a *professional consensus* yields information that is only as good as the memories and experiences of the people who are surveyed, so it is important to cast a wide net and interview people with diverse backgrounds.

The fifth and final strategy is to *examine the samples from other studies of mathematics textbooks*. A growing number of studies of mathematics textbooks have been published, and it may be helpful to consult these studies to see what textbooks were selected for a particular period of time and location, and to identify the methods that were used for sample selection. Caution must be applied, though, because some researchers do not list the books that are used in the sample (e.g., Dolev & Even, 2015), and other researchers may mistakenly claim that a single textbook or series is representative of all textbooks in a particular location and time.

Lessons Learned for Acquiring Textbooks

Once textbooks have been identified for inclusion in a sample, a researcher needs access to them. Personally, I prefer to *own a physical copy of the textbook*. Some publishers will provide textbooks requested by researchers. Many textbooks are available through online booksellers and auction sites. When a researcher owns a textbook, they can easily mark sections, pages, and tasks. Owning the textbook also means that the researcher can refer to it at a later time. Alternatively, if the books are not available for purchase or if the cost is prohibitively high, several universities have collections of mathematics textbooks.[2] Some collections are quite extensive. If the books are not available for inter-library loan, it may be worthwhile to visit the collection, and code within the library or digitally scan parts of the textbook for subsequent examination.

In a study where a team of researchers planned to examine five series of five textbooks each (Jones et al., 2015), purchasing physical copies was deemed to be too expensive. The researchers *contacted the publisher* to request access to electronic versions of the textbooks. After explaining the purpose of the research and the length of time needed for data collection, and then completing the necessary paperwork, the team was granted temporary online access to the textbooks.

CHOOSING COMPONENTS TO STUDY

Textbooks are organized into several components, with chapters being the primary and most common component. These chapters are often preceded by a Table of Contents, and perhaps a Preface. Chapters may be followed by an Index, Glossary, answers to selected problems, assessment resources, and other appendices. My research studies of mathematics textbooks have focused on the chapters, and not the other material. I refer to all pages in the chapters of the textbook as *instructional pages*. Other researchers may examine different components. For example, McCrory and Stylianides (2014) used the tables of contents and indexes of textbooks, "the published maps of each textbook" (p. 120), to search for reasoning-and-proving content within the textbook, attempting to mirror the practices of a student studying for a test or an instructor sequencing the topics of a course. In the remainder of this section, I describe my responses to challenges I faced when selecting chapters, pages, and tasks to include in the sample of a study.

Lessons Learned on Selecting Chapters

Several of my research studies have focused on the statistics and probability content of mathematics textbooks, so it was important to ensure that any chapters or sections on this specific content would be included in the sample. At the same time, prior experiences with textbooks revealed that some content related to statistics and probability may be included in other chapters, perhaps illustrating a connection among seemingly disjoint mathematical topics, or as a review. For example, one textbook series analyzed in Jones (2004) incorporated a review of prior topics within each lesson. If a lesson on the probability of an event preceded a lesson on geometry, then the set of exercises following the lesson on geometry would include an unrelated task requiring students to compute the probability of a compound event. Within this same study, another textbook contained probability tasks within a chapter focused on geometry. Students were asked to determine the likelihood of hitting a certain region on a dartboard by comparing the areas of figures.

A primary strategy for selecting chapters has been to *include all the chapters of the textbook in the sample* and examine each for the content of interest. This has proven to be beneficial; in some studies where all chapters were included in the sample (e.g., Jones, 2004; Jones & Basyal, 2019; Jones et al., 2015; Jones & Tarr, 2007), statistics and probability content was found in chapters on other topics. By including all textbook chapters in the sample, a researcher can identify more of the content of interest and see how it may be presented to students outside of the designated chapter. In other studies (e.g., Jones, 2014; Jones & Jacobbe, 2014), only the sections or chapters of the book that addressed statistics topics were examined. This was appropriate for the studies, but limited the conclusions that could be drawn from the analyses. In particular, extending the study of technology references in chapters related to statistics and probability (Jones, 2014) to the entire textbook (Jones et al., 2017) revealed that some textbooks addressed technology differently, depending on the topic of the chapter. Specifically, the same six textbooks were analyzed in both studies, and one measure was the count of technology references. The textbook with the most technology references within chapters on statistics and probability (Jones, 2014) moved down to fourth place when all chapters were considered (Jones et al., 2017). In studies like Jones (2014), the analysis, findings, and conclusions should be limited to the types of chapters in question. Overall, in studies that are not related to a mathematical strand, it is beneficial to include all textbook chapters in the sample.

Lessons Learned on Selecting Pages

In general, I recommend that researchers *include every page of the selected chapters in the sample of the study*. Other researchers of mathematics textbooks use different strategies, such as selecting all the pages in every other lesson from each chapter in a textbook (Bieda et al., 2014), or selecting a specific number of pages at random (Österholm & Bergqvist, 2013). Such methods are also reasonable. My purpose here is to illustrate why a researcher may want to include every page in the sample.

For many studies, my practice has been to examine every instructional page of a textbook. Essentially, this process is the same as conducting a census of those pages, where each instructional page in the population is included in the sample. Shortly after completing a study on the statistical content of U.S. elementary school textbooks (Jones et al., 2015), I worked with a colleague to conduct a post hoc study in which we used several page-sampling methodologies to estimate the proportion of pages containing statistics, and to estimate the number of tasks related to statistics within two textbook series (Jones & Jayawardena, 2019). In this study, there were five textbooks in each of the two series. We compared the census method (examining each page) to four other methods:

1. *Simple random sampling*: selecting pages for the sample without replacement using a random process;
2. *Systematic sampling*: including every other page (or every third page or tenth page, for example) in the sample;
3. *Stratified random sampling*: selecting pages for the sample at random from each chapter, so that the number of pages selected is proportional to the number of pages in the chapter;
4. *Cluster sampling*: selecting chapters without replacement using a random process and including all pages from the selected chapters in the sample.

One affordance of using a page sampling method other than a census is that it may save time. Bieda et al. (2014) stated that using a systematic sample "greatly enhanced the feasibility of the study" (p. 79). However, these affordances come with limitations regarding the level of certainty ascribed to the results. For example, Jones and Jayawardena (2019) found that none of the other four sampling methods consistently provided estimates close to the actual parameters determined by a census, stating, "this is likely due to the lack of homogeneity in the distribution of statistics tasks within textbooks, as textbooks are typically organized into content-based chapters" (p. 214). Jones and Jayawardena recommended further study about page sampling methods regarding textbook features that may be more homogeneously distributed throughout the textbook. In summary, I feel most comfortable including every instructional page in the sample, even if a census requires more time than another method.

Lessons Learned on Defining and Selecting Tasks

Once textbooks, chapters, and pages are identified for a sample, there is still the question about *what* on the page will be examined. Some researchers have examined geometric diagrams (Dimmel & Herbst, 2015) or definitions of terms (Usiskin et al., 2008). In my research studies, I have focused attention on the tasks (problems and exercises) that students are asked to complete. A focus on tasks brings about two related issues in textbook research: what constitutes a *task*, and which tasks will be included in the sample?

In my dissertation study, a task was defined as "an activity, exercise, or set of exercises in a textbook that has been written with the intent of focusing a student's attention on a particular idea" (Jones, 2004, p. 10). Under this definition, a set of eight numbered items related to the same context, such as determining probabilities of various events based on a spinner, would be considered a single task. A five-step sequence of questions related to a particular context, perhaps with parts labeled a, b, c, d, and e, would also be considered a single task. Although I originally thought that this definition would keep from overcounting the number of tasks, particularly within textbooks with several repetitive exercises, I sometimes found it difficult to divide a set of problems or exercises into tasks. For that reason, in later work, I adopted a slightly different definition: "The size and content of a task was determined by the textbook's marked divisions of an exercise or problem. The smallest marked division was considered a task" (Jones & Jacobbe, 2014, p. 6). That is, tasks are defined by the labeling sequence used by the textbook authors. Identifying tasks in this way has led to greater consensus among my research collaborators. Therefore, I now *identify tasks as the smallest labeled part of a problem, exercise, or activity.* Using this definition, the earlier examples in this paragraph would be counted as eight tasks and five tasks, respectively.

Once tasks have been defined, the researcher needs to determine which tasks to include in the sample. I suggest researchers *include every task in the sample*. By this, I mean that the researcher will look at each task on the selected textbook page, but not necessarily code each one. For example, in a study of technology references within textbooks (Jones et al., 2017), researchers looked at every task on a page, but only recorded and coded tasks that contained technology references. Similarly, in my dissertation study (Jones, 2004), I looked at every task on the page, but only included those that addressed probability content.

CONSIDERATIONS FOR NON-U.S. TEXTBOOKS

When analyzing textbooks from countries outside the U.S., it is important to consider whether the country has a set of approved textbooks for use in mathematics classrooms. In some countries, there may be only one official government-approved textbook series. Researchers examining textbooks from these countries should include the government-approved textbook in the sample. For example, in studies of the mathematics textbooks of Nepal (Basyal et al., 2022; Jones & Basyal, 2019), researchers included mathematics textbooks from the single government-developed and government-approved series. The structure and organization of Nepali mathematics textbooks is very similar to U.S. mathematics textbooks, with content divided into chapters and lessons, and each lesson containing some exposition, examples, and exercises. These features facilitated the application of sampling methods, used on U.S. textbooks, for research on Nepali textbooks. Furthermore, the research questions for the examination of Nepali textbooks were similar to those of studies with U.S. textbooks. Jones and Basyal (2019) examined every instructional page of the textbooks in the sample for statistical content, and Basyal et al. (2022) classified every mathematical task in three textbooks according to levels of cognitive demand. If researchers are examining textbooks that are organized differently than U.S. textbooks, these suggested sampling methods may not be appropriate.

SUMMARY

This chapter contains suggestions based on lessons I have learned from conducting content analyses of mathematics textbooks. I have examined books from different periods of history, and textbooks from both the U.S. and Nepal. Overall, I suggest that methods used for sampling textbooks, chapters, pages, and tasks be aligned with the study's research questions. The contrasting examples I provided from the work of others are presented as examples of quality research resulting from different sampling choices.

In summary, related to identifying textbooks for a sample, I make the following recommendations to researchers:

- Examine market research data to determine which textbooks are sold to schools. These data may be available for purchase, and the price can be high.
- Examine nationwide surveys and reports to identify textbooks used by mathematics teachers and how those books are used. Several reports, such as Banilower et al. (2018), are available for free and relate to mathematics instruction across all grade levels.
- Utilize lists of textbooks approved for adoption in various jurisdictions. Selecting textbooks from these lists allows a researcher to focus on students' opportunities to learn from within a particular district, state, or country.
- Talk to experts in the field to identify textbooks that may be worthy of examination. Their expertise can be valuable, particularly in cases where market research and survey data are unavailable.
- Examine the samples from other studies of mathematics textbooks. Be careful to cast a wide net and carefully examine the methodology of other studies.

Once textbooks have been identified for inclusion in the sample for a study, I recommend researchers obtain a physical copy of the textbook. Textbooks may be available for purchase from the publisher or other booksellers. Publishers may also send complimentary examination copies, depending on company policies. If it is not possible for researchers to own a physical copy, they may locate copies within libraries or contact the publisher to request access to an electronic version of the textbook.

Regarding which parts of the textbook to include in the sample of a study, I recommend researchers use the following methods based on my experiences with mathematical content analysis.

- Take a census of the instructional pages of the textbook; that is, include every instructional page of the textbook in the sample.
- Examine every task on the pages included in the sample.
- Define tasks as the smallest labeled part of a problem, exercise, or activity.

Individually, each of these three recommendations increases the likelihood that any particular feature of interest will be noticed. For example, by taking a census, a researcher examining the statistical content of a textbook will not overlook any pages containing statistics. Collectively, these three recommendations allow a researcher to conduct a comprehensive analysis of the mathematical content of the textbook. The findings can be presented with more certainty, and studies using these methods can be validated and replicated by other researchers. Even though these recommendations may come at the cost of more work and time for a researcher, I find this cost is balanced by an increased confidence in the findings. However, there are valid reasons for using different methodologies for choosing a sample, and other studies referenced in this chapter provide reasonable justifications for doing so.

As my relationship with mathematics textbooks has developed over my lifetime, I find that following these general guidelines allows me to characterize the mathematical content of textbooks in a way that I can share information with educators, teachers, and policymakers. My methods are certainly not the only way to conduct content analysis, and I am not suggesting that my methods are always the best for every research question. However, I believe that these suggestions will be helpful for those who wish to analyze the mathematical content of textbooks.

REFERENCES

Banilower, E. R., Smith, P. S., Malzahn, K. A., Plumley, C. L., Gordon, E. M., & Hayes, M. L. (2018). *Report of the 2018 NSSME+*. Horizon Research.

Basyal, D., Jones, D. L., & Thapa, M. (2022). Cognitive demand of mathematics tasks in Nepali middle school mathematics textbooks. *International Journal of Science and Mathematics Education, 21*(3), 863–879. https://doi.org/10.1007/s10763-022-10269-3

Bieda, K. N., Ji, X., Drwencke, J., & Picard, A. (2014). Reasoning-and-proving opportunities in elementary mathematics textbooks. *International Journal of Educational Research, 64*, 71–80. https://doi.org/10.1016/j.ijer.2013.06.005

Coxford, A. F., Fey, J. T., Hirsch, C. R., Schoen, H. L., Hart, E. W., Keller, B. A., Watkins, A. E., Ritsema, B. E., & Walker, R. K. (2001a). *Contemporary mathematics in context: A unified approach. Course 4 part A*. Everyday Learning.

Coxford, A. F., Fey, J. T., Hirsch, C. R., Schoen, H. L., Hart, E. W., Keller, B. A., Watkins, A. E., Ritsema, B. E., & Walker, R. K. (2001b). *Contemporary mathematics in context: A unified approach. Course 4 part B*. Everyday Learning.

Dimmel, J. K., & Herbst, P. G. (2015). The semiotic structure of geometry diagrams: How textbook diagrams convey meaning. *Journal for Research in Mathematics Education, 46*(2), 147–195. https://doi.org/10.5951/jresematheduc.46.2.0147

Dolev, S., & Even, R. (2015). Justifications and explanations in Israeli 7th grade math textbooks. *International Journal of Science and Mathematics Education, 13*, 309–327. https://doi.org/10.1007/s10763-013-9488-7

Greenberg, J., & Walsh, K. (2008). *No common denominator: The preparation of elementary teachers in mathematics by America's education schools*. National Center on Teacher Quality. https://www.nctq.org/publications/No-Common-Denominator:-The-Preparation-of-Elementary-Teachers-in-Mathematics-by-Americas-Education-Schools

Jones, D. L. (2004). *Probability in middle-grades mathematics textbooks: An examination of historical trends, 1957–2004* [Unpublished Doctoral dissertation, University of Missouri-Columbia].

Jones, D. L. (2014). The role of technology for learning stochastics in U.S. textbooks for prospective teachers. In K. Jones, C. Bokhove, G. Howson, & L. Fan (Eds.), *Proceedings of the International Conference on Mathematics Textbook Research and Development (ICMT-2014)* (pp. 269–274). University of Southampton.

Jones, D. L., & Basyal, D. (2019). An analysis of the statistics content in Nepali school textbooks. *Mathematics Education Forum Chitwan, 4*, 21–34. https://doi.org/10.3126/mefc.v4i4.26356

Jones, D. L., Brown, M., Dunkle, A., Hixon, L., Yoder, N., & Silbernick, Z. (2015). The statistical content of elementary school mathematics textbooks. *Journal of Statistics Education, 23*(3). https://jse.amstat.org/v23n3/jones.pdf

Jones, D., Hollas, V., & Klespis, M. (2017). The presentation of technology for teaching and learning mathematics in textbooks: Content courses for elementary teachers. *Contemporary Issues in Technology and Mathematics Teacher Education, 17*(1), 53–79.

Jones, D. L., & Jacobbe, T. (2014). An analysis of the statistical content in textbooks for prospective elementary teachers. *Journal of Statistics Education, 22*(3). https://jse.amstat.org/v22n3/jones.pdf

Jones, D., & Jayawardena, I. (2019). A comparison of page sampling methods for mathematics textbook content analysis. In S. Rezat, M. Hattermann, J. Schumacher, & H. Wuschke (Eds.), *Proceedings of the Third International Conference on Mathematics Textbook Research and Development* (pp. 209–214). Universitätbibliothek Paderborn.

Jones, D. L., & Tarr, J. E. (2007). An examination of the levels of cognitive demand required by probability tasks in middle grades mathematics textbooks. *Statistics Education Research Journal, 6*(2), 4–27. https://doi.org/10.52041/serj.v6i2.482

McCrory, R., & Stylianides, A. J. (2014). Reasoning-and-proving in mathematics textbooks for prospective elementary teachers. *International Journal of Educational Research, 64*, 119–131. https://doi.org/10.1016/j.ijer.2013.09.003

Meyer, M. R., & Langrall, C. W. (2008). *A decade of middle school mathematics curriculum implementation: Lessons learned from the Show-Me Project*. Information Age Publishing.

National Science Foundation. (2023a, March 6). *NSF Award Search: Award # 9714999—Show-Me Project: A National Center for Standards-based Middle School Mathematics Curriculum Dissemination and Implementation.* https://www.nsf.gov/awardsearch/showAward?AWD_ID=9714999

National Science Foundation. (2023b, March 6). *NSF Award Search: Award # 0333879—Center for the Study of Mathematics Curriculum.* https://www.nsf.gov/awardsearch/showAward?AWD_ID=0333879

Österholm, M., & Bergqvist, E. (2013). What is so special about mathematical texts? Analyses of common claims in research literature and of properties of textbooks. *ZDM Mathematics Education, 45*(5), 751–763. https://doi.org/10.1007/s11858-013-0522-6

Seeley, C. L. (2003). Mathematics textbook adoption in the United States. In G. M. A. Stanic & J. Kilpatrick (Eds.), *A history of school mathematics* (Vol. 2, pp. 957–988). National Council of Teachers of Mathematics.

Simba Information. (2022, October 12). *Mathematics/STEM market K–12: Teaching methods, traditional and digital materials used and needed, and market size.* https://www.simbainformation.com/Mathematics-STEM-Teaching-9587729/

Usiskin, Z., Griffin, J., Witonsky, D., & Willmore, E. (2008). *The classification of quadrilaterals: A study of definition.* Information Age Publishing.

Valverde, G. A., Bianchi, L. J., Wolfe, R. G., Schmidt, W. H., & Houang, R. T. (2002). *According to the book: Using TIMSS to investigate the translation of policy into practice through the world of textbooks.* Kluwer Academic Publishers.

Weiss, I. R. (1978). *Report of the 1977 national survey of science, mathematics, and social studies education.* Center for Educational Research and Evaluation.

Weiss, I. R., Banilower, E. R., McMahon, K. C., & Smith, P. S. (2001). *Report of the 2000 national survey of science and mathematics education.* Horizon Research.

ENDNOTES

1. Funding for the Show-Me Center began in 1997; funding for the Center for the Study of Mathematics Curriculum began in 2003.
2. Examples include, but are not limited to, the Curriculum Collection at the Sterling C. Evans Library at Texas A&M University, the Learning Resources Center in the School of Education at Baylor University, and the Blackwell History of Education Museum at Northern Illinois University.

CHAPTER 7

LESSONS LEARNED FROM CROSS-CULTURAL CURRICULAR RESEARCH
The Case of Proving-Related Activities

Hangil Kim
Chungnam High School, Korea

GwiSoo Na
Cheongju National University of Education, Korea

Eric Knuth
National Science Foundation, United States

Proving-related activities are fundamental to mathematical practice and play an important role in learning mathematics. Given the inextricable link between teachers' instructional practices and the curriculum materials they implement, the success of teachers' efforts to support students' learning to prove may thus depend on the nature of opportunities to engage students in proving-related activities provided in curricular materials. An international comparative study was designed to investigate the nature of proving-related tasks in a textbook from each of three countries: Korea, Singapore, and the United States. Results indicate that the majority of proving-related tasks are intended for students to test the truth of a conjecture and there is a scarcity of other types of proving-related activities. Issues in choosing the unit of analysis and challenges in analyzing differently designed curriculum materials, translating the language of a task, and methodological issues are discussed.

Keywords: comparative study; curriculum materials; proof; proving-related activities; textbook analysis

INTRODUCTION

The prominence and importance of proving-related activities has been recognized by researchers (Jeannotte & Kieran, 2017; Schoenfeld, 1994; Stylianides, 2009) and mathematicians (Alcock, 2010; Alcock & Inglis, 2008), as well as reflected in curricula across the globe (Department for Education [DoE], 2014; Ministry of Education [MoE], 2015; National Governors Association [NGA] Center for Best Practices & Council of Chief State School Officers [CCSSO], 2010). In stark contrast to its espoused prominence in school mathematics, teachers' and students' views about the nature and role of proof (Kim, 2021; Knuth, 2002a, 2002b) and their understanding about proof (Chazan, 1993; Senk, 1985) tend to be at odds with mathematicians' views on proof (Dawkins & Weber, 2017). To address issues in teaching and learning proof, researchers have investigated the nature of proving-related tasks available in textbooks (e.g., Bieda et al., 2014) given that textbooks are a primary source for teachers to prepare and implement daily instruction (Begle, 1973; Grouws et al., 2004). Moreover, textbooks have an impact on student learning of mathematics (Ball & Cohen, 1996; Fan, 2013; Stein et al., 2007) and can also promote teacher learning (Davis & Krajcik, 2005; Remillard & Bryans, 2004).

Investigating the nature of proving-related tasks in textbooks can offer insights into the nature of proving-related tasks, with cross-cultural examinations offering unique insights into the nature of proving-related tasks (Miyakawa, 2017). With a selection of Korean, Singaporean, and U.S. textbooks, we investigated the nature of proving-related tasks at Grade 8 (13–14 year-old students). The primary research question that guided this study was: *What is the nature of proving-related tasks in Grade 8 textbooks?* This question prompted us to investigate the conceptualization of proof and the kinds of activities identified as proving-related in the textbooks. We also confronted methodological issues, such as selecting which textbooks to analyze, determining the number of pages to analyze within each textbook, and deciding what constitutes a task. In this chapter, we report briefly on the results of the study, issues that surfaced while conducting the study, and lessons learned from addressing the issues.

BACKGROUND

In this section, we briefly summarize the extant literature on proving-related activities and cross-cultural research on textbooks. For subsequent discussions, it would be remiss not to make explicit our conceptualization of proof for this study, which rests on A. J. Stylianides's (2007) definition of proof as a mathematical argument with the following characteristics:

(1) uses statements accepted as true by the classroom community; (2) uses forms of mathematical reasoning understood by (or within the reach of) the classroom community; and (3) is presented in ways that are appropriate and understood by (or within the reach of) the classroom community.

Proving-Related Activities

Proving-related activities, such as developing, exploring, and proving conjectures (National Council of Teachers of Mathematics [NCTM], 2000), are fundamental to mathematical practices (Epstein & Levy, 1995) and play an important role in learning mathematics (NCTM, 1989, 1991; Schoenfeld, 1994). Accordingly, mathematics education scholars (e.g., Kim, 2022; Stylianides, Bieda, & Morselli, 2016; Stylianides, Stylianides, & Weber, 2016) and reform initiatives in the U.S. (e.g., NCTM, 2000; NGA & CCSSO, 2010) as well as in other countries (e.g., Australian Curriculum, Assessment and Reporting Authority, 2015; DoE, 2014; MoE, 2015) have increasingly called for proving-related activities to play a more central role in the mathematics experiences of all students. Yet, despite almost two decades of such calls, students continue to struggle learning to prove (e.g., Knuth et al., 2009; Reid & Knipping, 2010), and teachers, as well, struggle to facilitate the development of students' learning to prove (Stylianides et al., 2013).

Given the inextricable link between teachers' instructional practices and the textbooks they implement, the success of teachers' efforts to support students' learning to prove may thus depend, in large part, on the nature of opportunities to engage students in proving-related activities provided in textbooks (e.g., Bieda et al., 2014; Stylianides, 2009). Yet, few researchers have explicitly examined the nature of such opportunities in textbooks across countries. Accordingly, Miyakawa (2017) noted the importance and need for studies that allow us to "clarify different possibilities of the nature of proof to be taught and to make explicit what is implicit or taken for granted in other countries" (p. 38). To that end, the study[1] reported in this chapter investigates proving-related tasks in textbooks from three countries. On international studies of students' mathematics achievement (e.g., Mullis et al., 2016), two of the countries are high performing (Korea and Singapore), and one is middle performing (United States).

Curriculum Analysis: Framing and Units of Analysis

Curriculum research is often defined by three perspectives: *intended*, *implemented*, and *attained curriculum* (Schmidt et al., 1997; Valverde et al.,

2002). The curriculum perspective of interest in this study is the intended curriculum, that is, the actual written curriculum materials (i.e., textbook). As Stylianides's (2014) editorial suggested, textbook analysis possesses the potential to leverage reasoning and proving opportunities in the classroom because analyzing proving tasks sheds light on what a teacher might implement as part of classroom instruction. However, Stylianides also noted that research techniques for undertaking textbook analyses on proving need to be well developed in comparison to other techniques for research on student thinking or instructional practices. The literature on textbook analysis is relatively limited in scope, and studies are often framed by a focus on specific mathematical content and/or specific grade levels, and often vary in the units of analysis (e.g., Bergwall, 2021; Davis et al., 2014; Li, 2000).

In the case of studies focusing on specific content, for example, some researchers have examined geometry or specific geometric topics (e.g., Bergwall & Hemmi, 2017; Fujita & Jones, 2014; Otten et al., 2014), whereas others have examined algebraic topics such as exponents, logarithms, and polynomials (e.g., Thompson et al., 2012). Other researchers have examined specific content at particular grade levels, such as geometry (e.g., Bergwall, 2021) or algebra (Davis et al., 2014; Hong & Choi, 2014; Thompson et al., 2012) at the secondary school level, or various mathematical topics at the elementary-school level (e.g., Bieda et al., 2014; Charalambous et al., 2010). In the case of studies focusing on different units of analysis, Fujita and Jones (2014), for example, examined Japanese geometry textbooks and took as their unit of analysis *a block*: "a 'block' is taken to be one or more paragraphs united by a theme, or one figure or group of figures" (Valverde et al., 2002, p. 84). In contrast, in their textbook analysis, Zhang and Qi (2019) took as the unit of analysis tasks that had *a separate answer* appearing in the teacher's edition, thus identifying a task by the inclusion of a separate answer. Collectively, studies with such differing units of analysis led Cai and Cirillo (2014) to comment, "Greater methodological consistency could help support more generalized findings from analyses" (p. 137).

With respect to textbook analyses of proving-related activities, researchers have examined the types of tasks presented to students (e.g., Fujita & Jones, 2014; Otten et al., 2014; Thompson et al., 2012), the justifications expected from students (e.g., Bieda et al., 2014; Stylianides, 2009), and the frequency of tasks within particular content areas (e.g., Bieda et al., 2014; Hong & Choi, 2014) or throughout entire textbooks (e.g., Davis et al., 2014; Otten et al., 2014). For example, Stylianides (2009) examined students' textbooks focusing on the warrants provided for a mathematical claim, the frequency with which proving-and-reasoning related tasks appeared in the text, and the argument type (either proof or non-proof) anticipated by students. Researchers have also examined the textbook guidance offered for teachers to teach proving (e.g., McCory & Stylianides, 2014; G. J. Stylianides, 2007). G. J. Stylianides (2007), for example, analyzed the teacher's edition of a textbook to examine the nature of guidance offered to support teachers in their implementation of proving-related activities.

METHODS: CROSS-CULTURAL ANALYSIS OF PROVING-RELATED ACTIVITIES

In this section, we describe the methodology for this study. This includes our approach to selecting textbooks from Korea, Singapore, and the United States, and the methods we used to analyze proving-related tasks in the textbooks.

Data Sources

The tasks that are the focus of the analyses were drawn from textbook series from Korea, Singapore, and the United States. The three countries were chosen based on performance of students on international assessments (e.g., Trends in Mathematics and Science Study [TIMSS]). Korean and Singaporean students are typically among the highest performing students; U.S. students are typically "middle of the pack," performing behind many of their peers from other advanced industrial nations. We conjectured that such performance differences might be explained, at least in part, by differences in the textbooks themselves. We decided to focus on Grade 8 (13–14 year-old students) because national curriculum and standards guidelines (e.g., DoE, 2014; MoE, 2015; NCTM, 2000; NGA & CCSSO, 2010) for this grade level typically include an emphasis on engaging students in proving-related activities. The specific textbooks were chosen based on two considerations: the textbooks are currently being used in their respective countries; and the authors had access to each textbook. In addition, the U.S. textbook was selected as a "best case scenario" with respect to its potential for engaging students in proving-related activities, given the explicit attention to such activities by the authors of books in this series (Knuth et al., 2009; Stylianides, 2009). The three textbooks selected for inclusion in the study are as follows:

- Korea: *Middle School Mathematics 2* (Kang et al., 2020)
- Singapore: *Think! Mathematics* (Heng et al., 2021a, 2021b)
- United States: *Connected Mathematics Project* (Lappan et al., 2013a, 2013b, 2013c).

We chose to use the teacher's editions for analysis because these include answer keys and suggestions for tasks.

Data Analysis

The tasks analyzed in each textbook were selected through the following sequential process. First, a random number generator was used to select a page, excluding "non-lesson" pages (e.g., Table of Contents, Index, Glossary). Second, the set of exercises within the lesson that included the random number generated page were examined. Third, tasks were identified from the set of exercises in the given lesson, where a task was defined as an exercise or part of an exercise with separate answers (cf. Davis et al., 2014). For example, an exercise with three parts is shown in Figure 7.1; in this case, this exercise has three tasks.

Figure 7.1

A Lesson Exercise (with the corresponding answer key at the bottom)

Give the coordinates for each landmark location:		
(a) art museum	(b) hospital	(c) greenhouse
Answers: (a) (6, 1)	(b) (-6, -4)	(c) (-6, 0)

Source: Adapted from Lappan et al. (2013b, p. 176).

We followed this process for a 30% sample of the pages (excluding non-lesson pages), a sample size informed by prior textbook analyses (e.g., Bieda et al., 2014; Stylianides, 2009; Thompson et al., 2012).

In the first phase of coding, tasks were identified from among the lesson exercises. Each task was coded as proving-related (or not) and coded with respect to the proving-related activities in which students are expected to engage. Using the exercise in Figure 7.1 as an example, three tasks were identified, and each task was coded as *not* proving-related because the responses expected from students do not involve proving-related activities (e.g., formulating a conjecture, evaluating a proof, providing a justification).

Those tasks that were identified as proving-related were then coded to distinguish the type of proving-related activities anticipated by textbook authors based on the language of a task (e.g., show, explain, justify), the corresponding answer key, and any guidance offered to teachers for implementing the task. In coding the type of proving-related activities, we drew from descriptions of activities from prior research (e.g., Ellis et al., 2019; Knuth et al., 2009; Stylianides, 2009) as well as from reform documents (MoE, 2015; NCTM, 2000; NGA & CCSSO, 2010). The resulting coding scheme is shown in Table 7.1.

Table 7.1

Types of Proving-Related Activities

Type	Definition	Example (Task and Answer)
Testing the Truth of a Conjecture	An activity that requires a judgement about the truth of a conjecture.	*Task*: Toby said, if an even number is multiplied by an odd number, the product is odd. Do you agree? *Answer*: True.
Formulating a Conjecture	An activity that requires a conjecture be made based on the information provided.	*Task*: Consider the sum of each number sentence: $2 + 4$, $6 + 6$, and $4 + 8$. Make a conjecture about what you notice. *Answer*: I think the sum of two even numbers will always be even.
Providing a Proof	An activity that requires a general argument to justify the truth of a conjecture.	*Task*: Show that the sum of two even numbers is always even. *Answer*: Even numbers can be represented by pairs with no leftovers (for example, 6 has 3 pairs, while 7 has 3 pairs and 1 leftover). The sum of two even numbers will consist of both sets of pairs with no leftovers, thus the sum of two even numbers is always even.
Identifying Patterns	An activity that requires the identification of a pattern based on a finite set of data.	*Task*: Ryan receives $10 each week. He had $20 in the first week, $30 in the second week, and so on. How much will Ryan have after 10 weeks? *Answer*: Ryan has $10n + 10$ in week n. Thus, he will have $110 in week 10.
Evaluating a Justification	An activity that requires a judgement regarding the validity of an argument of a conjecture's truth.	*Task*: Cindy gave a proof for the following conjecture: A sum of two odd numbers is even. I checked $3 + 5$, $7 + 11$, and $15 + 47$ and each time the sum was even. *Answer*: Cindy's argument only shows that the sum will be even for three instances, but it does not show it will be true for all possibilities.

As an example of using this coding scheme, consider the task in Figure 7.2, which requires students to determine whether the given equation has a solution. In this case, the proving-related activity was identified as *testing the truth of a conjecture*. In coding a task posed as "determine if the equation has any solution" as *testing the truth of a conjecture*, we viewed the conjecture as "the equation has a solution" if this did not significantly change the intent of the task. (The issue of translating the language of a task will be revisited in a later section of this chapter.) Note also that based on the response expected from students of a proof to justify the conclusion "there is no solution" (as shown in the answer key), the task was also coded for the proving-related activity *providing a proof*.

Figure 7.2

A Task Coded for Testing the Truth of a Conjecture and Providing a Proof

Determine if the equation (below) has any solution.

$$\frac{\dfrac{1}{x^2 - 49} + \dfrac{2}{x - 7}}{\dfrac{4}{x + 7} - \dfrac{2}{x - 7}} = 1$$

Source: Adapted from Heng et al. (2021a, Chapter 6, p. 24).

RESULTS

As shown in Table 7.2, the number of tasks contained in the 30% sample of each textbook is considerably different in the Korean textbook (381 tasks) compared with the Singaporean and American textbooks (1,219 tasks and 1,131 tasks, respectively). The number of proving-related tasks is also quite different across the three textbooks, ranging from 29 (Korean textbook) to 117 tasks (American textbook); however, the proportion of proving-related tasks to total tasks has much less variation (5–10% across the three textbooks). Given the emphasis expressed by mathematics educators as well as in national curriculum standards for engaging students in proving-related activities, these results suggest the need for greater attention by curriculum developers. We return to this idea in the "Lessons Learned" section of this chapter.

We looked at the types of proving-related activities in which students are expected to engage as a function of textbook. Summary results are shown in Table 7.3.[2] The majority of proving-related activities involved tasks requiring students to test the truth of a conjecture, with such activities

Table 7.2

Summary Counts of Tasks and of Proving-Related Tasks for Three Eighth-Grade Textbooks From Three Countries

Textbooks	Number of Tasks	Number of Proving-Related Tasks
Middle School Mathematics 2 (Korea)	381	29 (8%)
Think! Mathematics (Singapore)	1,219	61 (5 %)
Connected Mathematics (United States)	1,131	117 (10%)

accounting for 55%, 77%, and 84% of the Korean, Singaporean, and American proving-related tasks, respectively. Students were expected to provide a proof in 48% and 25% of proving-related activities in the Korean and Singaporean textbooks, respectively, whereas none of the activities in the American textbook received this code. Only the American textbook included expectations for students to identify patterns and to formulate conjectures (9% of proving-related tasks).

Table 7.3

Summary Counts of Proving-Related Activities in Three Eighth-Grade Textbooks From Three Countries

Textbook	Identifying Patterns	Providing a Proof	Evaluating a Justification	Formulating a Conjecture	Testing the Truth of a Conjecture
Middle School Mathematics 2 (Korea)	0 (0%)	14 (48%)	0 (0%)	0 (0%)	16 (55%)
Think! Mathematics (Singapore)	0 (0%)	15 (25%)	0(0%)	0 (0%)	47 (77%)
Connected Mathematics (United States)	10 (9%)	0 (0%)	1 (0.1%)	10 (9%)	98 (84%)

Two examples of tasks, each intended to solicit two proving-related activities, are shown in Figure 7.3 and Figure 7.4. The first task (Figure 7.3), from the Korean textbook, requires students to first decide whether the conjecture (i.e., the triangles are congruent) is true, and then provide a reason for the decision. The answer key for this activity indicates a particular type of reason is expected—a proof. (The expectation for students to provide a proof may not be well conveyed to students given the prompt requesting them to provide a reason to a friend, which is another point we come back to in the *Lessons Learned* section of this chapter.) This activity was coded as *testing the truth of a conjecture* and *providing a proof*. In the second activity shown in Figure 7.4 from the U.S. textbook, students are expected to notice and identify a pattern (Part b), and then provide a conjecture based on the pattern noticed. This activity was coded as *identifying a pattern* and *formulating a conjecture*.

Figure 7.3

Sample Task Illustrating Anticipated Proving-Related Activity from the Korean Textbook

> (Translation from Korean to English): Decide the congruence of two right triangles with a leg of the same length and two angles (not the right angle) respectively equal to each other. Tell the reason to your friend.

Source: Adapted from Kang et al. (2020, p. 343).

Figure 7.4

Sample Task Illustrating Anticipated Proving-Related Activity from the U.S. Textbook

> Half circles have been drawn on the sides of the right triangle (at right). Find the area of each half circle. How are the areas of the half circles related to each other?

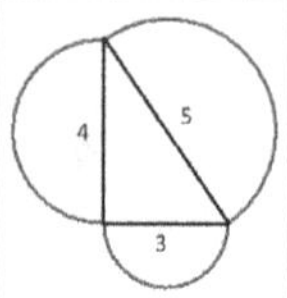

Source: Adapted from Lappan et al. (2013b, p. 216).

In general, there was lack of variety in the types of anticipated proving-related activities, with only two types of anticipated proving-related activities appearing in the Korean and Singaporean textbooks, and four types of anticipated proving-related activities appearing in the American textbook. In the American textbook, however, three of the four activities only occurred in a small proportion of the tasks (and one of these three activities, evaluating a justification, only once). Clearly, if textbooks are to provide students with opportunities to engage in various types of proving-related activities, then curriculum developers need to pay more explicit attention to providing such opportunities.

LESSONS LEARNED

In conducting this study, several issues surfaced with respect to the methodology and interpreting the results. In this section, we highlight each issue and briefly discuss the "lessons learned," as they may be useful for others interested in conducting curriculum analyses and, in particular, cross-cultural curriculum analyses.

Methodological Considerations

Several methodological considerations arose. These pertain to what constitutes a task, what tasks to count, mathematical content addressed, choice of student or teacher textbook edition, and meanings and interpretations of translations.

What Counts as the Unit of Analysis

In our review of prior curriculum studies, it was clear that the choice for unit of analysis varied across studies. In our case, we decided that a *task* was a problem presented to students that had its own distinct answer. In other words, a problem with three sub-problems was counted as three tasks. Such a choice leads to different counts for the number of tasks presented to students, relative to a study that may consider a problem with multiple sub-problems as a single task. Thus, the choice for the unit of analysis, a *task* in this case, has implications for interpreting results across studies and for making generalizations about results across studies.

What Problems to Consider

In addition to considerations for what constitutes the unit of analysis, choices were made about which problems to consider in the analysis. For example, in the *Connected Mathematics* (United States) textbook, every unit contains Application, Connection, and Extension (ACE) problems designed to provide additional learning opportunities for students. Are additional problems, like the ACE problems that are not included within an individual lesson, to be considered in the analysis? A related issue concerns additional problems accessible given the digital (or web-based) format of many textbooks. Both the *Middle School Mathematics 2* (Korea) and *Think! Mathematics* (Singapore) textbooks were also available in digital formats that provided additional problem sets for students; the *Connected Mathematics* textbook includes hyperlinks that direct students to additional materials (e.g., worksheets). Thus, the availability of additional problems external to the physical textbook adds another layer of complexity to considerations of what to include in textbook analyses.

Same Grade Same Content?

Even when conducted at the same grade level, in cross-cultural textbook analyses, the content as well as specific topics addressed may be very different. In the U.S., for example, much of the algebraic content in the middle-school grades (11–14 year-old students) might be considered prealgebra, whereas in other countries, like Korea and Singapore, the algebraic content might be considered more similar to first-year algebra.[3] Similarly, specific topic coverage in a textbook can vary across countries. For example, the equation of a circle with a center at the origin is addressed in the Grade 8 *Connected Mathematics* (United States) textbook; however, neither the Korean nor the Singaporean textbooks address the equation of a circle in Grade 8. Thus, the similarity (or difference) in content, as well as topics addressed, are important considerations for conducting international curricular studies.

Student Edition Versus Teacher Edition

The choice of textbook edition—student or teacher—can also lead to different conclusions about expectations for students. For example, the task shown in Figure 7.2, on its own (as shown in the student edition of the textbook), might lead us to have coded it as *testing the truth of a conjecture* (i.e., a solution exists). However, in reviewing the teacher edition of the textbook, it was clear that the anticipated response from students was not only to determine if a solution exists (*test the conjecture*), but also to *prove that a solution does (or does not) exist*. In addition, the teacher edition of textbooks varied in the nature of guidance offered to teachers with respect to the tasks presented to students. For example, the *Think! Mathematics* (Singapore) teacher edition offers specific questions to pose for advanced learners as well as for struggling learners, and the *Middle School Mathematics 2* (Korea) teacher edition offers guidance about what teachers should pay attention to when evaluating students' work and what questions might be helpful to extend a student's understanding of a task. Thus, understanding what is expected from students as they engage with a task may vary depending on whether one considers the student or teacher edition of a textbook.

Language Translation and Meaning

Conducting cross-cultural curriculum analyses often requires both the translation of text and the interpretation of the translated text. In the case of the research reported here, two of the authors are native speakers of Korean, who translated Korean words into their English word equivalents. However, what counts as "equivalent" can be challenging; for example, the Korean verb prompt 설명하시오 may be translated as *describe*, *demonstrate*, or *explain*. In contrast, the Singaporean textbook, written in British English, did not require a translation into English. However, can one assume, for example, the verb prompt *explain* has the same meaning in the Singapore textbook as it does in the American textbook? Thus, consideration must be given to the cultural context when translating text as well as when giving meaning to the translated text.

Interpreting the Results

The results of this study also raised considerations for researchers conducting cross-cultural textbook analyses and curriculum developers. We now briefly discuss the lessons learned from these considerations.

What Counts as ...?

Earlier in this chapter, we noted a challenge with translating words into English and on deciding their English word equivalents, but there is an additional challenge related to interpretation of the translated words. For example, consider the use of verb prompts. In proving-related tasks, the Korean textbook uses the verb *explain* with the expectation that students will provide a proof, whereas the Singaporean textbook uses the verbs *explain*, *show*, and *prove* with the expectation that students will provide a proof. Our ascribing meaning to prompts such as these was based on the expectations inferred from answer keys provided in teacher editions of the textbooks. Yet, what constitutes an explanation is also dependent on the classroom norms in place (e.g., Yackel & Cobb, 1996), and serves to highlight potential differences between the intended curriculum and the implemented curriculum (Schmidt et al., 1997). Thus, decisions about what is expected from students as they engage with curricular tasks needs to take into account the challenges associated with language translation and meaning, while at the same time, considering cultural and classroom norms with respect to word meaning.

Curriculum Design

This study also raised several considerations with respect to curriculum design and students learning to prove. First, given recommendations that proving-related activities be a regular and consistent part of students' mathematics education, the results of this study suggest a need for greater attention to including tasks that provide students with opportunities to engage in more proving-related activities, as well as a variety of proving-related activities. Second, although not addressed in our study, we wonder whether proving-related activities are distributed equally across the curriculum or concentrated in specific mathematics content within the curriculum. Finally, we also raise a question about the sequencing of student opportunities to engage in proving-related activities; in other words, are the opportunities intentionally designed to support students' transition from informal reasoning to more formal reasoning (i.e., proofs). Thus, curriculum designers need to take into consideration the number of opportunities for students to engage in proving-related activities, the distribution of these opportunities across the curriculum and across content areas, and the sequencing of these opportunities so they promote students' learning to prove.

CONCLUDING REMARKS

In this chapter, we reported brief results from a study that examined the nature of proving-related activities across middle-grades textbooks from three countries: Korea, Singapore, and the United States. In sharing lessons we learned from conducting the study, we hope that researchers interested in conducting textbook analyses, especially cross-cultural analyses, find useful insights about their own study design, analysis, and interpretation of results. We also noted lessons learned that might be helpful for curriculum developers with respect to students' learning concepts related to proof.

REFERENCES

Alcock, L. (2010). Mathematicians' perspectives on the teaching and learning of proof. In F. Hitt, D. Holton, & P. Thompson (Eds.), *CBMS issues in mathematics education* (Vol. 16, pp. 63–91). American Mathematical Society.

Alcock, L., & Inglis, M. (2008). Doctoral students' use of examples in evaluating and proving conjectures. *Educational Studies in Mathematics, 69*(2), 111–129. https://doi.org/10.1007/s10649-008-9149-x

Australian Curriculum, Assessment and Reporting Authority. (2015). *Australian curriculum: Mathematics*. Retrieved July 28, 2021, from https://australiancurriculum.edu.au/f-10-curriculum/mathematics/

Ball, D. L., & Cohen, D. K. (1996). Reform by the book: What is—or might be—the role of curriculum materials in teacher learning and instructional reform? *Educational Researcher*, 25(9), 6–14. https://doi.org/10.3102/0013189X025009006

Begle, E. G. (1973). Some lessons learned by SMSG. *The Mathematics Teacher*, 66(3), 207–214. https://doi.org/10.5951/MT.66.3.0207

Bergwall, A. (2021). Proof-related reasoning in upper secondary school: Characteristics of Swedish and Finnish textbooks. *International Journal of Mathematical Education in Science and Technology*, 52(5), 731–751. https://doi.org/10.1080/0020739X.2019.1704085

Bergwall, A., & Hemmi, K. (2017). The state of proof in Finnish and Swedish mathematics textbooks—Capturing differences in approaches to upper-secondary integral calculus. *Mathematical Thinking and Learning*, 19(1), 1–18. https://doi.org/10.1080/10986065.2017.1258615

Bieda, K. N., Ji, X., Drwencke, J., & Picard, A. (2014). Reasoning-and-proving opportunities in elementary mathematics textbooks. *International Journal of Educational Research*, 64, 71–80. https://doi.org/10.1016/j.ijer.2013.06.005

Cai, J., & Cirillo, M. (2014). What do we know about reasoning and proving? Opportunities and missing opportunities from curriculum analyses. *International Journal of Educational Research*, 64, 132–140. https://doi.org/10.1016/j.ijer.2013.10.007

Charalambous, C., Delaney, S., Hui-Yu Hsu, H., & Mesa, V. (2010). A comparative analysis of the addition and subtraction of fractions in textbooks from three countries. *Mathematical Thinking and Learning*, 12, 117–151. https://doi.org/10.1080/10986060903460070

Chazan, D. (1993). High school geometry students' justification for their views of empirical evidence and mathematical proof. *Educational Studies in Mathematics*, 24(4), 359–387. https://doi.org/10.1007/BF01273371

Davis, E. A., & Krajcik, J. S. (2005). Designing educative curriculum materials to promote teacher learning. *Educational Researcher*, 34(3), 3–14. https://doi.org/10.3102/0013189X034003003

Davis, J. D., Smith, D. O., Roy, A. R., & Bilgic, Y. K. (2014). Reasoning-and-proving in algebra: The case of two reform-oriented U.S. textbooks. *International Journal of Educational Research*, 64, 92–106. https://doi.org/10.1016/j.ijer.2013.06.012

Dawkins, P. C., & Weber, K. (2017). Values and norms of proof for mathematicians and students. *Educational Studies in Mathematics*, 95(2), 123–142. https://doi.org/10.1007/s10649-016-9740-5

Department for Education [DoE]. (2014). *Mathematics programmes of study: Key stage 4* (National curriculum in England). https://assets.publishing.service.gov.uk/government/uploads/system/uploads/attachment_data/file/331882/KS4_maths_PoS_FINAL_170714.pdf

Ellis, A. B., Ozgur, Z., Vinsonhaler, R., Dogan, M. F., Carolan, T., Lockwood, E., Knuth, E., & Zaslavsky, O. (2019). Student thinking with examples: The criteria-affordances-purposes-strategies framework. *Journal of Mathematical Behavior*, 53, 263–283. https://doi.org/10.1016/j.jmathb.2017.06.003

Epstein, D., & Levy, S. (1995). Experimentation and proof in mathematics. *Notices of the AMS, 42*(6), 670–674.

Fan, L. (2013). Textbook research as scientific research: Towards a common ground on issues and methods of research on mathematics textbooks. *ZDM Mathematics Education, 45*(5), 765–777. https://doi.org/10.1007/s11858-013-0530-6

Fujita, T., & Jones, K. (2014). Reasoning-and-proving in geometry in school mathematics textbooks in Japan. *International Journal of Educational Research, 64*, 81–91. https://doi.org/10.1016/j.ijer.2013.09.014

Grouws, D. A., Smith, M. S., & Sztajn, P. (2004). The preparation and teaching practices of U.S. mathematics teachers: Grades 4 and 8. In P. Kloosterman & F. Lester (Eds.), *The 1990 through 2000 mathematics assessments of the National Assessment of Educational Progress: Results and interpretations* (pp. 221–269). National Council of Teachers of Mathematics.

Heng, C., Yeo, J., Lee, S., Seng, T., Lai, W., & Hilong, W. (2021a). *Think! Mathematics: 2A*. Shing Lee Publisher.

Heng, C., Yeo, J., Lee, S., Seng, T., Lai, W., & Hilong, W. (2021b). *Think! Mathematics: 2B*. Shing Lee Publisher.

Hong, D., & Choi, K. (2014). A comparison of Korean and American secondary school textbooks: The case of quadratic equations. *Educational Studies in Mathematics, 85*(2), 241–263. https://doi.org/10.1007/s10649-013-9512-4

Jeannotte, D., & Kieran, C. (2017). A conceptual model of mathematical reasoning for school mathematics. *Educational Studies in Mathematics, 96*(1), 1–16. https://doi.org/10.1007/s10649-017-9761-8

Kang, O., Kwon, E., Hwang, H., Jeon, D., Noh, J., Woo, H., Yoon, S., Lee, H., Yoo, S., Yoon, H., Hong, C., & Jung, K. (2020). *Middle School Mathematics 2*. Donga Publication.

Kim, H. (2021). Problem posing in the instruction of proof: Bridging everyday lesson and proof. *Research in Mathematical Education, 24*(3), 255–278. http://doi.org/10.7468/jksmed.2021.24.3.255

Kim, H. (2022). Secondary teachers' views about proof and judgements on mathematical arguments. *Research in Mathematical Education, 25*(1), 65–89. https://doi.org/10.7468/jksmed.2022.25.1.65

Kim, H. (2023). *Investigating the nature of opportunities for proving-related activities in Korean, Singaporean, and United States secondary textbooks* [Doctoral dissertation, University of Texas at Austin]. https://repositories.lib.utexas.edu/bitstream/handle/2152/120498/KIM-DISSERTATION-2023.pdf

Knuth, E. J. (2002a). Teachers' conceptions of proof in the context of secondary school mathematics. *Journal of Mathematics Teacher Education, 5*, 61–88. https://doi.org/10.1023/A:1013838713648

Knuth, E. J. (2002b). Secondary school mathematics teachers' conceptions of proof. *Journal for Research in Mathematics Education, 33*(5), 379–405. https://doi.org/10.2307/4149959

Knuth, E. J., Choppin, J., & Bieda, K. (2009). Middle school students' production of mathematical justifications. In D. Stylianou,, M. Blanton, & E. Knuth (Eds.), *Teaching and learning proof across the grades: A K–16 perspective* (pp. 153–170). Routledge.

Lappan, G., Phillips, E., Fey, J., & Friel, S. (2013a). *Thinking with mathematical models*. Prentice Hall.

Lappan, G., Phillips, E., Fey, J., & Friel, S. (2013b). *Looking for Pythagoras*. Prentice Hall.

Lappan, G., Phillips, E., Fey, J., & Friel, S. (2013c). *Growing, growing, growing*. Prentice Hall.

Li, Y. (2000). A comparison of problems that follow selected content presentations in American and Chinese mathematics textbooks. *Journal for Research in Mathematics Education*, *31*(2), 234–241. https://doi.org/10.2307/749754

McCory, R., & Stylianides, A. J. (2014). Reasoning-and-proving in mathematics textbooks for prospective elementary teachers. *International Journal of Educational Research*, *64*, 119–131. https://doi.org/10.1016/j.ijer.2013.09.003

Ministry of Education [MoE]. (2015). *2015 Korean revised mathematics curriculum*. Ministry of Education.

Miyakawa, T. (2017). Comparative analysis on the nature of proof to be taught in geometry: The cases of French and Japanese lower secondary schools. *Educational Studies in Mathematics*, *94*(1), 37–54. https://doi.org/10.1007/s10649-016-9711-x

Mullis, I. V. S., Martin, M. O., Foy, P., & Hooper, M. (2016). *TIMSS 2015 International Results in Mathematics*. Retrieved from Boston College, TIMSS & PIRLS International Study Center website: http://timssandpirls.bc.edu/timss2015/international-results/

National Council of Teachers of Mathematics [NCTM]. (1989). *Curriculum and evaluation standards for school mathematics*.

National Council of Teachers of Mathematics. (1991). *Professional standards for teaching mathematics*.

National Council of Teachers of Mathematics. (2000). *Principles and standards for school mathematics*.

National Governors Association [NGA] Center for Best Practices & Council of Chief State School Officers [CCSSO]. (2010). *Common core state standards for school mathematics*. http://www.thecorestandards.org/Math

Otten, S., Males, L. M., & Gilbertson, N. J. (2014). The introduction of proof in secondary geometry textbooks. *International Journal of Educational Research*, *64*, 107–118. https://doi.org/10.1016/j.ijer.2013.08.006

Reid, D. A., & Knipping, C. (2010). *Proof in mathematics education: Research, learning and teaching*. Sense.

Remillard, J. T., & Bryans, M. B. (2004). Teacher's orientations toward mathematics curriculum materials: Implications for teacher learning. *Journal for Research in Mathematics Education*, *33*(5), 352–388. https://doi.org/10.2307/30034820

Schmidt, W. H., McKnight, C. C., & Raizen, S. A. (1997). *A splintered vision: An investigation of U.S. science and mathematics education*. Kluwer.

Schoenfeld, A. H. (1994). What do we know about mathematics curricula? *Journal of Mathematical Behavior*, *13*(1), 55–80. https://doi.org/10.1016/0732-3123(94)90035-3

Senk, S. (1985). How well do students write geometry proofs? *The Mathematics Teacher*, *78*(6), 448–456. https://doi.org/10.5951/MT.78.6.0448

Stein, M. K., Remillard, J., & Smith, M. S. (2007). How curriculum influences student learning. In F. K. Lester, Jr. (Ed.), *Second handbook of research on mathematics teaching and learning* (pp. 319–370). National Council of Teachers of Mathematics.

Stylianides, A. J. (2007). Proof and proving in school mathematics. *Journal for Research in Mathematics Education, 38*(3), 289–321.

Stylianides, A. J., Bieda, K. N., & Morselli, F. (2016). Proof and argumentation in mathematics education research. In A. Gutiérrez, G. C. Leder, & P. Boero (Eds.), *The second handbook of research on the psychology of mathematics education* (pp. 315–351). Sense.

Stylianides, G. J. (2007). Investigating the guidance offered to teachers in curriculum materials: The case of proof in mathematics. *International Journal of Science and Mathematics Education, 6*(1), 191–215. https://doi.org/10.1007/s10763-007-9074-y

Stylianides, G. J. (2009). Reasoning-and-proving in school mathematics textbooks. *Mathematical Thinking and Learning, 11*(4), 258–288. https://doi.org/10.1080/10986060903253954

Stylianides, G. J. (2014). Textbook analyses on reasoning-and-proving: Significance and methodological challenges. *International Journal of Educational Research, 64*, 63–70. https://doi.org/10.1016/j.ijer.2014.01.002

Stylianides, G. J., Stylianides, A. J., & Shilling-Traina, L. N. (2013). Prospective teachers' challenges in teaching reasoning-and-proving. *International Journal of Science and Mathematics Education, 11*(6), 1463–1490. https://doi.org/10.1007/s10763-013-9409-9

Stylianides, G. J., Stylianides, A. J., & Weber, K. (2016). Research on the teaching and learning of proof: Taking stock and moving forward. In J. Cai (Ed.), *Compendium for research in mathematics education* (pp. 237–266). National Council of Teachers of Mathematics.

Thompson, D. R., Senk, S. L., & Johnson, G. J. (2012). Opportunities to learn reasoning and proof in high school mathematics textbooks. *Journal for Research in Mathematics Education, 43*(3), 253–295. https://doi.org/10.5951/jresematheduc.43.3.0253

Valverde, G. A., Bianchi, L. J., Wolfe, R. G., Schmidt, W. H., & Houang, R. T. (2002). *According to the book: Using TIMSS to investigate the translation of policy into practice through the world of textbooks.* Springer Science & Business Media.

Yackel, E., & Cobb, P. (1996). Sociomathematical norms, argumentation, and autonomy in mathematics. *Journal for Research in Mathematics Education, 27*(4), 458–477. https://doi.org/10.2307/749877

Zhang, D., & Qi, C. (2019). Reasoning and proof in eighth-grade mathematics textbooks in China. *International Journal of Educational Research, 98*, 77–90. https://doi.org/10.1016/j.ijer.2019.08.015

ENDNOTES

1. The study is from the first author's dissertation (Kim, 2023).

2. Percentage sums may exceed 100% due to a proving-related task being coded for multiple proving-related activities.
3. This observation is based on the third author's experience conducting algebra-related research at the middle-grades level for over 25 years.

CHAPTER 8

LESSONS LEARNED IN CREATING MOSAICS TO PORTRAY THE ALGEBRA STRAND OF U.S. HIGH-SCHOOL TEXTBOOKS

Mary Ann Huntley
Cornell University

Maria S. Terrell
Cornell University

Nicole L. Fonger
Syracuse University

By describing the methodology and selected findings from an in-depth content analysis of the algebra strand of six U.S. high-school textbook series, we provide a context to discuss the lessons learned and solutions to challenges we encountered along the way. The metaphor of mosaics guided our development of timeline plots, which are introduced as a novel way to represent the breadth, sequence, and depth of topics covered in each textbook series.

Keywords: algebra; curriculum; high school; mathematics; textbook analysis

The broad question motivating this study was: *What is high-school[1] algebra in the United States (U.S.), as defined by textbooks?* We consider the algebra strand of the school curriculum to include symbolic algebra and functions. The impetus for this study was threefold. First, textbooks and other materials used during classroom instruction have a strong influence on teachers' practice and students' opportunity to learn mathematics (e.g., Begle, 1973; Fan & Kaeley, 2000; Floden, 2002; Schmidt et al., 1997; Thompson et al., 2012). Second, algebra plays a central role in the school mathematics curriculum (e.g., Committee on Prospering in the Global Economy of the 21st Century, 2007; Gamoran & Hannigan, 2000; Katz, 2007; Moses & Cobb, 2001; RAND Mathematics Study Panel, 2003). Third, there is large variation in the content and nature of algebraic activity in Grades K–12 mathematics classes across schools in the U.S. and in other countries (Kendal & Stacey, 2004). As stated by Yerushalmy and Chazan (2002), "school algebra is a complicated curricular arena to describe, one that is undergoing change" (p. 725). Taken together, these factors convinced us to develop an approach to conduct an in-depth analysis of the algebra strand of several high-school textbook series.

We knew from the outset of the project that we wanted to examine two main dimensions used for textbook analysis: *content*, which is the subject matter within a field of study; and *cognition*, which includes the sets of cognitive processes expected of students as they engage with the content. Before we began coding textbooks, we realized that other information could easily be obtained during our coding, including whether problems were set in a real-world context, and whether solving problems would likely involve the use of technology or manipulatives. Our focus on these curriculum variables motivated the following three research questions:

1. What is the content, including the breadth, sequence, and depth of topics covered?
2. What sets of behaviors are expected of students as they engage with the content?
3. To what extent are problems set in real-world contexts, and to what extent are tools (technology and manipulatives) expected to be used (per the authors' intentions) to solve problems?

GENESIS OF THE STUDY

As originally conceived, this textbook analysis study was the first phase of a larger U.S. National Science Foundation (NSF)-funded project involving a cross-curricular comparative study of the effects of various approaches to high-school algebra. Due to numerous factors, the nature of the project changed and some planned phases could no longer be carried out. At this point, the scope of the textbook analysis project widened, and we decided to rigorously apply coding taxonomies to *all* items from the narrative and homework sections of four textbook series. This approach is more comprehensive than other researchers' methodologies that involved analyzing portions of textbooks, such as specific problem sets (e.g., Brown et al., 2013; Sherman et al., 2016), certain pages (e.g., Rivers, 1990), tables of contents (e.g., Chávez et al., 2009; Chávez et al., 2010), or identified learning goals (e.g., American Association for the Advancement of Science [AAAS], 1999). Midway through coding the textbooks for this study, we obtained supplemental funding from the NSF, allowing us to code two additional series, bringing the total to six textbook series.

The idea of *mosaics* served as the guiding metaphor for the project. Each tile in a mosaic conveys a small piece of information. It is only when all tiles are put together that a picture emerges. The analogous situation for this project is that our coding of several dimensions of every item in six textbook series resulted in an enormous dataset, with each individual code carrying a small bit of information. Yet, as one looks across the codes for all items in a textbook, one can see a distinct *portrait of algebra* for each textbook series.

In the remainder of this chapter, we discuss the methodology for the study, decisions we made along the way, rationales for our choices, and lessons learned. We restrict our presentation of results to one aspect of the content analysis to illustrate how we constructed timeline plots (i.e., mosaics) to address the first research question.

CODING THE TEXTBOOKS (GATHERING THE TILES)

Although we were not conducting an evaluation study, our research design was strongly influenced by the report from a committee of the National Research Council (2004) that provides guidance for researchers who are evaluating mathematics textbooks. This report identifies elements that an effective evaluation study needs to encompass. Our research design was also informed by other researchers' methodological approaches to textbook analysis (e.g., AAAS, 1999; Adams et al., 2000; Chávez et al., 2010; Senk et al., 2008). In this section, we discuss how we selected textbooks to include in the study, and how we obtained the unit of analysis, analytic frameworks, and coding procedures.

Selecting and Obtaining Textbooks

Wanting to document the wide variation of approaches to high-school algebra, we cast a broad net when selecting textbooks to include in the study. We chose some books that are organized in a *subject specific fashion*, with each content strand appearing in a separate course, and others that are *integrated*, with all content strands appearing in each course. Of those textbooks that are subject specific, we chose some that underwent extensive field-testing during development, and others that were commercially generated. We chose some textbooks that received funding for their development from the NSF. For each type of textbook, we chose those with the greatest market penetration when the study began in 2008. Data about market penetration were obtained through conversations with textbook publishers, textbook writers, and reports (e.g., Weiss et al., 2001). The six textbook series that were selected for this study are listed in Table 8.1.

Table 8.1

Textbooks Included in the Study

Textbook Series	Book Titles/Units	Publisher	Year
INTEGRATED[a]			
Core-Plus Mathematics Program (CPMP)	Course 1 Course 2 Course 3	Glencoe/ McGraw-Hill	2008–2009
Interactive Mathematics Program (IMP)	Year 1 Year 2 Year 3	Key Curriculum Press	2008–2009
SUBJECT-SPECIFIC—EXTENSIVE FIELD-TESTING DURING DEVELOPMENT			
Center for Mathematics Education (CME)	*Algebra 1* *Algebra 2*	Pearson	2009
Discovering Mathematics (DM)	*Algebra* *Advanced Algebra*	Key Curriculum Press	2007, 2004
University of Chicago School Mathematics Program (UCSMP)[b]	*Algebra* *Advanced Algebra*	Wright Group/ McGraw-Hill	2008–2009
SUBJECT-SPECIFIC—COMMERCIALLY GENERATED			
Glencoe	*Algebra 1* *Algebra 2*	Glencoe/ McGraw-Hill	2008

[a] The NSF program solicitation from which CPMP and IMP were subsequently developed required instructional materials resulting from these grants to be integrated (NSF, 1991).

[b] Although the UCSMP textbooks are subject specific, there is some integrated content (e.g., geometry in *Advanced Algebra*, and statistics in *Algebra*).

For the integrated textbook series (CPMP and IMP), only those units within the first three years that have a major focus on algebra were coded. These units were identified through discussions with the textbook series' authors. For the subject-specific textbook series, the entire *Algebra 1*

(CME, Glencoe)/*Algebra* (DM, UCSMP) and *Algebra 2* (CME, Glencoe)/ *Advanced Algebra* (DM, UCSMP) books were coded. At the onset of the study, we hypothesized that each of the six textbook series represents a distinct approach to algebra. Citations for the six textbook series that were analyzed are provided in the Appendix. It was relatively easy to obtain textbooks free of charge from publishers who were eager to see their books mentioned in research publications. But one must be careful in communicating to publishers that there is no guarantee that the research findings will cast a positive light on the instructional materials. We have found that publishers who stand behind their materials are willing to assume this risk.

We wanted to code all items that students have an opportunity to solve in the student's text. In the U.S., it is common for the student version of a textbook to be embedded within a corresponding teacher's edition. We chose to code from the teacher's editions of the books. In this way, we saw the problems for students and had insight into the authors' intended solution strategy for the problems.

Identifying the Unit of Analysis

The first research question calls for a fine-grained examination of the content of textbooks, including the breadth, sequence, and depth of topics covered. In the U.S., it is common for textbooks to be divided into chapters, with further subdivisions into lessons. Each lesson typically begins with some narrative, which is an expository text with worked-out examples and some problems/activities/investigations for students to solve or explore. This is followed by a set of exercises (homework problems) for students to complete. We wanted to gather as much information as possible about students' opportunities to learn the content, so we chose a textbook item as the unit of analysis. In other words, every part of a multi-part question (e.g., 1a, 1b, 1c) was assigned a set of codes. Every textbook item in the narrative and homework problems of each textbook series was coded, with two exceptions. We did not code worked-out examples in narrative portions of textbooks because students' cognitive processes may be reduced if they rely on the solutions when solving such problems. We also did not code UCSMP end-of-chapter projects because these are optional. If a textbook series included chapter reviews or tests within the student book, these were also coded. Hence, with rare exception, we coded every mathematical problem in each textbook series that was available for students to solve. This approach to textbook analysis—in which *every* problem was examined in both the narrative and exercises portions of the books—is unprecedented, and contrasts with the approach by Jones (this volume), who calls for a census approach looking at every page and all tasks within a sample.

Choosing Analytic Frameworks

To address the first two research questions that involve examining the content and cognitive behavior of textbook items, we decided to use existing analytic frameworks, instead of developing our own. Using frameworks that were established and widely used by other researchers allowed us to focus our energy on other aspects of the project. Moreover, we anticipated that using existing frameworks might provide an entry point for readers of our research who are already familiar with the frameworks.

Content Framework

The *Survey of Enacted Curriculum (SEC) K–12 Mathematics Taxonomy* (Wisconsin Center for Education Research [WCER], 2007) was used to code the mathematical content of the six textbook series. As shown in Figure 8.1, the SEC taxonomy consists of 16 general areas, and each of these 16 general areas is divided into 4–19 specific topics. In Figure 8.1, the subcodes in the three content domains of the taxonomy that are most relevant to this study—basic algebra, advanced algebra, and functions—are provided. The algebra and advanced algebra content areas of the taxonomy reflect a symbolic approach to algebra (e.g., the use of symbols as variables, the emphasis on writing equivalent expressions, and solving equations in one and two variables).

We choose to use the WCER (2007) content framework because of its straightforward listing of mathematical topics in the Grades K–12 curriculum. This taxonomy has a small grain size. Other frameworks we considered lacked specificity, and some conflated instructional approach with content. For example, consider code 602 in the WCER taxonomy, which is "systems of equations." The corresponding standard for Grade 12 in the framework for the National Assessment of Educational Progress (National Assessment Governing Board, 2008) includes not only this content but also methods of solution: "Solve (symbolically or graphically) a system of equations or inequalities and recognize the relationship between the analytical solution and graphical solution" (National Assessment Governing Board, 2008, p. 36).

We attempted to retain the intention of the codes in the original taxonomy, yet we treated the taxonomy as a living document, revising it as new issues arose. In our adapted SEC content taxonomy, there are 217 topic codes, including 16 that we added for this project within six (of the 16 different) taxonomic categories:

- Measurement (Units);
- Basic Algebra (Polynomial Expressions);

Figure 8.1

Coding Categories in the Content Analysis Taxonomy

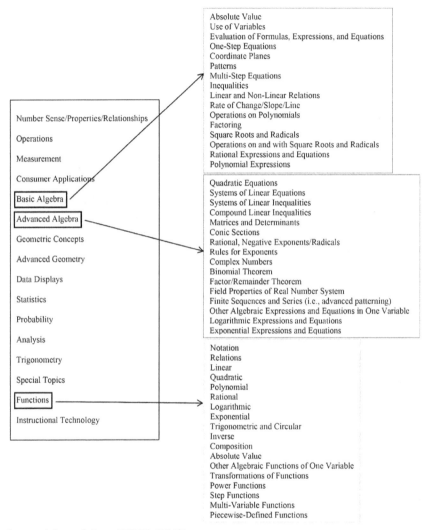

Source: Adapted from WCER (2007).

- Advanced Algebra (Finite Sequences and Series, Other Algebraic Expressions and Equations in One Variable, Logarithmic Expressions and Equations, Exponential Expressions and Equations);
- Statistics (Distributions, Experimental Design);
- Special Topics (History of Mathematics);

- Functions (Absolute Value Functions, Other Algebraic Functions of One Variable, Transformations of Functions, Power Functions, Step Function, Multi-Variable Functions, Piecewise-Defined Functions).

We added extensive annotations to the SEC Taxonomy (WCER, 2007) to illustrate how coders were to apply content codes to textbook items. We found it helpful to annotate with specific mathematics problems, sometimes with illustrative examples being actual problems within the textbooks we were coding. Other annotations consisted of notes to coders to facilitate consistent application of the content codes. For example, in Table 8.2 we provide a portion of the original and annotated taxonomies corresponding to conic sections (code 606) and rational, negative exponents/radicals (code 607) within the advanced algebra portion of the SEC Taxonomy (WCER, 2007).

Consistent with the approach of the developers of the SEC Taxonomy (WCER, 2007), we assigned one, two, or three content codes to each textbook item. See Table 8.3 for an illustration of applying one, two, or three codes to textbook items.

The content taxonomy is a coarse sieve. For example, with code 606 (conic sections), no distinction is made between different types of conic

Table 8.2

Portions of the Original and Annotated Content Taxonomy

Code	Portion of Original Content Taxonomy	Portion of Annotated Content Taxonomy
606	Conic Sections	Conic Sections (This includes equations of the form $Ax^2 + By^2 + Cx + Dy + E = 0$; i.e., circles, hyperbolas, parabolas, ellipses) e.g., (UCSMP Adv. Alg. Lesson 12-2 #3) Consider the circle with equation $x^2 + y^2 = 34^2$. What is the radius of the circle? <u>Note</u>: If the problem asks for the equation of the axis of symmetry, directrix, or asymptotes, do not also code 510 (Rate of Change/Slope/Line) because it is an assumed prerequisite.
607	Rational, Negative Exponents/Radicals	Rational, Negative Exponents/Radicals (excludes square roots unless the notation is $x^{\frac{1}{2}}$; apply this code if the problem focuses on notation) e.g., Evaluate $25^{\frac{-1}{2}}$, $16^{\frac{1}{4}}$, a^0, $(-4)^{-5}$ e.g., Write $6^{\frac{1}{5}}$ and $a^{\frac{n}{m}}$ in radical form. e.g., Simplify $121^{\frac{1}{2}}$, $125^{\frac{2}{3}}$

Source: Based on WCER (2007).

Table 8.3

Textbook Items With One, Two, and Three Content Code(s) Assigned

Textbook Item	Content Code(s)
3. Sketch each inequality on a number line. a. $x \leq -5$ b. $x > 2.5$ c. $-3 \leq x \leq 3$ d. $-1 \leq x < 2$	508 (Inequalities)
4. Consider the inequality $y < 2 - 0.5x$. a. Graph the boundary line for the inequality on axes scaled from -6 to 6 on each axis.	508 (Inequalities) 510 (Rate of Change/Slope/Line)
4. b. Determine whether each given point satisfies $y < 2 - 0.5x$. Plot the point on the graph you drew in 4a. Label the point T (true) if it is part of the solution or F (false) if it is not part of the solution region. i. $(1, 2)$ ii. $(4, 0)$ iii. $(2, -3)$ iv. $(-2, -1)$	503 (Evaluation of Formulas, Expressions, and Equations) 508 (Inequalities) 510 (Rate of Change/Slope/Line)

Note: Items are from *Discovering Algebra* (Lesson 5.6, p. 316).

sections, or evaluating, setting up, or solving equations involving conic sections, or graphing conic sections. We found it helpful to use a different color font for our annotations on the taxonomy to easily distinguish them from the original wording. This also helped focus our attention on the clarifying examples and advice to coders.

Cognitive Behavior Framework

To examine how the mathematical content within textbooks is intended to be experienced by students, we chose to use the categories of cognitive processes outlined in the *TIMSS Advanced 2008 Assessment Framework* (Garden et al., 2006).[2] This framework consists of three types (or domains) of cognitive behaviors expected of students as they engage with mathematical content in textbooks: *knowing*, *applying*, and *reasoning*. As shown in Table 8.4, each of these three domains is further divided into sub-categories, with a list of descriptors to identify the types of cognitive processes associated with each subcategory.

We assigned each textbook item a single cognitive code: K (knowing), A (applying), or R (reasoning). We originally attempted to code at a finer-grained level, namely, the cognitive processes associated within each cognitive domain. However, the process of coming to agreement across coders on which code to apply was extremely time consuming. The three broad domains of cognitive behavior were sufficient for our purposes; we reasoned that, if it was desired, secondary analysis of the data could dive deeper into the cognitive behavior codes at a finer grain size.

As with the content taxonomy, it was our goal to retain the intention of the codes in the original TIMSS cognitive behavior taxonomy. We made

Table 8.4

Overview of the TIMSS Cognitive Behavior Taxonomy

	Knowing	Applying	Reasoning
Definition	The facts, procedures, and concepts students need to know	The ability of students to make use of this knowledge to select or create models and solve problems	The ability to use analytical skills, generalize, and apply mathematics to unfamiliar or complex contexts
Cognitive Processes	Recall Recognize Compute Retrieve Recount	Select Represent Model Solve Routine Problems	Analyze Generalize Synthesize/Integrate Justify Solve Non-routine Problems

Source: Based on Garden et al. (2006).

extensive annotations to illustrate the codes and to ensure reliable application across coders. To illustrate use of the cognitive behavior taxonomy, Figure 8.2 contains problems from the UCSMP textbook series for each of the three cognitive domains. The content code for each of these problems is conic sections.

Figure 8.2

Sample Items Coded Within Each Cognitive Domain (Knowing, Applying, and Reasoning)

Knowing

Consider the circle with equation $x^2 + y^2 = 34^2$. What is the radius of this circle?

(UCSMP *Advanced Algebra*, p. 808)

Applying

Write an equation for two circles that are concentric and centered at the origin.

(UCSMP *Advanced Algebra*, p. 808)

Reasoning

Is there a maximum possible value for the eccentricity of an ellipse? Why or why not?

(UCSMP *Advanced Algebra*, p. 829)

Coding Real-World Context, Technology Use, and Manipulatives

We used a binary code to indicate which items were set in a real-world context. We did not code an item as having real-world context if the context was purely mathematical, such as the example shown in Figure 8.3.

Figure 8.3

A Problem that was Not Coded as Having Real-World Context

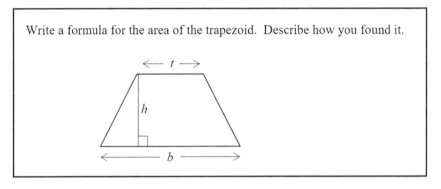

We used binary codes to indicate which items involved students using calculators[3] or computers, and which items indicated use of physical objects (manipulatives). We coded an item as involving technology or use of a manipulative only if we thought the authors' intent was for the problem to be solved using these tools, whether or not they were actually necessary to solve the problem. Indications that a problem required use of technology included a screenshot in the teacher or student solutions, examples in the narrative of solving similar problems using technology, and operations on large or small numbers (e.g., five or more digits, decimals with tenths or smaller). We acknowledge that in some cases the decision to code an item as involving the use of technology was high inference. In UCSMP, for example, there are no symbols in the books to indicate students should use technology, as the focus is on students determining when they might be appropriate or needed. Also, some items that might necessitate the use of calculators for a first course in algebra can often easily be done without technology in a second course in algebra.

Coding Procedures

A small, but diverse team of five people coded the textbooks. All have strong content knowledge and extensive experience in Grades K–12 mathematics education. When the textbooks were coded, two members of the team were university faculty—one with a doctorate in mathematics education and a master's degree in applied mathematics (Huntley), and one with a doctorate in pure mathematics and experience working with Grades K–12 teachers (Terrell). One other coder was, at the time, a postdoctoral research fellow at a large research university and has prior high-school teaching experience (Fonger). The two other coders were high-school classroom teachers with strong content knowledge, as evidenced by a bachelor's degree in mathematics and a master's degree in either mathematics or mathematics education, and extensive teaching experience (with more than 10 years at the high school level).

The lead researcher (Huntley) wrote a coding manual, which was used in a 2.5-day coder training session before coding began. Each coding team consisted of a teacher and either a university mathematics educator or mathematician. Teams were formed to avoid any perceived conflicts of interest (such as a textbook author being a dissertation advisor to a coder). Coding teams stayed intact for an entire textbook, but often were reformed for coding other books based on a coder's availability. Each coder worked separately to apply codes to items in the textbooks, which were entered into a spreadsheet called a *coding template*. As shown in Figure 8.4, the set of codes applied to each textbook item included up to three content codes (Topic1, Topic2, Topic3),[4] one cognitive behavior code (CogBeh),[5] and binary codes for use of calculator/computer (C), manipulatives (M), and context. There was also space for a coder to write a note. This was often used to document the rationale for applying a certain code or set of codes to an item.

Figure 8.4

Portion of the Coding Template

Rater:								
Book:								
Date:								
Item #	Topic1	Topic2	Topic3	CogBeh	C	M	Context	Comment

The coding pair discussed any differences in their assigned codes, and through a process of negotiation came to an agreement on the set of codes to apply to each item. All codes—content, cognitive behavior, technology, manipulatives, and context—were negotiated to reach 100% agreement across the two coders. Although we did not measure inter-rater reliability of our coding, pairs of coders came to 100% agreement on the set of codes assigned to every item across the six textbook series. This underscores the strength of our approach and the reliability of the results. We maintained consistency in coding, and minimized coding fatigue and drift, through frequent conversations across coding teams. As necessary, recoding was done when a clarification was made to the content taxonomy that changed the meaning of a code. A limitation is that we did not measure the extent to which coders remained consistent in their coding over time.

Practicing teachers were invited to serve on the team because of the researchers' strong conviction that teachers are valued research partners. Teachers' insights into the realities of classroom instructional practice give authenticity and grounding to research data. Throughout the project, the voices of teachers carried equal weight in discussions with researchers. Although teacher professional development was not one of the goals of the project, it was an unanticipated consequence of involving teachers as research partners. After coding over 40,000 textbook items, one teacher said that being involved in the project helped her become more mindful of her problem selection and its influence on students' opportunity to learn (Mayer et al., 2019). Although a reflective person by nature, being involved in the project was the impetus for her delving more deeply into the reasons her students might give incorrect answers to questions: "Are students' errors due to 1506 or is it 515? Or is it as fundamental as 511 (factoring) or 601 (solving quadratics)?" (Mayer et al., 2019, p. 384). She became more aware of the embedded content within problems and began planning her lessons more purposefully. As department chair, her experience with this research team influenced her discussions with novice teachers in her school. She spoke about the powerful impact that being involved in coding the textbooks had on her practice:

> Speaking as one of the math teachers on the research team, I felt that having those conversations (with a mathematician and mathematics educators) about reconciling the different codes was perhaps the best professional development I have ever been party to. Those talks also caused me to reexamine my practice in ways that I had not previously considered … the far-reaching consequences for me and my students are immeasurable. I am a better practitioner because of it. (Mayer et al., 2019, p. 383)

ANALYZING AND REPRESENTING THE DATA (FORMING MOSAICS)

Altogether, 63,174 textbook items were coded across the six textbook series. A statistician used the software package *Mathematica* (Wolfram Research Inc., Versions 8 [2010] and 9 [2012]) to tabulate and explore the data, and we used descriptive, rather than inferential statistics, to report the data. Returning to the metaphor of mosaics, we now describe the process of transforming the coding data (i.e., tiles) into results (i.e., mosaic portraits of algebra).

To respond to the first research question,[6] we needed to develop a way to represent the density, distribution, and sequencing of content across each textbook series. This was an enormous challenge given the large number of items that were coded. The statistician with whom we consulted devised a way to plot topic codes as a function of "time." We call these representations of the data *timeline plots*. A separate timeline plot was developed for each textbook series. Examples of timeline plots are shown in Figure 8.5a and Figure 8.5b, corresponding to the textbook series CME and CPMP, respectively.[7]

In a timeline plot, "time" is represented as the sequentially coded items in the textbook. Reading a plot from left to right for a subject-specific textbook series, such as CME (Figure 8.5a), the content progresses from the first page to the last—the dots are at the exact places where that content occurs in the textbook, with each dot representing a content code applied to an item within the textbook. Reading a plot from left to right for an integrated textbook series, such as CPMP (Figure 8.5b), the units that were coded are displayed in the timeline plots using the same ordering of the units as in the textbooks themselves,[8] and units that were not coded are omitted from the plots. Timeline plots use transparency; that is, lone dots are faint, and as they overlap, the image gets darker. It is not intended for people to see clearly the position of each dot. Rather, the aim is for the timeline plots to reveal differences in density, distribution, and sequencing of content topics within a textbook series. The categories of content (listed in Figure 8.1) appear along the left-hand margin of a timeline plot, beginning with *number sense/properties/relationships* at the bottom, and progressing upwards to *instructional technology*. The dashed vertical lines denote the beginning and end of the textbooks within a series.

We receive a consistent response when we share a timeline plot with a colleague: "There was an error in transmission; there's just a bunch of dots on the page. Please resend." People are unaccustomed to seeing data represented in this fashion. And the tendency, we think, is for people to want to focus on each individual dot within a timeline plot, instead of the image as a whole. The timeline plot, as a whole, is what we consider to be the *portrait of algebra* for a textbook series. These portraits are the mosaics!

A textbook's timeline plot reflects the authors' stance on algebra. For example, *functions* are introduced approximately halfway into the first CME book, and content related to functions is emphasized more heavily in the

Figure 8.5a

Mosaic Timeline Plot for CME

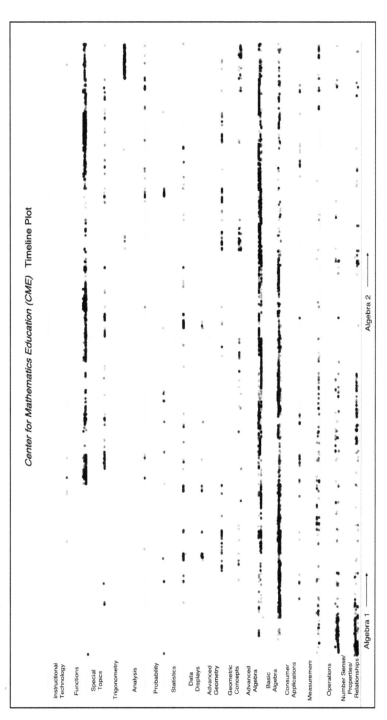

Figure 8.5b

Mosaic Timeline Plot for CPMP

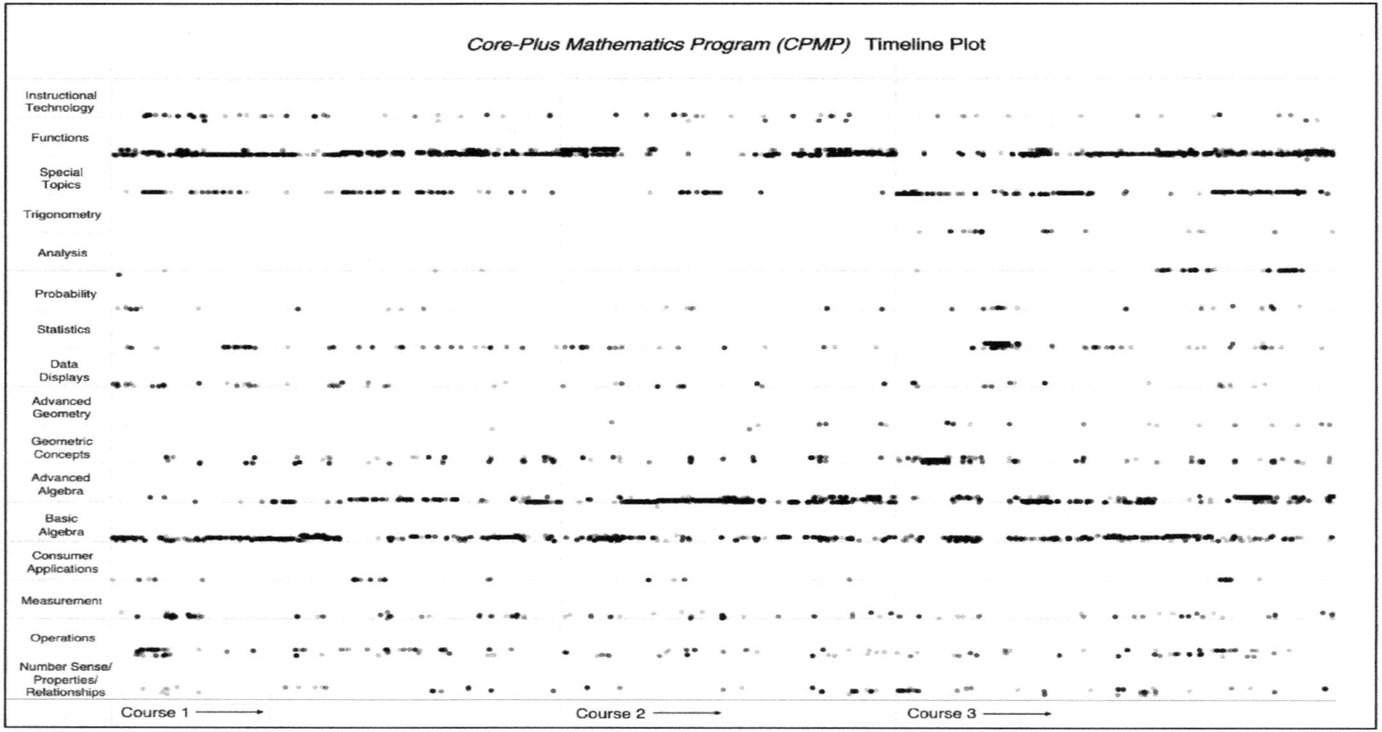

second book in the series, as evidenced by the heavier concentration of dots corresponding to functions to the right of the second vertical dashed line (Figure 8.5a). In this way, the timeline plot suggests that the authors of CME believe that students' work with symbolic algebra (taxonomic categories of *Basic Algebra* and *Advanced Algebra* in the timeline plots) is foundational to their work with functions. By contrast, the timeline plot for CPMP shows content related to functions being introduced early in the first unit, concurrent with symbolic algebra (Figure 8.5b). We infer that the authors of CPMP believe that exposure to functions should proceed in tandem with students learning symbolic algebra. Moreover, there is a high density of codes related to functions throughout the CPMP textbook series, underscoring the authors' focus on a functions approach (e.g., Nemirovsky, 1996) to algebra and algebraic thinking.

Another example of timeline plots reflecting textbook authors' perspectives on algebra is illustrated in the CME timeline plot (Figure 8.5a). The CME authors develop students' algebraic reasoning from the concept of number (especially early in the series), and topics in measurement are prominent throughout the CME books. In this way, the CME timeline plots bring into focus the authors' developing the ideas of algebra through the content domains of number and measurement.

Portions of the timeline plots corresponding to the taxonomic categories *Basic Algebra*, *Advanced Algebra*, and *Functions* are shown in Figures 8.6a–c for all six textbook series to allow readers to clearly see similarities and differences in presentation of algebra content across the series. For example, as with CME, the UCSMP authors also introduce functions about halfway through the first book in the series (Figure 8.6c). As another example, in general there is a higher density of dots for the content category *basic algebra* in the portion of the plots earlier in the subject-specific textbook series, and the density of dots in this content category decreases as one progresses through the books (Figure 8.6a). Similarly, there is a lower density of dots for the content category *advanced algebra* in the beginning portion of the plots for the subject-specific books, and the density of dots in this category increases as one progresses through the books (Figure 8.6b).

Further information about the study and results, including involving an author of each of the textbook series in a set of tasks to introduce them to our research methods and confirm that our application of codes reflected the objectives embedded in their textbooks, is included in Huntley et al. (2024). In that paper, we also share results from analysis of the other curriculum variables we examined (cognitive behavior of textbook items, as well as calculator use, context, and manipulatives).

Figure 8.6a

Basic Algebra Portion of the Mosaic Timeline Plots for All Textbook Series

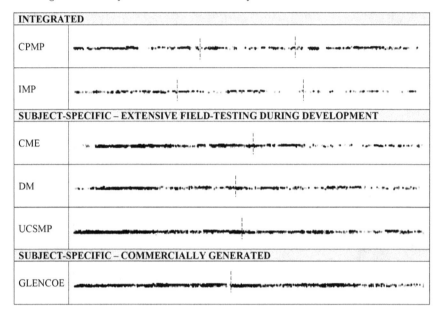

Figure 8.6b

Advanced Algebra Portion of the Mosaic Timeline Plots for All Textbook Series

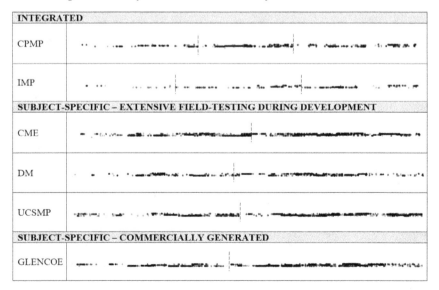

Figure 8.6c

Function Portion of the Mosaic Timeline Plots for All Textbook Series

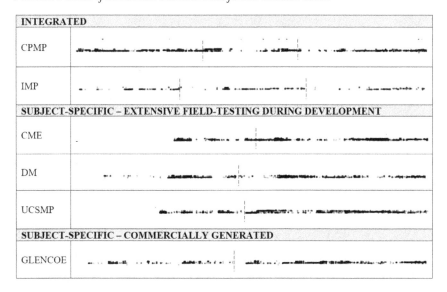

DISCUSSION

In this chapter we have shared our methodology, and rationale for our methodological choices in analyzing the algebra strand of six U.S. high-school mathematics textbook series. We have illustrated how using the metaphor of mosaics helped conceptualize the development of textbook-specific portraits of algebra. Throughout this chapter, we have provided advice and lessons learned, which are summarized in Figure 8.7.

One goal of this project is to provide novel representations of contrasting approaches to algebra. We have made concerted efforts *not* to pit one textbook series against another. We are careful to use a neutral tone when reporting results. Of course, the real test of any textbook series is twofold: What opportunities to learn algebra are provided to students through a particular approach to algebra, and how does this influence their learning of algebra? Although this question is beyond the scope of this project, we hope our problem-by-problem analysis of textbook items and resulting *mosaics of algebra* are useful to a variety of audiences (e.g., teachers, textbook selection committee members, policymakers, researchers). We urge caution in using and interpreting the results, being mindful that describing the nature of algebra, as represented in textbooks, cannot be captured by a single narrative or representation of the data.

Figure 8.7

Lessons Learned From Coding of Algebra Content in Six Textbook Series

---
- Use a metaphor (e.g., mosaic) to help guide thinking about a complex idea.
- Draw from existing analytic frameworks instead of developing new ones.
- Provide extensive annotations to the analytic frameworks, using a different color font and specific problems from the textbooks being coded.
- Have a diverse team of coders, each with strong content knowledge and extensive K–12 experience.
- Engage in frequent discussions across the teams of coders to achieve 100% agreement on the set of codes to apply to each textbook item.
- Be flexible in light of circumstances changing the scope of the project.
- Be open to recoding when it is deemed necessary.
- Be patient, as coding all items within a textbook can become tedious, and results are not able to be generated quickly.
- Anticipate questions from colleagues about reading and interpreting novel representations of results.

ACKNOWLEDGMENTS

This research was supported by a grant from the U.S. National Science Foundation (Award 090131). Any opinions, findings, and conclusions or recommendations expressed in this material are those of the author(s) and do not necessarily reflect the views of the National Science Foundation. We dedicate this chapter to Jennifer Mayer, one of our teacher collaborators, who recently passed away.

REFERENCES

Adams, L., Tung, K. K., Warfield, V. M., Knaub, K., Mudavanhu, B., & Yong, D. (2000). *Middle school mathematics comparisons for Singapore Mathematics, Connected Mathematics Program, and Mathematics in Context (including comparisons with the NCTM Principles and Standards 2000)* [Unpublished manuscript]. Report to the National Science Foundation.

American Association for the Advancement of Science. (1999). *Algebra textbooks: A standards-based evaluation.*

Begle, E. G. (1973). Some lessons learned by SMSG. *Mathematics Teacher, 66,* 207–214. https://doi.org/10.5951/MT.66.3.0207

Brown, J., Schiller, K., Roey, S., Perkins, R., Schmidt, W., & Houang, R. (2013). *Algebra 1 and geometry curricula: Results from the 2005 high school transcript mathematics curriculum study* (Report No. NCES 2013-451). U.S. Department of Education.

Chávez, O., Grouws, D., Tarr, J., Ross, D., & McNaught, M. (2009, April 13–17). *Mathematics curriculum implementation and linear functions in secondary mathematics: Results from the Comparing Options in Secondary Mathematics Project* [Paper presentation]. American Educational Research Association Annual Meeting, San Diego, CA.

Chávez, O., Papick, I., Ross, D., & Grouws, D. (2010, April 30–May 4). *The essential role of curricular analyses in comparative studies of mathematics achievement: Developing "fair" tests* [Paper presentation]. American Educational Research Association Annual Meeting, Denver, CO.

Committee on Prospering in the Global Economy of the 21st Century. (2007). *Rising above the gathering storm: Energizing and employing America for a brighter economic future*. National Academy Press.

Fan, L., & Kaeley, G. (2000). The influence of textbooks on teaching strategies: An empirical study. *Mid-Western Educational Researcher, 13*(4), 2–9.

Floden, R. E. (2002). The measurement of opportunity to learn. In A. C. Porter & A. Gamoran (Eds.), *Methodological advances in cross-national surveys of educational achievement* (pp. 231–236). National Academy Press.

Gamoran, A., & Hannigan, E. C. (2000). Algebra for everyone? Benefits of college-preparatory mathematics for students with diverse abilities in early secondary school. *Educational Evaluation and Policy Analysis, 22*, 241–254. https://doi.org/10.3102/01623737022003241

Garden, R. A., Lie, S., Robitaille, D. F., Angell, C., Martin, M. O., Mullis, I. V. S., Foy, P., & Arora, A. (2006). *TIMSS advanced 2008 assessment frameworks*. TIMSS & PIRLS International Study Center.

Huntley, M. A., Terrell, M. S., & Fonger, N. L. (2024). A content analysis of the algebra strand of six commercially available U.S. high school textbook series. *Education Sciences, 14*.

Katz, V. J. (Ed.). (2007). *Algebra: Gateway to a technological future*. Mathematical Association of America.

Kendal, M., & Stacey, K. (2004). Algebra: A world of difference. In K. Stacey, H. Chick, & M. Kendal (Eds.), *The future of the teaching and learning of algebra: The 12th ICMI Study* (pp. 329–346). Kluwer.

Mayer, J. M., Huntley, M. A., Fonger, N. L., & Terrell, M. S. (2019). Professional learning through teacher-researcher collaborations. *Mathematics Teacher, 112*, 382–385. https://doi.org/10.5951/mathteacher.112.5.0382

Moses, R. P., & Cobb Jr., C. E. (2001). *Radical equations: Math literacy and civil rights*. Beacon Press.

National Assessment Governing Board. (2008). *Mathematics framework for the 2009 National Assessment of Educational Progress*. https://www.nagb.gov/content/dam/nagb/en/documents/publications/frameworks/mathematics/2009-mathematics-framework.pdf

National Research Council. (2004). *On evaluating curricular effectiveness: Judging the quality of K–12 mathematics evaluations*. National Academies Press.

National Science Foundation [NSF]. (1991). *Instructional materials for secondary school mathematics: Program solicitation and guidelines.*

Nemirovsky, R. (1996). A functional approach to algebra: Two issues that emerge. In N. Bernarz, C. Kieran, & L. Lee (Eds.), *Approaches to algebra* (pp. 295–313). Springer. https://doi.org/10.1007/978-94-009-1732-3_20

RAND Mathematics Study Panel. (2003). *Mathematical proficiency for all students: Toward a strategic research and development program in mathematics education.*

Rivers, J. (1990, April 16–20). *Contextual analysis of problems in algebra I textbooks* [Paper presentation]. American Educational Research Association Annual Meeting, Boston, MA.

Schmidt, W. H., McKnight, C. C., & Raizen, S. A. (1997). *A splintered vision: An investigation of U.S. science and mathematics education.* Kluwer.

Senk, S. L., Thompson, D. R., & Johnson, G. J. (2008, July 6–13). *Reasoning and proof in high school textbooks from the USA* [Paper presentation]. International Congress on Mathematics Education, Monterey, Mexico.

Sherman, M. F., Walkington, C., & Howell, E. (2016). Comparison of symbol-precedence view in investigative and conventional textbooks used in algebra courses. *Journal for Research in Mathematics Education, 47*(2), 134–146. https://doi.org/10.5951/jresematheduc.47.2.0134

Thompson, D. R., Senk, S. L., & Johnson, G. J. (2012). Opportunities to learn reasoning and proof in high school mathematics textbooks. *Journal for Research in Mathematics Education, 43*, 253–295. https://doi.org/10.5951/jresematheduc.43.3.0253

Weiss, I. R., Banilower, E. R., McMahon, K. C., & Smith, P. S. (2001). *Report of the 2000 National Survey of Science and Mathematics Education.* Horizon Research.

Wisconsin Center for Education Research [WCER]. (2007). *SEC K–12 mathematics taxonomy.*

Wolfram Research, Inc. (2010). *Mathematica* (Version 8). [Computer software].

Wolfram Research, Inc. (2012). *Mathematica* (Version 9). [Computer software].

Yerushalmy, M., & Chazan, D. (2002). Flux in school algebra: Curricular change, graphing technology, and research on student learning and teacher knowledge. In L. English (Ed.), *Handbook of international research in mathematics education* (pp. 725–755). Erlbaum.

APPENDIX: BIBLIOGRAPHY OF TEXTBOOK RESOURCES

Benson, J., Cuoco A., D'Amato, N. A., Erman, D., Baccaglini-Frank, A., Golay, A., Gorman, J., Harvey, B., Harvey, W., Hill, C. J., Kerins, B., Kilday, D., Maurer, S., Palma, M., Saul, M., Sword, S., Thomas, B., Ting, A., & Waterman, K. (2009). *Center for Mathematics Education Project Algebra 1* [Teacher text]. Pearson.

Brown, S. A., Breunlin, R. J., Wiltjer, M. H., Degner, K. M., Eddins, S. K., Edwards, M. T., Metcalf, N. A., Jakucyn, N., & Usiskin, Z. (2008). *University of Chicago School Mathematics Project Algebra* (3rd ed.) [Teacher text]. Wright Group/McGraw-Hill.

Cuoco A., Baccaglini-Frank, A., Benson, J., D'Amato, N. A., Erman, D., Harvey, B., Harvey, W., Kerins, B., Kilday, D., Matsuura, R., Maurer, S., Sword, S., Ting, A., & Waterman, K. (2009). *Center for Mathematics Education Project Algebra 2* [Teacher text]. Pearson.

Fendel, D., Resek, D., Alper, L., & Fraser, S. (2009–2011). *Interactive Mathematics Program Years 1–3* (2nd ed.) [Teacher texts]. Key Curriculum Press.

Flanders, J., Lassak, M., Sech, J., Eggerding, M., Karafiol, P. J., McMullin, L., Weisman, N., & Usiskin, Z. (2010). *University of Chicago School Mathematics Project Advanced algebra* (3rd ed.) [Teacher text]. Wright Group/McGraw-Hill.

Hirsch, C. R., Fey, J. T., Hart, E. W., Schoen, H. L., & Watkins, A. E. (2008/2009). *Core-Plus Mathematics Program Courses 1–3* (2nd ed.) [Teacher texts]. Glencoe/McGraw-Hill.

Holliday, B., Luchin, B., Marks, D., Day, R., Cuevas, G. J., Carter, J. A., Casey, R. M., & Hayek, L. M. (2008). *Algebra 1* [Teacher text]. Glencoe/McGraw-Hill.

Holliday, B., Luchin, B., Marks, D., Day, R., Cuevas, G. J., Carter, J. A., Casey, R. M., & Hayek, L. M. (2008). *Algebra 2* [Teacher text]. Glencoe/McGraw-Hill.

Murdock, J., Kamischke, E., & Kamischke, E. (2007). *Discovering algebra: An investigative approach* (2nd ed.) [Teacher text]. Key Curriculum Press.

Murdock, J., Kamischke, E., & Kamischke, E. (2010). *Discovering advanced algebra: An investigative approach* (2nd ed.) [Teacher text]. Key Curriculum Press.

ENDNOTES

1. In the U.S., high school is generally Grades 9–12, corresponding to students of ages 15–19 years.
2. TIMSS stands for the Third International Mathematics and Science Study.
3. Only scientific and graphing calculators were mentioned in the textbooks we analyzed.
4. A topic code is a three- or four-digit number.
5. The cognitive behavior codes are K (knowing), A (applying), or R (reasoning).
6. The first research question is: What is the content, including the breadth, sequence, and depth of topics covered?
7. Timeline plots for the other four textbook series are in Huntley et al. (2024).
8. Recall that only the problems from units that the authors identified as having a major emphasis on algebra were coded for the integrated textbook series (CPMP and IMP).

SECTION III

CURRICULUM DESIGN AND DEVELOPMENT

CHAPTER 9

THE CURRICULUM RESEARCH FRAMEWORK

Supporting the Scientific Creation, Implementation, and Evaluation of Curricula

Julie Sarama and Douglas H. Clements
University of Denver

Curriculum development is often not a fully scientific endeavor. To address this concern when creating an early mathematics curriculum, we developed a framework for authoring and evaluating research-based curricula. The Curriculum Research Framework (CRF) *incorporates broad theoretical and methodological approaches, as well as attention to social values and equity issues. We describe our application of the CRF to the* Building Blocks *curriculum, emphasizing lessons learned. We argue that those designing curricula should ensure their work is scientifically based and evaluated. Those studying existing curricula should understand how they were developed and validated (or not) and that a comprehensive evaluation program involves more than student outcomes. We use the CRF to draw implications for research in development and evaluation projects. For each phase of the framework, we discuss how publishable research and curriculum development might occur, as well as what opportunities there may be for evaluation research alone. In all cases, we list useful methods.*

Keywords: curriculum; curriculum research framework; evaluation; learning trajectories; professional development

The development of both the *Building Blocks* preschool mathematics curriculum (Clements & Sarama, 1999, 2007a, 2013; Sarama & Clements, 2003) and the Curriculum Research Framework (CRF, Clements, 2002, 2007), on which all our research-and-development (R&D) is based, were motivated by our observation that curriculum development was often not a scientific endeavor and that it could and should so evolve (Battista & Clements, 2000; Clements & Battista, 2000). Our quarter-century journey lent credence to our optimism that a scientific approach was possible. Through this work, we extended and enriched our notions about how scientific endeavors must be conceived broadly across theoretical and methodological fields and must be complemented by other perspectives (Clements, 2007), such as historical (Kilpatrick, 1992), aesthetic (Eisner, 1998), or narrative (Bruner, 1986; James, 1892/1958). Societally-determined values and goals and equity issues (culture, ethnicity, race, language, disability) are also substantive components of any high-quality curriculum; no research or development program should ignore these components.

We have worked on all levels of curriculum, from the official curriculum to the learned curriculum (Burkhardt et al., 1990). Although researchers can study a curriculum, we believe the isolation of development from research harms both (cf. Clements, 2007; Clements & Battista, 2000; Lagemann, 1997; Lloyd et al., 2017; Sarama & Clements, 2008). Therefore, those designing any curriculum or curriculum component (e.g., an intervention or instructional unit) should ensure it is research-based *and* scientifically evaluated. To guide this research, we created an R&D framework.

THE CURRICULUM RESEARCH FRAMEWORK

Finding no clear, complete foundation for research-based curriculum development and evaluation when we began, we created the Curriculum Research Framework (CRF) based on a synthesis of our previous work (Clements & Sarama, 1999) as well as the work of others. Empirical research has supported the efficacy of the CRF for research and development (Clements, 2007; Doabler et al., 2019; Sarama & Clements, 2019b).

The CRF was designed (a) to develop a truly research-grounded *and* empirically-validated curriculum (the available or adopted curriculum, Burkhardt et al., 1990) and (b) to answer important questions of effect and conditions for three audiences. For practice, questions included: Is the curriculum effective, *and* when, where (under what conditions), and for whom? For policy, we asked: Are the goals important, what are the effects and effect sizes for learning of students and teachers, *and* what support is needed in various conditions or contexts? For theory, the queries were: Why is the curriculum effective or not, what processes were responsible, how were they connected to the theoretical bases, *and* why do specific sets of conditions alter effectiveness?

To answer these questions, the CRF has 10 phases in three categories. The first category, *A Priori Foundations*, starts with reviews of research in Phases 1–3. The second category, *Learning Trajectories*, invests substantial time and effort in developing research-based paths of learning and teaching (Phase 4). The third category, the *Evaluation of Enacted Curriculum*, progresses through multiple phases of formative evaluation (Phases 5–8) and summative evaluation (phases 9–10). Phase 5 is market research with adults. Phase 6 involves working with individual or small groups of children, Phase 7 with small groups of closely collaborating teachers and their classes, and finally, with larger numbers of new teachers in Phase 8. The summative evaluations also start small and expand. Qualitative research and development pervade the first 8 of the 10 phases; although the two summative phases emphasize quantitative analyses, qualitative research is always included.

There are three reasons that learning trajectories (LT) play a foundational role in all CRF phases. First, the purpose of learning trajectories is to support curriculum development (*curriculum* is from the Latin word for racecourse or path). Extending Marty Simon's (1995) original construct, we conceptualized

> learning trajectories as descriptions of children's thinking and learning in a specific mathematical domain [topic], and a related, conjectured route through a set of instructional tasks designed to engender those mental processes or actions hypothesized to move children through a developmental progression of levels of thinking, created with the intent of supporting children's achievement of specific goals. (Clements & Sarama, 2004, p. 83)

That is a long definition, reflecting the fact that creating LTs is a complicated endeavor.

The second reason for learning trajectories' central role is that they are asset-based; that is, they begin with children's strengths and what they know and can do. They are the mathematics of children (Baratta-Lorton, 1976; Steffe, 2017)—the thinking of children as it naturally unfolds. The LT approach does not "break down" mathematics into pieces taught in isolation, but rather, *builds up* mathematics from what the child knows and can do. Rigorous experiments have supported the effectiveness of the learning trajectory approach (e.g., Baroody et al., 2022; Sarama & Clements, 2021; Sarama, Clements, Baroody et al., 2021).

The third reason for learning trajectories' central role in the CRF is that they allow *teachers* to understand and build children's mathematics. With an understanding of developmental progressions, teachers can ensure that the goals and activities outlined are within the developmental capacities of children. In addition, their understanding that each level provides a natural building block to the next level can motivate and support their use of the critical strategy of formative assessment.

The main lesson learned in this complex project is that *scientific curriculum development and evaluation are both possible and valuable but require a comprehensive framework*. The remainder of this chapter is organized using the three broad categories of the framework, encompassing the 10 phases. Within each category, we describe each phase, as well as the research opportunities, questions, and methods within that category, providing examples from the *Building Blocks* curriculum developed on the basis of this framework (Sarama & Clements, 2019c). Although a comprehensive program would involve all phases, valid R&D can be conducted in a subset.

A PRIORI FOUNDATIONS: REVIEWING RESEARCH

The first category corresponds with the focus on *research-based* curricula. Although the phrase is often used freely and even carelessly to imply comprehensive research evaluation, we use *research-based* to mean what it says—that the curriculum is *grounded in existing research* (or claims to be). The phases in this category involve reviewing research and drawing implications for curriculum development. The questions concern subject matter content, cognition and learning, and pedagogy. A criterion for an equitable literature review is the inclusion of different populations, such as English language learners, different ethnic groups, those with Individual Educational Plans, and other underrepresented populations.

Phase 1: Subject Foundations

In Phase 1, research is used to identify mathematics content that is generative in students' development of future mathematical understanding, developmentally appropriate, and interesting to students in the target population. Sources include national and international standards (see National Governors Association and Council of Chief State School Officers [NGA/CCSSO], 2010), suggestions by mathematicians (e.g., Howe, 2011, 2014; Wu, 2011), and mathematics education research, including content predicting later achievement (Nguyen et al., 2016; Rittle-Johnson et al., 2018; Verdine et al., 2017). We have found it important to be cautious with correlational studies that relate variables but do not experimentally test a curriculum or other intervention (Bailey et al., 2018). These research-oriented strategies constitute parts of comprehensive content analyses (National Research Council, 2004) that can lead to important publications, but that are too rare in the field.

Phase 2: Cognitive Foundations

In Phase 2, developers similarly review theories and studies on students' thinking and learning. For example, to ground our curriculum projects, we created the hierarchical interactionalism theory from a synthesis of empiricism, (neo)nativism, and interactionalism theories (Sarama & Clements, 2009a). In addition, there has been a burgeoning of research on (a) students' understanding and learning of mathematics (e.g., Cai, 2017; Clements & Sarama, 2021), (b) learning opportunities and equity (e.g., Civil, 2016), and (c) curriculum development and research (e.g., Clements et al., 2020; Lloyd et al., 2017; Stein et al., 2007). Such publications can be consulted to guide future R&D or evaluation efforts.

Phase 3: Pedagogical Foundations

In Phase 3, empirical findings on the creation of specific types of instructional activities that are educationally effective and motivating are reviewed as general guidelines for the creation of instructional tasks and pedagogical strategies (e.g., Baroody et al., 2022; Frye et al., 2013; Sakakibara, 2014), including their relationships to assessments (Penuel & Shepard, 2016). The goal is to establish basic guidelines, not yet specific activities.

Designing, Conducting, and Publishing Research Reviews

Research syntheses, when they exist, are good resources for research reviews in this category. When they do not yet exist, the topic invites original research publications. For cognitive and pedagogical foundations, we assume that no single theory, body of research, or methodology can satisfactorily answer the question "What is best?" There are logistical limitations, of course, and a grounding of such issues in cultural values (Clements, 2007). Strict theoretical or methodological consistency may limit what can be garnered from research. More promising is the integration of issues and findings that researchers and teachers from diverse philosophical positions experience and report.

LEARNING TRAJECTORIES (PHASE 4)

This major category and phase led to another main lesson learned: *Use learning trajectories* (LT). We knew we wanted to make the curriculum about children's mathematics (Carpenter & Fennema, 1992; Steffe & Tzur, 1994). We built the LTs slowly and methodically.

Designing, Studying, and Publishing on Learning Trajectories

To begin, we reviewed empirical research on children's thinking and learning in each targeted topic to ascertain if there are existing developmental progressions that can be used or if they can be meaningfully built. For example, does the literature suggest a "natural" or relatively universal development progression of levels of thinking and learning that can guide assessment and teaching (e.g., Clements & Sarama, 2021; Maloney et al., 2014; National Research Council, 2009; Sarama & Clements, 2009a)? If not, a research review may suggest a developmental progression; a synthesis of this literature is a contribution worthy of publication. In our work, we usually had to integrate separate studies, often with children of different ages, using different tasks with different populations. To create the developmental progressions, we began by aligning results from the studies using students' ages as a guide, yielding a rough nascent draft of a progression. Then a series of cross-sectional clinical interviews, using tasks designed to elicit pertinent concepts and processes, helped us examine students' knowledge of the content topic, including conceptions, strategies, intuitive ideas, and informal strategies used to solve problems. From these, we hypothesized mental objects (e.g., concepts) and actions (processes) that define each level of thinking, the specification of which allows a high degree of precision. To ensure that the LTs were asset (i.e., strength) based (Celedòn-Pattichis et al., 2018), we reviewed and conducted research involving students from diverse communities, while at the same time making sure the LTs remained consistent with our theory and LT approach.

Learning trajectories are not simply "educated guesses" but rather, they are based on empirically supported developmental progressions. The benefit for the teacher is to have a well-formed and specific set of expectations about children's ways of learning and a likely pace along a learning path that includes central, worthwhile ideas. For example, the composition of geometric figures is a basic geometric competence; actions of creating and then iterating geometric figures and higher-order geometric units (composite figures) in the context of constructing patterns, measuring, and computing are established bases for mathematical understanding and analysis (Clements et al., 1997; Clements et al., 2001). The developmental progression moves from a lack of competence in composing or matching shapes to outlines (see *Separate Shapes Actor: Foundations* at LearningTrajectories.org for elaboration and videos), then to using individual shapes to make a picture but with each shape playing a unique semantic role (*Piece Assembler*), then combining shapes—initially through trial and error (e.g., *Picture Maker*), and then finally to concatenating shapes to form a component of a picture but not necessarily conceptualizing these creations as geometric shapes. At the *Shape Composer* level, children have developed competence with angles and anticipatory imagery and thus compose shapes to create superordinate shapes intentionally. (For details on other learning trajectories, see Clements & Sarama, 2021; Sarama & Clements, 2009a.)

Developmental progressions establish the basis for instruction. Tasks from teaching experiments may be refined and extended to concretize the external objects and actions that mirror the hypothesized mental actions-on-objects of the target level. For example, objects may be shapes or connecting cubes, and actions might be creating, copying, uniting, disembedding, and hiding individual and composite units. In addition, contexts, narratives, and teaching strategies can give these objects and actions meaning. These ideas are extended by finding activities for each level in the literature. For all topics, we suggest broad literature reviews, seeking research evidence (e.g., Clarke et al., 2002; Clements & Sarama, 2021; Maloney et al., 2014; Sakakibara, 2014) and the wisdom and creativity of expert practice (e.g., Cross et al., 2012; MacDonald & Shumway, 2016; Raisor & Hudson, 2018) to identify activities and interactions shown as effective in promoting the learning of students to achieve each level by encouraging children to construct the concepts and processes that define it.

Returning to the shape composition example, a leading instructional task that is also motivating requires children to solve outline puzzles with shapes off and on the computer. (See complete lessons and videos at LearningTrajectories.org.) The objects are shapes and composite shapes, and the actions include creating, duplicating, positioning (with geometric motions), combining, and decomposing. Both tasks on and off the computer require actions on these objects that correspond to each level in the learning trajectory. For example, tasks for *Piece Assemblers* scaffold children's growing ability to match shapes without, and then with, sides touching. Those for the *Picture Maker* level increasingly require that children compose shapes, and those for *Shape Composer* have more extensive areas in which angles must be addressed. Teachers are guided to use the cognitive model of each level for formative assessment, identifying children's level of thinking and supporting them in developing understanding at the subsequent level. There is ample opportunity for student-led, student-designed, open-ended projects in each set of activities.

Research continues with teaching experiments or design experiments of appropriate portions of the LTs, using repeated, short cycles of testing and revision, so that each LT describes (a) what students are able to do, (b) what they are not yet able to do but should be able to learn, and (c) *why*—that is, how they think at each level and how they learned these levels of thinking. Such research develops, refines, or validates LTs that need increasing attention from researchers. Of course, activities and teaching strategies are continually under revision and open to more effective approaches, as are all aspects of LTs. The LTs cannot capture all that is needed for high-quality teaching, such as the application of teachers' knowledge of children's culture and individual social and emotional needs. They should, however, illustrate and embody the *type* of activity hypothesized to be appropriate and efficacious for students at a given level of the developmental progression.

What is discussed here is *research and development* (R&D). More outlets for such work are available, and both direct descriptions of the R&D on specific LTs and retrospective analyses of the work can make substantive contributions to instructional theory. In addition, detailing relationships between the developmental progression and the instructional tasks and strategies can fill gaps in the "science of education" that were created historically because such work was not produced and published.

Lessons Learned on Combining Learning Trajectories Into a Coherent Curriculum

Many other issues must be carefully considered as the entire curriculum is developed in its first complete form (Clements & Sarama, 2008; Sarama & Clements, 2019a). We describe six such issues and the corresponding lessons.

Considerations of Grain Size

Although we respect and learn from projects that have used large grain sizes with success, such as the growth points by Australian colleagues (e.g., Clarke et al., 2002), we believe developmental progressions are best designed with a detailed grain size to support curriculum development and teaching. That is, our vision was to have each level represent a qualitatively distinct pattern of thinking, but one that is a reasonable goal for the next curricular step within children's Zone of Proximal Development (Vygotsky, 1935/1978).

Our results with this smaller grain size have been successful; each time a new activity or substantial modification of an activity occurs within the curriculum, it signals a goal of a new level of thinking. For example, we found that teachers could easily master the developmental progression for subitizing (rapid recognition of the numerosity of sets without counting), given the reasonable number of levels and the nature of the advance from perceptual to conceptual subitizing (seeing parts and synthesizing them into a whole). However, other topics were more complex. For those, we have made an innovation that was motivated by the concern that teachers became lost within the progression and may have forgotten the "big picture." We maintained the fine grain size for teaching but helped teachers see the "forest and trees" of each *major* LT, where trees were each level of the LT's developmental progression and the "forest" was the broad category that provides the "big picture" of the major phases.

Levels of Development That Are Not Curriculum Goals or Do Not Sound Asset-Based

We made two mistakes. Our original LTs included all developmental levels, including those that were not instructional targets. For example, research has identified a level at which children count each set to compare the number of objects in sets. However, if the set with fewer items has physically larger objects, they will maintain that the set with larger objects has *more*. Despite its developmental veracity, some teachers thought they should *teach* children the behavior of that level, even though the results were mathematically incorrect. Thus, we eliminated all such non-instructional levels as levels and combined the description of that behavior into the *prior* level (e.g., "May count the sets, but still claim the set with larger objects has more even if it has fewer items").

Extending LTs to infants and toddlers, we noticed some levels did not show that we always build on children's strengths. For example, our earliest levels often had names such as *Pre-Composer*. This gave the impression that a child had no competencies on which to build. So we changed this name to *Same Thing Comparer: Foundations*.

Number of Learning Trajectories in Professional Development

We needed LTs for all topics but soon decided to spend most professional development (PD) hours on a handful of essential topics, such as counting, subitizing, comparing and sequencing numbers, arithmetic, and describing and composing geometric shapes. Shorter sessions were given on other topics when the curriculum focused on those topics, including length measurement and patterning. As a result, less PD time was given to other topics. However, we included simple short instructions and then curriculum-provoked reminders regarding sorting and classifying, spatial orientation (other spatial skills were embedded throughout), multiplication, disembedding geometric figures, area, volume, and angle. This approach was consistent with the recommendation in the *Standards for Mathematics Teacher Preparation* (Association of Mathematics Teacher Educators, 2017) that teachers learn one or two LTs deeply at first and then have resources to learn others.

Integrating Learning Trajectories

Due to the primacy of young children's development (this may not apply to LTs for older students), we decided that topic-specific LTs should be interwoven rather than taught in separate curricular units for five reasons. First, children's mathematics learning is continuous and incremental (Clements & Sarama, 2007b; Siegler, 1996), and consistent exposure to activities is important to children's learning (Fuson, 1988). Second, the progressions for each LT cover years of development, so compressing the learning into a unit of instruction would be inappropriate. Third, preschool is a substantial period of cognitive development, with vast individual differences; therefore, distributing opportunities to learn topics across the year will be more effective for more children. Fourth, at all ages, distributed practice yields better recall and retention (Cepeda et al., 2006; Rohrer & Taylor, 2006). Fifth, interweaving may facilitate mutual reinforcement between LTs (Clements & Sarama, 2007b).

As an example, early learning of subitizing supports the development of a critical (and oft-neglected) level of thinking in the counting trajectory in which children recognize the last counting word indicates how many in the counted set. That is, if children count a group of objects, "One, two, three," and immediately recognize a group as containing three objects via subitizing, their understanding of the cardinality of the last counting word is facilitated. Conversely, establishing that level of thinking in counting supports the development of higher levels of subitizing, especially conceptual subitizing.

Testing this approach, we conducted a randomized-trials design to compare the effectiveness of *Building Blocks* to a business-as-usual curriculum (effect size, 1.07) and also to a different preschool mathematics curriculum that did not interweave topics but used a unit approach (effect size, .47). Although confounded by other curricular differences, the results suggest we learned the right lesson, namely *learning trajectories should be integrated rather than taught separately* (Clements & Sarama, 2008).

Another lesson learned is that *the structure of the media in which LTs are described can inadvertently isolate them*. We originally designed each LT to be embedded in a matrix in which each column was a LT and levels were aligned developmentally across LTs. We did not abandon that format, but publishers forced us away from it. Publishers of the *Building Blocks* curriculum and our books did not like the structure of our matrices and separated them. In retrospect, we gained a focus on one LT at a time, which helped teachers avoid feeling overwhelmed by the amount of new information. However, we lost the interconnectedness in the original print materials. Instead, we slowly introduced such connections in subsequent professional development sessions.

Scripted Curriculum or Not?

Educators frequently champion the individual teacher's interpretation and even their creation of the curriculum (e.g., John et al., 2018). In contrast, research suggests that implementing systematic, scientifically-based practice is more effective than private, idiosyncratic practice (Raudenbush, 2009). This is not to say teachers should deliver a *scripted* curriculum; indeed, such an approach contradicts the CRF's use of LTs in the service of formative assessment and its insistence on adaptation to local conditions. Instead, focusing on the shared scientific base is an effective and efficient way to improve education for children. Although not scripted, *Building Blocks* does include detailed instructions. So, a lesson learned is that *teachers uncomfortable or new to teaching mathematics might follow curriculum instructions closely. Others can and should adapt them, often substantially, and this should be encouraged by the curriculum and related professional development.*

Such scientifically-grounded shared practice is, somewhat paradoxically, *more* likely than private practice to generate creative contributions. Teachers will contribute modifications of effective practice that are already shared, and thus understood and more easily adopted, and that in turn will be accessible to discussion and further scientific investigation. We found it crucial to help teachers differentiate between productive adaptations and lethal mutations to the intervention (Brown & Campione, 1996). In a similar vein, the introduction of uncoordinated interventions had to be continually monitored for the deleterious effects it may have on the intervention.

Equity

We previously stated that "equity issues (culture, ethnicity, race, language, disability) are substantive components of any high-quality curriculum." Accordingly, all funded projects in *Building Blocks* included a main equity justification based on research that suggested racial differences were minor when socioeconomic status (SES) was controlled (Entwisle & Alexander, 1992; see also Payne & Biddle, 1999). However, SES and ethnic inequities have different origins, structures, and implications (Valant & Newark, 2016).

Therefore, *developers must consider all sources of inequity*. Curricula should address culture, at the very least, including resources for culturally responsive teaching. Adaptations for full inclusion of children with disabilities must be present as well (Clements, Vinh et al., 2021).

In summary, for all phases, it is important to emphasize that equity must be considered. Do existing studies include marginalized groups? Does any original planned research? Convenience samples are usually inadequate and inappropriate. Similarly, do reviews and original work include funds of knowledge from various cultures (Civil, 2002; Moll et al., 1992; Presmeg, 2007)? As one example, developing students' spatial abilities through art and design (e.g., tilings or tessellations), as well as puzzles (e.g., tangrams), can be based on art universally found in cultures (Danesi, 2009). Such activities are engaging and can increase students' mathematical self-efficacy (Casey et al., 2011; Cheng & Mix, 2012). Other issues are underdeveloped and under-researched, such as the relative lack of Spanish versions of curricula and the absence of curricula and resources in other languages (Park et al., 2017). Many other issues need R&D, such as the tension between creating local, specific, culturally responsive curriculum, and curricula that are adaptable but also can be widely disseminated.

EVALUATION OF ENACTED CURRICULUM

There are many phases in the last category, which involve *evaluating* the curriculum that has been developed. This first involves *formative* evaluation, and then *summative* evaluation, to assess the appeal, usability, and effectiveness of an instantiation of the curriculum.

Phase 5. Market Research

Market research is commercially-oriented research about customers, what they want, and what they will buy. The CRF does not ignore it but keeps it public and shared, contrary to usual commercial procedures (Jobrack, 2011).

Phase 6. Formative Research—Individual or Small Group

Pilot testing with individuals or small groups of students evaluates a particular activity or section of the curriculum. Early interpretive work evaluates components using a mix of model-testing and model-generation strategies, including group-based teaching experiments (Blanton et al., 2015; Lamberg & Middleton, 2009; Simon et al., 2016). The following questions are asked in this phase: What meaning do students give to the curriculum objects and activities? Does the enacted curriculum develop the hypothesized mental actions-on-objects or how they are intended to be instantiated? What elements of the teaching and learning environment contribute to student learning? In an intensely iterative phase, the evaluation and redesign cycle should occur in quick succession, possibly as often as every day or two. Field notes are best combined with video or audio recordings. Effectiveness of the enacted curriculum must be checked at every phase, maintaining links to the hypothesized theories and models.

Phase 7. Formative Research in a Single Classroom

Teachers may agree with the curriculum's goals and approach, but their *implementation* may not be aligned with the developer-researchers' vision (Sarama et al., 1998), so methods such as those of Phase 7 are used to determine the meaning that various curricular materials have for teachers and students. This phase studies curricular enactment with two interrelated methods. One method is classroom-based teaching experiments to help track and evaluate student learning, with the goal of making sense of the curricular activities as individual students experience them (Blanton et al., 2015; Lamberg & Middleton, 2009; Simon et al., 2016; Stephan et al., 2001). The other method involves the entire class, which is observed for information concerning the usability and effectiveness of the curriculum. Ethnographic participant observation (Spradley, 1980) allows the researcher to record both the teacher's and students' actions to ascertain how the materials are used, how the teacher guides students, and how the students act and learn.

A lesson we already knew but repeatedly relearned is that *teachers are at the core of successful enactment and must be provided resources and support.* Through the subsequent phases, we continued to add resources within and supplement the nascent *Building Blocks* curriculum.

Phase 8. Formative Research: Multiple Classrooms

Researchers observe multiple classrooms from diverse communities to gauge the effectiveness and usability of the curriculum with teachers new to the curriculum (cf. Drake et al., 2014). Observational instruments, whether general or specific fidelity measures, are often used to focus and direct these observations. In our view, fidelity instruments should go beyond simple compliance fidelity to include fidelity to the vision of the curriculum (Clements & Sarama, 2000/2018; Clements et al., 2015). Complementing these data collection efforts are data from teacher questionnaires and interviews. Researchers also investigate whether teacher characteristics, community characteristics, or contexts link to different enactments (Lloyd et al., 2017). Much is known about professional development's impact on teaching practices, but oftentimes little is known regarding the impact of professional development on children's learning (Markussen-Brown et al., 2017). We need to achieve both. Therefore, well-documented studies in Phases 7 and 8 can make substantial contributions to the literature (Lloyd et al., 2017).

Phase 9. Summative Research on a Small Scale

This phase evaluates what can be achieved with typical teachers under realistic circumstances (cf. Burkhardt, 2006). Again, in multiple diverse classrooms (about 4 to 10), experimental or quasi-experimental designs using standardized measures as pretests and posttests of learning are often implemented in conjunction with, and to complement, methodologies previously described (e.g., Clements & Sarama, 2007c; Doabler et al., 2014). Surveys of teacher participants also may be used to compare data collected before and after they have enacted the curriculum. The interpretive and survey data triangulate whether curricular supports are viewed as helpful by teachers and other caretakers and whether their teaching practices have been influenced. Have teachers *changed* previous approaches to assessing and teaching mathematics? Perhaps most important, do they understand their students' learning using learning trajectories?

In our work, a series of studies were conducted to evaluate and refine the *Building Blocks* curriculum and its implementation (Clements & Sarama, 2008; Clements et al., 2011; Clements et al., 2013; Sarama & Clements, 2009b; Sarama et al., 2008). Research on the curriculum showed a strong effect on achievement and an increase in the rate of children's development through the levels (Dumas et al., 2019). The children using *Building Blocks* outperformed those in the control group on a measure of oral language (Sarama et al., 2012). Other studies indicated that work with *Building Blocks* develops executive function competencies, such as response inhibition, attention shifting, and updating working memory (Clements, Sarama, Layzer et al., 2020). Studies such as these (Bierman et al., 2008) suggest general educational benefits (Mattera et al., 2021).

Phase 10. Summative Research at a Large Scale

The CRF is not complete until evaluations are conducted on a large scale. Such research should use a broad set of instruments to assess the impact of the implementation on participating children, teachers, program administrators, and parents, and relate this to the fidelity of teachers' implementation across diverse contexts (cf. Clements & Sarama, 2000/2018; Clements et al., 2011; Remillard et al., 2009; Sarama et al., 2008). The goal should be to measure and analyze the critical variables: (a) *contextual variables*, such as urban/suburban/rural; type of program; class size; teacher characteristics; student/family characteristics; and (b) *implementation variables*, such as engagement in professional development opportunities; fidelity of implementation; leadership, such as principal leadership, as well as support and availability of resources, funds, and time; peer relations at the school; "convergent perspectives" of the developer-researchers, school administrators, and teachers in a cohort; and incentives used. For example, a randomized experiment assesses the average impact of being exposed to the curriculum. Because not all intervention classrooms will implement the program with equal veracity, a separate series of analyses would relate outcome measures with a set of target contextual and implementation variables.

Simultaneously, qualitative methods are used within that structure. The combination of quantitative and qualitative analyses is critical. Quantitative methodologies provide experimental results, garnered under conditions distant from the developer-researchers, that are useful in and of themselves in that they can generate political and public support. Such methodologies also can uncover unexpected and subtle interactions not revealed by qualitative investigations. Qualitative methodologies, however, are just as important. Curriculum research includes interpreting interactions within diverse groups of individuals. It seeks to understand individual students' understanding and learning and how these change in the context of, and as a result of, the interactions among teachers and students around a specific curriculum. In a triangulation context, qualitative research may validate or invalidate quantitative results more so than the reverse (Russek & Weinberg, 1993).

Studies of *Building Blocks* had other important research lessons learned. First, African American children in the control averaged lower gains than other control children; in *Building Blocks* classrooms, African American children averaged higher gains than other children (Clements et al., 2011) that persisted into fifth grade. *Curricula built on LTs may be particularly effective in increasing the expectations for African American children as teachers see their untapped potential for learning mathematics* (Schenke et al., 2017). Both qualitative (Clements et al., 2011) and quantitative (Clements et al., 2023; Clements et al., 2011; Schenke et al., 2017) studies indicated that understanding children's learning and development, and increasing expectations of what they could learn, accounted for those positive results.

Second, a lesson learned is to *study domains to which high-quality mathematics curricula may transfer*. Although not consistent across all studies, the discussion and thinking promoted in *Building Blocks* contributed to preschoolers' expressive oral language (Sarama et al., 2012). That is, in an assessment in which children retell a story, students using *Building Blocks* scored significantly higher than children using the control curriculum on the number of key vocabulary words they used, their use of grammatically complex sentences, their willingness to reproduce narratives independently, and their inferential reasoning. A vital feature of the *Building Blocks* curriculum is talking about mathematics. For example, when a child identifies a square from a set of shapes, a *Building Blocks* teacher asks, "How do you know it's a square?" At the beginning of the school year, young children often answer such a question with simplistic responses, such as, "Because it looks like a square" or "I thought it in my head." As the curriculum progresses and children experience repeated invitations to explain their thought processes, even young children are supported in their ability to give accurate, reasoned responses, such as, "Because it has four sides and all right angles." In another study, contrary to our hypotheses, combining the *Building Blocks* and *Tools of the Mind* curricula did not increase executive function (EF), but focusing on the mathematics in *Building Blocks* alone did (Clements, Sarama, Layzer et al., 2020; Dong et al., 2021; see also, Weiland & Yoshikawa, 2013). Thus, a curriculum in a specific domain may be a valuable vehicle for improving both teacher practices and subsequently children's executive function (EF) (as in *Building Blocks* or Research-based, Developmentally Informed Head Start [REDI]) (Mattera et al., 2021, p. 14).

The third research lesson learned from studies of *Building Blocks* is that *its use positively affected teachers who showed higher fidelity scores two (Clements et al., 2015) to six years (Sarama et al., 2016) after the end of the project*. These teachers encouraged and supported children's discussions of mathematics and the use of formative assessment.

A New Approach: Rigorous Studies of the Basic Assumptions Used in *Building Blocks*

We took another approach by evaluating the individual *Building Blocks* LTs (BBLT). Multiple experiments rigorously evaluated whether instruction based on LTs for early mathematics is significantly more efficacious than alternatives *closely matched* to the *Building Blocks* curriculum but without one or two of the main characteristics of LTs that distinguish their application to curriculum and teaching from alternative pedagogical approaches. The first assumption is that instruction should move children from their present level to the next higher level and continue in this manner until the instructional goal is reached. A competing approach posits that it is more efficient and mathematically rigorous to teach the target level immediately by providing accurate definitions and demonstrating accurate mathematical procedures, potentially obviating the need for slower movement through each level. The second assumption of an LT approach is that there is a definite sequence of such levels of learning and teaching determined by research-based developmental progressions and that instruction is more efficacious if it builds each level in turn.

Methodological challenges, such as spillover in the teachers from one condition to the other, were daunting. However, most studies with adequate fidelity showed that the BBLT approach produced significantly better learning than the comparison condition (Baroody et al., in press; Clements, Sarama, Baroody, & Joswick, 2020; Clements et al., 2019; Clements, Sarama et al., 2021; Sarama et al., 2021). However, with the studies on patterns, the LT intervention resulted in significant learning but only for some objectives (Baroody et al., 2021). Lower levels of the patterning LT may facilitate learning higher levels but may not be necessary prerequisites. Indeed, some topics may not have strong sequential developmental progressions.

SUMMARY OF LESSONS LEARNED FROM RESEARCH-BASED CURRICULUM DEVELOPMENT AND RESEARCH

This brief chapter could but introduce the landscape of constructing LTs, using them to guide the development of a curriculum, and rigorously testing them. The references provided in this chapter, with additional supportive references in the Appendix, guide engagement with development and research in each phase of the process.

Table 9.1 contains a summary of the lessons learned in developing the *Building Blocks* curriculum and described throughout the discussion of the phases and issues within the three categories of the Curriculum Research Framework, including scale-up studies and other applications of the framework. A synthesis of curriculum development, classroom teaching, and research in mathematics education is necessary to contribute to a better understanding of mathematical thinking, learning, and teaching, and to progressive change in mathematics curricula. Curriculum development projects that produce rich tasks, together with authentic settings in which to test them, provide fertile ground for researchers to test innovative ideas. Without concurrent research, curriculum developers and teachers will miss opportunities to learn about critical aspects of students' thinking and the particular features of software, curricula, and teaching actions that engender mathematical development. We believe curriculum research can help ameliorate these problems (Clements et al., 1997; Schoenfeld, 1999). Traditional research is conservative; it studies *what is* rather than *what could be*. When research is an integral component of the design process, when it helps uncover and invent models of children's thinking and builds these into a creative curriculum, then research moves to the vanguard in innovation and reform of education.

Curriculum research complements other forms of research. It directly contributes to practice, as curricula significantly influence practice (Lloyd et al., 2017; Whitehurst, 2009). Across different methods and within them, there are iterative cycles, which must work to proceed and reveal weaknesses if they do not work (avoiding confirmation bias, the tendency to look for and accept evidence that you hope to see (Greenwald et al., 1986)), and thus, offer tests that are both more frequent and more trustworthy than tests in most other approaches to educational research.

Table 9.1

Lessons Learned in Creation, Implementation, and Evaluation of Curricula

Lesson Learned	Implications
Basing curricula on scientific research and development practices is both feasible and desirable.	Using the multiple phases in the proposed Curriculum Research Framework (CRF) will help developers improve curricula and contribute to curriculum research.
	The education community, including policymakers and academe, should support and require research-based curriculum development with explicated methods and results, improving curricula, research, and the public's opinion of educational research.
The Curriculum Research Framework is a viable guide to research-based and research-validated curricula development and evaluation.	Achieving the goals of CRF requires multiple methods, including qualitative and quantitative methodologies.
	Assessments must be equitable and valid for all children and all outcomes, including those to which the curricula may transfer.
Even if multiple methods are used, they are inadequate if they are all from a priori foundations phases. At best, this produces research-based rather than research-validated curricula.	Curricula should be produced using all appropriate applicable CRF phases; the more used, the better.
Learning trajectories are an effective core of a curriculum development project.	Apply or develop learning trajectories' three components: goal, developmental progression, and correlated instructional activities. Although we and others have developed definite learning trajectories for different mathematical topics, any curriculum development project should attend to the broad, comprehensive consideration of the goals or standards, to the asset-based approach of building on children's thinking and learning patterns, and to linking all teaching to these two.
	Consideration should be given to grain size, appropriate levels of development, number, integration (e.g., interweaving), structuring of LTs, degree, and type of guidance, and especially a comprehensive view of equity.

(Table continued on next page)

Table 9.1 (Continued)

Lessons Learned in Creation, Implementation, and Evaluation of Curricula

Lesson Learned	Implications
Subtle differences in activities can enhance or sabotage effectiveness.	Critical research must be conducted throughout the development process.
Teachers are at the core of successful enactment and must be provided resources and support in *educative* curricula.	Developers should support professional development, scale up of interventions, and systemic change.
High-quality curriculum development and research take substantial time.	Funding agencies should reconsider time frames and funding requirements for curriculum research.

ACKNOWLEDGMENT

This research was supported by the Institute of Education Sciences (IES), U.S. Department of Education, through grants R305K05157 and R305A110188 and also by the National Science Foundation (NSF), through grants ESI-9730804 and REC-0228440. The opinions expressed are those of the authors and do not represent the views of the IES or NSF.

REFERENCES

Association of Mathematics Teacher Educators [AMTE]. (2017). *AMTE standards for mathematics teacher preparation.* https://amte.net/standards

Bailey, D. H., Duncan, G. J., Watts, T. W., Clements, D. H., & Sarama, J. (2018). Risky business: Correlation and causation in longitudinal studies of skill development. *American Psychologist, 73*(1), 81–94. https://doi.org/10.1037/amp0000146

Baratta-Lorton, M. (1976). *Mathematics their way.* Addison-Wesley. https://www.center.edu/MathTheirWay.shtml

Baroody, A. J., Clements, D. H., & Sarama, J. (2022). Lessons learned from 10 experiments that tested the efficacy and assumptions of hypothetical learning trajectories. *Education Sciences, 12,* 195. https://doi.org/10.3390/educsci12030195

Baroody, A. J., Eiland, M. D., Clements, D. H., & Sarama, J. (in press). Does a learning trajectory facilitate the learning of early cardinal-number concepts? *Journal of Mathematical Behavior.*

Baroody, A. J., Yilmaz, N., Clements, D. H., & Sarama, J. (2021). Evaluating a basic assumption of learning trajectories: The case of early patterning learning. *Journal of Mathematics Education, 13*(2), 8–32. https://doi.org/10.26711/007577152790071

Battista, M. T., & Clements, D. H. (2000). Mathematics curriculum development as a scientific endeavor. In A. E. Kelly & R. A. Lesh (Eds.), *Handbook of research design in mathematics and science education* (pp. 737–760). Erlbaum.

Bierman, K. L., Nix, R. L., Greenberg, M. T., Blair, C. B., & Domitrovich, C. E. (2008). Executive functions and school readiness intervention: Impact, moderation, and mediation in the Head Start REDI program. *Development and Psychopathology, 20*(3), 821–843. https://doi.org/10.1017/S0954579408000394

Blanton, M., Brizuela, B. M., Gardiner, A. M., Sawrey, K., & Newman-Owens, A. (2015). A learning trajectory in 6-year-olds' thinking about generalizing functional relationships. *Journal for Research in Mathematics Education, 46*, 511–558. https://doi.org/10.5951/jresematheduc.46.5.0511

Brown, A. L., & Campione, J. C. (1996). Psychological theory and the design of innovative learning environments: On procedures, principles, and systems. In R. Glaser (Ed.), *Innovations in learning: New environments for education* (pp. 289–325). Erlbaum.

Bruner, J. (1986). *Actual minds, possible worlds*. Harvard University Press.

Burkhardt, H. (2006). From design research to large-scale impact: Engineering research in education. In J. Van Den Akker, K. P. E. Gravemeijer, S. McKenney, & N. Nieveen (Eds.), *Educational design research* (pp. 133–162). Routledge.

Burkhardt, H., Fraser, R., & Ridgway, J. (1990). The dynamics of curriculum change. In I. Wirszup & R. Streit (Eds.), *Developments in school mathematics around the world* (Vol. 2, pp. 3–30). National Council of Teachers of Mathematics.

Cai, J. (Ed.). (2017). *Compendium for research in mathematics education*. National Council of Teachers of Mathematics.

Carpenter, T. P., & Fennema, E. H. (1992). Cognitively guided instruction: Building on the knowledge of students and teachers. *International Journal of Educational Research, 17*(5), 457–457. https://doi.org/10.1016/S0883-0355(05)80005-9

Casey, B. M., Dearing, E., Vasilyeva, M., Ganley, C. M., & Tine, M. (2011). Spatial and numerical predictors of measurement performance: The moderating effects of community income and gender. *Journal of Educational Psychology, 103*(2), 296–311. https://doi.org/10.1037/a0022516

Celedòn-Pattichis, S., Borden, L. L., Pape, S. J., Clements, D. H., Peters, S. A., Males, J. R., Chapman, O., & Leonard, J. (2018). Asset-based approaches to equitable mathematics education research and practice. *Journal for Research in Mathematics Education, 49*(4), 373–389.
https://doi.org/10.5951/jresematheduc.49.4.0373

Cepeda, N. J., Pashler, H., Vul, E., Wixted, J. T., & Rohrer, D. (2006). Distributed practice in verbal recall tasks: A review and quantitative synthesis. *Psychological Bulletin, 132*, 354–380. https://doi.org/10.1037/0033-2909.132.3.354

Cheng, Y.-L., & Mix, K. S. (2012). Spatial training improves children's mathematics ability. *Journal of Cognition and Development, 15*(1), 2–11.
https://doi.org/10.1080/15248372.2012.725186

Civil, M. (2002). Everyday mathematics, mathematicians' mathematics, and school mathematics: Can we bring them together? In J. N. Moschkovich & M. E. Brenner (Eds.), *Everyday and academic mathematics in the classroom (Journal for Research in Mathematics Education Monograph Number 11)* (pp. 40–62). National Council of Teachers of Mathematics.

Civil, M. (2016). STEM learning research through a funds of knowledge lens. *Cultural Studies of Science Education, 11*(1), 41–59. https://doi.org/10.1007/s11422-014-9648-2

Clarke, D. M., Cheeseman, J., Gervasoni, A., Gronn, D., Horne, M., McDonough, A., Montgomery, P., Roche, A., Sullivan, P., Clarke, B. A., & Rowley, G. (2002). *Early Numeracy Research Project final report*. Department of Education, Employment and Training, the Catholic Education Office (Melbourne), and the Association of Independent Schools Victoria.

Clements, D. H. (2002). Linking research and curriculum development. In L. D. English (Ed.), *Handbook of international research in mathematics education* (pp. 599–636). Erlbaum.

Clements, D. H. (2007). Curriculum research: Toward a framework for "research-based curricula." *Journal for Research in Mathematics Education, 38*(1), 35–70.

Clements, D. H., & Battista, M. T. (2000). Designing effective software. In A. E. Kelly & R. A. Lesh (Eds.), *Handbook of research design in mathematics and science education* (pp. 761–776). Erlbaum.

Clements, D. H., Battista, M. T., Sarama, J., & Swaminathan, S. (1997). Development of students' spatial thinking in a unit on geometric motions and area. *The Elementary School Journal, 98*, 171–186.

Clements, D. H., & Sarama, J. (1999). *Preliminary report of Building Blocks—Foundations for mathematical thinking, Pre-Kindergarten to Grade 2: Research-based materials development* (NSF Grant No. ESI-9730804). State University of New York at Buffalo.

Clements, D. H., & Sarama, J. (2018). *Building Blocks fidelity of implementation*. University of Buffalo, State University of New York/University of Denver. (Original work published 2000)

Clements, D. H., & Sarama, J. (2004). Learning trajectories in mathematics education. *Mathematical Thinking and Learning, 6*, 81–89. https://doi.org/10.1207/s15327833mtl0602_1

Clements, D. H., & Sarama, J. (2007a). *Building Blocks—SRA Real Math Teacher's Edition, Grade PreK* [Computer software]. SRA/McGraw-Hill.

Clements, D. H., & Sarama, J. (2007b). Early childhood mathematics learning. In F. K. Lester, Jr. (Ed.), *Second handbook of research on mathematics teaching and learning* (Vol. 1, pp. 461–555). Information Age Publishing.

Clements, D. H., & Sarama, J. (2007c). Effects of a preschool mathematics curriculum: Summative research on the *Building Blocks* project. *Journal for Research in Mathematics Education, 38*(2), 136–163.

Clements, D. H., & Sarama, J. (2008). Experimental evaluation of the effects of a research-based preschool mathematics curriculum. *American Educational Research Journal, 45*(2), 443–494. https://doi.org/10.3102/0002831207312908

Clements, D. H., & Sarama, J. (2013). *Building Blocks, Volumes 1 and 2*. McGraw-Hill Education.

Clements, D. H., & Sarama, J. (2021). *Learning and teaching early math: The learning trajectories approach* (3rd ed.). Routledge. https://doi.org/10.4324/9781003083528

Clements, D. H., Sarama, J., Baroody, A. J., & Joswick, C. (2020). Efficacy of a learning trajectory approach compared to a teach-to-target approach for addition and subtraction. *ZDM Mathematics Education, 52*(4), 637–648. https://doi.org/10.1007/s11858-019-01122-z

Clements, D. H., Sarama, J., Baroody, A. J., Joswick, C., & Wolfe, C. B. (2019). Evaluating the efficacy of a learning trajectory for early shape composition. *American Educational Research Journal, 56*(6), 2509–2530. https://doi.org/10.3102/0002831219842788

Clements, D. H., Sarama, J., Baroody, A. J., Kutaka, T. S., Chernyavskiy, P., Joswick, C., Cong, M., & Joseph, E. (2021). Comparing the efficacy of early arithmetic instruction based on a learning trajectory and teaching-to-a-target. *Journal of Educational Psychology, 113*(7), 1323–1337. https://doi.org/doi.org/10.1037/edu0000633

Clements, D. H., Sarama, J., Layzer, C., & Unlu, F. (2023). Implementation of a scale-up model in early childhood: Long-term impacts on mathematics achievement. *Journal for Research in Mathematics Education, 54*(1), 64–88. https://doi.org/10.5951/jresematheduc-2020-0245

Clements, D. H., Sarama, J., Layzer, C., Unlu, F., & Fesler, L. (2020). Effects on mathematics and executive function of a mathematics and play intervention versus mathematics alone. *Journal for Research in Mathematics Education, 51*(3), 301–333. https://doi.org/10.5951/jresemtheduc-2019-0069

Clements, D. H., Sarama, J., Spitler, M. E., Lange, A. A., & Wolfe, C. B. (2011). Mathematics learned by young children in an intervention based on learning trajectories: A large-scale cluster randomized trial. *Journal for Research in Mathematics Education, 42*(2), 127–166. https://doi.org/10.5951/jresematheduc.42.2.0127

Clements, D. H., Sarama, J., & Wilson, D. C. (2001). Composition of geometric figures. In M. Van den Heuvel-Panhuizen (Ed.), *Proceedings of the 25th Conference of the International Group for the Psychology of Mathematics Education* (Vol. 2, pp. 273–280). Freudenthal Institute.

Clements, D. H., Sarama, J., Wolfe, C. B., & Spitler, M. E. (2013). Longitudinal evaluation of a scale-up model for teaching mathematics with trajectories and technologies: Persistence of effects in the third year. *American Educational Research Journal, 50*(4), 812–850. https://doi.org/10.3102/0002831212469270

Clements, D. H., Sarama, J., Wolfe, C. B., & Spitler, M. E. (2015). Sustainability of a scale-up intervention in early mathematics: Longitudinal evaluation of implementation fidelity. *Early Education and Development, 26*(3), 427–449. https://doi.org/10.1080/10409289.2015.968242

Clements, D. H., Vinh, M., Lim, C.-I., & Sarama, J. (2021). STEM for inclusive excellence and equity. *Early Education and Development, 32*(1), 148–171. https://doi.org/10.1080/10409289.2020.1755776

Cross, D. I., Adefope, O., Mi Yeon, L., & Pérez, A. (2012). Hungry for early spatial and algebraic reasoning. *Teaching Children Mathematics, 19*(1), 42–49. https://doi.org/10.5951/teacchilmath.19.1.0042

Danesi, M. (2009, April 24). *Puzzles and the brain*. Retrieved October 11, 2018, from https://www.psychologytoday.com/us/blog/brain-workout/200904/puzzles-and-the-brain

Doabler, C. T., Clarke, B., Fien, H., Baker, S. K., Kosty, D. B., & Cary, M. S. (2014). The science behind curriculum development and evaluation: Taking a design science approach in the production of a tier 2 mathematics curriculum. *Learning Disability Quarterly*, *38*(2), 97–111. https://doi.org/10.1177/0731948713520555

Doabler, C. T., Clarke, B., Firestone, A. R., Turtura, J. E., Jungjohann, K. J., Brafford, T. L., Sutherland, M., Nelson, N. J., & Fien, H. (2019). Applying the Curriculum Research Framework in the design and development of a technology-based tier 2 mathematics intervention. *Journal of Special Education Technology*, *34*(3), 176–189. https://doi.org/10.1177/0162643418812051

Dong, Y., Clements, D. H., Sarama, J., Dumas, D. G., Banse, H. W., & Day-Hess, C. A. (2021). Mathematics and executive function competencies in the context of interventions: A quantile regression analysis. *The Journal of Experimental Education*, *90*(2), 297–318. https://doi.org/10.1080/00220973.2020.1777070

Drake, C., Land, T. J., & Tyminski, A. M. (2014). Using educative curriculum materials to support the development of prospective teachers' knowledge. *Educational Researcher*, *43*(3), 154–162. https://doi.org/10.3102/0013189X14528039

Dumas, D. G., McNeish, D., Sarama, J., & Clements, D. H. (2019). Preschool mathematics intervention can significantly improve student learning trajectories through elementary school. *AERA Open*, *5*(4), 1–5. https://doi.org/10.1177/2332858419879446

Eisner, E. W. (1998). The primacy of experience and the politics of method. *Educational Researcher*, *17*(5), 15–20. https://doi.org/10.3102/0013189X017005015

Entwisle, D. R., & Alexander, K. L. (1992). Summer setback: Race, poverty, school composition, and mathematics achievement in the first two years of school. *American Sociological Review*, *57*, 72–84.

Frye, D., Baroody, A. J., Burchinal, M. R., Carver, S., Jordan, N. C., & McDowell, J. (2013). *Teaching math to young children: A practice guide*. National Center for Education Evaluation and Regional Assistance (NCEE), Institute of Education Sciences, U.S. Department of Education. https://ies.ed.gov/ncee/wwc/PracticeGuide/18

Fuson, K. C. (1988). *Children's counting and concepts of number*. Springer-Verlag. https://doi.org/10.1007/978-1-4612-3754-9

Greenwald, A. G., Pratkanis, A. R., Leippe, M. R., & Baumgardner, M. H. (1986). Under what conditions does theory obstruct research progress? *Psychological Review*, *93*, 216–229.

Howe, R. (2011). Three pillars of first grade mathematics. *The De Morgan Journal*, *1*(1), 3–16.

Howe, R. (2014). Three pillars of first grade mathematics, and beyond. In Y. Li & G. Lappan (Eds.), *Mathematics curriculum in school education* (pp. 183–207). Springer.

James, W. (1958). *Talks to teachers on psychology: And to students on some of life's ideas*. Norton. (Original work published 1892)

Jobrack, B. (2011). *Tyranny of the textbook: An insider exposes how educational materials undermine reforms*. Rowman & Littlefield.

John, M. S., Sibuma, B., Wunnava, S., Anggoro, F., & Dubosarsky, M. (2018). An iterative participatory approach to developing an early childhood problem-based STEM curriculum. *European Journal of STEM Education, 3*(3), 07. https://doi.org/10.20897/ejsteme/3867

Kilpatrick, J. (1992). A history of research in mathematics education. In D. A. Grouws (Ed.), *Handbook of research on mathematics teaching and learning* (pp. 3–38). Macmillan.

Lagemann, E. C. (1997). Contested terrain: A history of education research in the United States, 1890–1990. *Educational Researcher, 26*(9), 5–17. https://www.jstor.org/stable/1176271

Lamberg, T. D., & Middleton, J. A. (2009). Design research perspectives on transitioning from individual microgenetic inteviews to a whole-class teaching experiment. *Educational Researcher, 38*, 233–245. https://doi.org/10.3102/0013189X09334206

Lloyd, G. M., Cai, J., & Tarr, J. E. (2017). Issues in curriculum studies: Evidence-based insights and future directions. In J. Cai (Ed.), *Compendium for research in mathematics education* (pp. 824–852). National Council of Teachers of Mathematics.

MacDonald, B. L., & Shumway, J. F. (2016). Subitizing games: Assessing preschool children's number understanding. *Teaching Children Mathematics, 22*(6), 340–348.

Maloney, A. P., Confrey, J., & Nguyen, K. H. (Eds.). (2014). *Learning over time: Learning trajectories in mathematics education*. Information Age Publishing.

Markussen-Brown, J., Juhl, C. B., Piasta, S. B., Bleses, D., Højen, A., & Justice, L. M. (2017). The effects of language-and literacy-focused professional development on early educators and children: A best-evidence meta-analysis. *Early Childhood Research Quarterly, 38*, 97–115. https://doi.org/10.1016/j.ecresq.2016.07.002

Mattera, S. K., Rojas, N. M., Morris, P. A., & Bierman, K. (2021). Promoting EF with preschool interventions: Lessons learned from 15 years of conducting large-scale studies. *Frontiers in Psychology, 12*, 1–20. https://doi.org/10.3389/fpsyg.2021.640702

Moll, L. C., Amanti, C., Neff, D., & Gonzalez, N. (1992). Funds of knowledge for teaching: Using a qualitative approach to connect homes and classrooms. *Theory into Practice, 31*, 132–141.

National Research Council. (2004). *On evaluating curricular effectiveness: Judging the quality of K-12 mathematics evaluations*. Mathematical Sciences Education Board, Center for Education, Division of Behavioral and Social Sciences and Education, The National Academies Press. https://nap.nationalacademies.org/catalog/11025/on-evaluating-curricular-effectiveness-judging-the-quality-of-k-12

National Research Council. (2009). *Mathematics learning in early childhood: Paths toward excellence and equity*. The National Academies Press. https://doi.org/10.17226/12519

National Governors Association and Council of Chief State School Officers [NGA/CCSSO]. (2010). *Common core state standards*. http://thecorestandards.org/

Nguyen, T., Watts, T. W., Duncan, G. J., Clements, D. H., Sarama, J., Wolfe, C. B., & Spitler, M. E. (2016). Which preschool mathematics competencies are most predictive of fifth grade achievement? *Early Childhood Research Quarterly*, *36*, 550–560. https://doi.org/10.1016/j.ecresq.2016.02.003

Park, M., O'Toole, A., & Katsiaficas, C. (2017). *Dual language learners: A national demographic and policy profile*. Migration Policy Institute. https://eric.ed.gov/?id=ED589153

Payne, K. J., & Biddle, B. J. (1999). Poor school funding, child poverty, and mathematics achievement. *Educational Researcher*, *28*(6), 4–13.

Penuel, W. R., & Shepard, L. A. (2016). Assessment and teaching. In D. H. Gitomer & C. A. Bell (Eds.), *Handbook of research on teaching* (5th ed., pp. 787–850). American Educational Research Association. https://doi.org/https://doi.org/10.3102/978-0-935302-48-6_12

Presmeg, N. C. (2007). The role of culture in teaching and learning mathematics. In F. K. Lester, Jr. (Ed.), *Second handbook of research on mathematics teaching and learning* (Vol. 1, pp. 435–458). Information Age Publishing.

Raisor, J. M., & Hudson, R. A. (2018). Bottle caps as prekindergarten mathematical tools. *Teaching Children Mathematics*, *24*(6), 370–377. http://www.jstor.org/stable/10.5951/teacchilmath.24.6.0370

Raudenbush, S. W. (2009). The *Brown* legacy and the O'Connor challenge: Transforming schools in the images of children's potential. *Educational Researcher*, *38*(3), 169–180. https://doi.org/10.3102/0013189X09334840

Remillard, J. T., Lloyd, G. M., & Herbel-Eisenmann, B. A. (2009). *Mathematics teachers at work: Connecting curriculum materials and classroom instruction*. Routledge.

Rittle-Johnson, B., Fyfe, E. R., & Zippert, E. (2018). The roles of patterning and spatial skills in early mathematics development. *Early Childhood Research Quarterly*, *46*(1), 166–178. https://doi.org/10.1016/j.ecresq.2018.03.006

Rohrer, D., & Taylor, K. (2006). The effects of overlearning and distributed practise on the retention of mathematics knowledge. *Applied Cognitive Psychology*, *20*, 1–16.

Russek, B. E., & Weinberg, S. L. (1993). Mixed methods in a study of implementation of technology-based materials in the elementary classroom. *Evaluation and Program Planning*, *16*, 131–142.

Sakakibara, T. (2014). Mathematics learning and teaching in Japanese preschool: Providing appropriate foundations for an elementary schooler's mathematics learning. *International Journal of Educational Studies in Mathematics*, *1*(1), 16–26.

Sarama, J., & Clements, D. H. (2003). *Building Blocks* of early childhood mathematics. *Teaching Children Mathematics*, *9*(8), 480–484.

Sarama, J., & Clements, D. H. (2008). Linking research and software development. In G. W. Blume & M. K. Heid (Eds.), *Research on technology and the teaching and learning of mathematics: Cases and perspectives* (Vol. 2, pp. 113–130). Information Age Publishing.

Sarama, J., & Clements, D. H. (2009a). *Early childhood mathematics education research: Learning trajectories for young children*. Routledge. https://doi.org/10.4324/9780203883785

Sarama, J., & Clements, D. H. (2009b, April 13-17). *Scaling up successful interventions: Multidisciplinary perspectives*. [Paper presentation] American Educational Research Association, San Diego, CA.

Sarama, J., & Clements, D. H. (2019a). The Building Blocks and TRIAD projects. In P. Sztajn & P. H. Wilson (Eds.), *Learning trajectories for teachers: Designing effective professional development for math instruction* (pp. 104–131). Teachers College Press.

Sarama, J., & Clements, D. H. (2019b). From cognition to curriculum to scale. In D. C. Geary, D. B. Berch, & K. M. Koepke (Eds.), *Cognitive foundations for improving mathematical learning* (Vol. 5, pp. 143–173). Academic Press. https://doi.org/10.1016/B978-0-12-815952-1.00006-2

Sarama, J., & Clements, D. H. (2019c). Research and curricula. In K. R. Leatham (Ed.), *Designing, conducting, and publishing quality research in mathematics education*. (*Research in Mathematics Education Series*) (pp. 61–83). Springer. https://doi.org/10.1007/978-3-030-23505-5_5

Sarama, J., & Clements, D. H. (2021). Long-range impact of a scale-up model on mathematics teaching and learning: Persistence, sustainability, and diffusion. *Journal of Cognitive Education and Psychology, 20*(2), 112–122. https://doi.org/10.1891/JCEP-D-20-00005

Sarama, J., Clements, D. H., Baroody, A. J., Kutaka, T. S., Chernyavskiy, P., Shi, J., & Cong, M. (2021). Testing a theoretical assumption of a learning-trajectories approach in teaching length measurement to kindergartners. *AERA Open, 7*(1), 1–15. https://doi.org/10.1177/23328584211026657

Sarama, J., Clements, D. H., & Henry, J. J. (1998). Network of influences in an implementation of a mathematics curriculum innovation. *International Journal of Computers for Mathematical Learning, 3*, 113–148.

Sarama, J., Clements, D. H., Starkey, P., Klein, A., & Wakeley, A. (2008). Scaling up the implementation of a pre-kindergarten mathematics curriculum: Teaching for understanding with trajectories and technologies. *Journal of Research on Educational Effectiveness, 1*(1), 89–119. https://doi.org/10.1080/19345740801941332

Sarama, J., Clements, D. H., Wolfe, C. B., & Spitler, M. E. (2016). Professional development in early mathematics: Effects of an intervention based on learning trajectories on teachers' practices. *Nordic Studies in Mathematics Education, 21*(4), 29–55.

Sarama, J., Lange, A., Clements, D. H., & Wolfe, C. B. (2012). The impacts of an early mathematics curriculum on emerging literacy and language. *Early Childhood Research Quarterly, 27*(3), 489–502. https://doi.org/10.1016/j.ecresq.2011.12.002

Schenke, K., Nguyen, T., Watts, T. W., Sarama, J., & Clements, D. H. (2017). Differential effects of the classroom on African American and non-African American's mathematics achievement. *Journal of Educational Psychology, 109*(6), 794–811. https://doi.org/10.1037/edu0000165

Schoenfeld, A. H. (1999). Looking toward the 21st century: Challenge of educational theory and practice. *Educational Researcher, 28*, 4–14.

Siegler, R. S. (1996). *Emerging minds: The process of change in children's thinking*. Oxford University Press.

Simon, M. A. (1995). Reconstructing mathematics pedagogy from a constructivist perspective. *Journal for Research in Mathematics Education*, 26(2), 114–145. https://doi.org/10.2307/749205

Simon, M. A., Placa, N., & Avitzur, A. (2016). Participatory and anticipatory stages of mathematical concept learning: Further empirical and theoretical development. *Journal for Research in Mathematics Education*, 47(1), 63–93. https://doi.org/10.5951/jresematheduc.47.1.0063

Spradley, J. P. (1980). *Participant observation*. Holt, Rhinehart & Winston.

Steffe, L. P. (2017). Psychology in mathematics education: Past, present, and future. In E. Galindo & J. Newton (Eds.), *Proceedings of the 39th annual meeting of the North American Chapter of the International Group for the Psychology of Mathematics—Synergy at the crossroads: Future directions for theory, research, and practice* (pp. 27–56). Hoosier Association of Mathematics Teacher Educators. https://www.pmena.org/pmenaproceedings/PMENA%2039%202017%20Proceedings.pdf

Steffe, L. P., & Tzur, R. (1994). Interaction and children's mathematics. *Journal of Research in Childhood Education*, 8(2), 99–116.

Stein, M. K., Remillard, J. T., & Smith, M. S. (2007). How curriculum influences student learning. In F. K. Lester, Jr. (Ed.), *Second handbook of research on mathematics teaching and learning* (Vol. 1, pp. 319–369). Information Age Publishing.

Stephan, M., Cobb, P., Gravemeijer, K. P. E., & Estes, B. (2001). The role of tools in supporting students' development of measuring conceptions. In A. Cuoco (Ed.), *The roles of representation in school mathematics* (pp. 63–76). National Council of Teachers of Mathematics.

Valant, J., & Newark, D. A. (2016). The politics of achievement gaps: U.S. public opinion on race-based and wealth-based differences in test scores. *Educational Researcher*, 45(6), 331–346. https://doi.org/10.3102/0013189x16658447

Verdine, B. N., Golinkoff, R. M., Hirsh-Pasek, K., & Newcombe, N. S. (2017). Links between spatial and mathematical skills across the preschool years. *Monographs of the Society for Research in Child Development*, 82(1), 7–30. https://doi.org/10.1111/mono.12280

Vygotsky, L. S. (1978). *Mind in society: The development of higher psychological processes*. Harvard University Press. (Original work published 1935)

Weiland, C., & Yoshikawa, H. (2013). Impacts of a prekindergarten program on children's mathematics, language, literacy, executive function, and emotional skills. *Child Development*, 84(6), 2112–2130. https://doi.org/10.1111/cdev.12099

Whitehurst, G. J. (2009). *Don't forget curriculum*. Brown Center on Education Policy, The Brookings Institution.

Wu, H.-H. (2011). *Understanding numbers in elementary school mathematics*. American Mathematical Society.

APPENDIX: ADDITIONAL SUPPORTIVE REFERENCES

Barnes, E. M., & Stephens, S. J. (2019). Supporting mathematics vocabulary instruction through mathematics curricula. *Curriculum Journal, 30*(3), 322–341. https://doi.org/10.1080/09585176.2019.1614470

Baroody, A. J. (2016). Curricular approaches to connecting subtraction to addition and fostering fluency with basic differences in grade 1. *PNA, 10*(3), 161–191.

Björklund, C. (2018). Powerful frameworks for conceptual understanding. In V. Kinnear, M. Y. Lai, & T. Muir (Eds.), *Forging connections in early mathematics teaching and learning* (pp. 37–53). Springer. https://doi.org/10.1007/978-981-10-7153-9_3

Brown, R. D. (Ed.). (2018). *Neuroscience of mathematical cognitive development: From infancy through emerging adulthood*. Springer. https://doi.org/10.1007/978-3-319-76409-2

Cai, J., Morris, A., Hohensee, C., Hwang, S., Robison, V., & Hiebert, J. (2017). Clarifying the impact of educational research on learning opportunities. *Journal for Research in Mathematics Education, 48*(3), 230–236. https://doi.org/10.5951/jresematheduc.48.3.0230

Choppin, J. M., Roth McDuffie, A., Drake, C., & Davis, J. (2020). The role of instructional materials in the relationship between the official curriculum and the enacted curriculum. *Mathematical Thinking and Learning, 24*(2), 123–148. https://doi.org/10.1080/10986065.2020.1855376

Clarke, B. A. (2008). A framework of growth points as a powerful teacher development tool. In D. Tirosh & T. Wood (Eds.), *Tools and processes in mathematics teacher education* (pp. 235–256). Sense.

Clarke, D. M., Cheeseman, J., Clarke, B., Gervasoni, A., Gronn, D., Horne, M., McDonough, A., Montgomery, P., Rowley, G., & Sullivan, P. (2001). Understanding, assessing and developing young children's mathematical thinking: Research as a powerful tool for professional growth. In J. Bobis, B. Perry, & M. Mitchelmore (Eds.), *Numeracy and beyond* (24th Annual Conference of the Mathematics Education Research Group of Australasia, Vol. 1) (pp. 9–26). Mathematics Education Research Group of Australasia.

Cobb, P., Confrey, J., diSessa, A., Lehrer, R., & Schauble, L. (2003). Design experiments in educational research. *Educational Researcher, 32*(1), 9–13. https://doi.org/10.3102/0013189X032001009

Delgado, C. (2009). *Development of a research-based learning progression for middle school through undergraduate students' conceptual understanding of size and scale*. [Unpublished Doctoral dissertation, The University of Michigan].

Donegan-Ritter, M. M., & Zan, B. (2017). Designing and implementing inclusive STEM activities for early childhood. In C. Curran & A. J. Peterson (Eds.), *Handbook of research on classroom diversity and inclusive education practice* (pp. 222–249). IGI Global. https://doi.org/10.4018/978-1-5225-2520-2.ch010

Franke, M. L., Kazemi, E., & Battey, D. (2007). Mathematics teaching and classroom practice. In F. K. Lester, Jr. (Ed.), *Second handbook of research on mathematics teaching and learning* (Vol. 1, pp. 225–256). Information Age Publishing.

Freiman, V. (2018). Complex and open-ended tasks to enrich mathematical experiences of kindergarten students. In F. M. Singer (Ed.), *Mathematical creativity and mathematical giftedness: Enhancing creative capacities in mathematically promising students* (pp. 373–404). Springer International Publishing. https://doi.org/10.1007/978-3-319-73156-8_14

Garon-Carrier, G., Boivin, M., Lemelin, J.-P., Kovas, Y., Parent, S., Séguin, J., Vitaro, F., Tremblay, R. E., & Dionne, G. (2018). Early developmental trajectories of number knowledge and math achievement from 4 to 10 years: Low-persistent profile and early-life predictors. *Journal of School Psychology, 68*, 84–98. https://doi.org/10.1016/j.jsp.2018.02.004

Geary, D. C., Berch, D. B., & Koepke, K. M. (Eds.). (2019). *Cognitive foundations for improving mathematical learning* (Vol. 5). Academic Press.

Hiebert, J. C., & Grouws, D. A. (2007). The effects of classroom mathematics teaching on students' learning. In F. K. Lester, Jr. (Ed.), *Second handbook of research on mathematics teaching and learning* (Vol. 1, pp. 371–404). Information Age Publishing.

Hill, H. C., Blunk, M. L., Charalambous, C. Y., Lewis, J. M., Phelps, G. C., Sleep, L., & Ball, D. L. (2008). Mathematical knowledge for teaching and the mathematical quality of instruction: An exploratory study. *Cognition and Instruction, 26*, 430–511.

Inchaustegui, Y. A., & Alsina, A. (2020). Learning patterns at three years old: Contributions of a learning trajectory and teaching itinerary. *Australasian Journal of Early Childhood, 45*(1), 14–29. https://doi.org/10.1177/1836939119885310

Larson, R. S., Dearing, J. W., & Backer, T. E. (2017). *Strategies to scale up social programs: Pathways, partnerships and fidelity*. The Wallace Foundation. wallacefoundation.org

May, H., Gray, A., Sirinides, P., Goldsworthy, H., Armijo, M., Sam, C., Gillespie, J. N., & Tognatta, N. (2015). Year one results from the multisite randomized evaluation of the i3 scale-up of reading recovery. *American Educational Research Journal, 52*(3), 547–581. https://doi.org/10.3102/0002831214565788

McCoy, D. C., Salhi, C., Yoshikawa, H., Black, M., Britto, P., & Fink, G. (2018). Home- and center-based learning opportunities for preschoolers in low- and middle-income countries. *Children and Youth Services Review, 88*, 44–56. https://doi.org/10.1016/j.childyouth.2018.02.021

McGonigle-Chalmers, M., & Kusel, I. (2019). The development of size sequencing skills: An empirical and computational analysis. *Monographs of the Society for Research in Child Development, 84*(4), 1-202. https://doi.org/10.1111/mono.12411

McMahon, K. A., & Whyte, K. (2020). What does math curriculum tell us about continuity for PreK–3? *Curriculum Journal, 31*(1), 48–76. https://doi.org/10.1002/curj.8

Meaney, T. (2018). Mathematics curricula: Issues of access and quality. In M. Jurdak & R. Vithal (Eds.), *Sociopolitical dimensions of mathematics education: From the margin to mainstream* (pp. 171–189). Springer International Publishing. https://doi.org/10.1007/978-3-319-72610-6_10

Mulligan, J. T., Prescott, A., Mitchelmore, M. C. (2004). Children's development of structure in early mathematics. In M. J. Høines & A. B. Fuglestad (Eds.), *Proceedings of the 28th Conference of the International Group for the Psychology of Mathematics Education* (Vol. 3, pp. 393–400). Bergen University College.

Mulligan, J., Verschaffel, L., Baccaglini-Frank, A., Coles, A., Gould, P., He, S., Ma, Y., Milinković, J., Obersteiner, A., Roberts, N., Sinclair, N., Wang, Y., Xie, S., & Yang, D.-C. (2018). Whole number thinking, learning and development: Neuro-cognitive, cognitive and developmental approaches. In M. G. Bartolini Bussi & X. H. Sun (Eds.), *Building the foundation: Whole numbers in the primary grades: The 23rd ICMI study* (pp. 137–167). Springer International Publishing. https://doi.org/10.1007/978-3-319-63555-2_7

Murata, A. (2004). Paths to learning ten-structured understanding of teen sums: Addition solution methods of Japanese Grade 1 students. *Cognition and Instruction, 22*, 185–218.

Obersteiner, A., Reiss, K., & Heinze, A. (2018). Psychological theories in mathematics education. *Journal für Mathematik-Didaktik, 39*(1), 1–6. https://doi.org/10.1007/s13138-018-0134-3

Pasnak, R. (2017). Empirical studies of patterning. *Psychology, 8*(13), 2276–2293. https://doi.org/10.4236/psych.2017.813144

Radford, L. (2018). The emergence of symbolic algebraic thinking in primary school. In C. Kieran (Ed.), *Teaching and learning algebraic thinking with 5- to 12-year-olds: The global evolution of an emerging field of research and practice* (pp. 3–25). Springer. https://doi.org/10.1007/978-3-319-68351-5_1

Sack, J. J. (2013). Development of a top-view numeric coding teaching-learning trajectory within an elementary grades 3-D visualization design research project. *The Journal of Mathematical Behavior, 32*(2), 183–196. https://doi.org/10.1016/j.jmathb.2013.02.006

Sarama, J., Clements, D. H., Wolfe, C. B., & Spitler, M. E. (2012). Longitudinal evaluation of a scale-up model for teaching mathematics with trajectories and technologies. *Journal of Research on Educational Effectiveness, 5*(2), 105–135. https://doi.org/10.1080/19345747.2011.627980

Sari, E. F., Nugraheni, N., & Kiptiyah, S. M. (2019). The implementation of blended learning based realistic mathematics education in mathematics teaching. *International Journal of Innovation, Creativity and Change, 5*(5), 353–361. https://www.scopus.com/inward/record.uri?eid=2-s2.0-85082874544&partnerID=40&md5=6f4b4fbea22b8421932ac60235df05b8

Stein, M. K., & Kaufman, J. H. (2010). Selecting and supporting the use of mathematics curricula at scale. *American Educational Research Journal, 47*(3), 663–693.

Sullivan, P. (2011). *Teaching mathematics: Using research-informed strategies*. Australian Council for Educational Research.

Sztajn, P., Confrey, J., Wilson, P. H., & Edgington, C. (2012). Learning trajectory based instruction: Toward a theory of teaching. *Educational Researcher, 41*, 147–156. https://doi.org/10.3102/0013189X12442801

Vamvakoussi, X., Christou, K. P., & Vosniadou, S. (2018). Bridging psychological and educational research on rational number knowledge. *Journal of Numerical Cognition, 4*(1), 84–106. https://doi.org/10.5964/jnc.v4i1.82

Veraksa, A. N., Aslanova, M. S., Bukhalenkova, D. A., Veraksa, N. E., & Liutsko, L. (2020). Assessing the effectiveness of differentiated instructional approaches for teaching math to preschoolers with different levels of executive functions. *Education Sciences, 10*(7), 1–16. https://doi.org/10.3390/educsci10070181

Wakabayashi, T., Andrade-Adaniya, F., Schweinhart, L. J., Xiang, Z., Marshall, B. A., & Markley, C. A. (2020). The impact of a supplementary preschool mathematics curriculum on children's early mathematics learning. *Early Childhood Research Quarterly, 53,* 329–342. https://doi.org/10.1016/j.ecresq.2020.04.002

Wilson, P. H. (2009). *Teachers' uses of a learning trajectory for equipartitioning* [Unpublished Doctoral dissertation, North Carolina State University].

Wright, R. J., Stanger, G., Stafford, A. K., & Martland, J. (2006). *Teaching number in the classroom with 4-8 year olds*. SAGE.

Ye, B. (2018). Developments and changes in the primary school mathematics curriculum and teaching material in China. In Y. Cao & F. K. S. Leung (Eds.), *The 21st century mathematics education in China* (pp. 107–125). Springer Berlin Heidelberg. https://doi.org/10.1007/978-3-662-55781-5_6

Zielinski, C. D. (2018). *A "re-envisioned" instruction model: Minimizing students' learning difficulties in mathematics* [Unpublished Doctoral dissertation, University of Michigan-Dearborn]. https://deepblue.lib.umich.edu/bitstream/handle/2027.42/140755/Zielinski%20Final%20Dissertation.pdf?sequence=1&isAllowed=y

CHAPTER 10

LESSONS LEARNED IN DESIGNING AN EFFECTIVE EARLY ALGEBRA CURRICULUM FOR GRADES K–5

Maria Blanton
TERC

Angela Murphy Gardiner
TERC

Ana C. Stephens
University of Wisconsin–Madison

Rena Stroud
Merrimack College

Eric Knuth
U.S. National Science Foundation

Despina Stylianou
City College of New York–CUNY

We describe lessons learned in designing an early algebra curriculum to measure early algebra's impact on children's algebra readiness for middle grades. The curriculum was developed to supplement regular mathematics instruction in the U.S. for Grades K–5. Lessons learned centered around the importance of several key factors, including using conceptual frameworks to design the components of our curriculum, treating early algebra content as a set of core algebraic thinking practices across several core content domains, and using a staged approach to curriculum design research. We also learned the importance of recognizing and addressing gaps in (a) the assessment tools available for measuring growth in students' learning, (b) the empirical research base for early algebra, and (c) support for teachers around professional learning in early algebra. We found curriculum design to be a complex and expensive process that requires careful pacing and deliberation.

Keywords: curriculum design; early algebra; elementary grades; frameworks; learning progressions

THE NEED FOR AN EARLY ALGEBRA CURRICULUM

In recent decades, views on teaching and learning algebra in school mathematics have changed deeply. Prompted by algebra's historical status as a gateway to academic and economic success (Schoenfeld, 1995), scholars have argued that developing children's informal notions about algebraic ideas beginning in kindergarten would better prepare them for success in formal algebra in later grades. Such a substantial change in school mathematics has significant costs that include "deep curriculum restructuring, changes in classroom practice and assessment, and changes in teacher education—each a major task" (Kaput, 2008, p. 6). This raises fair questions about the impact that algebraic thinking in the elementary grades (referred to here as *early algebra*) might have on children's algebra readiness for middle grades. In essence, as we have asked elsewhere (Blanton et al., 2019), does early algebra matter?

In considering the tension around costs and benefits of early algebra, over a decade ago we began a program of research to better understand the impact of early algebra. We wanted to know what difference a comprehensive, research-based approach to instruction around early algebraic concepts across elementary grades might make in children's algebraic thinking as they entered middle grades. Our immediate challenge was clear: Elementary school programs[1] were not yet equipped to develop children's algebraic thinking in the deep, systemic way we felt was needed to fairly measure the impact of early algebra.

First, elementary teachers are critical to the success of algebra reforms, yet they have not historically been provided with sufficient professional learning opportunities to develop classroom instruction that fosters the rich and connected kinds of algebraic thinking that constitute early algebra (e.g., Greenberg & Walsh, 2008; Kaput & Blanton, 2005). Second, existing curricula have not adequately addressed early algebraic concepts and practices in a manner that focuses on important ways of thinking algebraically. Even now, widely used mathematics curricula for elementary grades too often treat algebra as a collection of "things to do" (e.g., solve an equation) rather than as a set of practices or habits of mind, such as generalizing mathematical relationships or justifying claims about general relationships.

Second, mathematics curricula have not always addressed core algebraic concepts with sufficient depth. Consider the concept of mathematical equivalence. The equal sign symbolizes an equivalence relation indicating two mathematical objects are equivalent (Jones et al., 2012) and should be understood *relationally* in elementary grades, meaning the two expressions in an equation have the same value. Yet many students view this symbol *operationally*, as a prompt to perform the computation indicated in the expression to the left of the equal sign (Jacobs et al., 2007).[2] Research suggests that students' operational misconceptions about the equal sign are present as early as kindergarten (Blanton, Otalora Sevilla et al., 2018) and persist in later elementary grades (e.g., Stephens et al., 2013). In a recent randomized study, we found that 80% of Grade 3 students exhibited operational misconceptions about the equal sign, even when their mathematics instruction used curricula aligned with the *Common Core State Standards for School Mathematics* (National Governors Association Center for Best Practices & Council of Chief State School Officers [NGA Center & CCSSO], 2010). These operational misconceptions have been found to persist into middle grades and negatively impact students' success solving algebraic equations (e.g., Knuth et al., 2006). Taken together, these studies suggest that understanding the meaning of the equal sign requires sustained attention over many years. Yet, well-designed curricula largely address this concept relationally—if at all—only in Grade 1, likely in alignment with its treatment in Grade 1 by the *Common Core*.

As this research illustrates, it is difficult to adequately prepare elementary students for algebra with curricula that do not deeply address core algebraic concepts through a sustained, coordinated approach across elementary grades. Thus, if we were to fairly understand early algebra's impact, we needed to develop an early algebra curriculum for Grades K–5 that could supplement regular mathematics curricula with a rigorous, research-based approach to developing children's algebraic thinking. In response to this need, we developed *LEAP: Learning through an Early Algebra Progression* (Blanton et al., 2021a–c; Blanton, Gardiner, Stephens, & Knuth, 2022a–c) as a supplemental curriculum for Grades K–5.[3]

THE EARLY ALGEBRA *LEAP* CURRICULUM

In this section, we first provide a brief overview of the *LEAP* curriculum and its particular components. We then describe findings from our investigations into its effectiveness. We consider a rigorous study of the impact of the *LEAP* curriculum—not just its design—to be an essential part of the development process.

Overview of the *LEAP* Curriculum

Our development of the *LEAP* curriculum has progressed in phases over the last 15 years, beginning with development of the Grades 3–5 curriculum, followed by development of the Grades K–2 curriculum. The *LEAP* curriculum consists of 18–20 lessons per grade level—60-minute lessons in Grades 3–5 and 35-minute lessons in Grades K–2—that are taught throughout the school year. Lessons begin with a *Jumpstart* to review concepts in previous lessons, and in Grades K–2 include a *Launch* to introduce the lesson focus. These are followed by small-group investigations (*Explore and Discuss*) in which students explore algebraic ideas and share their mathematical thinking. Lessons conclude with a *Review and Discuss* to summarize key ideas and formatively assess students' thinking. To illustrate, we have provided samples of these lesson components in the Appendix using Grade 1 lessons on the equal sign.

We designed the curriculum using a spiral approach so that students could continually revisit algebraic concepts and practices year-to-year, refining their understanding using increasingly sophisticated concepts, representations, and ways of thinking. Lessons are scaffolded with teacher questioning strategies to foster rich mathematical conversations around students' algebraic thinking. They focus on the use of meaningful problem contexts and—particularly in the early grades—concrete and visual tools linked to abstract representations to help students develop mathematical meaning for algebraic ideas.

Does the *LEAP* Curriculum Work?

Curriculum development should extend beyond simply designing instructional materials to examining whether the materials are effective. Given the lack of research using rigorous, experimental designs to evaluate the effect of curricular approaches on students' mathematical achievement (Clements, 2002; U.S. Department of Education, 2008), particularly with underserved student populations (Clements, 2007), we wanted to know whether the *LEAP* curriculum was effective for students across all demographics and academic abilities. Evidence that *LEAP* made a significant difference in students' understanding of core algebraic concepts and practices when taught as part of students' regular instruction would also provide a measure of early algebra's impact. In what follows, we share evidence to date of *LEAP*'s effectiveness.

We recently conducted a large-scale, randomized study of the intervention's effectiveness in Grades 3–5, where the intervention was taught by classroom teachers as part of their regular mathematics instruction. (See also Stylianou et al., this volume.) The study was conducted in 46 elementary schools using a demographically diverse sample of students from urban, rural, and suburban populations. To improve fidelity of implementation (FOI), throughout the intervention teachers were provided with professional development focusing on early algebra knowledge, their understanding of students' early algebraic thinking, and how tasks and instruction could support the development of this thinking. An analysis of teachers' FOI showed a significant positive relationship between components of teachers' implementation and students' performance (Stylianou et al., 2019).

We found that students who were taught the intervention as part of regular instruction significantly outperformed their peers who received only regular instruction (Blanton et al., 2019). At each of Grades 3–5, significant differences were found in students' knowledge of algebraic concepts and practices (i.e., item correctness) as well as their use of algebraic strategies to solve tasks. These significant differences persisted at the end of Grade 6, one year after the intervention ended (Stephens, Stroud et al., 2021).

Further, in a comparison of a subset of treatment and control schools where almost all students were from underserved communities (e.g., 100% low socioeconomic status [SES], over 90% students of color), treatment students also significantly outperformed control students at each of Grades 3–5 in both item correctness and use of algebraic strategies (Blanton et al., 2019). Thus, our findings suggest that students, regardless of demographic, are better prepared for algebra upon entering middle grades if they are taught the *LEAP* curriculum as part of regular instruction. This evidence pointed to *LEAP* as a promising curricular approach from which we might better understand the impact of early algebra.

LESSON LEARNED: FRAMEWORKS, FRAMEWORKS, FRAMEWORKS

Curriculum design—even for curricula focused on a particular strand such as early algebra—is a complex process. We are not curriculum designers by training, so the last 15 years have taught us much about designing effective curricula, even as we recognize that there is yet much to learn. Among the lessons learned, some have been serendipitous in nature in that we came to value a particular approach or lens used that we did not fully appreciate "in the moment."

One such aspect of our curriculum development that has become most valuable to us is that it has been frameworks driven. Frameworks provided critical scaffolding around how we designed our curriculum, how we addressed early algebra content in the curriculum, and the different stages in our curriculum design research. The use of frameworks was not always as explicit or intentional at the start of our work as it came to be as our work matured. We learned to appreciate how frameworks helped focus and stabilize our process and define more clearly the steps we needed to take. In what follows, we look at the critical ways in which frameworks informed our work.

A Learning Progressions Approach as a Framework for Designing *LEAP*

Learning progressions have become increasingly important as a research tool because of their ability to inform the design of standards, curricula, assessments, and instruction (Daro et al., 2011). From the start of our work, we adopted a *learning progressions approach* (e.g., Shin et al., 2009; Stevens et al., 2009) in which our curricular design involved the development of several core components (e.g., Clements & Sarama, 2004; see also Fonger et al., 2018): (1) empirically-derived learning goals around early algebra content; (2) grade-level instructional sequences (referred to here as the *LEAP* curriculum) designed to address these learning goals; (3) validated assessments to measure students' understanding of core algebraic concepts and practices as they advance through the instructional sequences; and (4) progressions that specify increasingly sophisticated levels of thinking students exhibit about algebraic concepts and practices in response to an instructional sequence.

This approach, which appealed to us in part because it already had significant traction in educational research, provided a critical over-arching framework that helped us think more systematically about the design process. It pushed us past simply pulling together interesting algebra tasks we might assimilate into some type of sequence and, instead, slowed down the design process to focus our attention on first building the scaffolding (i.e., learning goals) from which tasks and lessons could emerge in a coordinated way. That is, a learning progressions approach helped define our first objective—the development of grade-level learning goals—which, in turn, helped us build a more coherent and connected foundation for our curricular content than we might otherwise have built.

The learning progressions approach also helped us think about what we should consider in developing learning goals. Because learning progressions prioritize empirical research on children's thinking around specific content domains (Baroody et al., 2004), this approach directed us to first analyze empirical research on children's understanding of core algebraic concepts and what we might expect regarding their algebraic thinking at particular grades. We then reviewed available national and state curricular frameworks and standards for their treatment of algebra in Grades K–8, including *Principles and Standards for School Mathematics* (National Council of Teachers of Mathematics [NCTM], 2000), the *Curriculum Focal Points* (NCTM, 2006), and later, the *Common Core State Standards for Mathematics* (NGA Center & CCSSO, 2010). We also analyzed a range of instructional materials for Grades K–8 used in the U.S. for their treatment of algebra content; these included standard textbooks as well as those with a more investigative approach. Additionally, we considered mathematical perspectives on the sequencing of algebra content by examining formal algebra textbooks at both secondary and postsecondary levels. From these analyses, we looked for coherency between empirical research on algebra learning in Grades K–8 and benchmarks of algebra learning advocated in curricula and state and national frameworks, keeping in mind that research would likely be "ahead" of learning standards. We then synthesized our findings to develop grade-level learning goals that would form the backbone of our curriculum. A learning progressions approach helped us be more systematic and intentional in our analysis than we might otherwise have been. Moreover, it kept empirical research on children's algebraic thinking at the forefront of our design. With our learning goals in place, we could then develop grade-level instructional sequences, which are the heart of the *LEAP* curriculum, along with assessments to measure students' learning as they advanced through these sequences.

A Conceptual Framework for Early Algebra

As our work progressed, we came to appreciate how the design of our curriculum and assessments flowed from our learning goals, underscoring for us the importance of the learning progressions approach we used to develop these goals. Moreover, as we developed learning goals, a central question for us was: "How should we characterize the algebra that we want young children to learn?" There is a *lot* of algebra content present in learning standards, curricula, and research, and we needed a way to organize this content.

Members of our research team had experience in early algebra research prior to our curriculum design work, and from this experience they brought views on the nature of early algebra that aligned with Kaput's (2008) widely acknowledged conceptual analysis of algebra content. Kaput's content analysis of algebraic thinking involves two *core aspects*: (1) making and expressing generalizations using increasingly formal and conventional symbol systems; and (2) acting on symbols within an organized symbolic system through an established syntax, where conventional symbol systems available for use in elementary grades are interpreted broadly to include "[variable] notation, graphs and number lines, tables, and natural language forms" (p. 12). From these two core aspects, we identified *four essential algebraic thinking practices* that defined part of our conceptual framework for early algebra content (Blanton et al., 2011; Blanton, Brizuela et al., 2018): generalizing, representing, justifying, and reasoning with mathematical structure and relationships.

Kaput (2008) further argued that these core aspects occur across three *key strands*:

1. Algebra as the study of structures and systems abstracted from computations and relations, including those arising in arithmetic (algebra as generalized arithmetic) and quantitative reasoning.
2. Algebra as the study of functions, relations, and joint variation.
3. Algebra as the application of a cluster of modeling languages both inside and outside of mathematics. (p. 11)

Early algebra research has matured around several core content areas relative to these key strands. In developing our conceptual framework, we parsed these key strands around *three predominant domains of early algebra research*: (1) generalized arithmetic; (2) equivalence, expressions, equations, and inequalities; and (3) functional thinking. We see domains (1) and (2) as aligned with Kaput's key strand (1), and domain (3) aligned with strands (2) and (3).

The four algebraic thinking practices and three content areas where these practices can occur defined our conceptual framework for early algebra content. This framework was critical because it helped organize all our curricular content around *algebraic ways of thinking*. That is, rather than viewing curricular content as a set of things to do (e.g., solving equations), we viewed it through the lens of developing thinking practices or habits of mind.

Tasks and lessons were created with an eye towards how well they attended to these practices within the different content domains. We asked ourselves questions such as the following:

- What (and where) were opportunities for generalizing?
- What kinds of representations could be used and how could tasks build representational fluency across different forms, such as words, drawings, tables, graphs, and variable notation?
- How could tasks build students' capacity for developing strong arguments to justify general mathematical claims?
- What kinds of tasks promoted reasoning with generalizations?

Questions such as these helped us design content systematically and comprehensively around algebraic thinking practices. Irrespective of grade level, we thought about how each lesson might support students in developing their capacity to *generalize*, to *represent* their generalizations in different ways, to *justify* their claims with strong mathematical arguments, and to *reason with* the generalizations they built. To illustrate how we thought about lesson content within this conceptual framework, in Table 10.1 we show selected curricular content themes within the four algebraic thinking practices and their occurrence across the Grades K–5 *LEAP* curriculum.

We had not explicitly defined this conceptual framework *a priori*, although, given our prior research, it was already an underlying lens in our thinking. A clarifying moment in our design work involved making the explicit connection between the need for a way to organize algebra content and our existing way of thinking about early algebra content. We were fortunate that this framework already informed our thinking about algebraic content, and the design process helped solidify our understanding. The lesson learned here concerns the value of having a specific framework in mind for identifying curricular content prior to the design process.

We also found it helpful that the framework was organized around ways of thinking rather than specific content. This helped us connect content *across* Grades K–5. For example, the practice of generalizing arithmetic relationships about operations on evens and odds, how to develop viable arguments for these relationships, and how to use these relationships as building blocks to reason in novel situations, was a learning thread developed with increasing complexity across each grade, K–5. We distinguish this from an approach that addresses specific content related to even and odd numbers but misses the opportunity to develop mathematical arguments for general claims about relationships in this class of numbers.

A Framework for Curriculum Research

As we have discussed, we began our curriculum development by adopting a learning progressions approach as a framework to guide our thinking about the necessary components (e.g., learning goals, instructional

Table 10.1

Occurrence of Selected Curricular Content Within Algebraic Thinking Practices (ATP)

ATP	Curricular Content Themes within Algebraic Thinking Practices (ATP)	GRADE					
		K	1	2	3	4	5
Generalize	**Develop generalizations about**						
	Properties of operations	X	X	X	X	X	X
	Sums of evens/odds	X	X	X	X	X	X
	Rules for growing patterns	X					
	Functional relationships between two quantities	X	X	X	X	X	X
Represent	**Represent generalizations about structure/ relationships with words**						
	Expressions & Equations	X	X	X	X	X	X
	Properties of operations/arithmetic relationships	X	X	X	X	X	X
	Functions	X	X	X	X	X	X
	Represent generalizations about structure/ relationships with variables						
	Expressions & Equations		X	X	X	X	X
	Properties of operations/arithmetic relationships			X	X	X	X
	Functions			X	X	X	X
	Represent generalizations about functional relationships with graphs				X	X	X
	Represent relationships between quantities as equations in non-standard forms (i.e., $a = a$, $a = b + c$, and/or $a + b = c + d$)	X	X	X	X	X	X
Justify	**Develop representation-based arguments for**						
	Mathematical claims about specific but uncounted cases	X	X	X			
	General mathematical claims (generalizations)		X	X	X	X	X
	Identify best arguments for general claims			X	X	X	X
Reason	**Develop a relational view of '='**						
	Determine if equations in non-standard form (e.g., $a = a$, $a = b + c$, $a + b = c + d$) are true or false	X	X	X	X	X	X
	Find missing values in equations in non-standard form (e.g., $a = a$, $a = b + c$, $a + b = c + d$)	X	X	X	X	X	X
	Explicitly identify properties of operations to justify computational work		X	X	X	X	X

sequences, assessments) and how to best design these components in a way that prioritized empirical research on children's algebraic thinking. Along the way, we clarified our conceptual framework for early algebra content based on our already existing views on what it meant to think algebraically. A third type of framework used in our work concerns the research process itself. The benefit of a framework that outlines the components of a research process for developing a curriculum and testing its effectiveness had not occurred to us before our work began. We simply kept moving forward in what felt like obvious "next steps." Looking back, we are better able to articulate our process.

As we prepared to report on the effectiveness of our Grades 3–5 curriculum, we relied on Clements's (2007) Curriculum Research Framework (CRF) as a particularly relevant way to (retrospectively) characterize the stages of our work (see Sarama & Clements, this volume). The CRF consists of three phases: (1) *a priori foundations*, (2) *learning model*, and (3) *evaluation*. The *a priori foundations* phase involves identifying subject matter and relevant research in teaching and learning to inform the innovation's design. *Learning model* involves the design and sequencing of lesson content to reflect empirical models of children's thinking. The *evaluation* component involves the use of multiple methodologies to evaluate the appeal, usability, and effectiveness of an innovation.

We see the stages of our work as aligned with the CRF's phases for developing research-based curricular innovations (Blanton et al., 2019). Our conceptual framework for early algebra derives from Kaput's (2008) subject matter analysis of the content and practices of algebra. This, in conjunction with our analysis of empirical research on children's algebraic thinking, state and national learning standards and frameworks, existing K–8 regular mathematics curricula, as well as the canonical development of algebra as a mathematical discipline (Battista, 2004), provided the *a priori foundation* (Clements & Sarama, 2008) for the design of our intervention.

As with the *learning model* phase (Clements, 2007), the grade level instructional sequences in our curriculum were designed as conjectured routes whose sequencing was based on known or hypothesized progressions in children's thinking about our targeted subject matter domain (algebraic concepts and practices), with tasks sequenced to advance students' knowledge along a progression (Blanton et al., 2019). For example, research on the development of children's understanding of the equal sign as a relational symbol suggests the use of several types of tasks to challenge children's operational thinking (Rittle-Johnson et al., 2011). These include tasks where children are asked to define the equal sign, determine if equations are true or false, and find missing or unknown values in an equation. The sequencing of equation types is also important. For instance, layering in the use of operations in an equation (e.g., from a simple equation with no operations, to an equation with operations only to the left or right of the equal sign, to a complex equation with operations on both sides of the equal sign) can help scaffold students' relational thinking about the equal sign. We designed instructional sequences across Grades K–5 to account for this type of empirical research in grade appropriate ways.

Finally, as with the CRF's *evaluation* phase, our evaluation initially involved the use of design studies to test our instructional sequences. We subsequently used quasi-experimental cross-sectional and longitudinal studies to examine the curriculum's potential. More recently, we used a large-scale randomized study in Grades 3–5, with a follow-up retention study in Grade 6, to examine the curriculum's effectiveness. Having recently completed our design work and initial efficacy testing for the Grades K–2 portion of *LEAP*, our next goal is to conduct an effectiveness study of *LEAP* for Grades K–2.

The stages of our work for the Grades 3–5 design and how this aligns with the CRF is shown in Table 10.2. In retrospect, we see value in using a framework, such as the CRF, to guide the research process in curriculum development. Designing curricula is lengthy, complex, and expensive. A framework such as the CRF can help organize and sequence the design process and the acquisition of funding needed to support it.

LESSON LEARNED: ADDRESSING GAPS

Early algebra is a relatively new area, so there were gaps in key areas that we needed to consider in developing an early algebra intervention. In addition to the lack of a full, comprehensive curricular approach to early algebra, there were essentially no valid assessments that could adequately measure growth in students' early algebraic knowledge. Moreover, while there is a growing early algebra research base, there were gaps—particularly within early elementary grades—concerning learning progressions in children's algebraic thinking. Finally, elementary teachers have limited opportunities for professional development around children's algebraic thinking, so we wanted to take this into account through our curriculum design. In what follows, we describe how we addressed these areas.

Addressing Gaps in Assessments

As noted earlier, more rigorous experimental studies that show whether curricular approaches are effective in increasing students' mathematical

Table 10.2

Alignment of Our Development Process with the Curriculum Research Framework[a]

Stage of Work (Grades 3–5)	Alignment with Phases of CRF
Identification of subject matter (early algebra concepts and practices) through analysis of empirical research, national and state learning frameworks, regular K–8 mathematics curricula, and disciplinary knowledge; analysis of the alignment of these products with our conceptual framework for algebra	*A priori foundation*
Initial design of Grades 3–5 components: learning goals, grade level instructional sequences (curriculum), assessments	*Learning model*
Design studies for preliminary testing of the curricular design	*Evaluation*
Small-scale, quasi-experimental cross sectional and longitudinal studies of the curriculum's potential and feasibility	
Large-scale, cluster randomized trial (CRT) study of the curriculum's effectiveness in Grades 3–5	
Retention study of the curriculum's effectiveness in middle grades (Grade 6)	

[a] The Curriculum Research Framework (CRF) is based on Clements (2007).

achievement are needed (e.g., U.S. Department of Education, 2008). Robust measures of learning are essential for such studies. When developing a curriculum, it is important to consider whether there are adequate, validated measures that might be used to assess its effectiveness. Existing standardized assessments might be used, but one should consider if they are sensitive enough to measure concepts the curriculum addresses. In our case, there was a significant gap in available measures around early algebra. Even existing state standardized assessments are not calibrated closely enough to the algebraic concepts and practices *LEAP* addresses at a given grade level. *LEAP* accelerates algebraic development, so its content is sometimes beyond what students typically encounter in a given grade. This made typical standardized assessments for elementary grades an inadequate measure of *LEAP*'s effectiveness. Moreover, standardized assessments are sometimes only available in upper elementary grades, leaving assessment gaps in the earlier grades. To measure *LEAP*'s effectiveness (and to help us understand the impact of early algebra), we needed assessments that could closely measure growth in understanding of the particular algebraic concepts and practices we hoped the *LEAP* curriculum would foster in students' thinking.

Here again, our use of a learning progressions approach was fortuitous, as one of its key features is the design and use of assessments that measure growth along instructional sequences. This primed our thinking to be intentional about assessment development from the start of the design process. Our early algebra framework helped us think conceptually about the content of assessment items, as we had with the curriculum itself. That is, rather than just selecting different types of ubiquitous algebra tasks for our assessments, we thought about assessment design in terms of our conceptual framework around algebraic thinking practices. For example, did *LEAP* assessments measure students' capacity to generalize or to represent generalizations in different ways? Did they measure students' capacity to justify mathematical claims with strong (i.e., non-empirical) arguments? In essence, we learned that it is as important to think about assessment design in terms of the conceptual framework as it is the curriculum itself. Like curriculum development, designing and validating assessments can be complex. We came to appreciate the importance of devoting considerable energy and resources to developing good, validated assessments from the start of the design process.

Addressing Gaps in Research on Students' Thinking

Another area for which it is important to consider whether gaps might exist is the (empirical) research base on students' thinking about curricular content. We started our research on the impact of early algebra and developing the *LEAP* curriculum in Grades 3–5 because there was a much more robust and stable early algebra research base in this grade domain than in Grades K–2. While our work in Grades 3–5 progressed, we simultaneously began thinking about learning progressions around early algebraic concepts and practices for young (Grades K–2) learners in anticipation of extending our work to these earlier grades.

Gaps in the empirical research base for Grades K–2 led to several research projects in which we built on the early algebra research base concerning students' relational understanding of the equal sign (Blanton, Otalora Sevilla et al., 2018; Stephens et al., 2022; Stephens, Veltri Torres et al., 2021), generalizing functional and arithmetic relationships (Blanton et al., 2015; Ucles et al., 2020), using and interpreting variable and variable notation (Blanton et al., 2017; Brizuela et al., 2015; Veltri Torres et al., 2019; Ventura et al., 2021), and developing arguments for general mathematical claims (Blanton et al., 2022). This foundational research was critical in our later development of Grades K–2 learning goals, instructional sequences (curriculum), and assessments. It also points to the complexity of curriculum development. More than simply pulling tasks together—even if done in a cohesive and meaningful way—curriculum design needs to incorporate a relevant and robust research base. In our case, this required us to slow down and help build the research base needed to design the curriculum.

Addressing Gaps in Teachers' Professional Learning Opportunities

Designing a curriculum in a way that supports teachers' learning is not a new idea. In recent years, researchers have advocated for the development of *educative* curricula (e.g., Davis & Krajcik, 2005) that "incorporate elements that are meant specifically for *teacher* learning" (Stein et al., 2007, p. 334). Early algebra is in a particularly challenging position regarding teachers' access to professional learning opportunities. Elementary teachers are critical to the success of algebra reforms, yet a disproportionately large number of preservice and inservice elementary teachers have deeply rooted anxieties about mathematics (Battista, 1986; Haycock, 2001)—particularly algebra—that can impact the confidence with which they teach children (Bursal & Paznokas, 2006). Thus, despite early algebra now being part of the discourse of reform, many elementary teachers still need significant support in understanding how to routinely integrate early algebra into their daily instruction. At the same time, district priorities for teacher professional development often elevate literacy over mathematics (e.g., Bassok & Rorem, 2014; Wrabel et al., 2015), making it challenging for teachers to get the support they need, particularly in algebra—mathematical content they likely did not imagine teaching in elementary grades. The scarcity of resources for professional learning around teaching mathematics, along with anxieties elementary teachers might hold about algebra, can imperil teachers' ability to build early algebra-rich classrooms and impede reforms that introduce algebra in the elementary grades.

We designed early algebra lessons with these challenges in mind by considering how we might frame content in the curriculum as if it was *all* teachers saw. What if the curriculum was their sole professional learning opportunity? How could we design it in an educative way, to support their understanding of early algebra content, students' thinking about that content, and instructional practices that support children's algebraic thinking? There were several ways we addressed these questions. Each lesson contains a section called *Addressing Common Difficulties*. For example, as shown in Figure 10.1, this section contains insights from empirical research about challenges students might face in thinking about the concept of mathematical equivalence and how teachers might address these challenges.

Figure 10.1

Addressing Common Difficulties Example from Grade 1: Lesson 1.5 True or False? Comparing Towers

---

Addressing Common Difficulties

Students with an operational view of the equal sign will get "stuck" with equations that have operations on either side of the equal sign. They may reason that an equation such as $8 + 2 = 5 + 5$ is false because $8 + 2 \neq 5$, ignoring the "+ 5" on the right of the equation. Similarly, they might reason that $8 + 2 = 10 + 3$ is true, since $8 + 2 = 10$. Using tools such as cubes can help students see that either expression in an equation may have operations (or no operation at all). Cube towers provide a concrete way to reason about a relationship apart from its abstract representation (equation).

Students with an operational view of the equal sign sometimes add all numbers in an equation that doesn't make sense to them. Similarly, when building towers, they will combine towers representing each side of an equation into a single, large tower. It is important to encourage students to compare the heights of towers representing the expressions in an equation because this helps them to think about the two expressions as quantities to be compared, not combined.

Source: Blanton et al. (2022b, p. 40). Reprinted with permission of Didax.

Each lesson also contains a *Teaching Support* section that provides insights around general practices to support students' thinking about a particular concept or practice. An example is shown in Figure 10.2, again for the concept of mathematical equivalence.

Each lesson also includes a rationale for task designs so that teachers can better understand the purpose of a task in developing students' understanding of particular content. A sample is shown in Figure 10.3, taken from a Grade 1 lesson on equivalence.

Additionally, each lesson includes specific questions in boldface for teachers to ask students that can help scaffold or pace mathematical conversations. Teacher questions are followed by descriptions of what teachers might expect to learn about students' thinking. See the sample *Launch* in the Appendix for an illustration of teacher questions.

To be clear, we do not claim that these aspects of lessons are unique to the *LEAP* curriculum or that they represent all the professional support teachers need. We highlight these aspects of *LEAP* to illustrate a lesson we learned, namely, that *the curriculum is strengthened when its design considers how to support teacher, not just student, learning*. The challenges of implementing early algebra heightened our concern about teachers' professional learning opportunities, prompting us to consider how we might use the curriculum's design features to help offset these challenges.

Figure 10.2

Teaching Support Example from Grade 1 Lesson 1.5 True or False? Comparing Towers

Teaching Support

Using Tools
This lesson explores comparing the heights of towers of cubes as a way to compare the values of expressions in an equation. While some students will be able to reason from the equation itself, using concrete tools to compare the expressions in an equation is an important way to visualize equations to determine if they are true or false. Using concrete tools can also help students think about the more complicated equations that can promote a relational understanding of the equal sign.

Developing a Relational Understanding of the Equal Sign
Do not be surprised if students still demonstrate an operational view of the equal sign. Examining true/false equations will help them continue to develop a relational understanding. Using equations where operations occur on both sides of an equation, where operations occur on only the right side of an equation, or where there are no operations at all, is important for challenging students' operational thinking.

Asking Questions and Listening to Students' Thinking
Lessons depend on rich mathematical discussions focused around asking students questions and listening to their thinking. Ask questions that guide students' thinking rather than telling them how to solve a particular problem. For example, the question "How do your towers represent the equation?" encourages students to explain how they can represent an expression in an equation. For the equation $9 = 6 + 4$, questions such as "How could we change your towers to make their heights the same?" and "How does this change your equation?" encourage students to think about connections between the concrete and symbolic representations.

Source: Blanton et al. (2022b, p. 41). Reprinted with permission of Didax.

There are other contextual issues that we did not consider, but that, in retrospect, are important to bear in mind here. We designed our curriculum for a print format rather than an interactive digital one. The latter requires a different way of developing content, and *LEAP* would have benefited had we considered both formats simultaneously. For example, a digital format would have given us remote learning options that have been particularly needed during the COVID-19 pandemic in recent years. Planning for different ways teachers might access the curriculum and how its design or

Figure 10.3

Rationale for Tasks Example from Grade 1: Lesson 1.5 True or False? Comparing Towers

> **Rationale for the Tasks**
> - The use of true/false equations in forms other than standard form encourages students to think relationally, rather than operationally, about the equal sign.
> - The use of concrete tools such as cubes can help students reason about abstract representations such as equations.

Source: Blanton et al. (2022b, p. 36). Reprinted with permission of Didax.

format can support students' *and* teachers' learning can improve a curriculum's feasibility.

CONCLUSION

When we first began this journey, our plan was to do *all* of the work around understanding early algebra's impact in one four-year funded research project. In retrospect, this was far too ambitious. Among other things, we did not appreciate how long it would take to simply develop the tools (e.g., curriculum, assessments) from which we could begin understanding the impact of early algebra. Despite making significant progress, we still have work to do. We learned that with curriculum design (as with research), it is essential to pace the work so that it can be thoughtfully carried out in clear, methodical steps. We learned that developing a curriculum can involve taking detours to first do the research needed to understand what the curricular content should be.

In closing, we frame what we have learned as a series of questions to be considered when engaging in curriculum development:

- What framework will guide the curriculur components needed (for example, learning goals, instructional sequences) and how are they designed?
- What framework will guide how the content of the curriculum is conceptualized?
- What framework will guide the research process, from the curriculum's initial design to establishing its effectiveness?

- What are gaps that need to be addressed in areas such as curricular tools, student cognition, and meaningful classroom implementation? For example:
  - Are there available assessments for measuring growth in children's thinking?
  - Are there gaps in the empirical research base on children's thinking about curricular content?
  - Are there gaps in the learning opportunities teachers will have around implementing the curriculum and, if so, how can the curriculum be designed in ways that support teacher learning?

We continue to be amazed at how young children learn and grow algebraically, including through their experiences with *LEAP*. We are optimistic that the *LEAP* curriculum will be implemented in schools in ways that impact students' opportunities for success in algebra and promote teachers' learning. Most importantly, it is promising that schools now have effective curricular options such as *LEAP* for improving students' algebraic thinking. As Jim Kaput might say, that is a happy story.

ACKNOWLEDGMENT

This research was supported by the U.S. Department of Education under IES Award # R305A170378 and the U.S. National Science Foundation under Award #1720129. Any opinions, findings, and conclusions or recommendations expressed in this material are those of the authors and do not necessarily reflect the views of the Department of Education or the National Science Foundation.

REFERENCES

Baroody, A. J., Cibulskis, M., Lai, M.-L., & Li, X. (2004). Comments on the use of learning trajectories in curriculum development and research. *Mathematical Thinking and Learning, 6*(2), 227–260.
https://doi.org/10.1207/s15327833mtl0602_8

Bassok, D., & Rorem, A. (2014). *Is kindergarten the new first grade? The changing nature of kindergarten in the age of accountability.* University of Virginia.
http://www.cde.state.co.us/sites/default/files/20_Bassok_Is_Kindergarten_The_New_First_Grade.pdf

Battista, M. (1986). The relationship of mathematical anxiety and mathematical knowledge to the learning of mathematical pedagogy by preservice elementary teachers. *School Science and Mathematics, 86*(1), 10–19.

Battista, M. (2004). Applying cognition-based assessment to elementary school students' development of understanding of area and volume measurement. *Mathematical Thinking and Learning, 6*(2), 185–204.

Blanton, M., Brizuela, B., Gardiner, A., & Sawrey, K. (2017). A progression in first-grade children's thinking about variable and variable notation in functional relationships. *Educational Studies in Mathematics, 95*(2), 181–202. https://doi.org/10.1007/s10649-016-9745-0

Blanton, M., Brizuela, B., Gardiner, A., Sawrey, K., & Newman-Owens, A. (2015). A learning trajectory in 6-year-olds' thinking about generalizing functional relationships. *Journal for Research in Mathematics Education, 46*(5), 511–558. https://doi.org/10.5951/jresematheduc.46.5.0511

Blanton, M., Brizuela, B., Stephens, A., Knuth, E., Isler, I., Gardiner, A., Stroud, R., Fonger, N., & Stylianou, D. (2018). Implementing a framework for early algebra. In C. Kieran (Ed.), *Teaching and learning algebraic thinking with 5- to 12-year-olds: The global evolution of an emerging field of research and practice* (pp. 27–49). Springer.

Blanton, M., Gardiner, A., Ristroph, I., Stephens, A., Stroud, R., & Knuth, E. (2022). Progressions in young learners' understandings of parity arguments. *Mathematical Thinking and Learning*. https://doi.org/10.1080/10986065.2022.2053775

Blanton, M., Gardiner, A., Stephens, A., & Knuth, E. (2021a). *LEAP: Learning through an early algebra progression (Grade 3)*. Didax.

Blanton, M., Gardiner, A., Stephens, A., & Knuth, E. (2021b). *LEAP: Learning through an early algebra progression (Grade 4)*. Didax.

Blanton, M., Gardiner, A., Stephens, A., & Knuth, E. (2021c). *LEAP: Learning through an early algebra progression (Grade 5)*. Didax.

Blanton, M., Gardiner, A., Stephens, A., & Knuth, E. (2022a). *LEAP: Learning through an early algebra progression (Grade K)*. Didax.

Blanton, M., Gardiner, A., Stephens, A., & Knuth, E. (2022b). *LEAP: Learning through an early algebra progression (Grade 1)*. Didax.

Blanton, M., Gardiner, A., Stephens, A., & Knuth, E. (2022c). *LEAP: Learning through an early algebra progression (Grade 2)*. Didax.

Blanton, M., Levi, L., Crites, T., & Dougherty, B. (2011). *Developing essential understanding of algebraic thinking for teaching mathematics in grades 3–5* (Essential Understanding Series). National Council of Teachers of Mathematics.

Blanton, M., Otalora Sevilla, Y., Brizuela, B., Gardiner, A., Sawrey, K., & Gibbons, A. (2018). Exploring kindergarten students' early understandings of the equal sign. *Mathematical Thinking and Learning, 20*(3), 167–201. https://doi.org/10.1080/10986065.2018.1474534

Blanton, M., Stroud, R., Stephens, A., Gardiner, A., Stylianou, D., Knuth, E., Isler, I., & Strachota, S. (2019). Does early algebra matter? The effectiveness of an early algebra intervention in grades 3–5. *American Educational Research Journal, 56*(5), 1930–1972. https://doi.org/10.3102/0002831219832301

Brizuela, B. M., Blanton, M., Sawrey, K., Newman-Owens, A., & Gardiner, A. (2015). Children's use of variables and variable notation to represent their algebraic ideas. *Mathematical Thinking and Learning, 17*, 1–30. https://doi.org/10.1080/10986065.2015.981939

Bursal, M., & Paznokas, L. (2006). Mathematics anxiety and pre-service elementary teachers' confidence to teach mathematics and science. *School Science and Mathematics, 106*(4), 173–179. https://doi.org/10.1111/j.1949-8594.2006.tb18073.x

Clements, D. H. (2002). Linking research and curriculum development. In L. D. English (Ed.), *Handbook of international research in mathematics education* (pp. 599–636). Lawrence Erlbaum.

Clements, D. H. (2007). Curriculum research: Toward a framework for "research-based curricula." *Journal for Research in Mathematics Education, 38*(1), 35–70.

Clements, D. H., & Sarama, J. (2004). Learning trajectories in mathematics education. *Mathematical Thinking and Learning, 6*(2), 81–89.

Clements, D. H., & Sarama, J. (2008). Experimental evaluation of the effects of a research-based preschool mathematics curriculum. *American Educational Research Journal, 45*, 443–494. https://doi.org/10.3102/0002831207312908

Daro, P., Mosher, F., & Corcoran, T. (2011). *Learning trajectories in mathematics: A foundation for standards, curriculum, assessment, and instruction.* https://files.eric.ed.gov/fulltext/ED519792.pdf

Davis, E. A., & Krajcik, J. S. (2005). Designing educative curriculum materials to promote teacher learning. *Educational Researcher, 34*(3), 3–14.

Fonger, N. L., Stephens, A., Blanton, M., Isler, I., Knuth, E., & Gardiner, A. (2018). Developing a learning progression for curriculum, instruction, and student learning: An example from mathematics education. *Cognition and Instruction, 36*(1), 30–55. https://doi.org/10.1080/07370008.2017.1392965

Greenberg, J., & Walsh, K. (2008). *No common denominator: The preparation of elementary teachers in mathematics by America's education schools.* National Council on Teacher Quality. https://www.nctq.org/publications/No-Common-Denominator:-The-Preparation-of-Elementary-Teachers-in-Mathematics-by-Americas-Education-Schools

Haycock, K. (2001). Closing the achievement gap. *Educational Leadership, 58*(6), 6–11.

Jacobs, V. R., Franke, M. L., Carpenter, T. P., Levi, L., & Battey, D. (2007). Professional development focused on children's algebraic reasoning in elementary school. *Journal for Research in Mathematics Education, 38*(3), 258–288.

Jones, I., Inglis, M., Gilmore, C., & Dowens, M. (2012). Substitution and sameness: Two components of a relational conception of the equals sign. *Journal of Experimental Child Psychology, 113*(1), 166–176.

Kaput, J. J. (2008). What is algebra? What is algebraic reasoning? In J. Kaput, D. Carraher, & M. Blanton (Eds.), *Algebra in the early grades* (pp. 5–17). Lawrence Erlbaum/Taylor & Francis Group.

Kaput, J. J., & Blanton, M. (2005). Algebrafying the elementary mathematics experience in a teacher-centered, systemic way. In T. Romberg, T. Carpenter, & F. Dremock (Eds.), *Understanding mathematics and science matters* (pp. 99–125). Lawrence Erlbaum.

Knuth, E. J., Stephens, A. C., McNeil, N. M., & Alibali, M. W. (2006). Does understanding the equal sign matter? Evidence from solving equations. *Journal for Research in Mathematics Education, 37*(4), 297–312.

National Council of Teachers of Mathematics [NCTM]. (2000). *Principles and standards for school mathematics.*

National Council of Teachers of Mathematics. (2006). *Curriculum focal points for prekindergarten through grade 8 mathematics: A quest for coherence.*

National Governors Association Center for Best Practices & Council of Chief State School Officers [NGA Center & CCSSO]. (2010). *Common core state standards for school mathematics.* https://www.thecorestandards.org/Math/

Rittle-Johnson, B., Matthews, P. G., Taylor, R. S., & McEldoon, K. L. (2011). Assessing knowledge of mathematical equivalence: A construct-modeling approach. *Journal of Educational Psychology, 103*(1), 85–104.

Schoenfeld, A. H. (1995). Report of Working Group 1. In C. Lacampagne, W. Blair, & J. Kaput (Eds.), *The algebra colloquium: Vol. 2. Working group papers* (pp. 11–18). U.S. Department of Education, Office of Educational Research and Improvement.

Shin, N., Stevens, S. Y., Short, H., & Krajcik, J. S. (2009, June 24–26). *Learning progressions to support coherence curricula in instructional material, instruction, and assessment design* [Paper presentation]. The Learning Progressions in Science (LeaPS) Conference, Iowa City, IA.

Stein, M. K., Remillard, J., & Smith, M. S. (2007). How curriculum influences student learning. In F. K. Lester, Jr. (Ed.), *Second handbook of research on mathematics teaching and learning* (pp. 319–370). Information Age Publishing.

Stephens, A., Knuth, E., Blanton, M., Isler, I., Gardiner, A., & Marum, T. (2013). Equation structure and the meaning of the equal sign: The impact of task selection in eliciting elementary students' understandings. *Journal of Mathematical Behavior, 32*(2), 173–182.

Stephens, A., Stroud, R., Strachota, S., Stylianou, D., Blanton, M., Knuth, E., & Gardiner, A. M. (2021). What early algebra knowledge persists one year after an elementary grades intervention? *Journal for Research in Mathematics Education, 2*(3), 332–348. https://doi.org/10.5951/jresematheduc-2020-0304

Stephens, A., Sung, Y., Strachota, S., Veltri Torres, R., Morton, K., Gardiner, A. M., Blanton, M., Knuth, E., & Stroud, R. (2022). The role of balance scales in supporting productive thinking about equations among diverse learners. *Mathematical Thinking and Learning, 24*(1), 1–18. https://doi.org/10.1080/10986065.2020.1793055

Stephens, A., Veltri Torres, R., Sung, Y., Strachota, S., Murphy Gardiner, A., Blanton, M., Stroud, R., & Knuth, E. (2021). From "You have to have three numbers and plus sign" to "It's the exact same thing": K–1 students learn to think relationally about equations. *Journal of Mathematical Behavior, 62.* https://doi.org/10.1016/j.jmathb.2021.100871

Stevens, S. Y., Shin, N., & Krajcik, J. S. (2009, June 24–26). *Towards a model for the development of an empirically tested learning progression* [Paper presentation]. The Learning Progressions in Science (LeaPS) Conference, Iowa City, IA.

Stylianou, D., Stoud, R., Cassidy, M., Knuth, E., Stephens, A., Gardiner, A., & Demers, L. (2019). Putting early algebra in the hands of elementary school teachers: Examining fidelity of implementation and its relation to student performance. In M. C. Cañadas, M. Blanton, & B. Brizuela (Eds.), *Early algebraic thinking*. Special Journal Issue for *Journal for the Study of Education and Development—Infancia y Aprendizaje, 42*(3), 523–569.

Ucles, R., Brizuela, B., & Blanton, M. (2020). Kindergarten and first-grade students' understandings and representations of arithmetic properties. *Early Childhood Education Journal, 50*, 345–356. https://doi.org/10.1007/s10643-020-01123-8

U.S. Department of Education. (2008). *Foundations for success: The final report of the National Mathematics Advisory Panel*. https://files.eric.ed.gov/fulltext/ED500486.pdf

Veltri Torres, R., Prough, S., Strachota, S., Stephens, A., Sung, Y., Gardiner, A., Blanton, M., & Knuth, E. (2019, April 5–9). *Describing the unknown: Moving toward variable notation and algebraic thinking in kindergarten* [Paper presentation]. The Annual Meeting of the American Educational Research Association, Toronto, Canada.

Ventura, A. C., Brizuela, B. M., Blanton, M., Sawrey, K., Gardiner, A., & Newman-Owens, A. (2021). A learning trajectory in kindergarten and first grade students' thinking of variable and use of variable notation to represent indeterminate quantities. *Journal of Mathematical Behavior, 62*, 100866. https://doi.org/10.1016/j.jmathb.2021.100866

Wrabel, S. L., Gottfried, M., & Polikoff, M. S. (2015, February). *Instructional groupings in the inclusive kindergarten classroom: A cross-cohort analysis*. http://citeseerx.ist.psu.edu/viewdoc/summary?doi=10.1.1.691.1696

APPENDIX

Sample Lesson Components From Grade 1 Leap Curriculum

Sample excerpts of lesson components (*Jumpstart, Launch, Explore and Discuss, Review and Discuss*) taken from Grade 1 lessons about the equal sign: Lessons 1.4 *Comparison Symbols* and 1.5 *True or False? Comparing Towers* (Blanton et al., 2022b. Reprinted with permission of Didax.)

Jumpstart

Complete the Jumpstart.

Display the equal sign symbol. Students should be familiar with it from Kindergarten **LEAP** lessons.

1. Do students describe the equal sign as meaning "the answer" or "the total"? This indicates they see it as an operational symbol and do not yet have a relational understanding. Look for descriptions such as the equal sign means "balance" or "the amounts on either side of the equal sign are the same."

2. Select students to share their equations. Notice whether students' examples are equations in standard form, with operations only to the left of the equal sign. Discuss why an equation with no operations, such as 8 = 8, is valid.

> **Jumpstart**
> 1. What does this symbol mean?
>   =
> 2. Give an example of how you would use it.

Launch

What does it mean for an equation to be true? Do students indicate an equation is true "if you get the right answer"? This indicates operational rather than relational thinking.

Is the equation 4 + 4 = 8 true? Is the equation 4 + 7 = 8 true? Display the equations for students. These will likely be easy for students to answer because the first equation is a doubles fact and both equations are in standard form.

Is the equation 10 = 6 + 4 true? Let's use cubes to think about whether 10 has the same value as 6 + 4. Display the equation. Review the meaning of the equal sign as a symbol that indicates two expressions have the same value. Using language such as "has the same value as" can reinforce this understanding.

With students, use cubes to build a 10-tower and a 6 + 4-tower. Use different colors for each addend to help students visualize the addends.

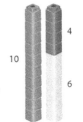

How can we use our towers to see if the equation is true? Do students compare the heights of the towers and notice they are the same? Discuss this strategy.

Is the equation 4 + 1 = 5 + 2 true? Do students think that the equation is false because "you can't have a plus sign after the equal sign"? This type of operational thinking is not uncommon at this point. Framing the question as "Is 4 + 1 the same amount as 5 + 2?" can help students think about the equation relationally.

Ask students how to build towers that represent the expressions in their equation (4 + 1 and 5 + 2). Notice whether they compare their heights. Relate their findings to the equation: 4 + 1 = 5, 5 + 2 = 7, and 5 ≠ 7, so the equation is false.

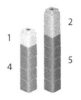

Explore and Discuss

Place students in partner pairs and give each group a set of cubes. It is helpful to represent a particular addend with the same color of cubes, so be sure students have a sufficient number of cubes of a given color.

Give students an equation and ask them to use the tower strategy discussed in the Launch to determine whether the equation is true or false.

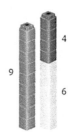

Display each equation, one at a time:

$9 = 6 + 4$

$3 + 2 = 1 + 4$

$8 = 3 + 2$

$3 + 4 = 7 + 2$

"I built a 9-tower and a 6 + 4-tower. When I held them up beside each other, the 9-tower was shorter. So, the equation is false."

After each equation, select students to share their strategies and the towers they built. Do students represent each expression in the equation with a tower? Do they compare their heights to determine if the equation is true or false? Keep in mind that equations with operations on either side are still new for students, and they might need additional support.

You might use additional equations to give students more practice. Be sure the addends are small enough that students can reasonably represent them with cubes. Use equations where operations are not only on the left side of the equal sign.

Thinking about Student Responses

Students who see the equal sign as an operational symbol might exhibit different misconceptions:

- They sometimes add all the numbers in an equation rather than compare the value of each side. With towers, these students might build a single tower showing all the addends (for example, a 9 + 6 + 4 tower for the equation 9 = 6 + 4) rather than compare the heights of the towers representing each expression in the equation.
- They may think an equation such as 4 + 1 = 5 + 2 is true because 4 + 1 = 5, essentially ignoring "+2" in the equation. Similarly, they may think an equation such as 3 + 2 = 1 + 4 is false, since 3 + 2 ≠ 1.
- They may see an equation such as 5 = 3 + 2 as "backwards" because there is no operation to the left of the equal sign.

 Review and Discuss

Describe the strategy we used today to find if an equation is true or false.

ENDNOTES

1. In the U.S., it is common for elementary grades to include Grades K–5 (ages 5–12).
2. Readers can refer to Figures 10.1 and 10.2 for discussion and examples of relational and operational understanding of the equal sign.
3. The *LEAP* Curriculum is available at https://www.didax.com/leap.

CHAPTER 11

LESSONS LEARNED FROM DECADES OF DESIGNING, RESEARCHING, AND REVISING A KINDERGARTEN TO GRADE 6 MATHEMATICS PROGRAM

Karen C. Fuson
Northwestern University

This chapter focuses on various lessons learned over a 30-year career in mathematics curriculum research. The lessons learned about intended curriculum center around work on the Learning Path Phases model for teaching mathematics topics through bridging from student methods to fluent methods. This includes an extended discussion of accessible and mathematically-desirable research-based methods supported by drawings and other visual supports, as well as A Nurturing Math Talk Community approach that supports engaging students in conversations needed for sense making. The lessons learned about teacher intended curriculum focus on the learn-while-teaching approach required by the use of new sense-making approaches. The implemented/enacted curriculum lessons involve a range of data collection methods, including observation of student methods. Lessons from the assessed curriculum include limitations of uncoordinated multiple levels of testing, none of which assess only grade-level goals or give data useful to teachers' approaches for reteaching. The learned curriculum lessons center on the power of errors in student work to improve a mathematics program or inform teachers to improve student learning.

Keywords: curriculum design; curriculum research; learning mathematics; Math Talk; teaching mathematics

I began my 30-year journey of developing a balanced mathematics curriculum with multiple goals: addressing issues brought up in the "math wars" (Klein, 2007; Schoenfeld, 2004); providing students with teaching based on how they think; and having 90% of Grade 3 students achieve grade-level goals in mathematics because it is difficult to catch up after Grade 3. I wanted to place mathematics topics at ambitious grade levels, similar to high-achieving countries (e.g., many Asian countries on international assessments), and find ways to support all students to meet these goals. So that results would be broadly applicable, I wanted to develop the program in English and in Spanish, and advance approaches that are applicable in high-poverty urban classrooms as well as in a range of other classrooms.

The journey began during a time of constructivism, emphasizing the value of student thinking. Based on this focus on student thinking, in the early 1990s, the U.S. National Science Foundation (NSF) funded the development of elementary mathematics programs with an innovative or investigative approach. As I studied and discussed these NSF-funded mathematics programs with their authors and teacher users and directed a 5-year longitudinal study of the early grades of one of them, I saw problems that needed to be addressed. These problems and the solutions I found are some of the lessons presented in this chapter. I have structured the chapter and lessons learned around the levels of curriculum as specified by Remillard and Heck (2014).

LESSONS LEARNED ABOUT THE OFFICIAL CURRICULUM

The official curriculum consists of national, state, or local standards and goals for what students should learn and on which they should be assessed (Remillard & Heck, 2014). These documents describe what to teach. Curriculum researchers and designers must be aware of the content of all pertinent official curriculum documents. For decades, the plethora of such documents in the United States, and the variation of topics and grade-level placement of topics across states, made curriculum design and research difficult and less generalizable than desired. For example, in the early years of my research, I had to design three different approaches to addition of fractions because this topic was either at Grade 3, 4, or 5 depending on the state (Reys, 2006). Before the *Common Core State Standards for School Mathematics* (CCSSM) (National Governors Association Center for Best Practices & Council of Chief State School Officers [NGA & CCSSO], 2010), teachers and all levels of mathematics educators spent great amounts of time deciding what to teach and when to teach it rather than considering how best to teach a given topic.

The CCSSM resulted in most states having the same mathematics standards for most topics, and there was general consistency in the major topics at each grade. Even some of the four states not adopting the CCSSM had similar standards developed by mathematics professionals in their state. As additional states have unadopted the CCSSM, their standards have remained relatively similar due to efforts of mathematics professionals in the state. But curriculum designers still had to decide the order of topics and the emphasis to give each topic, which can be thought of as the *mathematics learning path*, because these are not specified in the standards. My goal of having 90% of Grade 3 students achieve grade-level goals meant I had to decide on the most important grade-level topics, put them first in the curriculum, designate enough time on each so that all students could learn them, and have effective distributed practice materials to support learning and fluency. In this way, teachers who do not finish the program at least teach the most crucial grade-level content. This choice contrasted with the spiral approach used in one of the NSF-funded curricula in which a topic was revisited over and over. Some teachers in the longitudinal study of this curriculum indicated that this approach did not allow less-advanced students to learn grade-level content because such students did not have enough time on a topic. Some of these teachers asked me to use my research materials in their low-achieving classrooms, hoping that the concentrated time and different approach would work, which it did.

Deciding the most important topics was guided by knowledge of the structure of the CCSSM. Grades K, 1, and 2 focus on addition and subtraction including real-world situations in all grades. Then all types of word problems are introduced in Grades 1 and 2 together with a learning path of single-digit addition and subtraction solutions, as well as multidigit addition and subtraction representations and solutions. Representing and solving all types of word problems, similar to the range of problems solved in the Soviet Union (Stigler et al., 1986), is a significant advance for the United States. The variety of problem types, which include finding more difficult unknowns, enables students to develop algebraic understandings of situations and equations. This contrasts with the past focus in the U.S. on only the easiest unknowns. Some grade placements of content in the CCSSM are more advanced than have been typical in the U.S. but match placements of high-achieving countries. For instance, two-digit addition with regrouping is in Grade 1, and Grade 2 moves through multidigit addition and subtraction with regrouping for 3-digit numbers up to totals of 1,000. Students are to use concrete models or drawings and strategies based on place value, properties of operations, and/or the relationship between addition and subtraction, and relate their strategy to a written method. It takes the better part of a school year to support such understanding and fluency.

Grades 3 through 6 in the CCSSM focus heavily on multiplication and division of single-digit numbers and then of multidigit numbers, operations on fractions and decimal fractions, and some work on measurement and geometry beyond the first steps in Grades K, 1, and 2. Single-digit multiplication and division situations, patterns, and strategies are introduced and brought to fluency in Grade 3. This requires a great amount of time because patterns vary with the number and there are no general strategies as there are for single-digit addition and subtraction. Beginning the year with this topic provides enough time throughout the year for most students to obtain fluency.

The lack of repetition of topics in the CCSSM meant that teachers at specific grade levels were responsible for teaching particular topics. As teachers increasingly did so, the common practice of reviewing previous content until Christmas gradually lessened. For instance, as second-grade teachers began to teach all the multidigit addition and subtraction, third-grade teachers felt more comfortable beginning the year with multiplication. As a result, I saw movement from few teachers finishing a curriculum program at their grade level to many finishing all or most of a program. This leads to an important lesson: *Official curriculum that provides designers with clear non-repetitive standards at particular grades enables a curriculum to be designed with limited repetition across grades and supports teachers to teach with minimal repetition, as is done in many high-achieving counties.*

LESSONS LEARNED ABOUT THE INTENDED CURRICULUM

In many countries, the intended curriculum consists of the curriculum identified in textbooks (Remillard & Heck, 2014). For the past 30 years, my major research activity has been designing, writing, and researching intended curriculum for all major mathematics topics in kindergarten to Grade 6 and then also for prekindergarten. The curriculum evolved from research lessons, then to research units, then research units were formed into related grade-level learning paths, and later were research curricula revised to published books—*Math Expressions* (Fuson, 2004, 2009, 2011, 2013, 2018). In the following sections, I discuss some of the essential components in the design of the intended curriculum.

The Learning Path Phases Model: Bridging From Student Methods to Fluency

The intended curriculum that I designed had an overarching goal: to bridge from methods students invent and understand to research-based mathematically-desirable methods needed for fluency. Such bridges seemed to be lacking in the original NSF-funded programs, in which a large amount of time was spent on children's methods without moving most children forward to more desirable methods. Without these bridges, teachers would often teach "standard algorithms" right before a test, resulting in many students not understanding them. The *Learning Path Phases* model, a four-phase model of teaching, addresses this need (Fuson, 2009; Fuson et al., 2015; Fuson & Murata, 2007; Murata, 2008; Murata & Fuson, 2006).[1] Given the similarity between these four phases to those identified in Japanese classrooms, we named the phases with the terms used in Japan: *Guided Introducing*, *Learning Unfolding*, *Kneading Knowledge*, and *Maintaining Fluency and Relating to Later Topics* (Fuson et al., 2015; Murata, 2008).

In Phase 1, *Guided Introducing* of a topic, visual models, including math drawings, are introduced or developed by students while the teacher elicits students' prior knowledge. To solve problems, students use their own methods, which are often concrete, slow, and may have errors. Students share methods for solving problems and the teacher addresses errors when necessary.

In Phase 2, *Learning Unfolding*, the teacher uses a coaching and solve-and-explain classroom structure while students explain, discuss, and compare methods with drawings so that the mathematical aspects become explicit. The teacher introduces research-based methods and leads discussion to connect them to student methods. These solution methods are accessible to students but mathematically desirable enough to be standard methods. Phase 2 is the most crucial and often the longest phase as students become more fluent with mathematically desirable methods and the Phase 1 methods begin to disappear.

In Phase 3, *Kneading Knowledge,* students develop fluency with a chosen method from Phase 2, eventually solving problems without using a visual math drawing. In Phase 4, *Maintaining Fluency and Relating to Later Topics*, students maintain fluency and relate the current topic to other topics. As part of maintaining this fluency, on the back of each homework page, students have *Remembering*, a page with cumulative review, with short intervals between practice with initial feedback and then longer intervals as dictated by research (Rea & Modigliani, 1985).

These four phases are used in teaching each mathematics topic. For simple topics, these phases might be completed over one or two days. For more complex topics like multidigit computation, these four phases might require two weeks to months for all students to reach fluency. For all mathematics domains, continuity of concepts and of supportive visual representations through the grades is emphasized. More details can be found in the Teaching Progressions (karenfusonmath.com) for each mathematics domain in the CCSSM and in Fuson (2024).

This Learning Path Phases model provides strategies for teachers to differentiate instruction within the whole class. In Phases 1 and 2, students see the whole range of methods invented or used by students or introduced by the teacher or the mathematics program. These include methods with errors, slow methods early in the learning path, and methods that are accessible and mathematically desirable. Math drawings are used so that students can understand and discuss different methods so teachers can differentiate instruction *within the whole class* by helping everyone move forward on the learning path to a better method for *them*. Phase 2 has a consistent focus on using and explaining the more advanced methods. Students using less advanced methods are brought up to "good-enough" methods, not left to flounder with incorrect or slow methods.

This approach to designing a balanced program comes from research about two different approaches to teaching: dialogic and direct (Munter et al., 2015). These approaches are modifications of a reform/students-invent/student-directed approach and a traditional/teacher-telling/teacher-directed approach, opposite perspectives that figure prominently in the "math wars." Thus, the curriculum was developed with the goal of diffusing the "math wars" by including both perspectives in an appropriate order. Student invention is in Phases 1 and 2, and direct teaching of crucial mathematics is in Phases 2 and 3, where classroom discussion focuses on mathematics and connecting to efficient methods. That these approaches were clear to others is indicated by a study carried out by Munter et al. (2015). Nationally recognized experts from both the student-directed and teacher-directed perspectives met to find commonalities and differences. They examined and chose a mathematics program to use in research comparing the two teaching approaches. Both groups chose *Math Expressions*, the program I developed, as exemplifying their pedagogical approach.

The lessons learned in implementing the Learning Path Phases model are: *It is possible to create a math program that starts with students inventing but bridges to more mathematically desirable but still accessible methods for fluency. Furthermore, these characteristics can be recognized by experts favoring dialogic (students invent) or direct (teacher directed) teaching. Thus, programs which are predominantly dialogic or direct could add the other perspective in the appropriate places to satisfy the repeated national calls for both understanding and fluency.*

The Nurturing Math Talk Community

The four phases model emphasizes extensive student discussion of thinking, which requires a classroom environment that we called the "Nurturing Math Talk Community" (Fuson, 2009; Fuson et al., 2009; Hufferd-Ackles et al., 2004, 2015). Means of responsive assistance identified for reading by Tharp and Gallimore (1988) were useful to describe the assistance needed to build the Nurturing Math Talk Community (Fuson & Murata, 2007; Murata & Fuson, 2006). These responsive means are *engaging and involving*, *coaching* (model/show, instruct/explain, clarify, question, give feedback), and *managing*; the most important aspects of each of these are included in the four-phase model, but any means can be used in any phase. Importantly, students also use these means to assist their classmates. Initially, some students do so spontaneously and later all students can come to use these means of responsive assistance with teacher and student support.

During all aspects of Math Talk, the teacher builds, leads, and focuses the instructional conversation using the eight mathematical practices from the CCSSM. These eight practices can be paired and given names, resulting in this helpful sentence: *Teachers continually assist students to do* **math sense-making** *about* **math structure** *using* **math drawings** *to support* **math explaining**. Use of this sentence helped teachers remember all parts of the instructional conversations they were orchestrating in their classrooms. The bolded terms come from the following mathematical practices (MP):

- **math sense-making** summarizes MP1 "Make sense of problems and persevere in solving them" and MP6 "Attend to precision";
- **math structure** summarizes MP7 "Look for and make use of structure" and MP8 "Look for and express regularity in repeated reasoning";
- **math drawings** summarizes MP4 "Model with mathematics" and MP5 "Use appropriate tools strategically"; and
- **math explaining** summarizes MP2 "Reason abstractly and quantitatively" and MP3 "Construct viable arguments and critique the reasoning of others."

For Phases 1 and 2, the *Solve, Explain, Question, and Justify* activity structure was introduced so that students and the teacher could function well when students were explaining. In this structure, all students solve the same problem, usually with their own math drawing. Some students solve at the board and then two of them explain their methods, thus saving time because their methods are already on the board. Each of the two explaining students explains and relates their written method to the math drawing and then asks for questions and justifies the explanation or explains it further. Students at their desks use a large 12″ by 17″ whiteboard on which grade-level supports enable them to use math drawings and other visual supports. These boards provide empty space for student problem solving; if a teacher identifies an interesting solution, the student can explain it by bringing their whiteboard to the front of the classroom to share. This *Solve, Explain, Question, and Justify* structure arose partly as a solution to our observations that considerable time was wasted in many classrooms when explaining students went to the board and wrote out their method while other students waited.

The Nurturing Math Talk Community has proven effective with students from all backgrounds, especially students who are also learning English as they are learning mathematics. Math Talk with visual supports makes the math talk meaningful and increases English proficiency as well. Effectiveness was evidenced when outside state interviewers asked teachers in a school in which many students were identified as needing bilingual support how the second graders could explain 2-digit ungrouping for subtraction in English when they did not even know English words for parts of the body. The teachers explained that all students in their Math Talk classrooms were expected to make math drawings and explain their thinking in English. With considerable modeling and teacher help, as well as a belief that they could, these students learned to do so.

Some teachers are now using an approach called *Number Talks* for a part of their lesson time. Students must solve problems mentally. The teacher records the students' method, but students do not get a chance to record their thinking in written form. Also, usually the teacher does not extend student thinking to regular lessons. Fuson and Leinwand (2023) analyzed these and additional characteristics of Number Talks that are also limiting, and called for teachers to use a Nurturing Math Talk Community in regular lessons instead of Number Talks because students need extensive experience in representing and discussing their thinking.

Focusing on Phase 2

Phase 2 is a critical phase in the four phases model. It is where students move from their invented methods for solving problems to more sophisticated methods. Several aspects of this phase warrant further discussion.

The Importance of Math Drawings and Visual Supports

My classroom observations of other mathematics programs, and especially in my longitudinal study of one of the NSF-funded elementary curricula, indicated that some students discussed their thinking but often with nothing visible to help listeners understand their thinking, and some students rarely (or ever) described their thinking. In my research, we found that when students made mathematics drawings of situations and of quantities, these drawings helped their thinking and explanations, resulting in improved understanding by explainers and listeners. Therefore, designing visual supports and mathematics drawings for most topics in PK to Grade 6 became a major continuing research and design task. This became even more important when the CCSSM emphasized mathematics drawings related to written methods (Fuson & Li, 2014).

Mathematics drawings have many advantages over manipulatives. They can be drawn on a vertical class board so listeners can see them, or they can be drawn on an activity page or homework. They remain after they are drawn, so they can be an object of reflection and can be modified. They are inexpensive and easier to manage than manipulatives, and they do not get lost. Over the decades, classroom blackboards began to be filled with Word Walls and other items. For these reasons, I developed large (12″ by 17″) dry erase grade-level boards with visual supports useful for each grade and large workspaces for drawings and written notation.

Several visual learning supports were developed to help students build mathematical understandings. Examples of learning supports for understanding multidigit numbers are shown in Figure 11.1. The *Secret Code* cards (Figure 11.1a) show expanded notation when separated but show single-digits in each place when cards are placed on top of each other. Students visualize the zeroes hiding under each place. Montessori used cards like this over a hundred years ago (demonstrations of use of such cards by Montessori teachers can be watched at https://theglobalmontessorinetwork.org/resource/primary/number-cards-1-9000-english/ and https://montessorivivo-europe.com/en/home/134-number-cards-1-99999-3-sets.html). Small numbers on the top left and the place-value drawings on the back of each card were added by me. Students used English words (four hundred eighty-six) and regular hundreds/tens/ones words (four hundred eight tens six ones) for all multidigit work.

A large *Number Parade* (Figure 11.1b) poster in Kindergarten and Grade 1 shows numbers to 10 in 5-groups. Students make these horizontal (or sometimes vertical) 5-groups in math drawings of multidigit numbers because a value can be seen at a glance and also one can see how many more make ten for make-a-ten strategies. The 5-groups relate to the five fingers on a hand.

Nickel and dime strips (Figure 11.1c) use 5-groups and 10-groups to show numbers. These strips, which students find motivating, show pennies on one side and the dime or nickel on the other side. A dollar bill with 10 columns of 10 pennies with 5-groups within the columns shows 100 as 100 pennies and as 10 groups of 10, relating to the drawn vertical 10-sticks students make to show their computation method. The Money Flip Chart on the right has columns of 10 pennies or a dime or two nickels or one nickel that can be flipped down to show amounts up to 100. This is used in a Daily Routine as children build their understanding by counting to the number shown and seeing what those number words mean by the number of pennies organized in tens and ones. Children start counting the strips that show all pennies, and later move to the dime and nickel strips

224   K. C. FUSON

to practice those values. Each day one or more pennies are added to the chart. The sticky notes at the bottom that are used to cover the pennies not yet needed show the partners of 10 with the sticky notes there and the empty rectangles.

The 120 Poster (Figure 11.1d) has *columns* of 10 numbers instead of *rows* of 10 in the typical U.S. hundreds grid. These vertical columns support students to see the place value pattern that the first number is the same and the second number increases by one. The columns also relate to vertical drawn 10-sticks. This poster is used in Daily Routines with the Money Flip Chart (Figure 11.1c) to connect counting words, written numerals, and quantities in pennies and in the columns of 10 words. Children also say the number they have made, as shown in Secret Code Cards (Figure 11.1a).

Figure 11.1

Visual Learning Supports for Multidigit Understanding

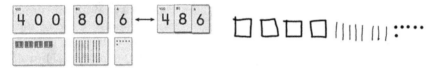

a. Secret Code Cards

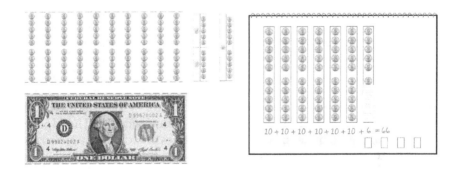

b. A Number Parade Poster

c. Nickel and Dime Strips for 5-Groups and 10-Groups

(Figure continued on next page)

Figure 11.1 (Continued)

Visual Learning Supports for Multidigit Understanding

	11	21	31	41	51	61	71	81	91	101	111
2	12	22	32	42	52	62	72	82	92	102	112
3	13	23	33	43	53	63	73	83	93	103	113
4	14	24	34	44	54	64	74	84	94	104	114
5	15	25	35	45	55	65	75	85	95	105	115
6	16	26	36	46	56	66	76	86	96	106	116
7	17	27	37	47	57	67	77	87	97	107	117
8	18	28	38	48	58	68	78	88	98	108	118
9	19	29	39	49	59	69	79	89	99	109	119
10	20	30	40	50	60	70	80	90	100	110	120

$66 = 60 + 6$

d. 120 Poster

Source: From *Math Expressions, Common Core*. Copyright ©2018 by Houghton Mifflin Harcourt Publishing Company. Reprinted by permission of the Publisher. All rights reserved.

Determining Mathematically-Desirable Methods and Standard Algorithms

A crucial clarification for multidigit computation is that the common methods taught in the U.S. are NOT "THE standard algorithms." There is no one "standard algorithm" that must be taught. Over the years, there have been many methods taught in the United States, and many methods are taught around the world. THE "standard algorithm" is a collection of different ways of writing a procedure that uses single-digit operations and concepts of place value (Common Core Writing Team, 6 March 2015; Fuson & Beckmann, 2012/2013). The accessible and mathematically-desirable methods used in Phase 2 are all standard algorithms, but they have features that are easier for students and do not stimulate errors.

In contrast, our research indicated that all the current common methods taught by some as "THE standard algorithm" have steps that mislead students and create errors (Fuson & Beckmann, 2012/2013). The terminology "THE standard algorithm" is unfortunate and has led to misunderstandings, but some members of CCSSM committees thought it was important to use. The terminology that should be used is "A standard algorithm."

To bridge between children's initial methods and the eventual mathematically-desirable methods used with fluency, several tasks needed to be completed and related to each other for each mathematical domain. These design and research tasks were:

- identify a learning path of student conceptual structures;
- analyze common mathematical words and written symbols to identify difficulties that students must overcome to understand the concepts and methods in the domain;
- identify accessible methods that are mathematically desirable; and
- analyze common written methods for ways in which they mislead students and create errors.

We discuss these tasks for multidigit computation because they are so crucial to students' mathematical success. Student conceptual structures were identified by observing, interviewing, teaching, consulting our research and the research of others, and by analyzing mathematical structures, words, and symbols to identify complexities that might lead to student difficulties in understanding. For example, the teen numbers are written as 1 then 4, but said with the four first as 14 (see Fuson, 1998, for a summary). Methods accessible to students were identified by first gathering a range of student invented methods from our research classrooms as well as from others' research. These student methods were analyzed, together with common written methods (including those taken to be "the standard algorithms"), for aspects that were easy or difficult, and for aspects that were mathematically desirable or conceptually misleading. The most promising methods were tested in classrooms to see how accessible they were to teachers and to students. This classroom research indicated that some students chose and worked better with one written method than another, so we chose two accessible methods to support this choice and variability (Fuson, 2020). These methods varied in the mathematical features that were highlighted, so having two methods enriched discussions and the conceptual structures students could build.

A first grader explaining the addition method *New Groups Below* (Figure 11.2a) shows her drawing beside the written method. Above each picture are two advantages of that method that do not apply to the typical standard algorithm shown in Figure 11.2c. In Figure 11.2b is a step-by-step diagram for 3-digit numbers, where the teen numbers for 14 ones and 13 tens are shown, showing how easy it is to add the numbers in each place and then add the one. The *New Groups Above* method, shown in Figure 11.2c, is often taken as "the standard algorithm" that must be taught. Disadvantages of this method are that you usually write the teen number in reverse

(6 then 1) and you cannot see the teen totals 16 and 14 because they are separated. Also, if you add the new one to the top number 8 to get 9 and then add 9 to the 5 below, you have to ignore two numbers that are there (1 and 8) and add 9 that you do not see. Also, some students say you are changing the problem by writing numbers on the top, and you actually are doing that. *Show All Totals* (Figure 11.2d) is accessible and mathematically desirable because it shows the place values ready to add to make the total and it can be done from the left, which many students prefer. The mathematically-desirable and accessible standard algorithms for multidigit subtraction, multiplication, and division are explained in Fuson and Beckmann (2012/2013), Fuson (2020), and in videos of students using these methods and video explanations of these methods on karenfusonmath.net under Classroom Videos.

Figure 11.2

Multidigit Addition Algorithms

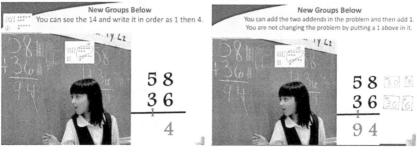

a. New Groups Below Method

b. Step-by-Step Math Diagram for Three-Digit Addition

c. The Typical Standard Algorithm

(Figure continued on next page)

Show All Totals

```
    1 8 9
+   1 5 7
  ─────────
    2 0 0
    1 3 0
      1 6
  ─────────
    3 4 6
```

d. The Show All Totals Algorithm

Source: From *Math Expressions, Common Core*. Copyright ©2018 by Houghton Mifflin Harcourt Publishing Company. Reprinted by permission of the Publisher. All rights reserved.

Math Drawings and Visual Supports for Many Mathematics Topics

Phase 2 math drawings and visual supports were important in teaching other math topics. Fractions were especially difficult because of the pressure to use number lines. Number lines are difficult for students because students look for things instead of lengths, and thus focus on the numerical labels and can be one number off. We found that pairing drawings of number bars with those of number lines helped students understand number lines and be able to use them in fraction computations. Drawings are especially important in understanding fraction equivalencies because the written methods show the written fractions getting larger but drawings are needed to show the unit fractions getting smaller (Fuson, 2019; Fuson & Beckmann, 2012).

For another example, our research at Grade 6 revealed confusions between fractions and ratios exacerbated by the misuse of fraction notation in initial ratio teaching and how writing ratios horizontally as *a:b* can reduce this confusion. Our approach set ratio tables and proportions within the multiplication table, initially showing in tables the multiplying column that links ratio columns, and had students solve factor puzzles made from proportions within the multiplication table (for details, see Abrahamson & Cigan, 2003; Fuson, 2024; Fuson & Abrahamson, 2005; Fuson & Beckman, 2012). This approach provided continuity from Grades 3 through 6 and facilitated access for low achievers who could understand, represent, and solve proportion problems. The use of length fraction bars and the number line throughout fraction lessons also provided consistency during these same grade levels (Fuson, 2019).

Summary

The discussion in this section has highlighted the importance of a learning phases model in the development of the intended curriculum. Specifically, we learned: *The curriculum needs to provide a bridge from student invented methods to accessible and mathematically-desirable algorithms that support fluency, and visual supports leading to math drawings made by students are helpful in providing this bridge. Analysis of computational methods and of mathematical symbols and words for their advantages and difficulties is also helpful. Finally, gathering methods students invent and researching whether they are accessible to other students and teachers is crucial because methods with fewer obstacles can be helpful to students in reducing errors and increasing understanding.*

LESSONS LEARNED ABOUT THE TEACHER INTENDED CURRICULUM

The teacher intended curriculum consists of those lessons and intentions of the teacher related to how they intend to implement the curriculum within their instructional environment (Remillard & Heck, 2014). We carried out frequent interviews with teachers to ascertain questions they had and misunderstandings they implemented and then adapted the curriculum to teacher thinking.

The Learn-While-Teaching Approach

For the teacher intended curriculum to be close to a designer's intended curriculum, teachers need to understand and use the teacher resources that support the mathematics program. The four Learning Path Phases, with their emphasis on new written methods and learning supports, meant that the Teacher Edition needed to have a *learn-while-teaching* approach because there would not be enough professional development time for teachers to learn everything before teaching it. Thus, a major design step was to put crucial material on the student page because some teachers rarely, and many teachers only occasionally, read the Teacher Edition. Furthermore, many teachers reported that they needed to learn mathematics with understanding because they had not done so in their own schooling. We used partially scripted lessons to help teachers with the new approaches, continually checking that these were helpful and not demeaning. One teacher said that using the lessons was like "having Karen Fuson sitting on my shoulder helping me."

I gathered extensive information from teachers in grade-level groups and weekly phone calls about lessons that needed fixing and ideas teachers had to improve lessons. These data often led to revisions to fix confusions or difficulties. There were many detailed revisions so that teachers could understand lessons better and move their intended curriculum closer to the intentions of the program.

We learned that two repeated features in the lessons helped teachers discuss errors and support Math Talk: *Puzzled Penguin* and *Math Talk in Action*. Because discussing errors made by a specific student made some teachers uncomfortable, the Puzzled Penguin feature asked students to help Puzzled Penguin correct and explain a common error (see Figure 11.3). These errors were collected by our extensive classroom research and from research studies and were previewed in the unit overviews that reviewed the approaches and learning supports to help teachers prepare for teaching a unit. Teachers were encouraged to discuss errors as they arose. Puzzled Penguin became a beloved classroom friend, and children were very supportive as they helped Puzzled Penguin understand an error and instead use a correct approach. Some teachers dressed up as Puzzled Penguin on Halloween. Puzzled Penguin helped errors become understood as normal and as a source of learning.

Figure 11.3

Puzzled Penguin in the Teacher Edition and on a Student Page

What's the Error?

MP3, MP6 Construct Viable Arguments/Critique Reasoning of Others | Puzzled Penguin Exercise 13 on Student Activity Book page 212 shows a common error children may make when learning 2-digit subtraction: neglecting to rewrite the tens after ungrouping a ten into 10 ones. Discuss with children what Puzzled Penguin did wrong. Children should cross out Puzzled Penguin's answer and find the correct answer.

- *What might have helped Puzzled Penguin?* If Puzzled Penguin had made a proof drawing, Puzzled Penguin would have seen that there should be 7 tens instead of 8 tens.

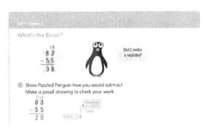

Source: From *Math Expressions, Common Core*. Copyright ©2018 by Houghton Mifflin Harcourt Publishing Company. Reprinted by permission of the Publisher. All rights reserved.

Note. Some text on the student page appears only in the Teacher Edition because it provides answers, as in the problem and drawing.

The *Math Talk in Action* feature supports teachers by giving examples of student talk to help teachers understand what students might say about a given topic. This feature also reminds teachers to elicit their students' thinking (see Figure 11.4).

Figure 11.4

An Example of a Math Talk in Action

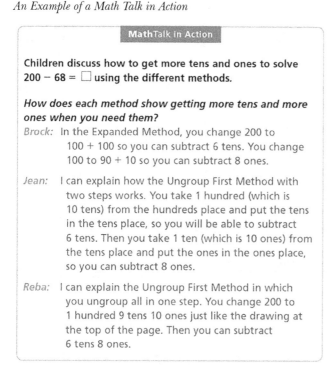

Source: From *Math Expressions, Common Core*. Copyright ©2018 by Houghton Mifflin Harcourt Publishing Company. Reprinted by permission of the Publisher. All rights reserved.

Many elementary teachers believe that they do not understand mathematics, and they feel damaged by their own mathematics education. In our work with teachers, we emphasized how none of us had learned mathematics with understanding, but that our research-based approaches would help teachers and their students understand mathematics. These encouragements, and especially the learn-while-teaching design efforts, were affirmed by pilot teachers in Chicago who said at a meeting, "Do you know what we call this program? We call it Math Therapy for Teachers because it is helping us understand math, and we never have before." An important lesson learned is: *Mathematics programs need to help teachers understand the mathematics. In other words, a math program focused on understanding for students also needs learning supports that help both teachers and students.*

Our observations indicated that many lessons in classrooms where another mathematics curriculum was being used were complex and required a lot of teacher organization of materials and students. This observation led to another lesson learned: *Keep materials to those absolutely essential and develop activity structures in which the mathematics can be varied so that student and teacher energy can focus on the mathematics.* Some examples of these are the *Math Talk*: *Solve, Explain, Question,* and *Justify* structure; the *Puzzled Penguin* activity; and the *Daily Routines* described above for K, 1, and 2 to learn complex grade-level concepts. The need for repetition is especially important in Kindergarten, where several working surfaces like the *Counting Mat* (see Figure 11.5) were developed so that children could use manipulatives to do many different related activities with the same materials. I continually simplified manipulatives and made all visual supports coherent across grades.

Figure 11.5

The Counting Mat Showing Activities Using Squares, Fingers, and Symbols to Add

Notice that the Counting Mat is turned lengthwise to hold all of the Number Tiles 1–10 at the top.

1. Get started.
2. Make a number.
3. Relate visual quantity to fingers, sounds, or body motions.
4. Practice visual imagery.
5. Describe arrangements.
6. Change arrangements.
7. Copy the arrangement of another person.
8. See partners of a number.
9. Use sounds to add and subtract.

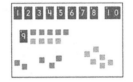

Encourage children to make 5-groups as one way to show a number. Children can also be creative in showing different ways to make a number.

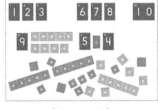

Source: From *Math Expressions, Common Core*. Copyright ©2018 by Houghton Mifflin Harcourt Publishing Company. Reprinted by permission of the Publisher. All rights reserved.

Levels in Math Talk Teaching

Our research identified four levels in teachers learning to elicit and support student explaining of mathematics (see Table 11.1, adapted from Hufferd-Ackles et al., 2004, 2015). These levels helped teachers move their intended curriculum closer to that intended in *Math Expressions*. Five components help teachers think about these crucial aspects of teaching:

teacher role, questioning, explaining mathematical thinking, mathematical representations, and building student responsibility within the classroom. The table was discussed by teachers in professional development sessions and during conference presentations. Teachers said that the components were helpful in thinking about their teaching and that the levels helped them visualize new ways of teaching. Videos of students explaining their thinking were also helpful in seeing that students can do such explaining (see the next section). A lesson learned is: *The importance of a framework, such as the table, helps teachers focus their thinking about the complex activities of teaching on a few crucial aspects of that complexity.*

Table 11.1

Levels of Math Talk Learning Community: Teacher and Student Action Trajectories

Levels of Math Talk Learning Community: Teacher and Student Action Trajectories					
Components of the Math Talk Learning Community				
Teacher Role	Questioning	Explaining mathematical thinking	Mathematical representations	Building student responsibility within the community
Overview of shifts among Levels 0–3: The classroom community grows to support students acting in central or leading roles and shifts from a focus on answers to a focus on mathematical thinking				
Shift from teacher as leader of conversation to students/ teacher as co-leaders.	Shift from teacher as questioner to students and teacher as questioners.	Students increasingly explain and articulate their math ideas.	Students increasingly explain their math thinking relying, as needed, on math drawings/ representations.	Students increasingly take responsibility for learning and evaluation of others and self. Math sense becomes the criterion for evaluation.
Level 0: Traditional teacher directed classroom with brief answer responses from students				
Teacher is at the front of the room and dominates conversation.	Teacher is only questioner. Questions serve to keep students listening to teacher. Students give short answers and respond to teacher only.	Teacher questions focus on correctness. Students provide short answer-focused responses. Teacher may tell answers.	Representations are missing or teacher shows them to students.	Students believe they need to keep ideas to themselves or just provide answers when asked.

(Table continued on next page)

Table 11.1 (Continued)

Levels of Math Talk Learning Community: Teacher and Student Action Trajectories

Levels of Math Talk Learning Community: Teacher and Student Action Trajectories				
Components of the Math Talk Learning Community				
Teacher Role	Questioning	Explaining mathematical thinking	Mathematical representations	Building student responsibility within the community
Overview of shifts among Levels 0-3: The classroom community grows to support students acting in central or leading roles and shifts from a focus on answers to a focus on mathematical thinking				
Level 1: Teacher beginning to pursue student mathematical thinking. Teacher plays central role in the Math Talk Community				
Teacher encourages sharing of math ideas and directs speaker to talk to the class, not to the teacher only.	Teacher questions begin to focus on student thinking and less on answers. Only teacher asks questions.	Teacher probes student thinking somewhat. One or two strategies may be elicited. Teacher may fill in an explanation. Students provide brief descriptions of their thinking in response to teacher probing.	Students learn to create math drawings to depict mathematical thinking.	Students believe their ideas are accepted by the classroom community. They begin to listen to each other supportively and to restate in their own words what another student said.
Level 2: Teacher modeling and helping students build new roles. Some co-teaching and co-learning begins as student-to-student talk increases. Teacher physically begins to move to the side or back of room.				
Teacher facilitates conversation between students, and encourages students to ask question of one another.	Teacher asks probing questions and facilitates some student-to-student talk. Students ask questions of one another with prompting from teacher.	Teacher probes more deeply to learn about student thinking. Teacher elicits multiple strategies. Students respond to teacher probing and volunteer their thinking. Students begin to defend their answers.	Students label their math drawings so others are able to follow their mathematical thinking.	Students believe they are math learners and that their ideas and the ideas of classmates are important. They listen actively so that they can contribute significantly.

(Table continued on next page)

Table 11.1 (Continued)

Levels of Math Talk Learning Community: Teacher and Student Action Trajectories

Levels of Math Talk Learning Community: Teacher and Student Action Trajectories					
Components of the Math Talk Learning Community					
Teacher Role	Questioning	Explaining mathematical thinking	Mathematical representations	Building student responsibility within the community	
Level 3: Teacher as co-teacher and co-learner. Teacher monitors all that occurs, still fully engaged. Teacher is ready to assist, but now in more peripheral and monitoring role (coach and assister).					
Students carry conversation themselves. Teacher only guides from the periphery of the conversation. Teacher waits for students to clarify thinking of others.	Student-to-student talk is student initiated. Students ask questions and listen to responses. Many questions ask 'why' and call for justification. Teacher questions may still guide discourse.	Teacher follows student explanations closely. Teacher asks students to contrast strategies. Students defend and justify their answers with little prompting from the teacher.	Students follow and help shape descriptions of others' math thinking through math drawings and may suggest edits in others' math drawings.	Students believe they are math leaders and can help shape the thinking of others. They help shape others' math thinking in supportive, collegial ways and accept the same.	

LESSONS LEARNED ABOUT THE IMPLEMENTED OR ENACTED CURRICULUM

The implemented or enacted curriculum consists of those aspects of the curriculum that actually occur in classrooms (Remillard & Heck, 2014). What occurs in classrooms is crucially important but is difficult to document. Early in our work, we did many classroom observations with various methods of recording, especially student methods and discussion. We did some video recording, although we found that we had little time to look at or use the video recordings to improve lessons or teaching because of the time it took to view them. Instead, in classrooms we looked at student thinking because we could tell if a teacher had been implementing a Math Talk community by how much and how well students explained their thinking and made drawings to show their thinking.

After teachers improved in their use of Math Talk, we recorded in a range of classrooms—mostly high poverty urban schools—to get examples of teaching to show teachers starting to use the program.[2] Our early efforts at video recording indicated how difficult the sound and light conditions are in many classrooms. Hence, we eventually used a non-profit documentary production company to do the recording. A variety of equipment was used to produce high-quality videos: teachers and student explainers wore microphones; a large mobile overhead microphone was used to capture Math Talk from classroom members; and two professional cameras were used to capture different perspectives. The filming crew said that filming in classrooms was the most difficult environment they had encountered. They were amazed by how well all of the students took to the intrusion of equipment and people, and carried on as if it was a normal school day. The lesson is: *Videos are existence proofs that students from backgrounds of poverty and those learning English can make mathematics drawings and explain their thinking. Teachers find these videos very helpful in forming mental images of new ways to teach. They can be difficult to obtain but are crucially important.*

LESSONS LEARNED ABOUT THE ASSESSED CURRICULUM

The assessed curriculum consists of the elements of the curriculum that appear on local (i.e., classroom), state, or national assessments (Remillard & Heck, 2014). A striking lesson from our research is that when we began 30 years ago, some districts had four levels of largely uncoordinated assessments: national or state tests, standardized tests, district tests, and school tests, and many districts had three such assessments. We found that school administrators, and thus teachers, were often more influenced by these assessments than by the official curriculum goals. Teachers spent a large amount of time focused on and preparing for these tests, continually feeling under pressure about student test performance. I felt that I had to include in the curriculum topics on national or state tests, but eventually I would only accept schools into our mathematics program who were willing to drop or modify school and district tests. Otherwise, teachers would cut and paste the program and interfere with its carefully designed learning paths, and thus reduce its potential to improve student learning.

"Standardized tests" have been used in the U.S. for decades, but they were not adequate measures of students' understanding of grade-level content. First, test makers added items beyond grade level to create a normal curve at each grade level so that half of the students will be above and half below the mean. Second, few or no test items were released, so it was difficult for teachers to know what to teach to prepare students for the test. There was no assurance that students were being tested on knowledge they had an opportunity to learn. Third, grade-level standards varied across states but a single version of a standardized test was used across states, so some students were tested on content they had not had an opportunity to learn. Not surprisingly, lower-income schools performed more poorly than higher-income schools. But even if you could raise scores, the test-makers could just add more difficult items to keep the normal curve. These tests have had pernicious effects on students and families, contributing to the perception that lower-income students and those with non-white skin are poor learners.

The CCSSM created common standards so that teachers could actually teach the same topics across many different states. Many of us who worked on the CCSSM hoped to help develop new kinds of tests that only contained items from the grade level content and that did not have many really difficult items; such changes would make it possible to know if students were learning grade-level content. I consulted on Smarter Balance and PARCC (Partnership for Assessment of Readiness for College and Careers), the new tests developed to assess the CCSSM. Eventually the test developers included items with above grade-level content. So, we still do not know how many students are learning grade-level goals.

Furthermore, assessments even now rarely give performance data for items and certainly do not show student work, which would be helpful in understanding student thinking. Thus, teachers do not receive useful information from assessments that could help them know what to reteach. Smarter Balance seems to give item information by student, which is what teachers need, but this is only available for interim assessments and not for the summative assessment. And the procedure is very slow to search for each item. What is needed is a grid of answers chosen for each item by individual students. This gives information valuable for teaching because it gives the errors students made by making certain choices.

In contrast, my unit tests for *Math Expressions* are grade-level tests designed for students to do well if they have learned the main content of the unit. There should be other tests that assess grade-level understanding in a reasonable way and that give easily accessible item error data for students.

As a final note on assessment, we had a distressing experience with grade-level pretests. Our grant reviewers had specified that we give grade-level or unit pretests and posttests. In the first year, we learned that pretests at the beginning of the year are cruel and useless. Students are not supposed to have learned grade-level content before they start that grade, so pretests on what they were going to learn that year were experienced as very difficult. Some students cried even though we said they were not expected to know what was on the test. For students to experience failure at the beginning of the school year was horrible. Thus, even though grants expected such pretests, we stopped giving them as part of our research and tried to persuade district administrators not to give them. *What one really needs to know is what students know at the end of the year, and students deserve a chance to learn that content during the year. Thus, fair assessments need to be focused only on grade-level content without tricky formats or content, and they need to provide results to teachers by student and include errors on items to help teachers know how to repair student partial understandings.*

LESSONS LEARNED ABOUT THE LEARNED CURRICULUM

The learned curriculum consists of what students actually master (Remillard & Heck, 2014). Student learning was our major goal. To consider what students had learned, we gathered and examined their homework, quizzes, and unit tests. We recorded wrong answers for each item so that we could modify the curriculum materials to reduce those errors. Many of these errors reflected partial but incomplete learning, so they were relatively easy for teachers to help students overcome. We created targeted one and two item quizzes given daily so we could track daily learning within a whole class and find errors that needed overcoming. We interviewed students on items gathered from other studies, and our students performed like East Asian children and much better than U.S. children in other programs (see Fuson, 2017; Fuson et al., 1997; Fuson & Smith, 2016). We did tutoring studies with low achievers to find learning issues and test how to overcome them (e.g., Fuson & Smith, 1995). We asked teachers how students were doing and what the major learning issues were. We used all these data to improve the program each year and to present research papers, but we rarely had enough time to write up results for journals. We tried repeatedly to get the students' results from state or national tests, but various issues interfered. Teachers would copy and give me their results, which were always considerably better than before my program, but I never could get systematic data of this kind.

We found in our years of giving many kinds of tests that not much variability is created by items with different numbers (e.g., Fuson, 2017), except for single-digit multiplication and division early in such learning. In testing, there is a huge emphasis on reliability, so repeated similar items are given. But we found this to be unnecessary for mathematics. Many tests are too long, and they could be shortened considerably because repeated items are not needed, especially if errors are the focus.

Wrong answers by item and by student provide formative and summative information about individual student thinking that can improve teaching and practice. In several schools we persuaded teachers to do what we had been doing: record the results of unit tests for each student by giving a check mark for correct answers and writing down any wrong answers. Because our unit tests focused on the main learning goals and were not tricky, this did not take much time. Teachers found the results informative in thinking about what confusions or small errors students were making. Teachers then could do a unit review targeted toward fixing the errors. Many teachers had not paid much attention to the errors before this experience, but they quickly saw how valuable this approach could be. If developers of mathematics programs wrote shorter unit tests designed to show student learning of crucial goals, recording only errors made by each student could become an effective common practice by teachers. This would also make it easy to give partial credit for some kinds of errors.

A major lesson learned is: *The field needs much more research at all grade levels with performance data published by item*. Over the years, it became almost impossible to collect control data because schools understandably did not want to take time away from learning. It also is very difficult to match samples; getting control data is expensive and time-consuming, with effort better spent on assessing learning of the intervention, for example, by interviewing students. I found the data in Stigler et al. (1990) invaluable. This international study had carefully designed first-grade and fifth-grade samples from Japan, Taiwan, and the United States with written and interview data published by item for each country. It would be helpful to have such data for all grade levels. Other studies that required interviewing students were also valuable, especially in identifying what children understood about tens and ones (see the tasks from various studies given in Fuson et al., 1997).

Publishing items with performance data would be useful, and it is easy to make similar items that are not released. When we found that only 26% of Grade 2 students were correct on a 2-digit problem with ungrouping on a standardized test at the end of the school year, this became an important target for improving such learning.

We also found that videos of students explaining their solution to important grade-level mathematics problems are valuable. In some schools, two videos a year for every student are put on secure websites. Parents are amazed and proud of their students, and we have found that parents finally understand the goals of the program as supporting students in understanding their mathematical thinking. I think these videos are more valuable than national or state tests, which mostly give useless data to compare students and districts, but do not give information about grade-level performance overall or about specific items. If our goal is for students to be able to explain their thinking, we need to collect and share data that reflects this goal.

CONCLUSION

The connected web of lessons learned across different levels of curriculum provides frameworks useful to program designers, mathematics coaches, and curriculum researchers. The Learning Path Phases model for teaching various topics emphasizes bridging from student methods to fluent methods with a Phase 2 extended discussion of accessible and mathematically-desirable research-based methods supported by mathematics drawings and other visual supports. A Nurturing Math Talk Community approach supports extensive conversations needed for sense making and deepening understanding. These sense-making approaches create the importance of a learn-while-teaching approach in a mathematics program to help teachers learn extensive new sense-making approaches. Related lessons learned are the limitations of current tests that include items beyond grade level, and the need for tests that assess only grade-level goals so district administrators can know how teaching is working. Tests should provide item data and student errors to teachers to guide reteaching. The posting on some district websites of videos of students explaining their thinking could be one phase of assessment that is consistent with the teaching goals of high-level standards like the CCSSM, specifically, the goal of having students understand mathematics and explain their thinking.

AUTHOR NOTE

Heartful thanks to all of the students, teachers, administrators, and parents who contributed to improving this mathematics program over the decades, to the many talented staff and graduate students who gathered and analyzed data to improve the program and shared ideas about all aspects of the program, to the small army of undergraduate work-study students who made and revised student pages in English and Spanish, and to the editors of Houghton Mifflin Harcourt for their dedication to implementing a new kind of math program.

REFERENCES[3]

Abrahamson, D., & Cigan, C. (2003). A design for ratio and proportion. *Mathematics Teaching in the Middle School*, *8*(9), 493–501.

The Common Core Writing Team. (2015, 6 March). *The NBT Progression for the Common Core State Standards*.

Fuson, K. C. (1998). Pedagogical, mathematical, and real-world conceptual-support nets: A model for building children's mathematical domain knowledge. *Mathematical Cognition*, *4*(2), 147–186.

Fuson, K. C. (2004, 2009, 2011, 2013, 2018). *Math expressions*. Houghton Mifflin Harcourt. [A 2026 update of the curriculum is underway and will be published by Heinemann, a subsidiary of Houghton Mifflin Harcourt.]

Fuson, K. C. (2009). Avoiding misinterpretations of Piaget and Vygotsky: Mathematical teaching without learning, learning without teaching, or helpful learning-path teaching? *Cognitive Development, 24*, 343–361. https://doi.org/10.1016/j.cogdev.2009.09.009

Fuson, K. C. (2017, April 2–8). *Kindergarten and grade 1 children living in poverty can learn the CCSS NBT concepts* [Paper presentation]. National Council of Supervisors of Mathematics Annual Conference, San Antonio, TX.

Fuson, K. C. (2019). Overcoming errors in fraction computation by emphasizing unit fractions, length drawings, and student explanations. *Universal Journal of Educational Research, 7*(8), 1663–1678. https://doi.org/10.13189/ujer.2019.070805

Fuson, K. C. (2020). The best multidigit computation methods: A cross-cultural cognitive, mathematical, and empirical analysis. *Universal Journal of Educational Research, 8*(4), 1299–1314. https://doi.org/10.13189/ujer.2020.080421

Fuson, K. C. (2024). *Math drawings and visual supports for sense making in Math Talk classrooms.* [Manuscript in preparation].

Fuson, K. C., & Abrahamson, D. (2005). Understanding ratio and proportion as an example of the apprehending zone and conceptual-phase problem-solving models. In J. Campbell (Ed.), *Handbook of mathematical cognition* (pp. 213–234). Psychology Press.

Fuson, K. C., Atler, T., Roedel, S., & Zaccariello, J. (2009). Building a nurturing, visual, math-talk teaching-learning community to support learning by English language learners and students from backgrounds of poverty. *New England Mathematics Journal, XLI*, 6–16.

Fuson, K. C., & Beckmann, S. (2012/2013). Standard algorithms in the Common Core State Standards. *National Council of Supervisors of Mathematics Journal of Mathematics Education Leadership, 14*(2, Fall/Winter), 14–30.

Fuson, K. C., & Beckmann, S. (2012, April 22–28). *Multiplication to ratio, proportion, and fractions within the Common Core* [Paper presentation]. National Council of Teachers of Mathematics Annual Conference, Philadelphia, PA.

Fuson, K. C., & Decker, R. S. (2017, April 2–8). *Learning cycles and mathematical practices in the classroom math talk community* [Paper presentation]. National Council of Supervisors of Mathematics Annual Conference, San Antonio, TX.

Fuson, K. C., & Leinwand, S. (2023). Building equitable Math Talk classrooms. *Mathematics Teacher Learning and Teaching, 116* (3), 164–173.

Fuson, K. C., & Li, Y. (2014). Learning paths and learning supports for conceptual addition and subtraction in the United States Common Core state standards and in the Chinese standards. In Y. Li & G. Lappan (Eds.), *Mathematics curriculum in school education* (pp. 541–558). Springer.

Fuson, K. C., & Murata, A. (2007). Integrating NRC principles and the NCTM process standards to form a class learning path model that individualizes within whole-class activities. *National Council of Supervisors of Mathematics Journal of Mathematics Education Leadership, 10*(1), 72–91.

Fuson, K. C., Murata, A., & Abrahamson, D. (2015). Using learning path research to balance mathematics education: Teaching/learning for understanding and fluency. In R. Cohen Kadosh & A. Dowker (Eds.), *Oxford handbook of numerical cognition* (pp. 1020–1038). Oxford University Press. [Also appears online in *Oxford Handbooks Online, July, 2014.*] https://doi.org/10.1093/oxfordhb/9780199642342.013.003

Fuson, K. C., & Smith, S. T. (1995). Complexities in learning two-digit subtraction: A case study of tutored learning. *Mathematical Cognition, 1*, 165–213.

Fuson, K. C., & Smith, S. T. (2016, April 13-16). *Children living in poverty can solve CCSS OA word problems* [Paper presentation]. National Council of Teachers of Mathematics Annual Conference, San Francisco, CA.

Fuson, K. C., Smith, S. T., & Lo Cicero, A. (1997). Supporting Latino first graders' ten-structured thinking in urban classrooms. *Journal for Research in Mathematics Education, 28*, 738–766.

Hufferd-Ackles, K., Fuson, K. C., & Sherin, M. G. (2004). Describing levels and components of a math-talk community. *Journal for Research in Mathematics Education, 35*(2), 81–116.

Hufferd-Ackles, K., Fuson, K. C., & Sherin, M. G. (2015). Describing levels and components of a math-talk learning community. In E. A. Silver & P. A. Kenney (Eds.), *More lessons learned from research: Volume 1: Useful and usable research related to core mathematical practices* (pp. 125–134). National Council of Teachers of Mathematics.

Klein, D. (2007). A quarter century of US 'math wars' and political partisanship. *Journal of the British Society for the History of Mathematics, 22*(1), 22–33. https://doi.org/10.1080/17498430601148762

Munter, C., Stein, M. K., & Smith, M. S. (2015). Dialogic and direct instruction: Two distinct models of mathematics instruction and the debate(s) surrounding them. *Teachers College Record, 117*(11), 1–32. https://doi.org/10.1177/016146811511701102

Murata, A. (2008). Mathematics teaching and learning as a mediating process: The case of tape diagrams. *Mathematical Thinking and Learning, 10*(4), 374–406.

Murata, A., & Fuson, K. (2006). Teaching as assisting individual constructive paths within an interdependent class learning zone: Japanese first graders learning to add using ten. *Journal for Research in Mathematics Education, 37*(5), 421–456.

National Governors Association Center for Best Practices & Council of Chief State School Officers [NGA & CCSSO]. (2010). *Common core state standards for school mathematics.* http://www.thecorestandards.org/Math

Rea, C. P., & Modigliani, V. (1985). The effect of expanded versus massed practice on the retention of multiplication facts and spelling lists. *Human Learning, 4*, 11–18.

Remillard, J. T., & Heck, D. J. (2014). Conceptualizing the curriculum enactment process in mathematics education. *ZDM Mathematics Education, 46*(5), 705–718. https://doi.org/10.1007/s11858-014-0600-4

Reys, B. (2006). *The intended mathematics curriculum as represented in state-level curriculum standards: Consensus or confusion?* Information Age Publishing.

Schoenfeld, A. (2004). The math wars. *Educational Policy, 18*(1), 253-286.

Stigler, J., Fuson, K. C., Ham, M., & Kim, M. S. (1986). An analysis of addition and subtraction word problems in Soviet and American elementary textbooks. *Cognition and Instruction, 3*, 153–171.

Stigler, J. M., Lee, S.-Y., & Stevenson. H. W. (1990). *Mathematical knowledge of Japanese, Chinese, and American elementary school children.* National Council of Teachers of Mathematics.

Tharp, R. G., & Gallimore, R. (1988). *Rousing minds to life.* Cambridge University Press.

ENDNOTES

1. Also, see the table in Fuson & Decker (2017).
2. See the 61 Classroom Videos on karenfusonmath.net.
3. References with Fuson in them and other relevant papers are available on my website, karenfusonmath.com, under Publications or Visual Presentations.

CHAPTER 12

LESSONS LEARNED FOR DEVELOPING AND ENACTING A DIGITAL COLLABORATIVE PLATFORM WITH AN EMBEDDED PROBLEM-BASED MATHEMATICS CURRICULUM

Alden J. Edson
Michigan State University

In this chapter I report on a design research project of a digital collaborative platform for an embedded problem-based mathematics curriculum—the Connected Mathematics Project (CMP). The goal of the project is to enhance the teaching and learning of mathematics that occurs in paper-and-pencil classrooms by leveraging the affordances of digital technologies in a digital classroom environment. In this chapter I share lessons learned for developing the digital collaborative platform for students and teachers, focusing on how the project team: (1) reimagined mathematics problems delivered in a digital collaborative platform, (2) supported a model of collaboration in the digital platform, and (3) provided students with just-in-time supports in the digital collaborative platform. To illustrate the lessons learned, I report on the iterative changes made to features of the digital collaborative platform based on analysis of project data and feedback from teachers and students.

Keywords: computer-supported collaborative learning; digital environments; mathematics learning and teaching; middle-grades mathematics; problem-based curriculum

INTRODUCTION

The evolution of digital technologies has led to numerous internet-based resources that are accessible to students and teachers. Although many digital resources include lessons and activities, many publishers and education software companies are releasing comprehensive programs, platforms, and systems designed to supplement or supplant print textbooks. These resources present researchers, teachers, and curriculum developers with new affordances for how students and teachers enact mathematics curricula in digital environments. Understanding the extent to which digital resources have the potential to change curriculum enactment compared with conventional classroom environments is critically important to improve the teaching and learning of mathematics.

The set of curriculum materials embedded in the digital collaborative platform discussed in this chapter is the Connected Mathematics Project's middle grades problem-based curriculum, *Connected Mathematics* (CMP) (Lappan et al., 2014; Phillips et al., 2025). Problem-based learning occurs as students "engage a problem without preparatory study and with knowledge insufficient to solve the problem, requiring that they extend existing knowledge and understanding and apply this enhanced understanding to generating a solution" (Wirkala & Kuhn, 2011, p. 1157). The focus on student thinking in problem-based curricula differs from *delivery mechanism* curriculum materials, in which students memorize facts and practice demonstrated procedures in a direct instruction classroom (Choppin et al., 2015; Roth McDuffie et al., 2018).

Three important principles related to the teaching and learning of mathematics underscore the design of the digital platform for CMP. First, at the onset of the project, the platform was intended to enhance (not replace) face-to-face mathematics instruction, particularly student-centered, inquiry-oriented teaching and learning of mathematics. The project team developed student and teacher features that leverage the affordances of digital technologies to amplify or transform paper-and-pencil environments without creating new challenges for students or teachers. In physical classrooms, teachers using the platform typically have desktop computers connected to projection systems and might also have access to a tablet or laptop to use when circulating the classroom. As in paper-and-pencil environments, when using the digital collaborative platform, students continued to work in small groups of two-to-four. Rather than students sharing devices in pairs or triplets, it was important that each student had access to their own laptop connected to the internet. This was to ensure access to student resources in the classroom. Because students worked collaboratively with others using a curriculum, the project team designed the platform for laptops so that screen size could support viewing the mathematics problem, students' individual work, and the work of others. Not every day was a "digital day." On non-digital days, students could upload photos of their paper-and-pencil work onto the digital platform. Some districts provide students with their own laptops, some teachers borrowed district carts for problems, and some teachers used classroom sets provided by the CMP. During the COVID-19 pandemic, students and teachers used video conference tools for small- and whole-group collaboration. Although development of the digital platform began prior to the pandemic, and minor changes were made to support the pandemic (e.g., a notification system was created to let students know they were connected to the internet), the team did not design features for the platform solely for online or remote learning and teaching of mathematics.

Second, the digital platform for CMP differed from other platforms by its emphasis on student and teacher collaboration around a mathematics curriculum. In problem-based mathematics classrooms like CMP, individual and collaborative learning is operationalized through the *Launch-Explore-Summarize* instructional model in both the print materials and the digital environment (Edson & Phillips, 2021; Lappan et al., 2004). In the *Launch* phase, the context and challenge of a problem situation is introduced to students who make predictions and ask clarifying questions. During the *Explore* phase, students work collaboratively to explore and solve the problem by gathering data, sharing ideas, looking for patterns, making conjectures, developing strategies, and creating arguments to support their reasoning and solutions, while the teacher monitors individual and group strategies and provides questions to guide students' thinking. In the *Summarize* phase, teachers facilitate discussion about the mathematical learning goal(s) of the problem while students share, discuss, and refine their strategies and conjectures, which reveal embedded mathematical understandings in the problem situation and connect to their prior knowledge. Although research suggests that computer-supported collaborative learning has even more prominent effects than traditional small-group work without digital devices (e.g., Jeong & Hmelo-Silver, 2016), our emphasis was strengthening individual and collaborative learning during all three instructional phases.

Third, because students and teachers have different roles in the enacted classroom, the design team streamlined users' experiences so that students and teachers share platform interface and functionality whenever possible. The curriculum problems visually appear the same for both students and teachers, yet teachers have access to additional support features. For example, teachers can access embedded student strategies and answers for each problem, and can access the teacher guide support to plan, enact, and reflect on the mathematics problem, which they can incorporate into their teaching and planning documents. High-level features of the platform for students and teachers are shown in Figure 12.1.

Figure 12.1

High-Level Features of the Digital Collaborative Platform for Students and Teachers

Student Features	Teacher Features
• Investigate problems using a new CMP STEM (Science, Technology, Engineering, and Mathematics) problem format that is embedded in the platform and supports students as doers, knowers, creators, and communicators of mathematics • Develop mathematics with a variety of tools (e.g., graphs, tables, text, drawing, images) in a safe and collaborative setting • Document, share, refine, publish, curate, and retrieve their work, moving back and forth between individual and shared spaces • Extend learning of mathematics concepts by transforming their work into sharable and retrievable classroom artifacts for future use and reflection in a mathematics learning log	• Access curriculum materials including student text, teacher guide, and problem solutions • Monitor evidence of student thinking from individuals, groups, or the entire class in real-time or after school • Select, highlight, edit, and incorporate individual student work into the class workspace • Create and send "just-in-time" supports to individual, group, and classes of students • Collaborate with teacher colleagues and school coaches to create resources, collaborate in the digital environment, and support each other in planning, teaching, and reflecting

After discussing the development of the digital collaborative platform, in the remainder of this chapter I report on lessons learned while developing a digital collaborative platform embedded within CMP. The design principles focused on (1) reimagining mathematics problems delivered in a digital collaborative platform, (2) supporting a model of collaboration in the digital platform, and (3) providing students with "just-in-time" supports in the digital collaborative platform. For each design principle, I share the context for three features of the digital collaborative platform that were iteratively developed, tested, and refined. To illustrate the lessons learned, I discuss iterative changes made to features of the digital collaborative platform based on analysis of project data and feedback from teachers and students.

THE DEVELOPMENT OF THE DIGITAL COLLABORATIVE PLATFORM

Development of the digital platform is conducted by a research team that is a partnership between Michigan State University and Concord Consortium, and includes people with expertise in educational research, curriculum development, educational technology, and mathematics education. The overarching methodological approach to the research is similar to the iterative process for design research, such as design studies (e.g., Edelson, 2002), design experiments (e.g., The Design-Based Research Collective, 2003), and developmental research (e.g., Richey et al., 2004). The goal of design research is to:

> Use the close study of learning as it unfolds within a naturalistic context that contains theoretically inspired innovations, usually that have passed through multiple iterations, to then develop new theories, artifacts, and practices that can be generalized to other schools and classrooms. (Barab, 2014, p. 151)

To create a digital collaborative platform embedded within the CMP curriculum materials, the team engages in an agile development process through rapid iterative cycles of designing, building, testing, and refining. The development process begins with the design of print problem-based curriculum materials; these are embedded into a digital collaborative platform for students; and soon thereafter, the team focuses on features in the platform for teachers that are linked to the student platform. The team uses conjecture mapping for each development phase (Sandoval, 2014). *Conjecture mapping* is a mechanism that makes explicit the relationships between the design of the platform, the enacted experiences within classrooms, and the outcomes associated with those experiences. The classroom testing is conducted with experienced mathematics teachers using the CMP curriculum materials from school districts in the midwestern and northeastern U.S. Because teachers enact seventh-grade CMP units at different times throughout the year, the team engages in multiple iterative cycles per year.

LESSON ONE: REIMAGING MATHEMATICS PROBLEMS FOR THE DIGITAL COLLABORATIVE PLATFORM

The first lesson focuses on connecting the mathematics problem with the student and teacher workspaces. In many classrooms, textbooks have been supplanted by consumable workbooks and internet-downloadable worksheets (Rohrer et al., 2020; Umriani et al., 2020). The design and format of written tasks are variable, as outlined in the following contrasts:

- Do the written tasks come across as blanks to fill in, or do they invite students to explore and solve problems involving important mathematics?

- Does the structure of tasks help provoke and develop student thinking over time, or do they support a model where students memorize facts and practice demonstrated procedures?
- Does the structure offer an assumption that subsequent work is based on problems of the same kind, or do students need to consider ways in which subsequent questions relate to each other?

The design of written tasks and their format is important for studying students' opportunities to learn mathematics and how they are taken up in classrooms by students and their teacher.

To enhance the teaching and learning of mathematics, the project team considered the task design and structure of the written mathematics problem. The team recognized that current written tasks–either those oriented toward student thinking or delivery mechanism—are structured around a conventional linear format (e.g., A1, A2, B1, B2, B3). Three considerations were critical in reimagining the task design and structure: (a) the research design work that went into prior versions of the print CMP materials (Edson et al., 2019); (b) new possibilities for how students (and their teachers) engage in and learn mathematics (Edson, 2016, 2017); and (3) the resources needed to support teachers' classroom practice.

Platform Design Feature: The CMP STEM Problem Format

The CMP STEM Problem format resembles the work that STEM professionals do to solve problems, build deep knowledge and skills, and meaningfully connect these solutions to inform the needs of society. The CMP STEM Problem feature consists of three connected components of the task structure: (1) *Initial Challenge*, which poses the mathematical challenge and provides open access for students in terms of access points, possible strategies, and nature of solutions; (2) *What If...?*, which unpacks the mathematical understandings and where students probe deeper at the mathematics by considering different situations when you change quantities, contexts, or strategies; and (3) *Now What Do You Know?*", which connects prior and future knowledge so students can self-assess and consolidate their learning of mathematics. For students, the task design promotes student engagement and learning as they collaborate to design solutions, make conjectures, offer critiques, and communicate their mathematical understandings. For teachers, the task design is more explicit in terms of the goals for mathematics problems, namely to (a) attend to the strategies students use to solve problems, (b) help students recognize the embedded or encoded mathematical understandings, and (c) connect these understandings to prior and future knowledge (Edson et al., 2019).

Figure 12.2 shows an example of the unit *Stretching and Shrinking: Developing Proportional Reasoning in the Context of Similarity* (*Scale Drawings*). In this problem, students investigate the effects that various algebraic rules have on a hat for Mug. In the *Initial Challenge*, students explore what happens to the location of a hat when numbers are added to or subtracted from each coordinate, versus multiplied which changes the size of the hat (as occurred in the prior problem). During the *What If...?* students distinguish rules that produce similar figures from rules that do not, recognizing the role that multiplication plays in scaled drawings or similarity relationships. The *Now What Do You Know?* not only provides an opportunity to discuss algebraic rules but also sets the stage for work on proportional relationships in the unit *Comparing and Scaling: Developing Proportional Reasoning in the Context of Number* (*Quantities*).

Connecting the Mathematics Problem and the Student Spaces

One lesson learned relates to the interconnectivity between the presentation of a mathematics problem and a student's individual and collaborative workspaces. In paper-and-pencil classrooms, curriculum materials and student notebooks are typically separate resources. In a digital environment, these resources can connect—students can access the problem and their workspaces simultaneously. In our development work, presentation of a mathematics problem progressed along a continuum from: (1) containing the problem and the notebook within the same section of the platform; to (2) separate sections for the problem and the notebook in the platform, with no interactive features between them; to (3) separate sections for the problem and the notebook, with interactive features between the sections. Figure 12.3 illustrates the placement of components of the digital platform at three development stages.

In the early development portion of Figure 12.3, tiles exist for students to access the various components. For example, student prompts and questions are embedded within some of the workspace tiles. This is not ideal for an inquiry-oriented classroom, as students need the authority and autonomy to generate their own work. Additionally, these separate tiles require students to navigate and manage multiple tiles on a screen, re-sizing and tracking as they move forward with the problem. From a task design perspective, there are limited features for presenting the problem, such as formatting text, providing labels for problem headers (*Initial Challenge*, *What If...?*, *Now What Do You Know?*). Due to a connection between the problem and the student notebook, students' notebook tiles are linked to each other so that students can access evidence of student thinking from each other.

Figure 12.2

Example of the CMP STEM Problem Format

Problem 2.2: Hats Off to the Wumps: Changing a Figure's Size and Location

Initial Challenge

Zack experiments with multiplying Mug's coordinates by different whole numbers to make other characters. Marta asks her uncle how multiplying the coordinates by a decimal or adding numbers to or subtracting numbers from each coordinate will affect Mug's shape. He gives her a sketch for a new shape (a hat for Mug) and some rules to investigate.

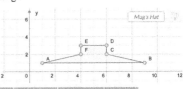

Point	Mug's Hat	Hat 1	Hat 2	Hat 3	Hat 4	Hat 5
	(x, y)	(x + 2, y + 3)	(x − 1, y + 4)	(x + 2, 3y)	(0.5x, 0.5y)	(2x, 3y)

- Look at each rule and predict what will happen to the hat with each rule.
- Test each rule. How does your result compare with your prediction?

What If…?

Situation A. Writing New Hat Rules

Several members of the computer club wrote different design criteria that would produce hats similar to the original Mug hat. What rule would you write for each design?

Syrah's Design

The side lengths are one third as long as Mug's hat.

Joe's Design

The side lengths are 1.5 times as long as Mug's hat.

Viola's Design

The hat is the same size as Mug's hat but has moved right 1 unit and up 5 units.

Kathy's Design

The image is in another quadrant of the graph.

Situation B. Raymond's Claim about Negative Numbers and Rules

Raymond's Claim

If you multiply each coordinate by a negative number, the image is similar but smaller.

Is he correct? Explain.

Situation C. Isaiah's Challenge: Putting the Hat on Mug

Isaiah's Challenge

I think it is possible to write a rule that will put Mug's hat on Mug. My group will work on finding the correct rule.

Is this possible? Why?

Now What Do You Know?

If the coordinate rule creates a similar figure, how can you use the rule to predict the side lengths of the image? The location of the image on the coordinate grid?

Source: Connected Mathematics Project (2023). (Reprinted with permission.)

Figure 12.3

Early, Mid, and Current Development of the Problem Presentation

Early Development

(Figure continued on next page)

Figure 12.3 (Continued)

Early, Mid, and Current Development of the Problem Presentation

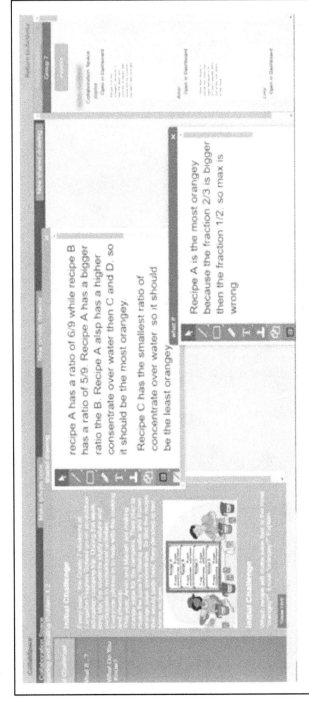

Mid Development

(Figure continued on next page)

Figure 12.3 (Continued)

Early, Mid, and Current Development of the Problem Presentation

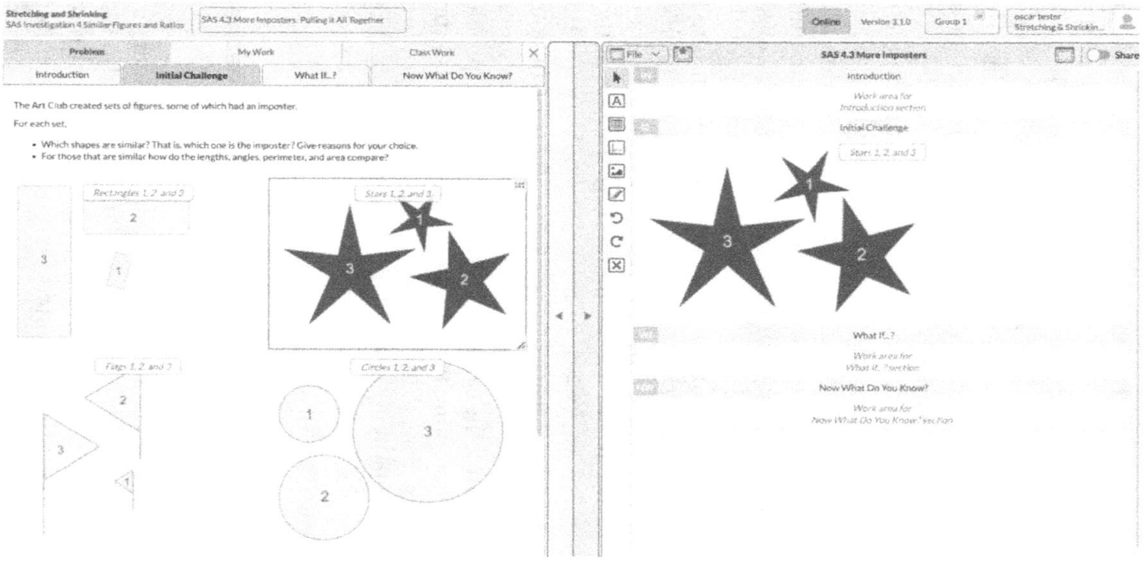

Current Development

Source: Connected Mathematics Project (2023). (Reprinted with permission.)

The middle portion of Figure 12.3 shows the mid development of the digital platform. In contrast to the early development, a separate tile contains the mathematics problem in its entirety. The formatting is more advanced than the early development; for example, photos can be displayed. This version allows students to access and navigate the entire problem. Students move tiles containing their work in the remaining space of the platform. However, students are not able to move content from the problem into their workspaces, and students' work is disconnected for each part of the problem. This is an important distinction because the three components are intentionally connected. As evidence of student thinking develops and grows over time, access to easily retrieve, copy, and modify is essential, as is accessing evidence of student thinking for the entire problem.

Figure 12.3 also shows the current version of the digital platform (at the time of publication of this chapter). In this version, students have a side-by-side view of the problem (shown on the left) and their student digital notebook (shown on the right). Students have access to important interactive features, such as the ability to click-and-drag aspects of the problem (e.g., text, representations, and photos) and move them into their notebook; re-size their screen to show only the mathematics problem or their notebook, resulting in a larger viewing screen for doing mathematics; and navigate to different parts of the problem using the tabs at the top, while maintaining the same workspace in their notebook. Despite connections being made between the mathematics problem and student notebooks, our development process is not over. A forthcoming change includes streamlining the problem so students can scroll through the entire problem at once, rather than viewing components separately. The team is also currently scaling up the number of units to include the entire seventh-grade curriculum. Changes in how problems are presented in the platform are likely to occur as mathematics topics, problem activities, and mathematics tools are written for the print materials.

Connecting the Mathematics Problem and the Teacher Dashboard and Workspace

Another lesson learned relates to connecting the mathematics problems to the teacher dashboard and workspace. In paper-and-pencil environments, a mathematics problem and a teacher's physical resources in a classroom are separate. In the United States, most curriculum resources for teachers are for planning, with suggestions for how teachers can enact the mathematics problems. The curriculum materials provide teaching resources, aids, and lab sheets that can be printed and used during class activities. Since the beginning stages of developing a digital platform, the project team linked the teacher dashboard to the student digital collaborative platform because the CMP curriculum materials reflect the understanding that teaching and learning are inextricably linked together (Edson et al., 2019). The team progressively modified the design of mathematics problems in the platform in ways that would support teachers in their planning, teaching, and reflecting on student thinking, including features so that teachers within a school district can collaboratively work together around planning, teaching, and reflecting on student work (Edson & Phillips, 2021).

As shown in Figure 12.4, teachers access problem solutions, strategies, and examples of student thinking as they plan their enactment of the mathematics problem. In the print curriculum, oftentimes solutions are minimal, contain limited elaborations, and are in a separate booklet or found in the back of the teacher guide materials. This is primarily due to publishers' page limit constraints. By contrast, in the digital platform, teachers can toggle solutions that are embedded within each component of a problem. Written as an informal discussion with teachers, these solutions provide explicit answers to mathematics problems that attend to the various strategies and understandings that are possible in CMP classrooms. And because the solutions are digital, teachers control the amount of content shown on their screen.

The team also developed a digital teacher space so teachers could access–in real-time–students' work for the entire class (see Figure 12.5). It was important that this teacher space be accessible on a computer or tablet so teachers could continue to circulate around the classroom. During class, it is often more difficult for teachers to access evidence of student thinking that is displayed on laptop screens, as compared with their paper-and-pencil work, because the screen may be angled or small, teachers need to be physically close to the screen, and the lighting in the room may limit visibility. Accessing evidence of student thinking was especially difficult during the COVID-19 pandemic when students and teachers were not in face-to-face settings. In the digital platform, teachers can observe evidence of student thinking for an entire group or zoom in to view one student's work. One of the challenges teachers often report to the design team is deciding when to initiate a whole-class summary discussion. To help teachers have a sense of student progress on mathematics problems, the team provides analytics on how many students in the class have work initiated for each component of a problem. For example, in Figure 12.5, the three icons in the vertical teacher space on the right under the word Progress show that 21 of 26 students have some work-in-progress for the *Initial Challenge*, 11 of 26 students for the *What If...?*, and 5 of 26 students for the *Now What Do You Know?* Another challenge that teachers experience in classrooms is identifying evidence of student thinking to highlight in a class discussion. By clicking on the *Initial Challenge*, *What If...?*, and *Now What Do You Know?*

icons on the right under *Progress*, the work of every student is automatically scrolled so the teacher can view all evidence of student thinking for that problem component, rather than manually scrolling through each individual student's workspace. Because all work shown in the teacher dashboard is synchronous and updates in-the-moment, the work displayed is current work in progress by students. Teachers can also navigate to the published work and see the version of work posted to the entire class.

As shown in Figure 12.6, the team has developed collaborative features of sharing and commenting so teachers can engage in planning, teaching, and reflection with their colleagues, synchronously or asynchronously. Because many teachers either plan their lessons by viewing the problems that their students see or using the suggestions in a teacher's guide, it is important to support teachers so they can discuss the curriculum, planning, teaching, and reflection documents, including the problems, teacher guide suggestions, embedded solutions, and student work on problems. These collaborative features of commenting and circulating documents allow teachers and coaches to have daily and ongoing conversations, support, and resources around planning, teaching, and reflection.

LESSON TWO: SUPPORTING A MODEL OF COLLABORATION IN THE DIGITAL PLATFORM

The second lesson relates to connecting individual and collaborative student notebooks in the digital platform. In paper-and-pencil classrooms, students use physical notebooks or paper to document their mathematical thinking. This may be challenging in small groups as students' work may be upside down to their groupmates. Students may also be working on different strategies, representations, or problem components. Some teachers create worksheets. However, in a digital environment, students create their own notebooks for exploring and solving problems. In the team's past experiences in mathematics classrooms, providing every student with laptops often shuts down discussions and stifles cooperation and collaboration. In thinking about ways to enhance enactment of CMP, the team positions collaboration through how it is operationalized in CMP classrooms (Edson & Phillips, 2021; Lappan et al., 2004). The Launch-Explore-Summarize instructional model frames how to create collaborative and discourse-based workspaces so that students can explore mathematics problems together while using individual laptops. To enhance the teaching and learning of mathematics, the project team considered the participant structure of students in small- and whole-groups. The team recognized that collaboration often has different interpretations depending on the specific educational setting. For example, collaboration can involve multiple people working

Figure 12.4

Teacher Feature: Embedded Problem Solutions, Strategies, and Student Thinking

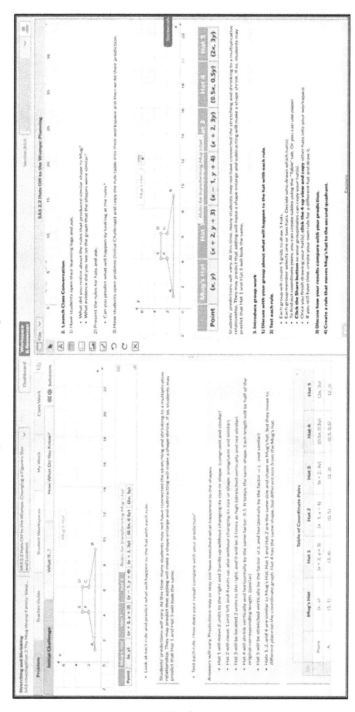

Source: Connected Mathematics Project (2023). (Reprinted with permission.)

Figure 12.5

Teacher Feature: Teacher Dashboard and Student Progress on Mathematics Problems

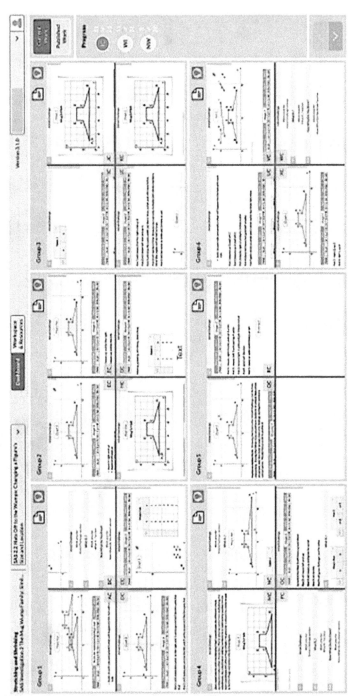

Source: Connected Mathematics Project (2023). (Reprinted with permission.)

Figure 12.6

Teacher Feature: Teacher Comments to Network Teachers and Coaches

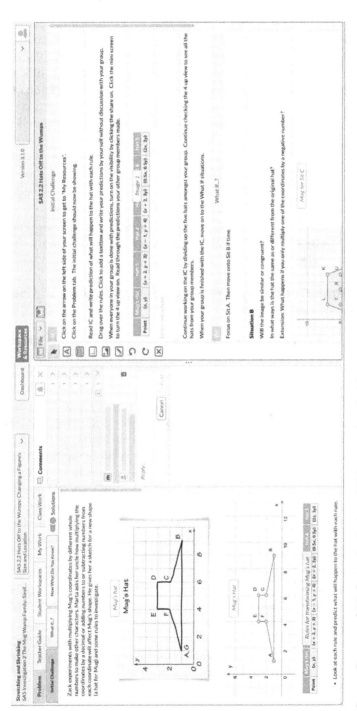

Source: Connected Mathematics Project (2023). (Reprinted with permission.)

together toward a shared learning goal, or it can be more cooperative with people working separately on a portion of a group product that is later combined into a single product (Jeong & Hmelo-Silver, 2016). These different interpretations have substantial implications for how technology supports collaborative learning. Here are sample questions the project team considered as it took up the challenge of determining a digital platform that connects individual and collaborative student notebooks:

- Do students share control over one tool in a workspace, like a marker on poster paper?
- How do collaborative workspaces relate to individual workspaces? Should individual workspaces exist?
- Whose work belongs to whom in different collaborative interpretations? What if a representation from a collaborative model gets deleted?

Platform Design Feature: Individual and Collaborative Student Notebooks

The digital notebooks contain several individual and collaborative features for students. Individually, students can access the CMP curriculum, and can click-and-drag text, graphs, and tables into their individual workspace (see Figure 12.7). Students can mark up or create new artifacts using tools such as text, tables, graphs, images, drawings, and more. Students can upload photos of their paper-and-pencil work. Students can embed work that is shared or published by others into their workspace and modify it. Students can create entries that span across problems and investigations, such as different strategies, key terms, *Summarize* notes, discussions, and mathematical reflections. Work is automatically saved on the platform.

Collaboratively, when students log onto the platform, they assign themselves to groups of four or are assigned by the teacher. Students access a "four-up" view (see Figure 12.8) where they access—in real time—the work of their groupmates, easily incorporating the work of others into their own. Figure 12.8 shows evidence of student thinking for the mathematics problem shown in Figure 12.7. As shown in Figure 12.8, the upper left-hand workspace (quadrant) is the students' individual workspace, and the three workspaces shown in the other quadrants are the individual workspaces from their group teammates. Students can discuss strategies and assign parts of the problem to individuals and then combine their strategies into a complete solution that can be shared with the class. Students in their small groups publish work to other groups and use it during whole-group discussions.

Figure 12.7

Curriculum-Embedded Individual Student Workspace

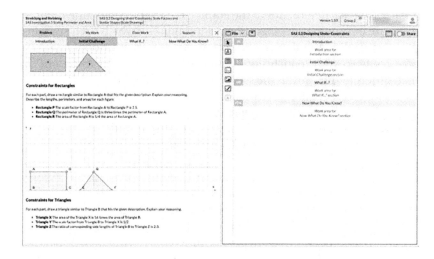

Source: Connected Mathematics Project (2023). (Reprinted with permission.)

Figure 12.8

Collaborative "Four-Up" View of Work

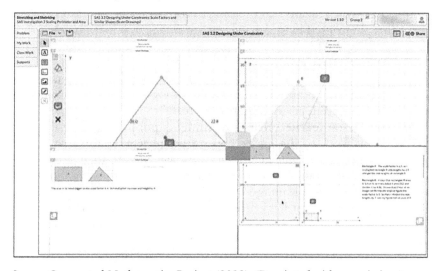

Source: Connected Mathematics Project (2023). (Reprinted with permission.)

Connecting Individual and Collaborative Student Notebooks

One lesson learned relates to the connectedness between individual and collaborative student notebooks. Early attempts—in which each student in a group controlled the same cursor on the screen, each student in a group shared control over the inclusion/exclusion of tiles, and each student controlled different tiles within the same workspace—did not work. For several reasons, we faced challenges in our attempts to develop collaborative student notebooks. For example, having multiple students controlling a single cursor mirrored the environment in which multiple students were sharing access to a single digital device, with only some students having access to the mathematics. Also, having multiple students with control created a group management concern, as multiple creations and entries appeared that easily distracted students from a particular solution strategy.

To address these concerns, the "four-up" view (Figure 12.8) was created to juxtapose a student's individual workspace with those of their other three group members. This means that students' individual workspaces are the same as their collaborative workspaces when they decide to collaborate; if they choose not to share their work, then other students cannot access their individual workspace. Typically, students ask questions within their group or provide time for others to work individually prior to collaborating. If all students in the group decide to share their work, they generally discuss their plan for moving forward before starting to share. Students also have the option for an individual workspace that is not connected to their groupmates, which means that their work is private from others, regardless of whether they opted in to sharing. An advantage of our approach is that students can co-opt work from their groupmates (who turn on sharing) into their individual space at any time without disrupting any work. As one student reports, "It feels good when a student uses my work to help them on their own problem because it feels like I'm doing something for someone else, even though I'm just doing my own work."

Another lesson learned relates to students publishing their individual or group work to the entire class. Student work published to the class has the same features of student work available to their group. For example, students can click-and-drag another student's work into their own workspaces. An advantage of this approach is that the teacher can also publish and co-opt work from students, providing support for the different ways *Launch* and *Summary* discussions are enacted in classrooms. In this way, teachers have features on the digital platform for structuring whole-group discussions, that is, the digital collaborative platform positions and leverages student work that can easily be used in whole-group discussions.

A forthcoming feature of the digital platform will allow students (and the teacher) to play back any work so that it can be viewed like a short video. This feature relies on sequences of the undo and redo features to provide a timeline of work. The unique aspect of this feature is that the timeline spans the entire workspace, not just individual tiles located within it. For example, by activating the playback feature, students and the teacher can see when the user moved from a table to a graph to their written response, which differs from separate playback features for each tile (graph, table, drawing, etc.). In essence, this provides a mechanism for students to unpack "finished" work done over time to highlight their underlying thinking processes. It also has the potential to support conversations on the strategies that are used or abandoned when exploring and solving problems. This is critical when students are confronted with new strategies and look for connections to the ways they currently think about a problem.

LESSON THREE: PROVIDING STUDENTS WITH JUST-IN-TIME SUPPORTS IN THE CMP DIGITAL PLATFORM

The third lesson focuses on connecting students with curriculum-provided and teacher-generated *just-in-time* supports (Novak et al., 1999), with the goal of moving a student's learning of mathematics forward in a classroom. In many curriculum materials, it is common for scaffolding to take the form of adaptive tutoring, with the goal of improving a student's speed, accuracy, and automaticity in performing routine mathematical procedures (Bakker et al., 2015). In these environments, learning mathematics becomes reduced to focusing on the performance of mathematical operations rapidly and correctly. In a problem-based curriculum, however, the goal is to develop a student's deep understanding and problem-solving ability. A balance is needed between providing an appropriate challenge for students and providing necessary supports, with the goal of students taking ownership of their learning (Edson, 2014, 2016). The CMP prompts to students that are embedded in the digital platform are called *just-in-time supports*.

Platform Design Feature: Just-in-Time Supports in the Digital Collaborative Platform

Drawing on the affordances of scaffolding that are static and given beforehand (e.g., Miyazaki et al., 2015; Schukajlow et al., 2015; Tropper et al., 2015) and scaffolding that is dynamic and used on-the-fly (e.g., Abdu et al., 2015), the team has developed two options for how specific supports are revealed as needed to students. One option is *curriculum-provided just-in-time supports*, which students select individually as they explore the problems. The other option is *teacher-generated just-in-time supports*, where teachers generate and send students prompts found in the student digital collaborative platform.

The curriculum-provided just-in-time supports are options that students can select on the digital platform. As shown in Figure 12.9, students can click on different shapes located at the bottom of their workspace. In Problem 2.2 (Hats Off to the Wumps, shown in Figure 12.2), examples of curriculum-provided supports include: (1) *Initial Challenge*—To predict what will happen to the hat with rules, recall the relationship between the rules to draw Zug, Lug, Bug, and Glug and its similarity in Problem 2.1[1]; (2) *Initial Challenge*–In Problem 2.1, how is a rule for adding or subtracting a number to the *x*- or *y*- coordinate different from a rule with multiplying a number?; and (3) *What If...?*—In *Initial Challenge* (of Problem 2.2), which rule is related to multiplying a number to the coordinates? What did the rule change in the hat? These prompts are only possible with the CMP STEM Problem format and the underlying learning progressions of the curriculum materials (Edson et al., 2019).

Figure 12.9

Curriculum-Provided Just-in-Time Supports

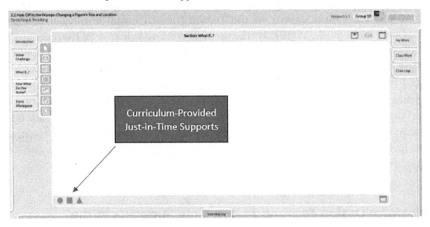

Source: Connected Mathematics Project (2023). (Reprinted with permission.)

As elaborated in Edson and Phillips (2021), the teacher-generated just-in-time supports are mechanisms where teachers can create prompts, questions, or comments, and send them to individuals, groups, or entire classes of students. For example, Figure 12.10 shows an example from the digital platform, showing how a teacher might create a message that can be sent to a group of students. The icons in the digital platform are in the dashboard so that teachers can monitor evidence of student thinking. The icons can also be found in the teacher workspace area, where they can access documents produced and published by students.

Figure 12.10

Teacher-Generated Just-in-Time Supports

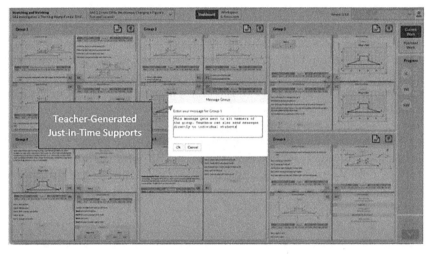

Source: Connected Mathematics Project (2023). (Reprinted with permission.)

Lessons Learned on the Curriculum-Provided Just-in-Time Supports

The curriculum-provided supports are no longer available on the digital collaborative platform. The prompts were offered as additional student learning opportunities when the mathematics problems were redesigned. The curriculum-provided supports were removed for three reasons. First, the redesign of the mathematics problems helped support students in their mathematics thinking. Second, curriculum developers cannot predict the specific supports needed to move students forward in solving a problem. Third, in a digital environment, students have access to a variety of other resources that can be leveraged as on-demand supports, including real-time access to the work of their groupmates, real-time access to published student work from other groups, mathematical representations moved from the curriculum into their notebooks to serve as a starting point for their thinking, and teacher-generated supports that can be released to students. As more features to support students are designed and developed for digital platforms, it is inevitable that earlier features may become superfluous over time. This lesson learned is consistent with a similar kind of support mechanism related to learner-controlled scaffolding at the high-school level that was not needed (Edson, 2014, 2016).

Lessons Learned on the Teacher-Generated Just-in-Time Supports

The teacher-generated just-in-time supports have three characteristics that are important in CMP classrooms: (a) teachers can compose them (like a document) using all the available tools (e.g., planning document, teaching document, student work) to customize a mathematics problem to meet an individual student's needs; (b) teachers can send supports to the entire class, to an entire group, or to an individual student; and (c) teachers can use them before, during, and after class. In CMP classrooms, we learned three different ways teachers use just-in-time supports.

First, based on a student's needs, teachers use just-in-time supports to create and customize in-the-moment interactions between themselves and their students using the robust set of tools. As one teacher explains:

> In the classroom, I am looking for kids who aren't doing anything, but I'm also looking for work that I want to highlight and talk to the whole group about. I'm not going to stop and have a "you should be doing" conversation with every other kid. I don't like doing that. But so it would be a great opportunity to be able to just shoot them a sticky note that says, "you know, hey, I noticed you didn't do much here. Do you have questions? Do you want to meet with me at lunch?"

Although teachers report that using the just-in-time supports during class is challenging if they do not have a tablet in the classroom, some have found this feature useful.

Second, teachers predominately create and send just-in-time supports when they monitor evidence of student thinking after class. In paper-and-pencil environments, it is not feasible for teachers to look at evidence of student thinking after class because students' notebooks leave the classroom with the students. In a digital environment, however, teachers can log into a digital platform to access student work, post comments on their solution strategies, and offer questions for students to reflect on their thinking.

Third, teachers can customize a mathematics problem delivered digitally to meet students' needs. Examples include questions for students to consider during the prompt, or additional/modified prompts and questions for students to assess their learning while they explore and solve problems. In practice, teachers prepare these activities before class and use them during lesson launches or formatively to assess student thinking at the end of class.

CONCLUSION

In this chapter, I have reported on lessons learned when developing a digital collaborative platform for students and teachers who use the Connected Mathematics Project curriculum (Phillips et al., 2025). The research team is unaware of other mathematics curricula that purposefully leverage digital technologies for high-level collaboration. The philosophy and design of the CMP materials prioritizes collaboration, which necessitated a careful examination of all possible resources that could be incorporated into the platform. The design principles of the digital collaborative platform focused on (1) reimagining mathematics problems, (2) supporting a model of collaboration, and (3) providing students with just-in-time supports. These design decisions led to changes in enacted experiences within CMP classrooms, resulting in enhanced teaching and learning of mathematics. Individual, collaborative, and classroom documents and artifacts created by students and their teachers are elevated in this digital platform. In this way, the resulting digital collaborative environment provides students and teachers with a more egalitarian environment than paper-and-pencil classrooms where conventional print materials drive enactment decisions.

ACKNOWLEDGMENTS

This work was supported by the National Science Foundation grants DRL-1620934, DRL-1620874, DRL-1660926, and DRL-2007842. Any opinions, findings, and conclusions or recommendations expressed in this material are those of the authors and do not necessarily reflect the views of the National Science Foundation.

REFERENCES

Abdu, R., Schwarz, B., & Mavrikis, M. (2015). Whole-class scaffolding for learning to solve mathematics problems together in a computer-supported environment. *ZDM Mathematics Education*, *47*(7), 1163–1178. https://doi.org/10.1007/s11858-015-0719-y

Bakker, A., Smit, J., & Wegerif, R. (2015). Scaffolding and dialogic teaching in mathematics education: Introduction and review. *ZDM Mathematics Education*, *47*(7), 1047–1065. https://doi.org/10.1007/s11858-015-0738-8

Barab, S. (2014). Design-based research: A methodological toolkit for engineering change. In R. K. Sawyer (Ed.), *The Cambridge handbook of the learning sciences* (2nd ed., pp. 151–170). Cambridge University Press.

Choppin, J., Roth McDuffie, A., Drake, C., & Davis, J. (2015). Curriculum metaphors in U.S. middle school mathematics. In T. G. Bartell, K. Bieda, R. Putnam, K. Bradfield, & H. Dominguez (Eds.), *Proceedings of the 37th annual meeting of the North American Chapter of the International Group for the Psychology of Mathematics Education* (pp. 65–72). Michigan State University.

Connected Mathematics Project. (2023). *The Connected Mathematics Project 4 Field-Test Materials*. Michigan State University.

The Design-Based Research Collective. (2003). Design-Based research: An emerging paradigm for educational inquiry. *Educational Researcher*, *32*(1), 5–8. https://doi.org/10.3102/0013189X032001005

Edelson, D. C. (2002). Design research: What we learn when we engage in design. *Journal of the Learning Sciences*, *11*(1), 105–121. https://doi.org/10.1207/S15327809JLS1101_4

Edson, A. J. (2014). *A deeply digital instructional unit on binomial distributions and statistical inference: A design experiment* [Unpublished Doctoral dissertation, Western Michigan University].

Edson, A. J. (2016). A design experiment of a deeply digital instructional unit and its impact in high school classrooms. In M. Bates & Z. Usiskin (Eds.), *Digital curricula in school mathematics* (pp. 177–193). Information Age Publishing.

Edson, A. J. (2017). Learner-controlled scaffolding linked to open-ended problems in a digital learning environment. *ZDM Mathematics Education*, *49*(5), 735–753. https://doi.org/10.1007/s11858-017-0873-5

Edson, A. J., & Phillips, E. D. (2021). Connecting a teacher dashboard to a student digital collaborative environment: Supporting teacher enactment of problem-based mathematics curriculum. *ZDM Mathematics Education*, *53*(6), 1285–1298. https://doi.org/10.1007/s11858-021-01310-w

Edson, A. J., Phillips, E., Slanger-Grant, Y., & Stewart, J. (2019). The Arc of Learning framework: An ergonomic resource for design and enactment of problem-based curriculum. *International Journal of Educational Research*, *93*(1), 118–135. https://doi.org/10.1016/j.ijer.2018.09.020

Jeong, H., & Hmelo-Silver, C. E. (2016). Seven affordances of computer-support collaborative learning: How to support collaborative learning? How can technologies help? *Educational Psychologist*, *51*(2), 247–265. https://doi.org/10.1080/00461520.2016.1158654

Lappan, G., Fey, J. T., Fitzgerald, W. M., Friel, S. N., & Phillips, E. D. (2004). *Getting to know Connected Mathematics: An implementation guide*. Pearson Prentice Hall.

Lappan, G., Phillips, E. D., Fey, J. T., & Friel, S. N. (2014). *Connected Mathematics 3* (Student Edition and Teacher Guide). Pearson.

Miyazaki, M., Fujita, T., & Jones, K. (2015). Flow-chart proofs with open problems as scaffolds for learning about geometrical proof. *ZDM Mathematics Education*, *47*(7), 1211–1224. https://doi.org/10.1007/s11858-015-0712-5

Novak, G. M., Patterson, E. T., Gavrin, A. D., & Christian, W. (1999). *Just-in-time teaching: Blending active learning with web technology*. Prentice Hall.

Phillips, E. D., Lappan, G., Fey, J. T., Friel, S. N., Slanger-Grant, Y., & Edson, A. J. (2025). *Connected Mathematics 4* (Student and Teacher Editions).

Richey, R. C., Klein, J., & Nelson, W. (2004). Developmental research: Studies of instructional design and development. In D. Jonassen (Ed.), *Handbook of research for educational communications and technology* (pp. 1099–1130). Lawrence Erlbaum.

Rohrer, D., Dedrick, R. F., Hartwig, M. K., & Cheung, C.-N. (2020). A randomized controlled trial of interleaved mathematics practice. *Journal of Educational Psychology*, *112*(1), 40–52. https://psycnet.apa.org/doi/10.1037/edu0000367

Roth McDuffie, A., Choppin, J., Drake, C., & Davis, J. (2018). Middle school mathematics teachers' noticing of components in mathematics curriculum materials. *International Journal of Educational Research*, *92*, 173–187. https://doi.org/10.1016/j.ijer.2018.09.019

Sandoval, W. (2014). Conjecture mapping: An approach to systematic educational design research. *The Journal of the Learning Sciences*, *23*(1), 18–36. https://doi.org/10.1080/10508406.2013.778204

Schukajlow, S., Kolter, J., & Blum, W. (2015). Scaffolding mathematical modelling with a solution plan. *ZDM Mathematics Education*, *47*(7), 1241–1254. https://doi.org/10.1007/s11858-015-0707-2

Tropper, N., Leiss, D., & Hänze, M. (2015). Teachers' temporary support and worked-out examples as elements of scaffolding in mathematical modeling. *ZDM Mathematics Education*, *47*(7), 1225–1240. https://doi.org/10.1007/s11858-015-0718-z

Umriani, F., Suparman, Hairun, Y., & Sari, D. P. (2020). Analysis and design of mathematics student worksheets based on PBL learning models to improve creative thinking. *International Journal of Advanced Science and Technology*, *29*(7), 226–237. http://sersc.org/journals/index.php/IJAST/article/view/9431.

Wirkala, C., & Kuhn, D. (2011). Problem-based learning in K–12 education: Is it effective and how does it achieve its effects? *American Educational Research Journal*, *48*(5), 1157–1186. https://www.jstor.org/stable/41306381

ENDNOTE

1. In Problem 2.1, students explore what rules will make similar shapes by considering coordinate rules. They draw a character Mug Wump and use the given rules to find other Wump characters: Zug ($2x$, $2y$), Lug ($3x$, y), Bug ($3x$, $3y$), and Glug (x, $3y$). Students compare rules and figure attributes and determine which family members are similar to Mug and which figures appear not to be similar and thus "imposters."

CHAPTER 13

STUDY OF TECHNOLOGY-BASED INQUIRY LEARNING

Automatic Analysis of Online Personal Feedback Processes

Raz Harel
University of Haifa and David Yellin College of Education

Michal Yerushalmy
University of Haifa, Israel

Studying inquiry-based learning often requires observations of complex learning situations in classroom, group, or individual settings. Analyzing the collected data can be a long process that is often restricted to small-scale research efforts. In this chapter, we describe a method of using automatically generated descriptive information to support the study of complex learning situations. We examined how students' participation in an elaborated feedback process, generated based on immediate online information, influences their inquiry processes. By redesigning textbook tasks into example-eliciting ones based on interactive diagrams, we developed an approach that supports automated analysis of students' personal example spaces. Through interactive exploration, student diagrams can be analyzed to identify their mathematical characteristics. Using the Seeing the Entire Picture (STEP) platform (Olsher et al., 2016) for the present study, we describe our method of analysis and application of the platform in two studies.

Keywords: elaborated feedback; example-eliciting task; student reflection; technology support for learning

INTRODUCTION

Using technology to provide feedback to support learning has been a challenge for educators since the 1970s. For example, Erlwanger (1973) demonstrated the complexities involved in supporting students' learning to use Individually Prescribed Instruction (IPI) in the mathematics curriculum. Currently, curriculum studies often use available systems to analyze students' work and produce a report for teachers or students. For example, Rezat (2021) studied digital mathematics textbooks that provide verification feedback and try-again information. ASSISTment (Kelly et al., 2013) is another widely used platform with the flexibility to study the use of feedback in autonomous learning situations quantitatively, including timely feedback on correctness and hints concerning mistakes (Roschelle et al., 2016). Systems of this type mainly collect answers to closed questions, multiple-choice questions, or numeric input, but the questions often do not lead to inquiry learning.

In the present chapter, we report on lessons learned from studies on feedback processes in which students interacted with technology tools for inquiry learning. We used the Seeing the Entire Picture (STEP) platform (Olsher et al., 2016), which generates descriptive textual information about students' work on an interactive example-eliciting task. Students receive a personalized elaborate report as they engage with the interactive diagram during their exploration. STEP can analyze the submitted work and automatically mark the identified characteristics of the submitted example to provide an elaborate report.

To enable the automatic analysis of students' inquiry processes using tools like STEP, it is necessary to design tasks that allow automatic analysis and to define the pedagogical-mathematical goals that can be assessed automatically. In this chapter, we share our experience and insights gained from implementing these principles with students in Israel.

THEORETICAL BACKGROUND

In this section, we first describe the types of tasks that lend themselves to inquiry processes via technology. Second, we describe the nature of students' personal feedback.

Assessing Students' Inquiry Processes Using Example-Eliciting Tasks

We explored inquiry processes in which students solve example-eliciting mathematical tasks in a digital environment. In *example-eliciting tasks*, students construct several examples for a given mathematical situation (Nagari-Haddif, 2017; Yerushalmy, 2020). Such tasks have certain advantages for the inquiry curriculum and automatic analysis. Constructing examples is essential for students to develop mathematical ideas (Zaslavsky & Zodik, 2014). By constructing examples using dynamic tools designed for exploration, students can expand their personal example space and conceptual understanding. Constructing examples of mathematical objects with specified properties without using known algorithms or procedures demands high-level exploration skills (Sangwin, 2013).

Recognition of the importance of example-eliciting tasks has led to studies in which researchers examined ways of designing such tasks (Nagari-Haddif, 2017; Sinclair et al., 2011). These studies have shown that creating and verifying examples can lead to the expansion of students' *personal example space*, which is the collection of examples that students generate in response to specific tasks within certain contexts (Watson & Mason, 2005). The personal example space can serve as a tool for explaining students' mathematical reasoning (Ayalon et al., 2017; Zaslavsky & Zodik, 2007). It can also enable studying students' concept images and concept definitions, as repeatedly demonstrated by Hershkowitz (1990). We offer our perspective on the use of these qualities of example spaces as part of curriculum research.

Personal Feedback

Personal feedback provided automatically to students, which allows them to reflect on their work, can serve as a fundamental tool to define pedagogical and mathematical goals amenable to automatic assessment. Traditionally, feedback has been treated as an object or information transferred from an agent, such as a teacher or digital environment, to the learner. Personal feedback plays two important roles: verification and elaboration. *Verification feedback* explicitly indicates the correctness of the student's answers. *Elaborated feedback* may explain why a response is correct or incorrect (Harel et al., 2022). Although verification feedback is the more common type, elaborated feedback is considered more effective to scaffold higher-order learning outcomes in mathematics (Van der Kleij et al., 2015).

From the variety of elaborated feedback, we studied the use of reported information that can function as a non-judgmental individual report to the student (Artigue & Trouche, 2021). This information goes further than an indication of correctness and addresses concrete features of the student's work in relation to the task (Butler, 1987; Corbett & Anderson, 2001). We classified such feedback as *elaborated attribute isolation*, which is one of six types of online elaborated feedback described by Shute (2008). We treated attribute isolation as a type of "elaborated feedback that presents information addressing central attributes of the target concept or skill being studied" (p. 160). As curriculum researchers, we emphasized the significance of providing students with a personal report that fosters a meaningful "inner" dialogue with themselves. Engaging in this dialogue helps students to reflect on their understanding and progress, providing valuable insights into their learning journey (Harel et al., 2022). That is, we adopted Carless's (2015) perspective that challenges the traditional feedback approach: "feedback is a dialogic process in which learners make sense of information from varied sources and use it to enhance the quality of their work or learning strategies" (p. 192). In the current chapter, we use the term *feedback* to describe the dialogic processes, and *report* to identify the information that technology can provide for use in such processes.

PREPARING THE FOUNDATIONS FOR AUTOMATIC ANALYSIS OF STUDENTS' EXPLORATION

In this section, we illustrate and explain two essential components required to support curricular studies using automatic analysis. The first describes principles that support the transformation of a textbook task into an example-eliciting task. The second focuses on organizing the desired pedagogical mathematical goals as a list of characteristics for assessment.

Redesigning a Task into an Example-Eliciting Task

The principal design feature of the STEP platform is that it analyses students' answers to example-eliciting tasks. Therefore, to conduct a meaningful assessment that can yield informative results, it is important to use tasks different from those commonly found in the school mathematics curriculum. For example, the following open motion task is an assessment item from the Israeli national final examination:

> Noga rode her bike from 8:00 A.M. to 12:00 P.M. and traveled 20 km. Draw a graph representing Noga's ride.

An answer that would typically earn the maximum score on a test would be a correctly drawn graph on a Cartesian system in which the student has successfully addressed the task requirements of distance and speed. Although a student may receive the maximum score for such an answer, the response may not necessarily provide educators with a clear picture of the student's understanding of the main goal of the task. For example, an answer that meets the full requirement may not provide valuable information about the student's concept image of *distance traveled* and whether the student conceptualized the total distance traveled during the ride as being identical to reading the end position of the ride (i.e., distance from the starting point). So, there is a need for a more thorough examination of their interactions with the mathematical objects. To achieve this objective, analysis of students' personal example spaces, rather than of a single example, should elicit a range of characteristics in their responses.

To transform the previous original motion task into an example-eliciting task, we modified the instructions by asking students to experiment with a given applet and construct different scenarios that meet the task and can be saved in the student's personal area on the STEP platform. Students submit three graphs instead of one. We also added a requirement that the graphs should be as different as possible. Asking students to submit three different supportive examples may expand their personal example space, challenge their preconceptions, and allow a more thorough analysis of the students' example space (Yerushalmy et al., 2017). This design principle encourages students to use different approaches and move beyond their comfort zone. Figure 13.1 displays a screenshot of the task interface.

Figure 13.1

Task Interface for Tasks in Study 1

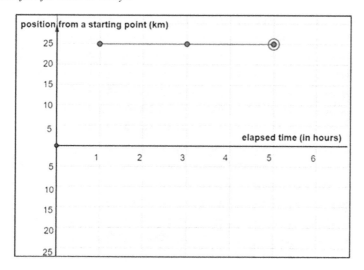

Characterizing the Pedagogical Mathematical Goals to Be Assessed Automatically

In this section, we present the additional component we used to analyze the students' example space, to demonstrate the potential of automatic analysis of student submissions for such example-eliciting tasks. Building upon Hershkowitz's (1990) studies on critical and non-critical characteristics, it is possible to assemble a list of characteristics that reflect the mathematical-pedagogical goals of the task, consisting of its critical and non-critical attributes.

We chose to design an elaborated personal report providing useful descriptive information to students. In the present study, we focused on the research use of automatically identified characteristics in the individual feedback reports provided to the students. Specifically, we examined the use of two key parameters: meeting task requirements; and the richness and variety of student examples.

To analyze the Noga riding task, we defined characteristics that were critical for the submissions to meet the task requirements. The *critical characteristics* included overall time, distance traveled, starting point, and endpoint. To analyze the variety of student examples, we considered additional characteristics that may or may not be present in each student's response to the task. These are the *non-critical characteristics* that include changing direction, incorporating stops during the ride, or constructing graphs with Point B that is below the horizontal axis. These non-critical characteristics of response support assessing insights into whether the example reflects a non-trivial ride. If none of the non-critical characteristics were present in any of the student's examples, we interpreted it as an indication of a narrow example space. The ability to automatically identify the critical and non-critical characteristics in student submissions enabled us to quantitatively analyze changes in the student's example space by designing a series of example-eliciting tasks. Below we show how we used these techniques in our studies.

AUTOMATED ANALYSIS OF STUDENTS' PERSONAL EXAMPLE SPACES

We conducted two studies using an activity similar to the one described in Yerushalmy et al. (2022). The first focuses on how information about critical and non-critical characteristics of tasks influences subsequent performance. The second focuses on how receiving immediate feedback during task exploration affects performance.

Study 1: Analyzing Effects of Information Given to Students About Critical and Non-Critical Characteristics

This study analyzed the effects of various components of the personal elaborated feedback and determined which ones can be assessed automatically. We analyzed the individual effects of critical and non-critical characteristics.[1] There were two similar example-eliciting tasks; each required students to construct and submit three different position versus time graphs that met the given story conditions:

- Task 1: Students were asked to construct a "Noga ride" in two intervals, creating a position versus time graph using *two* segments.
- Task 2: Students were to construct a "Noga ride" using *four* segments.

Students were encouraged to explore various situations before deciding which constructions to include in their submission. They were also allowed to reconstruct and change their decisions before submitting. Following each task, the participants could interact with the elaborated report. Because the students had to submit three examples, the elaborated personal report provided by STEP contained three lists side-by-side as seen in Figure 13.2.

Figure 13.2

Example of Elaborated STEP Report for Task 1

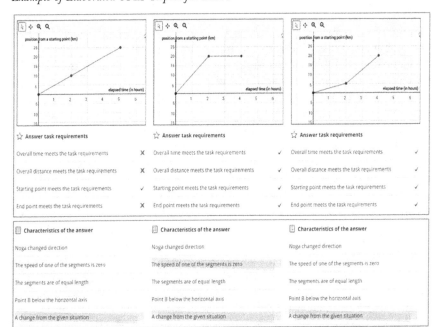

The critical characteristics are listed under the section titled *Answer task requirements* and the non-critical characteristics are listed under the section titled *Characteristics of the answer*. STEP marked the critical characteristics it identified as true with a check mark (✓) and those identified as not true with an × symbol. STEP highlighted in yellow (but shaded in this chapter) the non-critical characteristics it identified in the response. For example, in the student's submission shown in Figure 13.2, the student successfully addressed the task requirements in the examples appearing on the right and in the middle but did not meet three of the task requirements in the example on the left.

Nearly 100 students in grades 9-10 (ages 13 to 14) participated in the study, and were divided randomly into groups A and B. After submitting solutions to the first task, students in group A received an elaborated report that included only the critical characteristics; those in group B received an elaborated report that included only the non-critical characteristics. Students in both groups then completed the second task.

We used the STEP platform to collect students' submissions for both tasks and analyze the students' examples automatically. For each critical and non-critical characteristic, the STEP platform provided information on the number of students whose submissions included a given characteristic. The reports received by the students contained only critical or non-critical characteristics depending on their group membership, whereas the report received by the researchers included both types. Table 13.1 shows the percentage of submissions in which each critical characteristic appears in the first and second tasks for both groups of students.

According to Table 13.1, a lower percentage of submissions met the *distance* requirement in the second task than in the first one. We expected no improvement in meeting critical requirements by group B students, who received feedback only on non-critical characteristics. But the absence of improvement for group A students who did receive information on the critical characteristics came as a surprise. One explanation is that information to students about meeting task requirements (critical characteristics) is insufficient for elaboration toward understanding. Another explanation is that work on the second task, which required constructing a four-segment graph, was subject to a wider range of errors than constructing a two-segment graph.

Table 13.2 reports similar results to those of Table 13.1, but for the appearance of non-critical characteristics. The characteristic *Point B is below the horizontal axis* was not identified in submissions of the group A students and only negligibly in group B (5%). No significant improvement was found for the characteristic *Zero speed*. Both groups, however, improved on the *Change of Direction* characteristic between the first and the second task: Group A from 21% to 70% and Group B from 11% to 50%. To examine

Table 13.1

Percentage of Submissions in Which Each Critical Characteristic Appeared

Group	Characteristic	Percentage of submissions in which **critical** characteristics appeared	
		First Task	**Second Task**
Group A	Starting point	66	67
(n = 53)	Endpoint	58	70
	Distance	74	54
	Overall time	63	63
Group B	Starting point	76	82
(n = 44)	Endpoint	61	75
	Distance	89	65
	Overall time	64	72

Table 13.2

Percentage of Submissions in Which Non-Critical Characteristics Appeared

Group	Characteristic	Percentage of submissions in which the **non-critical** characteristics appeared	
		First Task	**Second Task**
Group A	Segments are of equal length	30	100
(n = 53)	Change from the given situation	100	100
	Point B is below the horizontal axis	0	0
	The speed of one of the segments is zero	8	19
	Noga changed direction	21	70
Group B	Segments are of equal length	27	98
(n = 44)	Change from the given situation	100	100
	Point B is below the horizontal axis	0	5
	The speed of one of the segments is zero	20	30
	Noga changed direction	11	50

whether the differences in the *Change of Direction* characteristic were significant, we conducted a Wilcoxon signed rank test on the means of the percentages of cases in which the characteristic appeared in the first and second tasks, revealing that the number of cases increased significantly between the first and the second task for Group A, $z = -5.099, p = .00$ and for Group B, $z = -3.441, p = .001$.

The findings suggest that when working on the first task before interacting with the elaborated report, most students assumed that Noga rode without stopping (i.e., no zero speed), did not change direction, and did not pass through the point where she started her ride (i.e., no point below the x-axis). Thus, students demonstrated a narrow example space consisting of examples that did not vary by non-critical characteristics. After receiving the elaborated report, students submitted more examples with the *Change of Direction* characteristic in which the total distance traveled and the distance from the start to the end position are not equal. Thus, although students still had difficulty creating examples that met the distance requirement, more students succeeded in expanding their personal example space.

The fact that both groups failed to show substantial improvement was contrary to our expectations and led to further investigation. We assumed that providing students with explicit information about the critical characteristics (Group A) would lead to an improvement in their ability to meet the task requirements. We also expected an enhancement in the variety of examples from students who received information that included the identification of non-critical characteristics (Group B).

To further analyze the results, we conducted several task-based interviews to understand students' expectations from the feedback. The first insight from these interviews, described by Harel et al. (2022) and Yerushalmy et al. (2022), is that students needed support and guidance to interpret the feedback effectively. Task-related interviews revealed that students faced difficulties using the elaborated report when they lacked advance knowledge about aligning their submissions with the task requirements.

The second insight concerns the words used to characterize the submitted examples. In the curriculum, students learned how to interpret motion stories from graphs, with a focus on understanding the values of points and their positions relative to the axes. In Study 1, our aim was to present the characteristics using the most comprehensible terminology based on students' existing knowledge; we were less attentive to maintaining consistency and uniformity of the terms of the characteristics. For instance, the characteristic *Point B is below the horizontal axis*, uses the terms of the Cartesian system to describe riding back through the starting point. Understanding and interpreting the elaborated information as being related to the ride probably carries an additional cognitive load. Therefore, we should have

Study of Technology-Based Inquiry Learning   283

formulated the text to align with the wording of the task, which we made a focus in Study 2.

Study 2: Redesigning the Activity and Elaborated Report So Critical Information Comes First

We realized that students were less likely to think about and elaborate on non-critical characteristics before receiving affirmation of their attempts to meet the task requirements. Hence, in Study 2, we designed the characteristics of the requirements to be interactively updated as textual information that appears within the environment of the dynamic diagrams. When the students worked on an example, whenever an element of the constructed graph met a requirement, the relevant characteristics of the report were highlighted. Figure 13.3 shows a snapshot of the results of a student's dragging during the graph construction of the first task.

Figure 13.3

Snapshot of Updated Textual Information

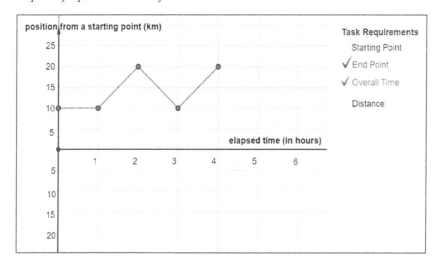

Designing the Elaborated Report: Articulating the Focus of the Characteristics

As in Study 1, speed, direction, and position were the focus of the work and the assessment. Figure 13.4 shows the revised list of characteristics included in the elaborated report, which are intended to articulate mathematical ideas and to refine the terms used to describe the phenomena that the students could interact with and which they could either reject or use to refine their submissions (Yerushalmy et al., 2022).

The first three characteristics are the critical ones. Unlike in Study 1, where each student received personal critical or non-critical information, the report in Study 2 included both critical and non-critical characteristics. Moreover, in Study 2, the characteristics were consistently formulated using the same words that appear in the text of the story and the task instruction. For example, the characteristic formulated in Study 1 as *Point B is below the horizontal axis* was formulated in Study 2 as *Noga passed through the city from which she left*.

Figure 13.4

List of Characteristics

Starting point meets the task requirements
Overall distance meets the task requirements
Overall time meets the task requirements

Non- critical Characteristics

Noga finished riding in the city from which she left
Noga passed through the city from which she left
Noga changed direction at least once
Noga stopped at least once
Noga rode at different speeds

Procedure

Twenty-nine students in grades 9–10 participated in Study 2 in a similar procedure to that of Study 1. We considered a student's submission to have met the task requirements if all three examples met the task requirements. To satisfy the characteristics *Noga finished riding in the city from which she left* and *Noga passed through the city from which she left*, students needed to construct an example in which there was a change in the direction of travel. Otherwise, the ride ended at point (4, 20). Therefore, we chose to determine *variety* based on the endpoint position. We considered students' submissions to be varied if not all three examples of graphs contained riding segments going in the same direction, and therefore not all three graphs ended at the same point (4, 20). Otherwise, we considered the submissions to lack variety.

This type of analysis is different from the one we conducted in Study 1. The analysis in Study 1 compared the percentage of appearances of each characteristic in the submissions of the first and second tasks. The analysis in Study 2 was based on connections between aspects of the task. To study which representation of the interactive information helped students engage with the elaborated feedback, we conducted an independent-samples t-test to compare meeting task requirements and the variety of students' examples before interaction with the feedback (first task) and after it (second task).

Findings

Ninety-seven percent of submissions for the first task and 96% for the second task met task requirements. The findings suggest that most students were able to cope with the task and meet its requirements in this design format. Table 13.3 presents the findings concerning the variety of the students' examples that met the task requirements.

Table 13.3

Variety of Submissions in Study 2

	Total submissions	Variety of the examples	
First task	$n = 28$	$n = 9$	32%
Second task	$n = 24$	$n = 10$	42%

According to an independent samples t-test, there was no significant difference between the variety in the first task ($M = .3$, $SD = .5$) and the second task ($M = .4$, $SD = .5$), $p > 0.05$. To gain more insight relative to the improvement in students' performance due to interaction with the elaborated information, we performed a separate calculation to determine the number of students whose variety improved between the first and second tasks (not reflected in Table 13.3). Of the 24 students who submitted the second task, 20 submitted examples lacking variety in their first submission, 5 of whom (25%) achieved variety in the second submission. Both studies demonstrated the feasibility of automatically analyzing students' example space using example-eliciting tasks and a list of critical and noncritical characteristics.

SUMMARY AND DISCUSSION

Our goal was to showcase the process of tracking and analyzing students' inquiry processes, specifically their work on example-eliciting tasks with the support of elaborated feedback. We investigated the automatic analysis of students' example space, focusing on two key parameters: meeting task requirements and the diversity of their examples. The elaborated feedback provided textual information to students about the critical and non-critical characteristics of the submitted tasks.

Based on the design of example-eliciting tasks and the mathematical conditions that characterized the answers, we were able to automate the analysis of student examples. STEP, the platform we used, played a decisive role in organizing and categorizing the information we sought to assess. Our findings demonstrate the value of platforms like STEP in analyzing and describing students' example spaces.

The studies produced two significant insights regarding designing such assessment studies. The first concerns the importance of maintaining consistent and uniform language in the feedback that aligns with the language used in the task. The second concerns students' preferences regarding the type of information they expect to see in the report. As a result of the findings of Study 1 and the task-related interviews, in which students expressed a desire for information about their task performance, in Study 2 we incorporated interactive updated information about critical characteristics available to students while working on the task. After submission, students had the opportunity to interact with the elaborated feedback, which contained both critical and non-critical characteristics.

Shute and Rahimi (2017) noted that complexity is one of the difficulties students face when engaging with elaborated feedback. We addressed the complexity of the elaborated report by presenting the critical characteristics first while working on the task. This approach allowed us to implement graded exposure of the fully elaborated report, containing both critical and non-critical characteristics after students had submitted their tasks. The interactive updated information served as a valuable tool to gradually introduce and clarify the comprehensive report, making sure that students were already acquainted with the critical characteristics. The findings of Study 2 provide evidence that the use of interactive updated information was beneficial in several ways. It helped students meet the task requirements effectively and encouraged them to expand the diversity of their examples. Additionally, it fostered student engagement with the elaborated feedback. The understanding achieved by students and the possibility of an automatic analysis of responses enable a broad assessment of comprehension during research and exploration.

The two studies involved different variations of the same task, but in the analyses we defined the *variety* parameter differently. In Study 1, we defined the submission of three examples as varied based on the number of non-critical characteristics identified in the three examples. In Study 2, we determined the parameter based on connections and interrelations between characteristics of the task. We sought to present methods of analysis that may be suitable for a wide range of tasks and feedback processes. In studies that use example-eliciting tasks where the characteristics of the task interrelate, a meta-analysis that considers the effects of interrelated characteristics could be of further interest. We do not favor a single method of analysis and show why different methods lead to different interpretations.

When the task characteristics are not explicitly defined, for example when using textbook problems for research purposes, it is possible to use another strategy, as demonstrated in Harel et al. (2022). This approach involves assembling a list of characteristics that include the following three categories: (a) basic characteristics that the designer found reasonable to expect in students' answers; (b) expected integrative characteristics, which are a combination of basic characteristics or initial directions for inquiry; and (c) characteristics that have the potential to challenge students' current perspectives. The basic characteristics are the equivalent of the critical task requirement characteristics, and the two other categories are the equivalent of the non-critical characteristics. This approach can be used to analyze and assess students' responses in various task scenarios.

We conclude by stressing the theoretical framework underlying the analyses. We used the theory of fundamental concepts (Hershkowitz, 1990; Tall & Vinner, 1981) to interpret the connection between the change in the example space of the students and their construction processes of relevant concepts. This theory and the method of analysis is supported by other theoretical approaches that value examples to be the important construct of inquiry learning. For example, Luz and Yerushalmy (2019) reported on an algorithmic analysis that supports the automatic online assessment of students' inquiry of geometry theorems using a dynamic geometry environment and analyzed the data based on variation theory (Marton et al., 2004). Harel et al. (2022) used the studies of Boero et al. (2007) to analyze students' conjecturing processes. The analysis methods we used in Studies 1 and 2 can also be implemented in activities that address a longer multiple-tasks unit and include other formats of example-eliciting tasks. We continue to develop and improve our methods of analysis, but we do not expect to obtain the same information from analysis of elaborated feedback data as we would from close observations of individuals. Our intention in devising ways to analyze larger samples was to present an additional perspective and possibly provide a deeper understanding of students' inquiry processes.

ACKNOWLEDGMENT

This research was supported by funding from the Israel Science Foundation (grant 147/18).

REFERENCES

Artigue, M., & Trouche, L. (2021). Revisiting the French didactic tradition through technological lenses. *Mathematics*, *9*(6), 629. https://doi.org/10.3390/math9060629

Ayalon, M., Watson, A., & Lerman, S. (2017). Students' conceptualizations of function revealed through definitions and examples. *Research in Mathematics Education*, *19*(1), 1–19. https://doi.org/10.1080/14794802.2016.1249397

Boero, P., Garuti, R., & Lemut, E. (2007). Approaching theorems in grade VIII: Some mental processes underlying producing and proving conjectures, and conditions suitable to enhance them. In P. Boero (Ed.), *Theorems in schools: From history, epistemology and cognition to classroom practice* (pp. 249–264). Sense.

Butler, R. (1987). Task-involving and ego-involving properties of evaluation: Effects of different feedback conditions on motivational perceptions, interest, and performance. *Journal of Educational Psychology*, *79*(4), 474–482.

Carless, D. (2015). *Excellence in university assessment: Learning from award-winning practice*. Routledge. https://doi.org/10.4324/9781315740621

Corbett, A. T., & Anderson, J. R. (2001). Locus of feedback control in computer-based tutoring: Impact on learning rate, achievement and attitudes. In J. Jacko & A. Sears (Eds.), *Proceedings of ACM CHI 2001 conference on human factors in computing systems* (pp. 245–252). Association for Computing Machinery Press.

Erlwanger, S. H. (1973). Benny's concept of rules and answers in IPI mathematics. *Journal of Children's Mathematical Behavior*, *1*, 7–26.

Harel, R., Olsher, S., & Yerushalmy, M. (2022). Personal elaborated feedback design in support of students' conjecturing processes. *Research in Mathematics Education*. https://doi.org/10.1080/14794802.2022.2137571

Hershkowitz, R. (1990). Psychological aspects of learning geometry. In P. Nesher & J. Kilpatrick (Eds.), *Mathematics and cognition* (pp. 70–95). Cambridge University Press.

Kelly, K., Heffernan, N., Heffernan, C., Goldman, S., Pellegrino, J., & Goldstein, D. S. (2013). Estimating the effect of web-based homework. In H. C. Lane, K. Yacef, J. Mostow, & P. Pavlik (Eds.), *International conference on artificial intelligence in education* (pp. 824–827). Springer.

Luz, Y., & Yerushalmy, M. (2019). Students' conceptions through the lens of a dynamic online geometry assessment platform. *Journal of Mathematical Behavior*, *54*, 100682. https://doi.org/10.1016/j.jmathb.2018.12.001

Marton, F., Runesson, U., & Tsui, A. B. M. (2004). The space of learning. In F. Marton & A. B. M. Tsui (Eds.), *Classroom discourse and the space of learning* (pp. 3–40). Lawrence Erlbaum Associates.

Maymon, L. (2020). *Two types of automated feedback to solutions of mathematics tasks by middle school students: A comparative analysis* [Unpublished Master's dissertation, University of Haifa].

Nagari-Haddif, G. (2017). Principles of redesigning an e-task based on a paper-and-pencil task: The case of parametric functions. In T. Dooley & G. Gueudet (Eds.), *Proceedings of the tenth conference of the European Society for Research in Mathematics Education* (pp. 3691–3698). Congress of the European Society for Research in Mathematics Education [CERME].

Olsher, S., Yerushalmy, M., & Chazan, D. (2016). How might the use of technology in formative assessment support changes in mathematics teaching? *For the Learning of Mathematics*, *36*(3), 11–18.

Rezat, S. (2021). How automated feedback from a digital mathematics textbook affects primary students' conceptual development: Two case studies. *ZDM Mathematics Education*, *53*, 1433–1445. https://doi.org/10.1007/s11858-021-01263-0

Roschelle, J., Feng, M., Murphy, R. F., & Mason, C. A. (2016). Online mathematics homework increases student achievement. *AERA Open*, *2*(4), 1–12.

Sangwin, C. J. (2013). *Computer aided assessment of mathematics*. Oxford University Press.

Shute, V. J. (2008). Focus on formative feedback. *Review of Educational Research*, *78*(1), 153–189.

Shute, V. J., & Rahimi, S. (2017). Review of computer-based assessment for learning in elementary and secondary education. *Journal of Computer Assisted Learning*, *33*(1), 1–19.

Sinclair, N., Watson, A., Zazkis, R., & Mason, J. (2011). The structuring of personal example spaces. *The Journal of Mathematical Behavior*, *30*(4), 291–303.

Tall, D., & Vinner, S. (1981). Concept image and concept definition in mathematics with particular reference to limits and continuity. *Educational Studies in Mathematics*, *12*(2), 151–169. https://doi.org/10.1007/BF00305619

Van der Kleij, F. M., Feskens, R. C. W., & Eggen, T. J. H. M. (2015). Effects of feedback in a computer-based learning environment on students' learning outcomes: A meta-analysis. *Review of Educational Research*, *85*(4), 475–511. https://doi.org/10.3102/0034654314564881

Watson, A., & Mason, J. (2005). *Mathematics as a constructive activity: The role of learner generated examples*. Erlbaum.

Yerushalmy, M. (2020). Seeing the entire picture (STEP): An example-eliciting approach to online formative assessment. In B. Barzel, R. Bebernik, L. Göbel, M. Pohl, H. Ruchniewicz, F. Schacht, & D. Thurm (Eds.), *Proceedings of the 14th international conference on technology in mathematics teaching—ICTMT 14: Essen, Germany, 22nd to 25th of July 2019* (pp. 26–37). Universität Duisburg-Essen. https://duepublico2.uni-due.de/receive/duepublico_mods_00070728

Yerushalmy, M., Nagari-Haddif, G., & Olsher, S. (2017). Design of tasks for online assessment that supports understanding of students' conceptions. *Zentralblatt für Didaktik der Mathematik*, *49*(5), 701–716.

Yerushalmy, M., Olsher, S., Harel, R., & Chazan, D. (2022). Supporting inquiry learning: An intellectual mirror that describes what it "sees." *Digital Experiences in Mathematics Education, 9*, 315–342. https://doi.org/10.1007/s40751-022-00120-3

Zaslavsky, O., & Zodik, I. (2007). Mathematics teachers' choices of examples that potentially support or impede learning. *Research in Mathematics Education, 9*(1), 143–155. https://doi.org/10.1080/14794800008520176

Zaslavsky, O., & Zodik, I. (2014). Example-generation as indicator and catalyst of mathematical and pedagogical understandings. In Y. Li, E. A. Silver, & S. Li (Eds.), *Transforming mathematics instruction: Multiple approaches and practices* (pp. 525–546). Springer. https://doi.org/10.1007/978-3-319-04993-9_28

ENDNOTE

1. The design of the activity and findings from Study 1 are based on research carried out as part of the master's thesis of Maymon (2020).

CHAPTER 14

NARROWING THE GAP BETWEEN SCHOOL MATHEMATICS CURRICULA AND CONTEMPORARY MATHEMATICS

Lessons Learned During a Four-Phase Curriculum R&D Study

Nitsa Movshovitz-Hadar
Technion—Israel Institute of Technology

Atara Shriki
Kibbutzim College of Education, Israel

Ruti Segal
Oranim College of Education, Israel

Boaz Silverman
The Center for Educational Technology, Israel

This chapter focuses on lessons learned from a four-phase implementation study of the Mathematics-News-Snapshots Project, a research-and-development project designed to bridge the gap between contemporary and high school mathematics by regularly interweaving Math-News-Snapshots in the curriculum. We describe challenges encountered before, during, and after the four phases of the study and how coping with these challenges contributed to its phase-to-phase evolution. Our results show that students gained insight into the nature of contemporary mathematics as a vivid, open-ended, application-rich discipline, and their teachers gained mathematical content for teaching and pedagogical content knowledge. The overarching lesson of the project is threefold: (1) laboratory-style small sample experimentation may obscure the difficulties when implemented in real-life educational systems; (2) curriculum development outcomes and encouraging implementation research results are not sufficient to make change happen because decision making and leadership are also needed; and (3) the gap between school mathematics curricula and contemporary mathematics takes ingenuity, vision, persistence, and courage to overcome.

Keywords: contemporary mathematics; innovative curriculum; reverse engineering study; student achievement-oriented curriculum study; teacher implementation-oriented curriculum study

IDENTIFYING A CRITIQUE-BASED NEED FOR CURRICULAR CHANGE

An observed gap of more than 200 years between contemporary mathematics and school mathematics curricula was the initial drive for the study that is the focus of this chapter. To determine this gap, we prepared (1) a list of mathematical areas that are commonly taught within the school mathematics curricula[1] (e.g., algebra, analytic geometry, calculus, Euclidean geometry, probability & statistics, sequences, series); and (2) a timeline in which specific relevant topics related to these areas were developed (Eves, 1983a, 1983b). We found that high school mathematics curricula do not typically involve topics developed beyond the early 19th century. We attribute this gap to the nature of mathematics, namely, it is open-ended, predominantly hierarchical, and mostly accumulative. For generations, mathematics has been a highly prolific discipline, with a vivid present and a promising future. Thousands of new results are published annually in hundreds of professional journals (American Mathematical Society, 2022); "... the overwhelming majority of works in this ocean contain new mathematical theorems and their proofs" (Sevryuk, 2006, p. 101). Newly created problems that previous findings yielded are constantly added to the plethora of yet unsolved problems that challenge mathematicians. As almost nothing becomes obsolete in the mathematical body of knowledge, it is ever-growing and expanding. Consequently, the gap between contemporary mathematics and school mathematics curricula is bound to widen unless someone intervenes.

We decided to invest our efforts in an attempt to narrow this gap, as school mathematics generally reflects neither the ever-growing nature of the field nor mathematicians' work to establish new results. Hence, students who graduate high school usually possess the wrong image of mathematics as a discipline in which all answers are known, leaving little room for further exploration. It is, therefore, likely that the number of students who choose to specialize in mathematics or mathematics-related disciplines will decrease, and society at the national level, particularly the economy, will pay the toll (cf. Kounine et al., 2008; Mathieson & Homer, 2022; Noyes & Adkins, 2017).

We believe this non-constructive, dead-end conception of mathematics is spread to the public and keeps the majority of people unmotivated or even disliking mathematics while unquestioningly admiring those who find mathematics intriguing. However, this situation does not agree with education laws in many countries that emphasize one primary goal of the education system—to open students' horizons to human culture and the intellectual achievements of humanity. Several attempts at changing public awareness of the true nature of mathematics are taking place, such as the Global Math Project led by James Tanton (https://www.globalmathproject.org), various activities at MoMath—the National Museum of Mathematics established in 2012 in New York City (U.S.) (https://momath.org/), and publications and talks by mathematicians to share their passion for mathematics with the public (e.g., Frenkel, 2013; Fry, 2015). Having identified a justifiable need to introduce a change in the official high school mathematics curriculum, we had a rationale for the initiative and our first lesson learned (see Lesson 14.1).

Lesson 14.1

Identifying Needs for Curricular Change

> **Lesson 14.1: Identifying a need for curricular change.**
>
> *Before embarking on curriculum research, it is necessary to identify a need for curricular change based on observation or solid critique of the present situation. Addressing such a need must involve (1) the development of an implementable solution that fits this need, and (2) the examination of its feasibility and impact on teachers and their students.*

ADDRESSING AN IDENTIFIED NEED FOR CURRICULAR CHANGE

Any attempt to narrow the gap between school mathematics curricula and contemporary mathematics faces a few hurdles, including (1) students' insufficient mathematical background, (2) systemic time constraints, (3) the demanding teaching load involved in preparing students for school-leaving exams, and (4) the need for specialized teacher training. The logic of curriculum studies would be to compare *several* strategies for coping with these challenges empirically (Pacheco, 2012). However, it seemed too demanding to develop more than one option and then run an empirical comparison study. Thus, our next step was to consider an optimal strategy based on previous knowledge of the circumstances.

One way to introduce school students to modern mathematics is to design learning modules on selected topics in contemporary mathematics. However, only very few topics in modern mathematics lend themselves to a module that is teachable in a single week, let alone a single day. Such a module typically interferes with teaching the mandatory curriculum or is offered as an after-school program to a select group of interested students. We sought a different solution, one that would be suitable for *any* high school student, not just the mathematically inclined, and one that could become an integrated part of the official mandatory curriculum in Israel. The approach we adopted was interweaving snapshots of mathematical news in teaching the compulsory curriculum (*Mathematics–News–Snapshot* [*MNS*]), as short intermezzos, with each taking at most one class period of 45 minutes. Introducing one *MNS* at a time during the school year would minimize interference with the flow of teaching the mandatory curriculum, thus addressing challenges (2) and (3) previously mentioned. To meet challenges (1) and (4), we devised the *MNS*s as short, concise "story-telling style" PowerPoint presentations, each conforming to a set of design principles developed according to our professional perception of best practices in mathematics education. The design principles fall into three categories, as seen in Figure 14.1.

Figure 14.1

Design Principles for the Mathematics–News–Snapshots

---

1. *Mathematical Content Design Principles*

    Each *MNS* presentation should include the following content:

    1.1 A mathematical result published in the last five decades, with possibly yet unsolved related problems;

    1.2 Concepts/definitions/theorems from the mathematics curriculum essential to make the news accessible and connected to students' presumed prerequisite knowledge;

    1.3 Relevant and essential extracurricular materials.

2. *Content about Mathematics and Mathematicians Design Principles*

    Each *MNS* presentation should include the following, as much as relevant:

    2.1 Elements reflecting the nature of mathematics, such as definition, problem posing and problem-solving, search for a pattern, generalization;

    2.2 Historical background relevant to the news, such as long-standing conjectures, failures and success in proving them, cultural background if appropriate;

    2.3 Details about the unique contribution of mathematicians involved in the piece of news or its history, as well as about the personal life of such mathematicians;

    2.4 Connections within mathematics (curricular or otherwise), to other disciplines, everyday life, or students' common interests.

(Figure continued on next page)

Narrowing the Gap Between School and Contemporary Mathematics

Figure 14.1 (Continued)

Design Principles for the Mathematics-News-Snapshots

---
3. *Pedagogy and Style-Related Design Principles*

Each *MNS* will be a PowerPoint presentation including:

3.1   The full text for the teacher to present the *MNS* in class, in the format of a Socratic dialogue promoting discourse and student involvement;

3.2   Animation and illustrations to demonstrate or represent mathematical ideas, concepts, and other elements;

3.3   A simple, concrete, and non-formal tone, to make the content accessible to high school students;

3.4   An annotated list for further reading;

3.5   A limited number of slides not exceeding 45-minute lesson duration, to address curricular pressures.

In developing each *MNS*, we considered the simile of "a glass bottom boat tour without getting wet," that is, we wish to introduce students to the world of a contemporary mathematics problem but without needing to explore the mathematics content in depth. Students can learn something about the mathematics (the metaphorical flora and fauna) but without having to "dive into the water" or know the details about every object visible through the glass. Figure 14.2 illustrates the typical structure of each *MNS* as students enjoy the glass-bottom boat tour.

Figure 14.2

The Underlying Structure of Each Mathematics-News-Snapshot

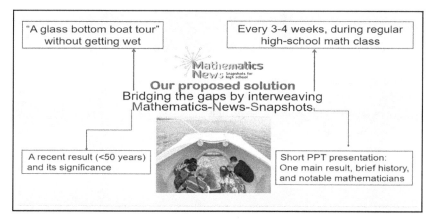

Source of the picture: Shutterstock.com. (Reprinted with permission.)

At this preliminary stage, we had one prototype, about Sudoku, that exemplified the rationale of our *MNS* and its realization through the three types of design principles. See Figure 14.3 for a summary of this prototype; for more details, see Movshovitz-Hadar (2008).

Figure 14.3

Summary of MNS Prototype About Sudoku

MNS Title: *Sudoku–Beyond Pastime*

MNS Summary:

The *MNS* focuses on the $3^2 \times 3^2$ Sudoku puzzle, its definition, the way Sudoku puzzles can be created, the level of difficulty and the number of solutions as related to the number of entries in the puzzle, and diverse uses for Sudoku arrays that influence the quality of life today. No prior familiarity with Sudoku is required.

The main questions addressed in this *MNS*:

- What is a Sudoku puzzle, and what is the challenge in solving it?
- How does one create a Sudoku puzzle?
- Who invented the riddle, and how did it become popular? What is the number of Sudoku puzzles (of order 9 × 9) having a unique solution?
- Is the puzzle easier to solve as the number of entries increases?
- In 2007, two Canadian mathematicians, Herzberg and Marty, presented an example that proves the existence of a Sudoku puzzle that has a unique solution with only 17 entries out of 81.
- Is there a Sudoku puzzle with *fewer than* 17 entries that has a unique solution?
- Must every Sudoku puzzle with *more than* 17 entries have a solution? And if there is, does it have to be unique?
- Is there a formula that allows one to solve *any* solvable Sudoku puzzle?
- Are there questions concerning Sudoku for which the answer is not yet known?
- Why are mathematicians interested in Sudoku puzzles of a higher order?
- What uses do Sudoku sets have in fields such as planning agricultural research, organizing a schedule, planning league and tournament competitions, disguising information and espionage, public health, and more?

Key issues related to the standard curriculum in this *MNS*:

- The natural numbers
- Naming large numbers
- The difference between a numeral and a number
- Elimination Existence proof
- Latin squares
- Rejecting a conjecture by a counterexample

Between 2009 and 2023, through collaboration between expert mathematicians and mathematics educators, we gradually developed, field-tested, and revised 27 *Mathematics-News-Snapshots*, all in the format of PowerPoint presentations. Their primary foci include, among others: Fermat's Last Theorem; the race for finding larger and larger primes; the Map Coloring problem; Goldbach's conjecture; random walks; the Möbius strip and its applications; the Cake Cutting problem; Catalan's conjecture; the Kepler conjecture; Benford's Law, a surprising discovery about the digit 1; Fibonacci numbers—myths and facts; the digits of π; soap bubbles; the Millennium problem—Is $P = NP$?; pentagonal tiling; the traffic paradox; the sofa moving problem; and mathematical matchmaking. Together, the *MNS*s provide a glimpse into contemporary mathematics and its applications, tailored to be accessible to high school students. We aim to have 30 *MNS*s to provide teachers with a sufficiently broad selection to interweave periodically in teaching the mandatory mathematics curriculum during the three high school years. All the *MNS*s include comments for the teacher in the notes section underneath the slides. They are accessible, in both English and Hebrew, through the project website at https://mns.org.il/.

Is there a way to give high school students a taste of modern mathematics, although they lack the mathematical background needed to delve into the topics in depth? Should we strive to offer high school teachers the means to do it? Or, is narrowing the gap between school mathematics curricula and contemporary mathematics a non-realizable dream? These were our initial doubts in 2007, before we embarked on the first phase of our study. Thirteen years later, upon completion of the fourth phase in Israel, we felt ready to call high school teachers worldwide to assume responsibility for narrowing the gap between contemporary and school mathematics by employing our *MNS*s (Movshovitz-Hadar, 2020). It took a four-phase study to establish the feasibility of incorporating *MNS*s into the mandatory curriculum and their impact on teachers and students. The road was strewn with pitfalls, leading to an important lesson learned (see Lesson 14.2).

Lesson 14.2

The Importance of Considering Practical Factors to Usage

> **Lesson 14.2: Making clear decisions about practical factors opens the door for a successful solution.**
>
> *Prior to any curriculum study that intends to address a recognized need for change in the school curriculum, one should consider (at least) three practical factors: the desired target audience; the form of the curriculum to be developed, including basic design principles; and the a-priori constraints that may endanger its implementation.*

THE FOUR-PHASE STUDY: AN OVERVIEW

Examining the role and impact of the *MNS*s as a vehicle for narrowing the gap between school mathematics curricula and contemporary mathematics was a combination of three types of curriculum study: (1) innovative idea-based curriculum development study (change-oriented); (2) student-centered curriculum study (achievement-oriented); and (3) teacher-centered curriculum study (implementation-oriented). At each subsequent phase, the project evolved from what had been designed due to challenges encountered in the previous phase. Phase I (2007/8-2010/11) was an action research project with one teacher and a few drafted *MNS*s. Phase II (2011/12-2012/13) was a clinical study of the implementation of the *MNS*s in two schools with the entire mathematics team. Phase III (2013/14-2014/15) was a scaled-up implementation study. Phase IV (2016/17-2018/19) was a three-year longitudinal study focused on the impact of the project on students. In the rest of this chapter, we describe the challenges faced in each phase, their effect on the transition from one phase to the next, and some lessons learned in each. To conclude, we specify our present insights, concerns, future plans, and over-arching lessons.

Phase I: Action Research, Its Challenges, and Lessons Learned

Following the preparatory steps, it was natural to start realizing the *MNS* idea through action research. We developed drafts of a few *MNS*s and had one teacher try them in a few classes, systematically reporting insights and obstacles. Fortunately, a high school teacher with a strong mathematics background applied for PhD study at the Technion and was captivated by the idea of the project. For three years, she drafted *MNS*s according to the design principles (Figure 14.1) and interweaved them one by one in her teaching of the mandatory curriculum in four high school classes in grades 10 and 11. This teacher's classes were matched to classes of four other teachers that served as comparison groups to assess progress in teaching the mandatory curriculum when interweaving *MNS*s into instruction.[2]

Typically, action research consists of various cycles (Coghlan & Brannick, 2005; Kemmis, 1997). In our case, it took place in two stages, each consisting of two cycles. Stage 1 was a design study whose first cycle was the development of first drafts for a few *MNS*s. The initial step in this cycle was a literature search for a *suitable* piece of news. Then, the design step took place according to the three categories of *design principles* (Figure 14.1).

Following a *field trial* of each *MNS* draft in one or two classes, it was revised based on observations made during the field trial. Updates were added if new related results had been published since the drafted version appeared.

Upon completion of the first cycle, it felt reasonable to examine the outcome retrospectively and substantiate the underlying characteristics of the *MNS*s as related to the design principles. So, the second cycle of Stage 1 was *a validation process* in the form of a *reverse engineering study*. We assigned five mathematics education experts a *reverse engineering task* to trace the design principles in the different slides of each *MNS*. Each expert received a detailed list of the design principles, a PowerPoint presentation file containing the slides, and a printed file with screenshots of each *MNS* slide. Adjacent to each screenshot was a table where the expert inserted the design principles of content and pedagogy attributed to each part of that slide. Figure 14.4 includes a sample of slides from the *MNS* on Kepler's conjecture as used for a reverse engineering task in a workshop for mathematics educators.[3]

The frequency occurrences of each uncovered design principle in each *MNS* were summarized to obtain the percentage of the total occurrence of every design principle per *MNS*, yielding the relative frequency distribution of design principles per *MNS*. A negotiation session took place to bring the five experts into agreement when discrepancies were found among their results. Following the validation process, it was possible to compare the inner composition of the *MNS*s to one another based on occurrences of the design principles in each and relate the results to other data collected during Stage 2 of the action research, described below. This task led to our third lesson.

Lesson 14.3

The Value of a Reverse-Engineering Study

> **Lesson 14.3: A reverse-engineering study sheds light on undercurrents in the development stage of innovative pedagogy.**
>
> *Given an innovative idea that one starts to implement according to some design principles, it is crucial to ensure that the development stage yields an outcome that reflects the rationale and the Design Principles, and that they were not overlooked throughout the process.*

Stage 2 of the action research was an implementation study, which also took place in two cycles. Cycle 1 involved the implementation of the same *MNS* in various ordinary senior high school classes. Cycle 2 involved the implementation of various *MNS*s in the same class.

Figure 14.4

Three Slides From a Reverse Engineering Task on the MNS Related to Kepler's Conjecture

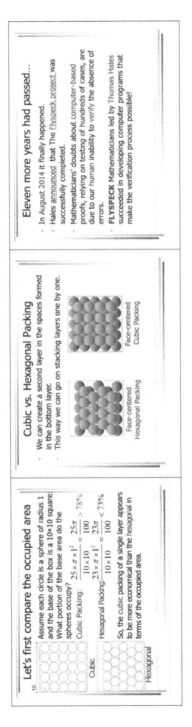

Quantitative and qualitative data collection took place before, during, and after these implementations (Amit et al., 2011). The action research summary from the implementing teacher's point of view was based on a structured journal portfolio documenting her daily experiences related to (i) curriculum issues, (ii) system issues, and (iii) budget issues, as seen in Figure 14.5.

Figure 14.5

Issues Faced by Teachers Implementing MNS as Part of an Action Research Study

---

Curriculum Issues

Teachers experience severe time constraints. Teachers are responsible for all students' success in school mathematics, no matter how difficult or boring a student finds mathematics. Every deviation from the curriculum and any extra-curricular activity is a threat to reaching this goal.

System Issues

Teachers and their classroom practice are subject to inspection (e.g., by the superintendent, principal, peer teachers, students, or parents). Any extra-curricular activity may invite criticism and often puts the teacher in a defensive position.

Budget Issues

Many schools are not yet equipped with the technical equipment (computers and projectors) required for implementation, nor are they able or willing to invest efforts to obtain such equipment. This puts the motivated teacher in a problematic situation. A partial solution is adapting the PowerPoint presentation to a frontal exposition using the blackboard. Implementation is highly demanding on the part of the teacher. Without appropriate compensation, they are unlikely to assume this challenge.

Despite the difficulties, for three consecutive school years, the action research teacher managed to interweave 7–8 *Mathematics-News-Snapshots* every year in her teaching of each grade 10 and grade 11 class, with no harm to "covering" the mandatory curriculum nor to students' achievement in mastering it. The action research indicated that: (1) a teacher who overcomes the obstacles will likely become a lifelong learner, constantly gaining new knowledge and possibly changing perceptions regarding contemporary mathematics, its history, and integration in mathematics education; and (2) students' benefits prove these efforts worthwhile. The teacher who conducted the action research summarized her personal experience by saying: "Now I am not just a math teacher who prepares her students for the school-leaving exams. I am the representative of the mathematics community in my classrooms" (Amit & Movshovitz-Hadar, 2011). These initial results led to Lesson 14.4.

Lesson 14.4

Design Principles Can Promote Pedagogy

> **Lesson 14.4: Hone the design principles to promote pedagogy.**
>
> *Each Snapshot demands a student's prolonged attention, much like a movie, a football game, or a concert. Students were asked to hold back and take notes while listening without interfering, which appeared valuable to their listening skills. During the action research, as the need to raise students' curiosity and maintain their attention for the duration of the exposition proved of utmost importance, we continually reshaped the design principles. In particular, the pedagogy of the presentation became consistently based on questions and answers in a Socratic dialogue mode. It also became clear that to support the teacher's self-confidence, the verbal part of the PowerPoint presentation in each MNS should be complete and self-contained rather than outlined. The teacher can then practice the oral presentation like an actor performing a play. (See examples in Figure 14.4.)*

During the first year of the action study, the school principal decided that the class size was too big and asked students to volunteer to switch to a different class. The parents protested, insisting that their children stay in that classroom because the other classroom teacher did not incorporate *MNS*s in the teaching of mathematics. In planning Phase II, we avoided objections to transfer among classes by getting the entire team of mathematics teachers in a school to participate in the study. This led to another lesson.

Lesson 14.5

Parents' Voices Are Important

> **Lesson 14.5: Consider the parent's voice in a curriculum study.**
>
> *The parents' attitudes toward an experimental change are influential. They are no less important than the teachers' attitudes or those of the students.*

Phase II: A Clinical Study, Its Challenges, and Lessons Learned

The availability of ten field-tested revised *MNS*s, and the proof that one teacher *was* capable of incorporating 7–8 of them yearly in her ordinary teaching while coping successfully with many difficulties, encouraged us to address the question: What would it take to have other teachers replicate her success? Obviously, the researcher/teacher of Phase I had some vested interest in the success of the project, as she was eager to earn her PhD. One could not expect every high school mathematics teacher to be as involved as she was in both the development and implementation of the *MNS*s. Phase II set out to prove that Amit's success in coping with the difficulties involved in interweaving 7–8 *MNS*s in the teaching of the mandatory curriculum every school year was replicable, leading to Lesson 14.6.

Lesson 14.6

Math News Snapshots and Professional Development Should Be Developed Jointly by Mathematicians and Mathematics Educators to Meet the Needs of Teachers

> **Lesson 14.6: New curricular material development is a joint venture.**
>
> *Polishing the existing MNSs and developing new ones must be carried out by expert mathematicians collaborating with expert mathematics educators independently and simultaneously with the design of professional development for teachers who would implement the MNSs according to their needs.*

In 2012, the Israel Science Foundation (ISF) provided a three-year grant (# 144/2011) to explore the feasibility of a broader implementation of the *MNS*. We sought the advice of the Vice-President (VP) in charge of the curriculum of Ort schools—a large, well-organized network of high schools in Israel. He liked the *MNS*s idea and suggested an intensive year-long clinical study with the entire team of mathematics teachers in two high schools.

The Ort VP selected two principals of comprehensive high schools in the ORT network to participate in the study's second phase. The principals were "turned on" and proud to be chosen. The next step was getting the schools' mathematics teams on board. We met the two principals, together with their mathematics department heads, to explain the project details and share a sample *MNS*. These were accepted with mixed feelings. Although they found the rationale captivating and the demonstrated *MNS* fascinating, serious concerns were expressed regarding teachers' ability to respond to students' questions related to topics in contemporary mathematics, most of which were new to the teachers. To address these concerns, another meeting was set with the entire mathematics team in each school, where teachers became acquainted with the idea that the field of mathematics has become so rich that there does not exist one comprehensive mathematics expert who knows everything in mathematics, as might have been true 100 years ago (Amit, 2016). We quoted the eminent logician and leader in mathematics education, Leon Henkin:

> One of the big misapprehensions about mathematics that we perpetuate in our classrooms is that the teacher always seems to know the answer to any problem that is discussed. This gives students the idea that there is a book

somewhere, with all the right answers to all the interesting questions, and that teachers know those answers. And if one could get hold of the book, one would have everything settled. That's so unlike the true nature of mathematics. (Steen & Albers, 1981, p. 89)

These preliminary meetings were attended by the principals, who encouraged the teachers by emphasizing their pride and trust in the mathematics team selected by the ORT administration to participate in this experiment. We summarize our lesson learned in Lesson 14.7.

Lesson 14.7

Importance of Developing Partnerships With Teachers

> **Lesson 14.7: It is imperative to establish mutual trust and partnership relationships with teachers.**
>
> *Teachers' feedback is crucial, and they should know that their feedback is seriously considered. To this end, we built proprietary software enabling the teachers' comments on the materials, as well as on their own performance in implementing them. The teachers' research diaries and written feedback became a vital source of evidence about the difficulties and successes of the moves that took place throughout the year. In addition, the partnership among the teachers in each school was significant to developing a teacher's community of practice, which allowed for one teacher to rely on another one to present the MNS in his/her class in case of absence.*

As evidence of the typical collaborative spirit that we were able to induce in each school, we note the initiative of one teacher in one of the schools to present her and her peers' experiences at the annual meeting of high school mathematics teachers in Israel, which we strongly supported. The school team members prepared it together, and she presented it (Orlovich, 2013). The presentation was very well received.

Phase II of our study was conducted in two high schools each year (31 teachers, grades 9–12), employing proprietary Wiki-like software deliberately designed to share feedback. A project staff member met weekly with each school team for a 90-minute professional development workshop, introducing one *MNS* every other week, and discussing its merits and its presentation to students. At the end of this phase, we found that teachers successfully coped with the difficulties they encountered. Each incorporated 6–7 *Snapshots* every school year in at least one classroom (674 high school students in all). They expressed satisfaction and reported students' enthusiasm as well as their own, even though the project was imposed on them by their school administration (Shriki et al., 2013). However, this allocation of resources led us to summarize another lesson.

Lesson 14.8

Scalability Concerns

> **Lesson 14.8: Limitations of a clinical study—the case of scalability.**
>
> *Assigning a project staff member to conduct weekly professional development meetings in each high school, which made the clinical study of Phase II successful, raised a scalability problem. So, towards the third and last year of the ISF grant, we decided to reach out to volunteer teachers throughout the country instead of continuing to work with a limited number of teachers in two schools whose participation resulted from their principal's decision.*

Phase III: A Scaled-Up Implementation Study, Its Challenges, and Lessons Learned

For the third phase, we sent a call for high school mathematics teachers to participate in a year-long professional development program in video conferencing format, planned similarly to the conduct of Phase II. To our great disappointment, only 17 teachers responded. Each teacher had at least seven years of high school mathematics teaching experience. We approached (in person) many high school mathematics teachers whom we knew to determine why they did not respond. We learned that our call was intimidating, as the targeted audience perceived the program as highly demanding. Many refrained from joining, given the amount of time they were expected to devote to familiarize themselves with the *MNS*s, allocate to integrating them in classroom instruction, and provide feedback through the proprietary software. In addition, there was apprehension involved in presenting topics about which they did not feel "at home." We did not offer compensation beyond intellectual satisfaction and official credit for professional development.

Phases II and III were accompanied by a formative evaluation study examining the experience of teachers and their students for one year. Our research questions involved teachers' and students' experiences. Phase III teachers were free to choose any class they wished for the experimental implementation of the *MNS*s. They were advised to: (a) select a senior high school grade (preferably 10th grade) to ensure students have a certain mathematical maturity to benefit from exposure to *MNS*s, as well as to enable a three-year duration of the study; (b) give an equal chance to all three streaming levels of the curriculum—Basic, Intermediate, and Advanced; and (c) designate another class in the same school that would receive the same parts of the standard curriculum, to serve as a control class for comparison of pre- and post-test achievement.

All participating students were given two questionnaires (pre- and post-interweaving *MNS*s) dealing with their level of agreement with a stated point of view (e.g., *Mathematics is a set of formulas and rules one has to memorize*), a statement about mathematics (e.g., *Mathematics began to be developed about 4,500 years ago*), and questions concerning their experience as mathematics students. The questionnaires asked participants to explain their answers.

A total of 1,074 high school students (from Phases II and III) answered the preliminary questionnaire, and 968 answered the summary questionnaire after being introduced to six or seven *MNS*s during one school year. The difference in numbers was mainly due to a failure in the electronic format in which students responded to the questionnaire, which could not keep track of nicknames some students adopted for the preliminary questionnaire, thus rendering it impossible to match students' responses to the questionnaires. The distribution of preliminary questionnaire responses was 30% in 9th grade, 45% in 10th grade, and 10% in 11th grade. (The rest were in 7th, 8th, or 12th grade.) Over the three streaming levels, 235 (22%) were Basic Level students, 251 (23%) were Intermediate Level, and 588 (55%) were Advanced Level. Both qualitative and quantitative methods were employed for analysis to provide a comprehensive understanding of the processes involved in the study (Caracelli & Greene, 1997). In Phase III, we added a control group comparison study.

Students reported that the *MNS*s changed their image of mathematics as a discipline, increased or deepened their mathematical knowledge, and developed their mathematical thinking. They also expressed emotional aspects, such as pleasure from mathematics and increased motivation (Movshovitz-Hadar, 2015).

For the sake of "research cleanliness," we were strict with the teachers and prohibited any changes in the *MNS*s, often to the teachers' dismay. They were only allowed to skip one slide here and there, or add an activity before or after the presentation. We required each teacher to participate in a demanding training program that included consecutive days of meetings each summer and other school vacations, regular video conference meetings during the school year, completing a feedback report after integrating each *Snapshot*, answering questionnaires, and participating in interviews. There were four similar questionnaires, one assigned at the beginning of the project's first year and the others at the end of each of the three years. The preliminary questionnaire consisted of Likert-type items (on a 1–5 scale) regarding teacher's perceptions about the nature of mathematics and the curriculum (e.g., *Mathematics, as it is currently taught in school, reflects mathematics as a discipline*) and the teachers' interest and motivation to expand their own knowledge about issues related to mathematics beyond the curriculum (e.g., *I read materials that deal with extra-curricular mathematics regularly*). Following each Likert-type item, the teachers had to justify their assigned ranking.

The three summary questionnaires were designed to trace teachers' perceptions regarding the process of integrating the *MNS*s during their participation in the project and the impact of the integration of *MNS*s on their students' perception of mathematics, as well as on the teachers' professional development. It included Likert-type statements (on a 1–5 scale) (e.g., *I/my students learned new things about the contribution of mathematics to science and society*). In addition, the questionnaires included yes/no items (e.g., *Was the integration of the MNSs a challenge for you? Yes/No. Please explain your choice*). As seen in Lesson 14.9, we may have overdone data collection. In retrospect, we should have learned from another curriculum study, Logic in Wonderland (Movshovitz-Hadar & Shriki, 2009, 2018a, 2018b), in which the pre- and post-questionnaires were part of the curriculum and served the learner for self-evaluation as well as the researchers for data collection.

Lesson 14.9

Learning to Empower Teachers and Not Restrict Them

Lesson 14.9: Research formalities should not restrict teachers, but rather empower them.

In a curriculum study, it is essential not to let the data collection take over the educational goals. In this study, we seemed to have overdone it. The requirements from the teachers, on top of the main assignment of integrating the MNSs into their ordinary teaching, appeared to be a burden. We failed to realize this in real-time. One solution is to make data collection tools an inherent part of the innovative project, so that the data collection does not become objectionable and oppressive.

At the end of this phase, we obtained satisfactory evidence of the difficulties teachers had to overcome to interweave the *MNS*s in their teaching for one school year and of the positive impact of *MNS*s on teachers. However, the effect on students was not sufficiently clear. Exposure to 6–7 *MNS*s for one year did not yield the impact on students we wanted to achieve. A major challenge that remained open was to determine the impact on students after experiencing the *MNS*s for three years.[4] It took over a year to receive a grant that enabled a longitudinal study and to get organized for Phase IV, reminding us of the importance of Lesson 14.10.

Lesson 14.10

Time Is Needed for Change to Manifest in Results

> **Lesson 14.10: Educational results take time and patience.**
>
> Although students in the experimental MNS groups expressed their enthusiasm, no significant difference concerning students' perceptions of mathematics as a discipline or as a profession was found between the experimental group and the comparison group after one year. Indeed, one cannot expect a truly significant effect after experiencing only six MNSs throughout one school year.

Phase IV: A Longitudinal Study, Its Challenges, and Lessons Learned

In the fourth phase, we aimed to characterize the impact of the *Mathematics-News-Snapshots* Project on high school students' perceptions of mathematics—both as a discipline and as a profession. More specifically, we asked how and to what extent high school mathematics students perceive the effect of their participation in the *MNS* project—after one, two, and three years of participation—on their image of mathematics as a discipline, their image of mathematics as a profession, and on their affections related to school mathematics. We asked: How do these differ from a similar group of students not exposed to the *MNS*s? In addition, we also studied the effect on the teachers who assumed responsibility for introducing their students to the *MNS*s. We examined how teachers perceived the impact of integrating *MNS*s for three years on the development of several components of their *Pedagogical Content Knowledge*, as originally defined by Shulman (1986), and on their *Mathematical Knowledge for Teaching* as defined by Ball and Bass (2003, 2009) and Ball et al. (2008). A competitive grant from the Israeli Ministry of Science supported this phase.

In this phase, we considered a new setup for teachers' professional development in three stages, one per year, which was necessary for a three-year study of the implementation of *MNS*s. As a preliminary stage, a two-day summer school was announced. This time, 61 teachers registered, 47 of whom had not taken part in a previous phase of our study and 14 who continued. This response, strikingly different from the one obtained to the call for participation in Phase III, made us happy. So much so that we unintentionally ignored two easily observable facts. First, only about one-third of the teachers with whom we worked in the past re-joined us; we realized it was due to the difficulties these teachers expressed in the previous phases. Second, the number of teachers who volunteered was almost the same as the number of high schools they came from—we had only a few cases of two teachers from the same school. As became evident later, this turned out to be an actual obstacle.

At the end of the summer school, the 42 applicants who showed up knew the challenges and demands expected for the three years ahead. They were asked to and 28 (66%) agreed to sign (1) a commitment for interweaving *MNS*s in at least one 9th or preferably 10th grade class in the first year and (2) an intent to continue with the same class for the entire three-year duration of the study.[5] Numerous teachers decided not to join us—some were unsure they would be assigned to 9th or 10th grade in the coming school year, and others were intimidated by the demands.

Although we did not face any teacher dropouts in the two previous phases, we anticipated some departures in the longitudinal study. It was not reasonable to assume that all 28 teachers who started the journey with 749 students would be able to persevere. However, our anticipation was less than the actual dropout reality. Only 13 teachers remained for the second year, and only 6 teachers followed their 10th grade class all the way to 12th grade and regularly exposed the same students to the various *MNS*s, one at a time, during the three years. In retrospect, this resulted from a mistake in the recruiting method. The total number of students exposed to the *MNS*s for three years (78) was much less than it could have been if we had invested in initially recruiting at least two or three teachers from the same school. In that case, it would have been possible for another teacher to carry on with a class whose teacher had left the school. Instead, students of a leaving teacher did not continue with us, and the student sample shrunk. Thus, we have Lesson 14.11.

Lesson 14.11

Teacher Recruitment is Crucial in a Longitudinal Study

> **Lesson 14.11: In planning a longitudinal study, one must consider conditions that may prohibit continuity.**
>
> *Recruiting teachers for a longitudinal study is better activated by calling the entire school team of a few schools, with the school principal's consent, rather than one or two teachers from many schools. This can help secure the participation of students whose teacher cannot stay for the duration of the study. It also has a bearing on the mutual support teachers can get from and give one another, thus contributing to the success of the desired curricular reform beyond the research issues.*

Between September 2016 and July 2019, four data collection instruments were employed to collect quantitative and qualitative data to address the research questions concerning the impact on students and on their teachers

(Bikner-Ahsbahs et al., 2015). Data were analyzed both qualitatively and quantitatively. The analysis comprised three parts. See Figure 14.6.

Figure 14.6

Data Collection and Analysis

Collection of Data

- Questionnaires were designed by borrowing from instruments used in previous phases. Each was constructed and validated by our team of mathematics education experts and administered at the beginning of the first year and the end of each school year.

- Student interviews (20 informal, semi-structured interviews) provided a direct route to students' responses (Kvale, 2007).

- Teacher feedback on their experiences was obtained through comments on the proprietary software and during meetings with the project staff.

- Teacher field notes and four questionnaires were answered by each teacher, at the beginning of the first year and the end of each school year.

Analysis of Data

- Identifying statements in each data source dealing with students' perceptions and affect regarding mathematics as a discipline and as a profession, as well as students' emotions for school mathematics.

- Utilizing a qualitative research paradigm, implementing a process of open and axial coding to identify the main categories and sub-categories in the identified statements in the student interviews (Corbin & Strauss, 2008). These categories were then used to classify data from student questionnaires and teacher feedback.

- Carrying out comparative analyses on the research questionnaires. We focused on student responses to questionnaires (both experimental and comparison groups) and used t-tests to compare them by grade and by level of study after one, two, and three years of participation.

The data indicated that regularly interweaving the *Snapshots* was advantageous for both students and teachers. Students gained insight into the true nature of contemporary mathematics as a vivid, open-ended, application-rich discipline, and their teachers gained improved mathematical knowledge for teaching and pedagogical content knowledge (Movshovitz-Hadar et al., 2019, 2021; Segal et al., 2019, 2021, 2023; Silverman et al., 2023). The accumulating results of the longitudinal study motivated us to translate all the existing *MNS*s into English during these years and start calling the international community's attention to them (Movshovitz-Hadar, 2019, 2020; Movshovitz-Hadar & Silverman, 2018). Requests from the international community yielded the translation of a few *Snapshots* into German, French, Italian, and Portuguese. In June 2020, the *Mathematics-News-Snapshot* Project received recognition as a site of the month on the Klein Project Blog of the International Commission on Mathematics Instruction.[6]

PRESENT DEVELOPMENTS AND FUTURE CHALLENGES

The education system in Israel is centralized, and it is expected that teachers will refrain from adopting the *MNS*s and introducing them regularly into instruction until the Israel Ministry of Education makes them a part of the mandatory requirement. Therefore, convincing the national superintendent of school mathematics and the national committee in charge of the mandatory school curriculum to include the *MNS*s as a part of the official curriculum is our next challenge. We presented a strong rationale and the results of our longitudinal study. Still, we were required to demonstrate the contribution of the *MNS*s to student achievement in school mathematics. For that to happen, we have recently considered a qualitative assessment of achievement from students' exposure to this new part of the curriculum, in the format of submitting an independent student's summary of the lessons learned from each *MNS*, so that its assessment can be added to each student's school leaving certificate. Without it, there is only a minimal chance that teachers will adopt the *MNS*s and introduce them regularly into instruction because they understandably prioritize topics that students must master to pass the matriculation exams. This has become even more evident than ever due to the effects of COVID-19 on education.

Soon after the completion of Phase IV of the study, we were ready to start disseminating the *MNS*s more relaxedly, without data collection and other associated research. In this spirit, in March 2020, we made all the 24 *MNS*s written and updated by that time freely and unconditionally available to every mathematics teacher on a designated website (https://mns.org.il).[7] Unfortunately, teachers and the education system as a whole became preoccupied with challenges presented by COVID-19 lockdowns. In fact, the pandemic halted the dissemination of the *MNS*s for two school years, as ultimate preference was given to closing gaps in implementing the mandatory curriculum. During these two years, however, we continued to develop new *MNS*s. Presently there are 28 *MNS*s on our website (https://mns.org.il); the 29th is in preparation and we are aiming for 30.

Recently, we also have been contemplating the idea that to narrow the gap between school mathematics curricula and contemporary mathematics, the *MNS*s might become a vehicle for making 21st century students more aware of 21st century competencies (Organization for Economic Cooperation and Development [OECD], 2018; United Nations Educational,

Scientific and Cultural Organization [UNESCO] Institute for Lifelong Learning, 2020). To this end, we have piloted a detailed analysis of one of the *MNS*s through the looking glass of these competencies and found the outcome supporting the idea. This will likely lead to Phase V of our study, which may focus on raising awareness of and imparting 21st century competencies simultaneously with narrowing the gap between school mathematics curricula and contemporary mathematics using the *MNS*s, and possibly also to acceptance of the *MNS*s by education authorities in Israel.

SUMMARY, DISCUSSION, AND CONCLUDING REMARKS

In this chapter, we shared lessons learned during our four-phase study dedicated to collecting pieces of evidence that support the following meta-mathematical statement about mathematics education: *Employing the pedagogy of interweaving Mathematical News Snapshots in high school mathematics teaching is an efficient strategy for narrowing the gap between contemporary mathematics and the school mathematics curricula. This 'wander-about' in the world of mathematics is intriguing enough to impress students, detailed enough to motivate them to do mathematics, and mind-opening enough to yield a desired image of mathematics as a vivid, creative, and ever-growing domain.* The reverse engineering analysis of the various *MNS*s provided evidence for a sequel to that statement: *The Mathematical News Snapshots are deeply rooted in the relevant history of mathematics, and the two go hand in hand in leading students to the perception of mathematics as a rich living discipline.* In phrasing these meta-mathematical statements, we follow Movshovitz-Hadar's (1993) viewpoint regarding mathematics education as a meta-mathematical discipline: "As such, statements about mathematics, with relevance to its learning, its understanding, its use, its teaching, and its communication belong to mathematics education" (p. 267). The proof for such meta-mathematical statements cannot take the form of deductive proof characteristic of mathematical statements. Instead, it takes the form of proof by supporting evidence commonly provided by empirical and social sciences (Amit & Movshovitz-Hadar, 2011; Malek & Movshovitz-Hadar, 2011; Movshovitz-Hadar, 1988, 2006).

There is a noticeable gap between the way mathematics is perceived by most education system graduates and professionals. Mathematicians and mathematics educators see mathematics as a creative, connected, vivid, and applicable domain. In contrast, many students perceive mathematics as an irrelevant and obsolete domain with nothing new left to discover (Boaler, 2016)—a far cry from its true nature. The study reported in this chapter explored the feasibility of providing high school students with a glimpse into the "ocean" of new mathematical results, to use Sevryuk's (2006) metaphor, and into the applicability of contemporary mathematics to the quality of life in the modern era. For instance, Sudoku has applications for cancer treatments; the Möbius Strip has applications for molecular chemistry.

Introducing students to contemporary mathematics by giving them a taste of it as a pleasurable experience is similar to taking students to the theater, to a concert, to a football game, to a dance performance, or to a painting exhibition so students can enjoy and appreciate the experiences, without expecting them to become professionals in these areas. To this end, we adopted frontal storytelling, often dramatical, pedagogy. This pedagogy goes hand-in-hand with the fact that students do not possess the background to delve into the topic in depth, and the teachers are there to give them glimpses into it. The *MNS*s were not meant to present an in-depth theory, but rather to guide "a wander about" in the world of contemporary mathematics.

We built the project on the premise that ALL high school students deserve to have the opportunity to learn about contemporary research in mathematics as part of the curriculum because mathematics is a significant part of human culture. Clearly, teachers are crucial to this endeavor. Without collaboration with teachers of real classrooms, no innovative idea has a chance, no matter how brilliant. We designed the *MNS*s to suit not only students who were the target audience but also teachers who face many challenges in implementing them. To this end, we provided teachers with full texts of *MNS*s for classroom presentations, expecting teachers to act them out, and decide independently to skip some parts or add some activities before or after the presentation. This strategy was chosen because we realized that mathematics teachers come to the teaching field with the mathematical knowledge they acquired as students at school and during their post-secondary teacher training studies at colleges or universities. The courses taught during the training usually include mathematics courses to deepen mathematical knowledge and courses dealing with the pedagogy of teaching mathematics while referring to the content presented in the mandatory curriculum. During their higher studies, teachers usually are not exposed to contemporary mathematical news.

The findings of Phase IV, the three-year longitudinal study (2016–2019), indicate that teachers perceived the interweaving of the *MNS*s as leverage for expanding and deepening their Mathematical Knowledge for Teaching (Ball et al., 2008). They become aware of the importance of exposing students to contemporary mathematical subjects beyond the curriculum to change their image of mathematics to a dynamic and constantly evolving field. They become aware of various connections between mathematics and its applications. For example, they were exposed to a rich expanse of examples illustrating the relationship between mathematics, science, and society, as well as between mathematics and the phenomena around them, which are part of teachers' Horizon Content Knowledge (HCK)

(Ball et al., 2008; Ball & Bass, 2009). Such knowledge includes a sense of the mathematical environment, major ideas and values of the discipline, and key discipline practices. The teachers developed their Specialized Content Knowledge (SCK) as well. As they expanded the repertoire of *MNS*s examples through which mathematical concepts can be presented, they were "making features of particular [mathematical] content visible" (Ball et al., 2008, p. 400), enabling the delivery of connected and relevant mathematics to their students. In addition, teachers recognized that interweaving the *MNS*s with their regular curriculum can be an opportunity to address students' fears of mathematics and expose them to examples that connect the development of mathematics and its applications in everyday life (Knowledge of Content and Student) (Ball et al., 2008).

Our multi-phase study evolved gradually. Completion of one phase raised the need for the next one. *This perpetual process in itself could be yet another typical aspect of curriculum studies.* We are presently looking at the potential *MNS*s may have for inducing 21st century competencies, hoping that continued production of new *MNS*s will become an integral part of the mathematics education system and inspire students to change their attitudes towards mathematics and become literate about the nature of mathematics as a prolific, open-ended, creative, and application-rich discipline. The overarching lesson is threefold, as expressed in Lesson 14.12.

Lesson 14.12

The Overarching Lesson

Lesson 14.12: The overarching lesson is threefold.

a. The examination of an innovative idea for a change in the curriculum should include considerations of the loaded and dynamic realities of the educational system. Laboratory-style small sample experimentation may not present the complete picture of the obstacles such a change may encounter when realized.
b. Curriculum development outcomes and research results, as impressive as they may be, do not suffice to make a system-wide change happen. We learned the hard way that no longitudinal curriculum study can be carried out successfully by teachers in real time without the school principal's collaboration, and without their inclusion in the mandatory curriculum. In a centralistic education system, decisions about the inclusion of new topics are in the hands of the superintendent of school mathematics and other government committees. Implementation of system-wide change takes leadership.
c. The gap between school mathematics curricula and contemporary mathematics takes ingenuity, vision, persistence, and courage to narrow.

REFERENCES

American Mathematical Society. (2022). *MathSciNet by the numbers*. https://mathscinet.ams.org/mathscinet/2006/mathscinet/help/about.html?version=2

Amit, A. (2016). *Is it still possible for a mathematician to be competent in most of the areas of fields of mathematics?* Quora. https://smg.quora.com/Is-it-still-possible-for-a-mathematician-to-be-competent-in-most-of-the-areas-or-fields-of-mathematics-1

Amit, B., & Movshovitz-Hadar, N. (2011). Design and high school implementation of Mathematical-News-Snapshots—An action research into 'Today's News is Tomorrow's History.' In B. E. Krongellner & C. Tzanakis (Eds.), *History and epistemology in mathematics education: Proceedings of the Sixth European Summer University* (pp. 171-184). Verlag Holzhausen GmbH/Holzhausen Publishing.

Amit, B., Movshovitz-Hadar, N., & Berman, A. (2011). Exposure to mathematics in the making: Interweaving Math News Snapshots in the teaching of high school mathematics. In V. Katz & C. Tzanakis (Eds.), *Recent developments on introducing a historical dimension in mathematics education* (pp. 91–102). Mathematical Association of America.

Ball, D. L., & Bass, H. (2003). Making mathematics reasonable in school. In J. Kilpatrick, W. G. Martin, & D. Shifter (Eds.), *A research companion to the Principles and Standards for School Mathematics* (pp. 27–44). National Council of Teachers of Mathematics.

Ball, D. L., & Bass, H. (2009). *With an eye on the mathematical horizon: Knowing mathematics for teaching to learners' mathematical futures* [Paper presentation]. The 43rd Jahrestagung der Gesellschaft für Didaktik der Mathematik, Oldenburg, Germany.

Ball, D. L., Thames, M. H., & Phelps, G. (2008). Content knowledge for teaching. *Journal of Teacher Education*, *59*(5), 389–407. https://doi.org/10.1177/0022487108324554

Bikner-Ahsbahs, A., Knipping, C., & Presmeg, N. (Eds.). (2015). *Approaches to qualitative research in mathematics education*. Springer.

Boaler, J. (2016). *Mathematical mindsets: Unleashing students' potential through creative math, inspiring messages and innovative teaching*. Jossey-Bass.

Caracelli, V. J., & Greene, J. C. (1997). Crafting mixed-method evaluation designs. *New Directions for Evaluation*, *74*, 19–32. https://doi.org/10.1002/ev.1069

Coghlan, D., & Brannick, T. (2005). *Doing action research in your own organization* (2nd ed.). SAGE.

Corbin, J., & Strauss, A. (2008). *Basics of qualitative research: Techniques and procedures for developing grounded theory* (3rd ed.). SAGE.

Eves, H. (1983a). *Great moments in mathematics (After 1650)* (1st ed., Vol. 7). Mathematical Association of America. http://www.jstor.org/stable/10.4169/j.ctt6wpwqx

Eves, H. (1983b). *Great moments in mathematics (Before 1650)* (1st ed., Vol. 5). Mathematical Association of America. http://www.jstor.org/stable/10.4169/j.ctt6wpwqx

Frenkel, E. (2013). *Love and math: The heart of hidden reality*. Basic Books.

Fry, H. (2015). *The mathematics of love: Patterns, proofs, and the search for the ultimate equation*. Simon and Schuster.

Kemmis, S. (1997). Action research. In J. P. Keeves (Ed.), *Educational research, methodology, and measurement: An international handbook* (2nd ed., pp. 173–179). Pergamon.

Kounine, L., Marks, J., & Truss, E. (2008). *The value of mathematics*. Reform.

Kvale, S. (2007). *Doing interviews*. SAGE.

Malek, A., & Movshovitz-Hadar, N. (2011). The effect of using transparent pseudo-proofs in linear algebra. *Research in Mathematics Education, 13*(1), 33–58. https://doi.org/10.1080/14794802.2011.550719

Mathieson, R., & Homer, M. (2022). "I was told it would help with my Psychology": Do post-16 Core Maths qualifications in England support other subjects? *Research in Mathematics Education, 24*(1), 69–87. https://doi.org/10.1080/14794802.2021.1959391

Movshovitz-Hadar, N. (1988). School mathematics theorems: An endless source of surprise. *For the Learning of Mathematics, 8*(3), 34–40. https://flm-journal.org/Articles/25D52D0AA169E7DE039E32AC027A45.pdf

Movshovitz-Hadar, N. (1993). The false coin problem, mathematical induction, and knowledge fragility. *Journal of Mathematical Behavior, 12*(3), 253–268.

Movshovitz-Hadar, N. (2006). What can mathematicians and mathematics educators communicate about? In Q. Douglas (Ed.), *Proceedings of the 3rd international conference on the teaching of mathematics at the undergraduate level* (ICTM). (CD Format, paper 900). Istanbul, Turkey.

Movshovitz-Hadar, N. (2008). Today's news are Tomorrow's history. Interweaving mathematical news in teaching high school math. In E. Barbin, N. Stehlikova, & C. Tzanakis (Eds.), *History and epistemology in mathematics education: Proceedings of the fifth European Summer University (ESU 5)* (pp. 535–546). Vydavatelsky Press.

Movshovitz-Hadar, N. (2015). Mathematics News Snapshots—Meeting the challenge of introducing senior high school students to contemporary mathematics. In *Proceedings of the 13th annual Hawaii International Conference on Education* (pp. 2209–2210). Hawaii International Conference on Education. ISSN#: 1541-5880. https://hiceducation.org/wp-content/uploads/proceedings-library/EDU2015.pdf

Movshovitz-Hadar, N. (2019, April 1–3). *Refreshing high school curricula with Mathematics-News-Snapshots* [Paper presentation]. National Council of Teachers of Mathematics Research Conference, San Diego, CA.

Movshovitz-Hadar, N. (2020). Is it our responsibility to introduce high school students to contemporary mathematics? *OnCore, The AATM—Arizona Association of Teachers in Mathematics Leadership Journal,* 66–73.

Movshovitz-Hadar, N., & Shriki, A. (2009). *Logic in Wonderland—Alice's Adventures in Wonderland* as the context of a course in logic for future elementary teachers. In B. Clarke, B. Grevholm, & R. Millman (Eds.), *Tasks in primary mathematics teacher education: Purpose, use and examples* (Vol. 4, pp. 85–103). Springer Mathematics Teacher Education Series.

Movshovitz-Hadar, N., & Shriki, A. (2018a). *Logic in Wonderland—an introduction to logic through reading Alice in Wonderland—Student's workbook*. World Scientific Publishing. https://www.worldscientific.com/worldscibooks/10.1142/10412

Movshovitz-Hadar, N., & Shriki, A. (2018b). *Logic in Wonderland—An introduction to logic through reading Alice in Wonderland—Teacher's guidebook*. World Scientific Publishing. https://www.worldscientific.com/worldscibooks/10.1142/10409

Movshovitz-Hadar, N., Segal, R., Shir, K., Shriki, A., & Silverman, B. (2021). A multi-stage attempt at narrowing the gap between contemporary mathematics and high school mathematics. In Y. Shimizu (Ed.), *Proceedings of Topic Study Group 37: Research on classroom practice at secondary level*. 14th International Congress on Mathematical Education.

Movshovitz-Hadar, N., Segal, R., Shir, K., Shriki, A., Silverman, B., & Zigerson, V. (2019). Bridging between school mathematics and contemporary mathematics: Turning a dream into reality. In A. Rogerson & J. Morska (Eds.), *Theory and practice: An interface or a great divide? The Mathematics Education for the Future Project—Proceedings of the 15th International Conference* (pp. 406–411). Maynooth University.

Movshovitz-Hadar, N., & Silverman, B. (2018, August 1–4). *Meeting the challenge of introducing senior high school students to contemporary mathematics* [Workshop]. The Annual Mathematical Association of America MathFest, Denver, CO. https://www.maa.org/sites/default/files/pdf/mathfest/2018/TechnionWorkshopSlides.pdf

Mullis, I. V. S., Martin, M. O., Gonzalez, E. J., & Chrostowski, S. J. (2004). *TIMSS 2003 International Mathematics Report: Findings From IEA's Trends in International Mathematics and Science Study at the Fourth and Eighth Grades*. TIMSS & PIRLS International Study Center, Boston College. https://timssandpirls.bc.edu/PDF/t03_download/T03INTLMATRPT.pdf

Noyes, A., & Adkins, M. (2017). *Rethinking the value of advanced mathematics participation*. Nuffield Foundation.

Oberman, J. (2018). Israeli mathematics curricula (K-12): A concise form based upon the Hebrew original published by the Ministry of Education. In N. Movshovitz-Hadar (Ed.), *K–12 mathematics education in Israel* (Vol. 13, pp. 405–430). World Scientific. https://doi.org/10.1142/10741

Organization for Economic Cooperation and Development [OECD]. (2018). *The future of education and skills: Education 2030*.

Orlovich, H. (2013). *Hevzekim BeOrt Holon: Havayot Mesaqrenot Mehasade* [*Math-news-Snapshots in Ort Holon: Intriguing experiences from the field*] [Paper presentation]. Annual Meeting of High School Mathematics Teachers, Shfayim, Israel. (Abstract published in the proceedings, p. 60.)

Pacheco, J. A. (2012). Curriculum studies: What is the field today? *Journal of the American Association for the Advancement of Curriculum Studies, 8*, 1–18. https://doi.org/10.14288/jaaacs.v8i0.187713

Segal, R., Shriki, A., Movshovitz-Hadar, N., & Silverman, B. (2019). Interweaving Mathematics–News–Snapshots as a facilitator for developing mathematical knowledge for teaching. In U. T. Jankvist, M. Van den Heuvel-Panhuizen, & M. Veldhuis (Eds.), *Proceedings of the Eleventh Congress of the European Society for Research in Mathematics Education* (pp. 3497–3504). European Research in Mathematics Education. https://hal.archives-ouvertes.fr/hal-02422748/document

Segal, R., Shriki, A., Silverman, B., & Movshovitz-Hadar, N. (2021). Interweaving Mathematics-News-Snapshots in class: Implications for teachers' Horizon Content Knowledge. In N. Buchholtz & C. M. Ribeiro (Eds.), *Proceedings of Topic Study Group 33: Knowledge in/for teaching mathematics at the secondary level*. 14th International Congress on Mathematical Education.

Segal, R., Shriki, A., Silverman, B., & Movshovitz-Hadar, N. (2023). *Detailed analysis of the results from a longitudinal study of interweaving Math News Snapshots in high school mathematics—The impact on the teachers* [Manuscript in preparation].

Sevryuk, M. (2006). Arnold's problems, book review. *Bulletin of the American Mathematical Society*, 101–109. https://doi.org/10.1090/S0273-0979-05-01069-4

Shriki, A., Movshovitz-Hadar, N., Amit, B., & Zohar, O. (2013, February 18–19). *Integrating snapshots from contemporary mathematics in high school teaching* [Paper presentation]. Jerusalem Conference for Research in Mathematics Education, Jerusalem, Israel.

Shulman, L. S. (1986). Those who understand: Knowledge growth in teaching. *Educational Researcher*, *15*(2), 4–14. https://doi.org/10.3102/0013189X015002004

Silverman, B., Shriki, A., Segal, R., & Movshovitz-Hadar, N. (2023). *Detailed analysis of the results from a longitudinal study of interweaving Math News Snapshots in high school mathematics—The impact on the students*. [Manuscript in preparation].

Steen, L. A., & Albers, D. J. (1981). Round and round at the round table. In L. A. Steen & D. J. Albers (Eds.), *Teaching teachers, teaching students: Reflections on mathematical education* (pp. 86–92). Birkhäuser Boston. https://doi.org/10.1007/978-1-4899-0427-0_12

United Nations Educational, Scientific and Cultural Organization [UNESCO] Institute for Lifelong Learning (UIL). (2020). *Embracing a culture of lifelong learning: Contribution to the futures of education initiative*. https://unesdoc.unesco.org/ark:/48223/pf0000374112

ENDNOTES

1. Sources about curricula in various countries can be found at https://files.eric.ed.gov/fulltext/ED535223.pdf, in the *TIMSS—Curriculum Frameworks for Mathematics and Science* (https://timssandpirls.bc.edu/timss1995i/timsspdf/tr2chap1.pdf, p. 6, Figure 12), and in TIMSS 2003 (Mullis et al., 2004). TIMSS is the Trends in Mathematics and Science Study.
2. At each grade level of high school (10th–12th grade, ages 16 to 18) in Israel, there are three different mathematics tracks called 3–4–5 units respectively, where students learn algebra, geometry, trigonometry, statistics, and precal-

culus to different levels of depth. For more details about the curriculum in Israel, see Oberman (2018).
3. This example is from a workshop conducted at ESU6—the Sixth European Summer University, that took participants through a reverse engineering process focusing on the Kepler Conjecture *MNS* following its presentation. The interested reader can find more practical details about this process from Amit and Movshovitz-Hadar (2011).
4. Yet another challenge was further development of new *MNSs* while revising, polishing, and updating the existing ones. Due to space limitations, we do not elaborate on this process in this chapter.
5. It is not common for high school teachers in Israel to continue with the same students for all three years. We were hoping that teachers would invest effort in staying with students.
6. http://blog.kleinproject.org/?p=4564
7. Advertisement of this site was through the Center for High School Teachers of Mathematics in Israel.

SECTION IV

RESEARCH ON THE TEACHER INTENDED CURRICULUM

CHAPTER 15

BROADENING CONCEPTUALIZATIONS OF THE INTENDED AND ENACTED CURRICULUM TO INFORM RESEARCH DESIGN

Emily Saxton
The Math Learning Center

Nicole Rigelman
The Math Learning Center and Portland State University

Corey Drake
The Math Learning Center

Kim Markworth
Unaffiliated

We outline a conceptual framework that supports our efforts to define our intended curriculum (Bridges in Mathematics) by identifying both core design features and essential elements. We discuss our conceptualization of the enacted curriculum, which intentionally prioritizes teacher responsiveness over fidelity of implementation. Together, these conceptualizations help us distinguish between implementations that are aligned with our intended curriculum from those that depart from our intended curriculum and they inform our research design. We share a suite of tools that measure the use of both the core design features, which are easily counted, and the essential elements, which generally require more complex measurement tools. Finally, we discuss how the measurement tools are designed to distinguish between enactments that are aligned and productive from those that depart from the intended curriculum in unproductive ways.

Keywords: curriculum fidelity; curriculum integrity; enacted curriculum; intended curriculum; productive adaptation

INTRODUCTION

The Math Learning Center's (MLC's) research agenda is driven not only by questions focused on whether our curriculum, *Bridges in Mathematics* (*Bridges*) (Math Learning Center, 2023), makes a difference for students, but also by questions focused on *for whom* and *under what circumstances* it makes a difference. We agree with recent calls for these latter questions to be prioritized in education research (National Academies of Sciences, Engineering, and Medicine, 2022). Attention to implementation is not new to research focused on curriculum effectiveness—it has long been deemed a necessary consideration for examinations of the impact of curriculum materials on student outcomes (Chval et al., 2009; National Research Council, 2004). However, implementation research that draws on the perspectives of both the curriculum developers—and their intentions for the curriculum—and the educators who enact the curriculum in their classrooms is still rare.

Many of the research questions we consider most important and pressing in our efforts to understand and continually improve *Bridges* curriculum materials rely on our ability to understand how our materials are used in classrooms. These implementation-focused research questions include the following:

- What variations in implementation do we observe across teachers?
- Do common implementation patterns emerge across teachers?
- How does teacher implementation change across successive years of implementation?
- What kinds of adaptations do teachers make to the written materials?
- Why do teachers make those adaptations?
- Are those adaptations aligned with the aims of the curriculum?

In pursuing answers to these questions, we first clarified how we conceptualized implementation because sound measurement relies on clear definitions of what we intend to measure (Wilson, 2005; Wolfe & Smith, 2007). This need for clear conceptualization pushed our organization to develop clear definitions of both the intended curriculum and the enacted curriculum (Remillard & Heck, 2014).

We first outline a conceptual framework we found useful in our efforts to clearly define our intended curriculum, which distinguishes between core design features and essential elements. *Core design features* are structural features of the curriculum (or the "what"), such as the variety of instructional formats, modalities, and embedded professional learning resources. *Essential elements* are instructional features of the curriculum (or the "how"), such as the nature of classroom discourse and the openness of instruction.

Next, we discuss our conceptualization of the enacted curriculum. We intentionally avoid ideas of fidelity of implementation because we do not expect that all teachers will implement the curriculum with strict adherence to the written materials. Instead, we hope that teachers will use the materials "in ways that are responsive to children's mathematical strengths, needs, and cultural and linguistic backgrounds to optimize children's learning in mathematics" (National Council of Teachers of Mathematics [NCTM], 2020, p. 40), and therefore enhance students' opportunities to access the mathematics in the materials and achieve the mathematical learning goals of the materials. This conceptualization of the enacted curriculum leads us to strive for ways to distinguish between implementations that are responsive to students and aligned with curricular aims (implementation with integrity) from those that depart from curricular aims.

Finally, we outline how these conceptualizations of the intended curriculum and the enacted curriculum inform our research design. We share a suite of tools that measure the use of core design features, which are easily counted with implementation logs or curricular annotation tools, and essential elements, which generally require complex measurement tools such as classroom observation protocols or teacher interviews. We conclude by discussing how the measurement tools in our research projects are designed to help distinguish between enactments that are aligned and productive from those that depart from and are unproductive to the overall aims of the intended curriculum (DeBarger et al., 2013; Remillard, 2005; Sherin & Drake, 2009). We believe the ability to distinguish between productive and unproductive adaptations will help us answer our research questions related to the *implementation circumstances* under which the *Bridges* curriculum makes a difference for students.

INTENDED CURRICULUM: DISTINGUISHING BETWEEN CORE DESIGN FEATURES AND ESSENTIAL ELEMENTS

MLC has a long history of developing and supporting the use of standards-based[1] curriculum materials. Central to MLC curriculum and ancillary materials is their use of faithful visual models, student investigation and problem solving, and rich mathematical discourse. As such, central to the design of the third edition of *Bridges* is attention to both what is taught (i.e., mathematics content and practice standards) and how it is taught (i.e., connections to NCTM's [2020] equitable and effective mathematics teaching practices). We view core design features of the curriculum as those present in the curriculum as written—the "what"; in contrast, the essential elements represent MLC's intended classroom implementation—the "how." The core design features and essential elements are summarized in Table 15.1 and more fully described in the remainder of this section.

Core Design Features

Bridges includes a purposeful variety of instructional formats and modalities as a core design feature, including three main curricular structures supporting student classroom learning: *Sessions*, *Work Places*, and *Number Corner*. The *Sessions* often begin with a problem posed by the teacher, followed by time for students to think independently, work for a period of time, and talk in pairs before reconvening to share and compare strategies and solutions as a whole class. For example, in Grade 2 Unit 5, students[2] first think-pair-share about patterns they notice in a pattern block sequence (Figure 15.1a). Next, student pairs build what they think the fifth arrangement will look like (Figure 15.1b). Then whole class discussions of additional arrangements follow, culminating in a discussion of how to generalize the pattern to the 100th arrangement (Figure 15.1c).

Figure 15.1

Example Session From Grade 2 Unit 5 of Bridges (3rd Edition)

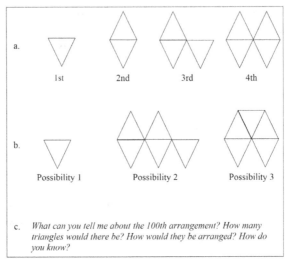

Source: Figure 15.1 is republished with permission of The Math Learning Center, from ©2023 *Bridges in Mathematics*, 3rd Edition, Grade 2 Teachers Guide. All Rights Reserved by The Math Learning Center.

Table 15.1

Core Design Features and Essential Elements of the Bridges in Mathematics Curriculum Materials

Core Design Features Characteristics and structures visible in the materials as written	**Essential Elements** Characteristics of classroom implementation aligned with the authors' intentions

• *Purposeful variety of instructional formats and modalities*
   o Three distinct formats:
     ▪ Sessions
     ▪ Work Places
     ▪ Number Corner
   o Instructional modalities:
     ▪ *Explicit integration of familiar routines*
     ▪ *Opportunities for technology integration*
	• *Active learning characterized by student-centered instruction*

• Good tasks
   o Standards aligned
   o Multiple entry points
   o Varied representations and tools
	• *Open-ended tasks promote reasoning, problem solving, and connections within and across representations*

• Faithful visual models
   o Positioned as supporting reasoning and sensemaking
   o Used throughout the learning process rather than abandoned in favor of abstraction
	• *Visual models further sensemaking and deepen conceptual understanding*

• Embedded professional learning
   o *Grade-level and unit overviews*
   o *Sidebars highlighting curricular alignment with equitable and effective practices*
   o *Teacher questions, sample student dialog, and discourse planners*
	• *Rich mathematical classroom discourse supports understanding and achievement*

• Attention to diversity, equity, and inclusion
   o Images represent classroom diversities
   o Contexts reflect students' lived experiences
   o Leverage and honor multiple competencies and student resources
	• *Positive mathematics identities reinforce deep mathematics learning*

Note: Features and elements shown in italics are those we measure through our instrument design and selection, described later in this chapter.

Work Places provide ongoing game-based practice with key skills and give teachers the opportunity to observe, assess, and provide appropriate challenge and support. An example Work Place from Grade 2 Unit 5 is *Close to 25¢*, where students draw from a card deck with values ranging from 4¢ to 25¢ and build the amount on each card with the goal of getting close to—but not over—25¢. This game provides students with opportunities to practice skip counting by 5s and adding sets of mixed coins using Money Value Pieces and 25-Frames.

Number Corner is a set of workouts (or lesson structures) that introduce, reinforce, and extend skills and concepts critical to grade-level instruction. For the Calendar Grid workout, every month has a calendar with daily markers and a related anchor chart where the class documents, observes, and predicts how repeating and growing patterns extend throughout the month. The Calendar Collector features a growing collection of various sorts (e.g., quantities of money, increments of time, data from experiments) and students have opportunities to estimate, count, measure, display, and make sense of the data. Another workout is the Daily Rectangle, where students construct quantities with magnetic tiles and notice attributes of numbers. For example, in September, students form "2-by" tile arrays to match the date. They come to discover that an even number of tiles can be seen as groups of two or two equal groups (doubles). They find the structure of an odd number of tiles is slightly different: when attempting to form a "2-by" array, there is an extra tile—groups of 2 plus 1—or the sum of two consecutive numbers (neighbors). Routines such as number talks and number strings are included in the Computational Fluency workout. Number Line activities develop skills with counting, skip-counting, and comparing and ordering whole numbers.

Another structural feature of *Bridges* is its standard format of eight units at each grade level with 20 lessons per unit. Taken together, *Bridges* students receive about 80 minutes of mathematics instruction per day, with *Sessions* and *Work Places* comprising 60 minutes and *Number Corner* another 20 minutes. Purposely embedded within the *Sessions* and *Number Corner* as a core design feature are familiar mathematical[3] and instructional[4] routines (herein referred to as "routines"). These routines are opportunities for teachers to elicit and use student thinking to support their instruction. Through a combination of action steps and sidebar commentaries in the Teachers Guide, teachers learn more about the multiple purposes of the recommended instructional moves (e.g., questions to ask, suggestions about thinking to highlight); when appropriate, sidebars also provide teachers the opportunity to integrate technology (i.e., virtual manipulatives and shareable math apps corresponding to the featured problem or routine).

Another core design feature of *Bridges* is good tasks. The tasks are standards aligned, allow for multiple entry points, and call for use of varied representations (i.e., verbal, physical, visual, symbolic, contextual). For many tasks, students have a choice of tools and representations. A related core design feature is the regular presence and use of physical and visual models so they can become faithful visual models. In this way, models are positioned as tools for deepening understanding for all students; their use is not relegated to lesson introduction or for students who may struggle.

Embedded in *Bridges* is teacher professional learning with, among other educative core design features, grade-level and unit overviews, sidebar commentary, and discourse resources. The unit overviews orient teachers to the focus of the *Sessions* and *Work Places* for the unit, provide mathematical background and purpose of related models and strategies, and convey how the work fits into the mathematical storyline for the grade level and beyond. The sidebars highlight connections to equitable and effective mathematics teaching practices (NCTM, 2020), allowing teachers to see the intentionality of the design decisions. In addition, *Sessions* include potential questions a teacher may ask, examples of likely student responses through sample dialog, and discourse planners.

A final core feature is attention to diversity, equity, and inclusion. To support students with seeing themselves in mathematics, the *Bridges* materials include images with diverse representations and problem contexts to reflect a broad range of students' lived experiences. Teachers are supported in seeing potential places and ways they might leverage students' competencies and resources (i.e., language, lived experiences, ways of knowing).

Essential Elements

The variety of instructional formats and modalities in *Bridges* materials supports teachers to engage students as active constructors of their learning. The intentional integration of instructional routines conveys the essential element of centering on students' mathematical thinking. Unlike many elementary mathematics technologies that position students to receive information, the MLC apps are interactive—mimicking the experience of physical models and supporting continued active learning.

Another essential element of *Bridges* is support for implementation of good tasks. Specifically, support includes instructional moves teachers can make to maintain the problematic and open-ended aspects of tasks while students work. When planning and implementing, teachers have access to (a) unit overviews that convey the intentionality of the within- and across-unit task sequencing, and (b) action steps and sidebars that encourage student reasoning and problem solving, support productive struggle, and press for connections within and across various representations. Physical and visual models are essential elements in *Bridges*, their regularized use further supports students' sensemaking and deepens conceptual understanding. Because models can powerfully represent mathematical ideas, they are purposefully integrated, revisited, and extended within and across grade levels, so they become faithful models for students.

Altogether, the educative features within *Bridges* communicate the importance of orchestrating meaningful mathematical discourse. The questions and dialog act as essential elements, conveying how teachers might engage students in meaningful discourse that strengthens students' understanding and achievement, while providing a vision for such practice.

When *Bridges* is implemented in alignment to MLC's intentions, where teachers leverage students' competencies and resources, students can come to see themselves as capable doers, knowers, and sense makers of mathematics. This student-centered enactment simultaneously supports the development of a positive mathematics identity and deeper learning.

Although there is overlap between the core design features and the essential elements, the distinction is important. If MLC were to only develop materials with the core design features in mind and not also the essential elements, teachers would be left to determine the rationale for the task sequences, the recommended instructional moves, or the purpose of a structure or interaction. By articulating the essential elements of *Bridges*, teachers are better equipped to make implementation decisions consistent with our intentions.

ENACTED CURRICULUM: ELEVATING RESPONSIVENESS OVER STRICT ADHERENCE

In our conceptualization of the enacted curriculum, we intentionally avoid fidelity of implementation because we do not expect all teachers to implement our curriculum with strict adherence to the written materials. We recognize that changes must be made as individual teachers and their students interact with each other around a set of curriculum materials; it is neither possible nor desirable for these interactions to be based on strict adherence to the words in the Teachers Guide. Although we reject the notion of implementation with fidelity, we are committed to conceptualizing, supporting, and measuring *implementation with integrity* (NCTM, 2020). This commitment reflects our hope that teachers will have both the freedom and the support to implement *Bridges* in ways that are (a) responsive to students' mathematical thinking, (b) supportive of the development of students' positive mathematics identities, and (c) consistent with the mathematical aims of the materials.

We define *implementation with integrity* as including two key components: responsiveness to students, and alignment with the mathematical goals of the curriculum. Each of these components can be observed and measured not only at the level of the lesson, but also across sequences of lessons or units. From this larger perspective, we define implementation with integrity as implementation that responds to and aligns with progressions of students' thinking and mathematical content over time. In prior work (Drake & Land, 2015), we identified *coordinated noticing*, attention to both the progression of students' mathematical thinking and to the progression of mathematical content underlying a given set of curriculum materials, as critical to implementation with integrity.

In conceptualizing implementation with integrity, it is important to consider which core design features and essential elements of the curriculum have been modified, how they are modified, and why they are modified. When developing our conceptualization of what does or does not count as implementation with integrity, we built on Ben-Peretz's (1990) notion of the *curriculum envelope*, which Sherin and Drake (2009) describe as "a range of adaptations that teachers might make while still maintaining the intended goals of a lesson" (p. 493). We also connected to DeBarger and colleagues' (2013) conceptualization of *productive adaptations* as those modifications that "emerge from teachers' careful observations of their classrooms (i.e., are evidence-based), are thus responsive to the demands of a particular classroom context, and are consistent with the core design principles and intentions of a curriculum intervention" (p. 301). Each of these conceptualizations, like ours, focuses on the need for teachers to respond to specific students and contexts while maintaining the mathematical goals of a lesson (or series of lessons) written for a general audience.

Based on these conceptualizations, certain adaptations might be considered to fall outside the curriculum envelope, and therefore, would not be consistent with our conceptualization of productive adaptation and implementation with integrity. These adaptations would include, for example, those that alter the mathematical goals of the lesson or sequence of lessons, those that lower the cognitive demand of a task, or those that limit student access to the concepts or activities of the lesson. In contrast, productive adaptations and enactments that support implementation with integrity are those that maintain the mathematical goals of the lesson (or series of lessons), maintain or increase the cognitive demand of the lesson, and increase access for all students to the mathematics. These adaptations can include purposeful use of number choice to respond to students' thinking, modifications of contexts to respond to students' cultural and community experiences, and inclusion of opportunities for students to develop and explain strategies for solving problems before a strategy is demonstrated by the teacher (Drake et al., 2015; Land et al., 2019).

HOW OUR INTENDED AND ENACTED CURRICULUM INFORM OUR RESEARCH DESIGN

These conceptualizations of the intended curriculum (the "what" and the "how" in our written materials) and the enacted curriculum (with our focus on productive adaptation and implementation with integrity) inform our research design because they act as our operationalization of, or clearly specified definition of, implementation. As such, our definitions of the intended curriculum and the enacted curriculum are a foundation for the development of a suite of measurement tools for our research projects. These measurement tools are designed to measure both the structural, easily counted core design features, and the complex, nuanced essential elements of instruction that, taken together, represent our intended curriculum.

Measurement Tools for Core Design Features

In our estimation, not all core design features of *Bridges* can be detected in a teacher's enactment of the curriculum because some core design features (e.g., whether tasks are *standards* aligned) are solely properties of the written materials (e.g., does the task as written align to the *standards*?). Other core design features, however, can be detected in a teacher's enactment of the curriculum, including (a) the extent to which the instructional formats and modalities in a given lesson are implemented in a classroom, and (b) the extent to which embedded professional learning resources are accessed by teachers. When measuring core design features, our goal is to understand how much of the curriculum was implemented and whether parts of lessons were adapted. Two measurement tools serve as our primary mechanism for understanding how these core design features are enacted in classrooms: the *Teachers Guide Annotation Tool* and *Unit Implementation Logs* (Table 15.2). Whenever possible, we leverage multiple measurement tools to gather implementation data on the same core design feature, because multiple data sources enrich our understanding of how teachers enact these core design features.

Teachers Guide Annotation Tool

Curricular annotation tools have been used in previous research studies to understand how teachers enact curriculum materials. For example, the Improving Curriculum Use for Better Teaching (ICUBiT) Project designed a set of tools that collectively measure how teachers interact with curriculum guides during planning (which they called the *Curriculum Reading Log*) and how much of the curriculum teachers implement (which they called

Table 15.2

Matrix of Core Design Features of Bridges Measured By Our Measurement Tools

	Measurement Tool	
Core Design Features	*Teachers Guide Annotation Tool*	*Unit Implementation Log*
Purposeful variety of instructional formats and modalities	X	X
Embedded professional learning	X	

the *Table of Contents Record*) (ICUBiT, 2022). Interestingly, these tools were used across five standards-based elementary mathematics curricula in the ICUBiT research project, which demonstrates these data collection methods were applicable to a variety of curricula.

In our research projects, we employ the *Teachers Guide Annotation Tool* to help us understand how teachers use the *Bridges in Mathematics* Teachers Guide during planning and implementation. Teacher annotations during planning are made in a designated color and in the left margin of the Teachers Guide (Figure 15.2). These give us insight into important questions:

- What portions of the Teachers Guide do teachers read when they are planning a unit or lesson?
- What parts of a lesson do teachers plan to use?
- What parts of a lesson do they plan to modify or skip?

Figure 15.2

Example Page of Teachers Guide Annotation Tool

Source: Figure 15.2 is republished with permission of The Math Learning Center, from ©2023 *Bridges in Mathematics*, 3rd Edition, Grade 2 Teachers Guide. All Rights Reserved by The Math Learning Center.

Subsequent annotations after implementation, which we ask teachers to make in a different color and in the right margin of the Teachers Guide, provide us with a data source that addresses:

- What portions of the Teachers Guide did teachers implement in their classrooms?
- Where did teachers make adaptations?

In terms of how this measurement tool helps us understand teacher use of core design features, we gain different insights during planning and implementation. During planning, the *Teachers Guide Annotation Tool* gives insights into whether teachers access the embedded professional learning structures (e.g., unit overviews, sidebars, teacher questions). We also gain insights as to which instructional formats (*Sessions*, *Work Places*, and *Number Corner*) and modalities (routines and technology) teachers plan to use during lessons. Through their annotations after implementation, teachers document whether and how they implemented the three curricular structures, instructional routines, and technological resources. Annotations also indicate which teacher questions were used during a lesson, provide margin notes documenting additional or alternative questions posed, or note the direction a subsequent student discussion took.

Unit Implementation Log

Implementation logs have been used in a variety of educational research studies to gather data that span different time scales and units of analysis. Researchers who use implementation logs most frequently measure implementation at time scales of a single day of instruction (Phelps et al., 2011; Rowan et al., 2004; Shechtman et al., 2019) or a week of instruction (Caswell et al., 2016; Riel et al., 2016; Stickles, 2011). Researchers also seek to understand implementation from different units of analysis, including instruction experienced by a single student (Rowan et al., 2004), by a representative sample of students (Rowan et al., 2009), or by a whole classroom of students (Caswell et al., 2016). These variations are appropriate given the varied goals of the respective studies, which may be focused on the role of implementation on student outcomes, creating generalizable records of implementation, or on understanding how teachers enact a written curriculum.

In our research, we are interested in understanding how teachers enact *Bridges* in their classrooms, thus we ask teachers to log their implementation weekly. To support teacher completion of logs, we email teachers each Friday when the log is open and specify a due date of the following

Monday. If the log was not yet completed by Monday, reminders are sent in the method of each teacher's choice (email, text, or phone call). The log explicitly asks teachers if they annotated their Teachers Guide that week; if teachers answer "no," survey skip logic routes them to a page that reminds them of (1) the importance of their annotations and (2) the instructions on how to annotate the Teachers Guide. Teachers receive a research stipend for the weekly log and for annotating the Teachers Guide. In terms of unit of analysis, we focus teacher reports on implementation at the whole-class level. These *Unit Implementation Logs* help us develop an understanding of the extent to which these core design features were enacted in each classroom (Table 15.3).

Table 15.3

Core Design Features of Bridges as Measured by the Unit Implementation Log

Core Design Feature Purposeful variety of instructional formats and modalities	Unit Implementation Log Question (Rating scale)
Distinct structure 1: *Sessions*	For each Session you've implemented, please pick the response that best describes your implementation. *(Omitted, Used as written, Adapted, Replaced)*
Distinct structure 2: *Work Places*	For each Work Place listed below, please choose all that apply for this week's instruction: *(Not played, Played as a class, Played independent of the teacher)*
Distinct structure 3: *Number Corner*	How many days this week did you teach Number Corner?
Explicit integration of instructional routines	For each instructional routine listed below, please select the response that best represents how often the routine was used this week: *(Not this week, Once, Two or three times, Four or five times)*
Opportunities for technology integration	Which MLC apps did you use this week [choose all that apply]: *(Yes or No for each particular app)* If you created any app share codes for your students this week, please share them here: *(Share code collected, if used)*

Measurement Tools for Essential Elements

The essential elements of *Bridges in Mathematics* are more complex and nuanced, so our measurement orientation necessarily shifts from quantifying enacted curriculum features toward determining whether enactments are aligned with the instruction intended in our curriculum. These essential elements are akin to what other curriculum effectiveness researchers have dubbed "consistency with the pedagogical orientation" of written curriculum materials (Chval et al., 2009). To determine whether teacher enactments of our curriculum are aligned with the essential elements, we use another set of measurement tools: a *Classroom Observation Protocol* and a *Teacher Interview Protocol* (Table 15.4). Here again, we leverage multiple measurement tools to gather implementation data on the same essential element to enrich our understanding of how well a teacher's enactment of an essential element aligns with the aims of our written curriculum.

Table 15.4

Matrix of Essential Elements of Bridges Measured By Our Measurement Tools

	Measurement Tool	
Essential Elements	Classroom Observation Protocol	Teacher Interview Protocol
Active learning characterized by student-centered instruction	X	X
Open-ended tasks promote reasoning, problem solving, and connections within and across representations	X	X
Visual models further sensemaking and deepen conceptual understanding		X
Rich mathematical discourse supports understanding and achievement	X	X

Classroom Observation Protocol

Classroom observation protocols are commonly used measurement instruments to understand how teachers enact curriculum (Chval et al.,

2009; Gujarati, 2011; Huntley, 2009). Unlike other measurement tools that rely on teacher-reported data, these tools utilize trained observers. They are arguably better suited than other measurement tools to get at the more nuanced, pedagogical aspects of implementation. When, for example, classroom discourse is of interest, there is little substitute for direct observation of classroom instruction. As a result, we view classroom observations as a valuable measurement tool for most of the essential elements of *Bridges*.

To select an observation protocol that aligned with our intended curriculum, MLC's research team reviewed eight existing observation protocols to determine whether each protocol would help us measure the essential elements of *Bridges*. The *Instructional Quality Assessment (IQA)* (Boston, 2012; Boston & Wolf, 2004, 2006; Junker et al., 2005) is our chosen observation protocol because the *IQA* rubrics have the closest correspondence to the essential elements we intend to measure in our research projects (Table 15.5).

Table 15.5

Essential Elements of Bridges Mapped onto the Instructional Quality Assessment Rubrics

Bridges in Mathematics Essential Elements	IQA Observation Protocol Rubrics
Active learning characterized by student-centered instruction	N/A (see table note)
Open-ended tasks promote reasoning, problem solving, and connections within and across representations	Potential of the task Implementation of the task *Both rubrics result in scores of 0–4*
Rich mathematical discourse supports understanding and achievement	Teacher's linking Students' linking Asking (teachers' press) Providing (students' responses) Student discussion following task Rigor of teacher's questions (Questioning) Mathematical residue *All rubrics result in scores of 0–4*

Note: There is no specific *IQA* rubric for the essential element focused on active learning. Our review determined that scores of 3 and 4 across the *IQA* rubrics position students to reason and problem solve and generally engage in meaningful mathematics work.

Teacher Interview Protocol

Researchers of teachers' use of mathematics curriculum materials have long used teacher interviews as a measurement tool to aid their understanding of how teachers interact with written curriculum during planning, enactment, and reflection. These interviews sometimes take place immediately before or after classroom observations or at key points in the overall study timeline. For example, Gujarati (2011) conducted debriefing interviews after classroom observations to clarify what was observed and gather data about teacher perceptions of the observed lesson. Sherin and Drake (2009) conducted both post-observation interviews and interviews at three key points throughout their study to shed light on how teachers interacted with curriculum before, during, and after implementation.

In our research, we leverage the other three data sources on implementation (the *Teachers Guide Annotation Tool*, *Unit Implementation Log*, and *Classroom Observation Protocol*) to tailor our semi-structured *Teacher Interview Protocol* to each teacher's enactment. Thus, our Teacher Interview Protocol is designed to gather *more* information about specific instances of a teacher's enactment of our curriculum and also pursue the *why* behind adaptations. Our semi-structured interview protocol is designed with a series of questions that center a specific instance of teacher enactment (e.g., a lesson that was adapted, omitted, or skipped). These specific instances are purposefully selected to gather additional data about one of the essential elements. For example, we may intentionally select a lesson that, as intended in the written material, was designed to promote *active learning characterized by student-centered instruction*, and then we ask a series of questions to understand how the teacher planned and implemented the lesson (Table 15.6).

Illustrative Example of How Data From Measurement Tools Create a Picture of Enactment

To illustrate how our measurement tools work together to create a rich picture of the enacted curriculum, we provide an example from one teacher's classroom from a recent multiple case study of teachers' enactments of Grade 2 Unit 5 from *Bridges* (3rd Edition) (Math Learning Center, 2023). The example draws data from the last sessions of Unit 5 (Session 3: *Pattern Block Sequences*, described in a previous section of the chapter, and Session 4: *More Growing Patterns*).

Core Design Feature

The written materials for Session 3 are designed to include a variety of instructional formats by calling for students to spend time on *Sessions* and *Work Places* and including the Think-Pair-Share instructional routine once in Session 3. We begin to have a picture of the enactment of this session

Table 15.6

Example Teacher Interview Questions Designed to Measure the Essential Elements of Bridges

Essential Elements	Teacher Interview Protocol

Active learning characterized by student-centered instruction
 	
Open-ended tasks promote reasoning, problem solving, and connections within and across representations
 	
Rich mathematical discourse supports understanding and achievement
	For (lesson title), you noted in your [unit log/Teachers Guide annotation] that you [omitted/adapted/replaced] this session. We'd like to learn more about this:
	
	a. When you first read this session during your planning, what did you think about it?
	b. During your planning, did you change the mathematical goals from those of the written materials? And, if so, why?
	c. Please describe what this session [or its replacement] looked like in your classroom. Follow up if adaptations were made:
	  1. Why did you make those changes?
	  2. Were these changes made in response to anything your students said or did during the lesson? If yes, please tell me more.
	d. Did your mathematical goals for your students change at any point during the lesson? And, if so, why?
	e. What, if anything, would you do differently in the future?

Visual models further sensemaking and deepen conceptual understanding	In what ways do you think visual models supported your students' development of conceptual understanding in this unit? Does a specific example come to mind that you could share? [If key visual models for the unit of instruction are not mentioned, ask specifically]: In what ways do you think [specific visual model] supported your students' conceptual understanding in this unit?

from the teacher's instructional log, where the teacher chose "adapted" as the best overall description of her implementation of Session 3 and indicated that she used the Think-Pair-Share routine once. We also know from the log that students in this classroom went to *Work Places* at some point that week. However, the *Teachers Guide Annotation Tool* sheds further light on the use of *Work Places* during this specific session because the teacher noted "not enough time" in the margins next to mention of sending students to *Work Places* in Session 3. Because Sessions 3 and 4 are the last two sessions of the unit and the only two sessions that were logged on the weekly log, we can conclude that students went to *Work Places* only during Session 4 that week. We also know from the log that students chose two specific *Work Place* games (*Close to 25¢* and *Beat you to $1*) and played the games independent of the teacher for less than 25 minutes. Taken together, the enactment data from the instructional log and the *Teachers Guide Annotation Tool* led us to conclude this core design feature (*purposeful variety of instructional formats and modalities*) was mostly implemented with integrity in Session 3.

Essential Elements

The written materials for Session 3 are intended to lead to active learning experiences for students who engage with Unifix cube and pattern block sequence tasks like the one described earlier in the chapter. Embedded questions in the Teachers Guide are designed to support teachers in implementing the essential element of *rich mathematical discourse to support understanding* by initially inviting students to make sense of the sequences through noting patterns and by discussing the generalization of the patterns as they grow. Our classroom observations then allow us to contrast intended teacher questions from the written materials with enacted teacher questions during the session (Table 15.7). For example, we see a loose alignment between the intended and enacted questions for Session 3 because the teacher asked students about the "100th sequence," rather than "100th arrangement," and provided some scaffolding with a restatement of the question ("*If I need three triangles for the 3rd, four for the 4th.... What would the 100th look like?*"). Later she followed up with a question that asked for procedural information ("*How many triangles would it have on the top? How many on the bottom?*") but did not ask questions that would encourage students to explain their reasoning as intended ("*How do you know?*"). Thus, student discussion for this task was rated as a level 2 on the *IQA* rubric (*Student Discussion Following the Task*); we concluded the enactment of this essential element was not aligned with the aims of the curriculum during Session 3.

For two other essential elements, however, observation and teacher interview data converge on a picture of teacher enactment of the last two sessions in Unit 5 that represent integrity of implementation. For the essential element of *active learning characterized by student-centered instruction*, we observed greater than 75% of students participating in small- or whole-group discussions during Sessions 3 and 4. The teacher interview data then reinforced our observations of active learning when the teacher shared that her students found the patterns in these sessions engaging and enjoyable:

They loved it. Oh my gosh! That was probably their favorite. They're always looking for patterns in *Number Corner* ... either the calendar itself or whatever the collection is, "what's the pattern?" And so, for them to just have four[5] straight days of patterns, they were in heaven. They loved it!

Table 15.7

Example Intended and Enacted Teacher Questions

Teacher Questions	
Intended from the materials	Enacted during the session
What patterns can you find in this sequence? What do you notice that repeats with each arrangement? What is different in each arrangement?	What do you notice about this shape?
What can you tell me about the 100th arrangement? How many triangles would there be? How would they be arranged? How do you know?	What would the 100th sequence look like? (Follow-up) If I need three triangles for the 3rd, four for the 4th...What would the 100th look like? (Follow-up) How many triangles would it have on the top? How many on the bottom?

The teacher also noted a progression in student confidence and understanding of patterns across the successive sessions, which culminated in student-created sequences in Session 4:

> And it showed when it came to making their own pattern. You could tell the kids were really like "Yeah, I got this!" and they just flew with making a pattern.

Finally, the combination of interview data and classroom observations of Session 4 again enriched our understanding of the enactment of the essential element of *open-ended tasks to promote reasoning, problem solving, and connections within and across representations*. The classroom observation of Session 4 allowed us to document how the teacher set up the task.

> I'm not going to give you a lot of directions today. You are going to make choices. You're going to make a growing pattern, not a repeating pattern like you did when you were in Kindergarten and First grade. I have pattern blocks, colored tiles, and Unifix cubes. You can work by yourself, or with a partner, or in a group of three. Three is the largest group. You are going to choose one bin and you are going to make your growing pattern. Make the 1st, 2nd, 3rd, and 4th, if you can. I will give you a sticky note to label your sequences.

This enactment was rated as a level 4 on the *IQA* rubric (*Implementation of the Task*). The variety of student-created sequences observed in this classroom further confirmed the openness of this task as enacted in Session 4 (Figure 15.3). We concluded this teacher implemented the Session 4 task in a manner that closely mirrors what was intended in the written materials.

Figure 15.3

Examples of Student-Created Sequences

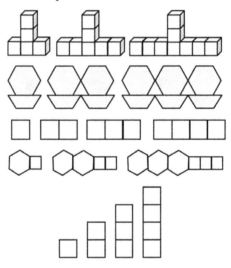

CONCLUSION

Curriculum researchers have lamented that the intentions behind a written curriculum are not always clear or explicit (Stein & Kim, 2009). This lack of clarity about the intended curriculum has at times limited the conclusions that can be drawn from research on curriculum use because it is difficult for researchers to move beyond agnostic descriptions of curriculum enactment without a clearly articulated intended curriculum. We also recognize that teacher enactment of curriculum is a complex practice involving many phases (planning, implementing, and reflecting) as well as decision points where teachers engage in a dynamic series of design choices based on their interactions with curriculum materials and their students (Brown, 2009; Lloyd et al., 2009). Thus, we consider the time spent defining what is intended in our curriculum and deciding what to prioritize when we examine the enactment of our curriculum as time well spent. It is through this conceptual work that we can be intentional in our measurement of curriculum use.

Together, our suite of measurement tools is designed to help us determine whether an enactment represents *implementation with integrity*, that is, implementations that are both responsive to students and aligned with curricular aims. Our measures of core design features (*Unit Implementation Log* and *Teachers Guide Annotation Tool*) help us identify moments in which teachers make the instructional design decision to adapt our written curriculum. Then our measures of essential elements (*Classroom Observation Protocol* and *Teacher Interview Protocol*) help us determine whether those adaptations fall within or outside the curriculum envelope. Making these determinations is no easy analytic task, but without those determinations, we have no sound basis for making decisions about future development efforts and no rigorous way of answering the question: Under what *implementation circumstances* does our curriculum make a difference for students?

REFERENCES

Ben-Peretz, M. (1990). *The teacher–curriculum encounter: Freeing teachers from the tyranny of texts*. State University of New York Press.

Boston, M. (2012). Assessing instructional quality in mathematics. *The Elementary School Journal, 113*(1), 76–104. https://doi.org/10.1086/666387

Boston, M., & Wolf, M. K. (2004, April 12–16). *Using the Instructional Quality Assessment (IQA) toolkit to assess academic rigor in mathematics lessons and assignments* [Paper presentation]. The Annual Meeting of the American Educational Research Association, San Diego, CA.

Boston, M. D., & Wolf, M. K. (2006). *Assessing academic rigor in mathematics instruction: The development of Instructional Quality Assessment Toolkit* (CSE Tech. Rep. No. 672). University of California, National Center for Research on Evaluation, Standards, and Student Testing (CRESST).

Brown, M. W. (2009). The teacher-tool relationship: Theorizing the design and use of curriculum materials. In J. T. Remillard, B. A. Herbel-Eisenmann, & G. M. Lloyd (Eds.), *Mathematics teachers at work: Connecting curriculum materials and classroom instruction* (pp. 17–36). Routledge.

Brown, S. I., & Walter, M. I. (2005). *The art of problem posing*. Psychology Press.

Caswell, L., Schwartz, G., Minner, D., Allexsaht-Snider, M., & Buxton, C. A. (2016). Using teacher logs to study project enactment and support professional learning in the LISELL-B project. In C. Buxton & M. Allexsaht-Snider (Eds.), *Supporting K–12 English language learners in science* (pp. 93–119). Routledge.

Chval, K. B., Chávez, Ó., Reys, B. J., & Tarr, J. (2009). Considerations and limitations related to conceptualizing and measuring textbook integrity. In J. T. Remillard, B. A. Herbel-Eisenmann, & G. M. Lloyd (Eds.), *Mathematics teachers at work: Connecting curriculum materials and classroom instruction* (pp. 70–84). Routledge.

DeBarger, A. H., Choppin, J., Beauvineau, Y., & Moorthy, S. (2013). Designing for productive adaptations of curriculum interventions. *Teachers College Record, 115*(14), 298–319.

Drake, C., & Land, T. J. (2015, April 16–20). *Ambitious curriculum use: Coordinating the use of curriculum materials with responding to children's mathematical thinking* [Paper presentation]. The Annual Meeting of the American Educational Research Association, Chicago, IL.

Drake, C., Land, T. J., Bartell, T. G., Aguirre, J. M., Foote, M. Q., Roth McDuffie, A., & Turner, E. E. (2015). Three strategies for opening curriculum spaces. *Teaching Children Mathematics 21*(6), 346–353.

Fosnot, C. T., & Dolk, M. (2001). *Young mathematicians at work: Constructing number sense, addition, and subtraction.* Heinemann.

Franke, M. L., Kazemi, E., & Turrou, A. C. (Eds.). (2018). *Choral counting and counting collections: Transforming the PreK–5 classroom.* Stenhouse.

Gujarati, J. (2011). From curriculum guides to classroom enactment: Examining early career elementary teachers' orientations toward standards-based mathematics curriculum implementation. *Journal of Mathematics Education at Teachers College, 2*(1), 40–46.

Humphreys, C., & Parker, R. (2015). *Making number talks matter: Developing mathematical practices and deepening understanding, grades 4–10.* Stenhouse.

Huntley, M. A. (2009). Brief Report: Measuring curriculum implementation. *Journal for Research in Mathematics Education, 40*(4), 355–362.

ICUBiT (Improving Curriculum Use for Better Teaching Project). (2022). *Tools.* https://icubit.gse.upenn.edu/tools.

Junker, B. W., Weisberg, Y., Matsumura, L. C., Crosson, A., Wolf, M., Levison, A., & Resnick, L. (2005). *Overview of the instructional quality assessment.* Regents of the University of California.

Kazemi, E., & Hintz, A. (2014). *Intentional talk: How to structure and lead productive mathematical discussions.* Stenhouse.

Land, T. J., Bartell, T., Drake, C., Foote, M. Q., Roth McDuffie, A., Turner, E. E., & Aguirre, J. M. (2019). Curriculum spaces for connecting to children's multiple mathematical knowledge bases. *Journal of Curriculum Studies, 51*(4), 471–493. https://doi.org/10.1080/00220272.2018.1428365

Lloyd, G. M., Remillard, J. T., & Herbel-Eisenmann, B. A. (2009). Teachers' use of curriculum materials: An emerging field. In J. T. Remillard, B. A. Herbel-Eisenmann, & G. M. Lloyd (Eds.), *Mathematics teachers at work: Connecting curriculum materials and classroom instruction* (pp. 3–14). Routledge.

Math Learning Center. (2023). *Bridges in mathematics* (3rd ed.).

McCoy, A., Barnett, J., & Combs, E. (2013). *High yield routines, grades K–8.* National Council of Teachers of Mathematics.

National Academies of Sciences, Engineering, and Medicine. (2022). *The future of education research at IES: Advancing an equity-oriented science.* National Academies Press. https://doi.org/10.17226/26428

National Council of Teachers of Mathematics [NCTM]. (2020). *Catalyzing change in early childhood and elementary mathematics: Initiating critical conversations.*

National Research Council. (2004). *On evaluating curricular effectiveness: Judging the quality of K–12 mathematics evaluations.* National Academies Press.

Parrish, S. (2010). *Number talks: Helping children build mental math and computation strategies, grades K–5*. Math Solutions.

Phelps, G., Corey, D., DeMonte, J., Harrison, D., & Loewenberg Ball, D. (2011). How much English language arts and mathematics instruction do students receive? Investigating variation in instructional time. *Educational Policy, 26*(5), 631–662. https://doi.org/10.1177/0895904811417580

Remillard, J. T. (2005). Examining key concepts in research on teachers' use of mathematics curricula. *Review of Educational Research, 75*(2), 211–246.

Remillard, J. T., & Heck, D. J. (2014). Conceptualizing the curriculum enactment process in mathematics education. *ZDM Mathematics Education, 46*(5), 705–718. https://doi.org/10.1007/s11858-014-0600-4

Richardson, K., & Dolphin, S. (2020). *Number talks in the primary classroom*. Didax.

Riel, J., Lawless, K. A., & Brown, S. W. (2016). Listening to the teachers: Using weekly online teacher logs for ROPD to identify teachers' persistent challenges when implementing a blended learning curriculum. *Journal of Online Learning Research, 2*(2), 169–200.

Rowan, B., Camburn, E., & Correnti, R. (2004). Using teacher logs to measure the enacted curriculum: A study of literacy teaching in third-grade classrooms. *The Elementary School Journal, 105*(1), 75–101. https://doi.org/10.1086/428803

Rowan, B., Jacob, R., & Correnti, R. (2009). Using instructional logs to identify quality in educational settings. *New Directions for Youth Development, 2009*(121), 13–31. https://doi.org/10.1002/yd.294

Shechtman, N., Roschelle, J., Feng, M., & Singleton, C. (2019). An efficacy study of a digital core curriculum for grade 5 mathematics. *AERA Open, 5*(2), 1–20. https://doi.org/10.1177/2332858419850482

Sherin, M. G., & Drake, C. (2009). Curriculum strategy framework: Investigating patterns in teachers' use of a reform-based elementary mathematics curriculum. *Journal of Curriculum Studies, 41*(4), 467–500. https://doi.org/10.1080/00220270802696115

Shumway, J. (2011). *Number sense routines: Building numerical literacy every day in grades K–3*. Stenhouse.

Silver, E. A. (1994). On mathematical problem posing. *For the Learning of Mathematics, 14*(1), 19–28.

Stein, M. K., & Kim, G. (2009). The role of mathematics curriculum materials in large-scale urban reform: An analysis of demands and opportunities for teacher learning. In J. T. Remillard, B. A. Herbel-Eisenmann, & G. M. Lloyd (Eds.), *Mathematics teachers at work: Connecting curriculum materials and classroom instruction* (pp. 57–75). Routledge.

Stickles, P. R. (2011). Using instructional logs to study teachers' adaptation to curricular reform. *School Science and Mathematics, 111*(2), 39–46. https://doi.org/10.1111/j.1949-8594.2010.00059.x

Teacher Education by Design [TEDD]. (2014). *Mathematics instructional activities*. University of Washington. https://tedd.org

Whitin, D. J. (2006). Problem posing in the elementary classroom. *Teaching Children Mathematics, 13*(1), 14–18.

Whitin, P. (2004). Promoting problem-posing explorations. *Teaching Children Mathematics, 11*(4), 180–186.

Wilson, M. (2005). *Constructing measures: An item response modeling approach*. Lawrence Erlbaum Associates.

Wolfe, E. W., & Smith, E. V. (2007). Instrument development tools and activities for measure validation using Rasch models: Part I—Instrument development tools. *Journal of Applied Measurement, 8*(1), 97–123.

ENDNOTES

1. Here and throughout this chapter, standards-based refers to both mathematics content standards and Standards for Mathematical Practice.
2. Grade 2 students in the U.S. are typically ages 7–8.
3. Mathematical routines include number talks, dot talks, quick images, same and different, choral counting, and number strings (cf. Fosnot & Dolk, 2001; Franke et al., 2018; Humphreys & Parker, 2015; McCoy et al., 2013; Parrish, 2010; Richardson & Dolphin, 2020; Shumway, 2011; Teacher Education by Design [TEDD], 2014).
4. Instructional routines include think-pair-share, gallery walk, and the following discussion structures: open strategy sharing, compare and connect, what's best and why?, define and clarify, and problem posing (cf., Brown & Walter, 2005; Kazemi & Hintz, 2014; Silver, 1994; Whitin, D., 2006; Whitin, P., 2004).
5. The last four sessions in Grade 2 Unit 5 focus on sequences and growing patterns. Thus, the teacher's reference to "four straight days of patterns."

CHAPTER 16

OUR JOURNEY THROUGH DATA ANALYSIS TO DEVELOP A MODEL OF CURRICULAR REASONING

Dawn Teuscher
Brigham Young University

Shannon Dingman
University of Arkansas

Lisa Kasmer
Grand Valley State University

Travis Olson
University of Nevada-Las Vegas

Curriculum (e.g., standards, textbooks, instructional materials) is important and teachers make numerous decisions about curricular materials as they plan and enact lessons. Therefore, it is paramount to identify how teachers make decisions and their reasoning for the decisions. In this chapter, we share the process of identifying U.S. middle grades teachers' curricular reasoning and their decisions, which is complex as these occur within teachers' minds. We share intentional methodological choices made in selecting the official and intended curricula to gather the needed data to make teachers' decisions and reasoning transparent. We then describe the iterative data analysis process used to gain insight into teachers' curricular reasoning. Research typically describes the data analysis of a study as though it were a linear process. We share examples from our data analysis to paint a picture of the messy data analysis process that often is not reported in published research. We conclude with four recommendations for curriculum researchers as they engage in future data analysis.

Lessons Learned From Research on Mathematics Curriculum, pp. 347–366
Copyright © 2024 by Information Age Publishing
www.infoagepub.com
All rights of reproduction in any form reserved.

Keywords: curricular reasoning; curriculum; teachers' decisions

Teachers make many curricular decisions every day as they prepare, implement, and reflect on their lessons. These decisions are informed by various resources that teachers utilize, such as the *Common Core State Standards for Mathematics* (*CCSSM*) (National Governors Association Center for Best Practices and Council of Chief State School Officers [NGA & CCSSO], 2010), scope and sequence documents, mathematics textbooks, and learning progressions/trajectories. We define these resources as curriculum because they are a "collection of written, published, or otherwise available materials that teachers utilize to structure tasks or lessons" (Dingman et al., 2021, p. 268). We also view teachers as designers of curriculum as they engage in a participatory relationship with the curriculum (Brown, 2009; Remillard, 2005). As designers of curriculum, teachers use curricular reasoning, defined as "the thinking processes that teachers engage in as they work with curriculum materials to plan, implement and reflect on instruction" (Breyfogle et al., 2010, p. 308). Teachers incorporate their own beliefs, knowledge, backgrounds, and experiences as they make these curricular decisions.

This process of identifying teachers' curricular decisions and their reasoning for these decisions is complex as it occurs in teachers' minds; therefore, teachers' reasoning is not transparent to an observer. Over the past several years, our research team has worked to unpack teachers' reasoning for these decisions. Specifically, we study the thinking processes that middle-grades mathematics teachers use as they plan and enact instruction based on curricular resources. In this chapter, we share our journey through the data collection and analysis process to demonstrate that this journey is not linear, but iterative. We share intentional methodological decisions made to gather the needed data and then the iterative data analysis techniques used to gain insight into teachers' curricular reasoning for their decisions. Our intention is to share with other researchers the "messy middle" of the research process so others can learn from and recognize that this is part of the research work that we all conduct.

INTENTIONAL METHODOLOGICAL DECISIONS

To examine in-depth the construct of curricular reasoning, we made two intentional methodological choices as we planned our research study. We intentionally selected mathematical content that Grade 8[1] teachers were not familiar teaching, as this content had not previously been part of their curriculum. Through our previous analysis of *CCSSM* in relation to state standards, we identified numerous mathematical topics that were shifted to earlier grade levels in *CCSSM*. One prominent shift was the teaching of congruence through the lens of geometric transformations (Tran et al., 2016). We then intentionally selected a mathematics textbook unit aligned with the Grade 8 *CCSSM* geometric transformation content that students were expected to learn, yet the textbook was not written for Grade 8. Although many may not see these choices as important or impactful for the study, we suggest that researchers intentionally need to think about methodological decisions as part of their study design. We provide more details about how each of these decisions impacted our study and data analysis.

Selection of Mathematics Content

The purposeful methodological choice of selecting mathematics content that teachers were not teaching and that was sequenced or approached in a different way than was typical impacted our study of middle-school teachers' curricular reasoning. In *A Priority Research Agenda for Understanding the Influence of the Common Core State Standards for Mathematics* (Heck et al., 2011), researchers are encouraged to analyze new curriculum by "choosing content topics that vary from the currently predominant approaches ... such as new content topics, or content topics that are approached, sequenced, or connected to other topics differently" (p. 10). This recommendation was used as we built on our previous findings from analyses of standards documents and textbooks (Dingman et al., 2013; Teuscher et al., 2015; Teuscher et al., 2016; Tran et al., 2016) to select mathematics content that was new to teach following adoption of the *CCSSM*.

Our research team identified different content areas that we could have selected, but we purposefully selected geometric transformations in Grade 8. We share examples from *CCSSM* to illustrate how geometric transformations are different for Grade 8 students. Two *CCSSM* standards (NGA & CCSSO, 2010) for Grade 8 students within the geometry domain are:

- Verify experimentally the properties of rotations, reflections, and translations. (*CCSSM* 8.GA.1)
- Understand that a two-dimensional figure is congruent to another if the second can be obtained from the first by a sequence of rotations, reflections, and translations; given two congruent figures, describe a sequence that exhibits the congruence between them. (*CCSSM* 8.GA.2)

These standards are students' initial introduction to geometric transformations (rigid motions) and the connection to congruence. This represents a different approach from how congruence was typically introduced in prior textbooks. Most often, congruence was introduced through examination of congruent sides and angles of figures (Teuscher et al., 2015). However, in the *CCSSM* students first define congruence through rigid motions; then, in high school, students are expected to build on this definition of congruence through transformations. The following *CCSSM* high-school standards (NGA & CCSSO, 2010) demonstrate how students continue to use the definition of congruence from Grade 8 to expand their understanding of congruence.

- Use the definition of congruence in terms of rigid motions to show that two triangles are congruent if and only if corresponding pairs of sides and corresponding pairs of angles are congruent. (*CCSSM* HSG.CO.B8)
- Explain how the criteria for triangle congruence (ASA, SAS, and SSS) follow from the definition of congruence in terms of rigid motions. (*CCSSM* HSG.CO.B8)

This selection of mathematics content was critical to identify teachers' decisions and their curricular reasoning. Because geometric transformations were new to Grade 8 and the *CCSSM* approach of connecting transformations to congruence was different from previous state standards, we hypothesized that teachers' decisions and reasoning would be more explicit for the research team to analyze. Schoenfeld (2011) suggests that the decision-making process is more apparent for non-routine decisions. Therefore, the Grade 8 teachers were asked to make non-routine decisions by using an unfamiliar curriculum that was aligned with the new standards to capture both their decisions and their reasoning.

Selection of Mathematics Textbook Unit

A second methodological choice was selecting a mathematics textbook unit aligned with the new approach to teaching geometric transformations. Although there were new textbooks available for teachers after *CCSSM* was released, none of the textbooks at Grade 8 aligned well with the approach recommended in the *CCSSM* standards (Teuscher et al., 2015; Tran et al., 2016). Therefore, we began looking more broadly to identify a textbook unit that would be aligned with the geometric transformation standards in Grade 8. We selected a high-school geometry textbook, *University of Chicago School Mathematics Project* (*UCSMP*) *Geometry* (Third Edition) (Benson et al., 2009), that approached connecting transformations to congruence as described in *CCSSM*. This textbook selection allowed us to provide teachers with a resource that was aligned with the new standards. However, because the textbook was written for high-school geometry students, it contained content that was not required for middle-school students. Having some geometry content in the high-school textbook that was aligned and other geometry content not aligned to the Grade 8 standards allowed teachers to make explicit decisions about usage and describe their reasoning for these decisions as we collected data. We expected that the middle-school teachers across our project would make different curricular decisions from each other and that their reasoning might be different as well. We found that having the teachers use an unfamiliar textbook unit and focus on standards that were new to them and their students placed them in a non-routine situation. As a result, they were able to identify and share with the research team their reasoning in explicit ways.

Example of Intentional Methodological Choices

To illustrate how the mathematics content and curriculum choices were helpful, we share various examples of teachers' decisions and their reasoning. All teachers in our study had to decide whether to introduce students to the definition that a reflection line is the perpendicular bisector for all points on the preimage and image. *UCSMP Geometry* includes this definition in the lesson materials; as it was a high-school textbook, we wondered whether teachers would choose to include it in Grade 8. Therefore, we asked teachers about their decision. Most of the middle-school teachers in our study, when first given the textbook unit, did not plan to include this definition as a major concept of reflections. Teachers' reasons for not including the definition of a reflection line were: "It doesn't include this in the *CCSSM*," or "This is too much depth, students do not need this to identify reflections." However, about half of the 17 teachers who participated in our study decided that the definition was important and they included the definition or parts of it in their lessons. These teachers reasoned that, "This is how you know that it is a true reflection," and "It is important for students to know how to create a reflection."

Intentionally selecting the mathematics content and providing a mathematics textbook unit aligned with the new standards were critical methodological choices for the research team to identify teachers' decisions and reasoning. We learned that not all teachers made the same mathematical decisions about whether to address the definition of a reflection, so our next step was to identify teachers' reasoning for these decisions. The analysis of teachers' reasoning helped our team make sense of their decisions and their reasoning.

MESSINESS OF DATA ANALYSIS

The research team's methodological choices placed teachers in a non-routine situation (i.e., defining congruence with geometric transformations) where they were asked to make their decisions and reasoning transparent. We then turned our attention to the messy work of data analysis. In most published research, it appears as though research is conducted and analyzed in a linear fashion with the results appearing through data analysis. We suggest that the data analysis process is not linear; rather, it is iterative in that researchers constantly return to the research questions and framework to modify, clarify, and add to their analysis for a deep understanding of the particular phenomenon being studied. In this section, we share important parts of this iterative process that helped our research team dig into the data.

Unit of Analysis

The first step in data analysis is to identify the unit of analysis to ensure consistent comparisons across all data. In curriculum research, this unit of analysis is often unclear. We had to identify what was and was not an instance of curricular reasoning. Without the unit of analysis being defined, we would not be able to compare curricular reasoning among the participating teachers. We found that the unit of analysis allowed us to break our data into manageable slices that we could process and code. The process of determining the unit of analysis is an iterative one; we modified our definition of the unit of analysis a number of times until we captured what we wanted to identify.

Our initial attempt at defining a unit of analysis was to identify teachers' decisions as they discussed the planning of their mathematics lesson. The research team anticipated there would be two types of decisions: mathematical and pedagogical. Although this seemed like an obvious starting point as we were studying teachers' reasoning for their decisions, we ran into two problems. First, as we began identifying the different decisions teachers made, we were coding every sentence. This split our data into pieces so small that we lost the larger context of teachers' reasoning for their decisions as they planned and reflected on their lessons. Second, the team struggled to separate teachers' decisions into mathematical or pedagogical categories. For example, our research team had many discussions about a specific teacher's decision to use a video clip from the Disney movie *Mulan*[2] in the launch of a lesson to highlight the idea of geometric reflections with her students. This was a decision that the teacher made as she planned her lesson, but the research team struggled with whether the decision was a mathematical decision, a pedagogical decision, or both a mathematical and pedagogical decision? As we discussed these issues, we considered whether identifying a teacher's decisions was the unit of analysis that would help the team make sense of her reasoning. The answer was no.

Our next attempt was to return to our theoretical framework on curricular reasoning using three ways that teachers reason when planning lessons, as identified by Roth McDuffie and Mather (2009). We determined that we would use these three ways to identify the unit of analysis. Although helpful, we again bumped into an issue. After coding a few videos, we had data coded as curricular reasoning; however, we had more data not coded for curricular reasoning. We wondered whether teachers were not using curricular reasoning during these times or whether we missed their curricular reasoning. As we reviewed the uncoded data, we found teachers were using curricular reasoning but not in ways aligned with the three initial ways of reasoning that we were using as our unit of analysis. We needed to capture all the ways in which teachers may reason about their decisions, not just those that had already been identified in prior research.

We went back to the drawing board and determined that we needed something different from our theoretical framework. We shifted to identifying *instances of curricular reasoning*, which we defined as *a teacher's decision about the mathematical trajectory or flow of the lesson*. Although this attempt focused on our research question related to the teacher's curricular decisions, it was too narrow and not as clear as we had anticipated. The interview protocols included questions where teachers would respond to their own mathematical thinking or knowledge, which were important ingredients to their decisions and reasoning about the mathematical trajectory or flow of the lesson. For example, teachers were asked in a pre-interview, "what does orientation of a figure mean within geometric transformations?" In most cases the teachers responded with either "it is the position of the image on the paper after it is transformed" or "it is the direction that the figure is labeled in comparison to the pre-image." These answers were important for us to understand the teachers' decisions and reasoning for content within the lesson, but they were excluded from the coding because the teachers were not discussing the mathematical flow or trajectory of the lesson. The research team agreed that these instances needed to be included to make sense of teachers' decisions and reasoning.

Our unit of analysis was once again refined to *an instance of curricular reasoning* where the importance of the mathematics of the lesson became explicit. We defined an instance of curricular reasoning to be when *a teacher indicates that they have arrived at a determination regarding their approach to instruction while considering aspects of their knowledge or practice <u>related to the mathematics of the lesson</u> (e.g., students, textbook materials, beliefs, lesson sequence, standards documents). Or a teacher expresses a thought, idea, or opinion formed* or makes a remark while considering an aspect of her or his knowledge or practice related to the mathematics of the lesson (e.g., students, textbook materials, beliefs, lesson sequence, standards documents). We now provide examples to demonstrate how one instance of curricular reasoning might have changed in our coding over time and to demonstrate the difference between an instance of curricular reasoning and one that was not curricular reasoning.

In the first example of curricular reasoning, Helen responds to a pre-interview question about how she used the UCSMP materials to plan her geometric reflections lesson:

> Textbooks make me uncomfortable ... and when I use a textbook, I use it for resources. I look at ... how do they sequence that. So, when I make decisions based off of a textbook, I look at sequence first and then I look at are the activities on my students' level, are those activities I can use, how can I rewrite that so it is on my students' level. At first I was very unsure of the sequence, but when I really dug in and looked at it, I was like, I get it now.

Originally, this instance was coded as three separate decisions: (1) Helen reviews the sequence within the textbook; (2) Helen reviews the student activity to see if it is at the appropriate student level; and (3) Helen was unsure whether to use the UCSMP materials sequence for geometric transformations. However, if these decisions are separated, then we lose the larger context of how Helen reasoned about using the UCSMP materials to plan her lesson. In this example, Helen arrived at a determination that the sequence within UCSMP, although different from how she had previously taught transformations, was something that made sense. In her interview, Helen explained that her understanding of why the UCSMP materials used the sequence of reflections, translations, and then rotations was why she decided to use the task from the UCSMP materials with her students.

The second example is one that was initially an instance of curricular reasoning, but became a non-example after the refinement of our definition. Helen explained how the students will work on their task during the lesson:

> The assignment I had for them had three different pieces to it, but I only put one page in front of them at a time. Because I knew if I gave them all three pages they would shut down immediately, "I can't do this," you know, the I can't would start coming in and so instead breaking it into ok, let's just get the material out and start there.

We initially coded this as an instance of curricular reasoning because it was a decision the teacher made in relation to the flow of the lesson; however, using the refined definition, this is not an instance of curricular reasoning. Helen made a decision to give students one page of the task at a time to keep students motivated. According to the refined definition, there is nothing in this decision related to the mathematics of the lesson; so, this instance was removed as curricular reasoning. As we coded the pre- and post-lesson interviews for instances of curricular reasoning, using the refined definition was much easier for us as well as the graduate students.

When we started this project, we had quite a bit of research experience. Identifying the unit of analysis was an iterative process of continually returning to our research questions and our theoretical framing of the research. The alignment among the research question, the theoretical framing, and the data analysis is an important characteristic of quality research (Leatham, 2019). A critical first step in analyzing our data was defining a unit of analysis with appropriate grain-size. Without this, the research team could not identify the specific curricular decisions teachers made with their reasoning attached.

Building on and Extending Other's Research

Findings from curriculum research suggest that curriculum (e.g., standards, textbooks) matters (Lloyd et al., 2017; Stein et al., 2007). High-quality curriculum is important for teachers for a variety of reasons: it provides a trajectory of the mathematics students will learn (e.g., Tarr et al., 2006); it provides learning experiences for teachers (e.g., Davenport, 2000); it provides types of questions that teachers can ask students (e.g., Boaler & Brodie, 2004); and it influences classroom instruction. Findings from curriculum research suggest that curriculum also matters for students. High-quality curriculum provides students with high cognitive demand tasks (e.g., Stigler & Hiebert, 2004); it provides students a focus on conceptual and procedural understanding (Stein et al., 2007); and it has been shown to improve student learning if tasks are implemented at a high level (Cai & Howson, 2013).

Cohen and Ball (1999) defined the instructional triangle in Figure 16.1a as "the interactions among teachers, and students, around educational materials" (p. 2). Our research team was focused on the curricular decisions a teacher makes, which we saw as a subset of the instructional decisions that Cohen and Ball used to connect the teacher, student, and educational materials (i.e., curriculum). With more research, Cohen et al. (2003) modified the instructional triangle, as shown in Figure 16.1b, suggesting that instructional decisions were interactions among teachers, students, and content. Although the second instructional triangle aligned with our study, we wondered why the authors made the modification from instructional materials to content. We also wondered if the authors defined content to include instructional materials or if content was something separate and removed from the interaction with teacher and student. At this point in our research, we took the change to content to include both content and instructional materials.

Figure 16.1a

Instructional Triangle

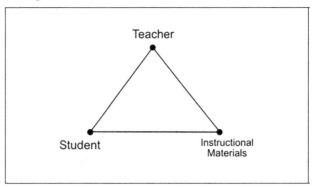

Source: Adapted from Cohen and Ball (1999, p. 3).

Figure 16.1b

Revised Instructional Triangle

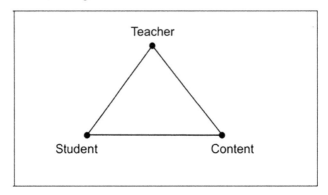

Source: Adapted from Cohen et al. (2003, p. 124).

In our research meetings, as we continued to discuss our coding of teachers' decisions and reasoning, we realized that assuming content included the instructional materials was problematic for our data analysis. We began to discuss the following questions:

- What is the difference between content and instructional materials?
- Should content and instructional materials be separated because they are interactions that happen among teachers and students?
- If content and instructional materials are different, do teachers see them as different?
- Are teachers reasoning about mathematics content and instructional materials differently?
- If there was a difference, how could we distinguish between the two and was it necessary to tease apart instructional materials and mathematics content?

To investigate these questions, we dug into our data to find multiple examples to illustrate differences between curriculum and content. For the purpose of this chapter, we share one example to illustrate the difference. Helen was a 19-year veteran Grade 8 mathematics teacher who taught mathematics without using a textbook for most of her career. Helen had typically taught geometric transformations in the following order: translations, reflections, and rotations. This sequence is fairly common in many textbooks; but because Helen did not use a textbook, we wondered why she taught the topics in this order. Helen's reason for teaching the transformations in this sequence was that translations are the easiest transformations; therefore, this sequence provides an easier access point for her students. Rotations are the most difficult transformations; therefore, she would teach these last after students had some understanding of the other transformations. Helen sequenced her lessons from easiest to hardest content to give her students access to the mathematics and build on their understanding of this content. Helen reasoned about the mathematical content she was teaching in relation to her students.

During our study, Helen was given both the teacher and student editions of the geometric transformation unit from *UCSMP Geometry* to use as she planned her unit. Helen noted the textbook unit had a different sequence than she had previously taught. In the textbook, reflections are introduced first, and then translations and rotations are built from reflections. Helen stated, "I didn't understand why they started this way." After reviewing and working through the lessons in the textbook unit, she decided to modify her lesson sequence to match the UCSMP sequence. Helen reasoned about her change in the lesson sequence:

> The UCSMP curriculum was aligned with the *Common Core State Standards for Mathematics* (*CCSSM*) that requires students to use sequences of transformations to identify if figures are congruent or similar.

In this example, the curriculum was a catalyst for her to think about the mathematics content differently than she had in the past 19 years. Without the UCSMP curriculum, Helen may not have changed her lesson sequence of geometric transformations, nor may she have made the connection between transformations and congruence.

After recognizing that all eight of our middle-school teachers who participated during the first year of data collection reasoned about the mathematics and curriculum in different ways, we decided to separate the content (mathematics) vertex in the instructional triangle into two elements (curriculum and mathematics). Figure 16.2 displays the Instructional Pyramid (Dingman et al., 2021) that models the four elements that interact as teachers make curricular decisions during the teaching process. As we continued to analyze our data, we saw the importance of keeping the curriculum (i.e., instructional materials) and mathematics (i.e., content) as distinct elements because often teachers thought about these two elements separately.

Making Sense of the Data and Extending Others' Research

After identifying our unit of analysis, we began coding teachers' curricular reasoning for their decisions. To begin the coding process, we used curricular reasoning aspects from Roth McDuffie and Mather (2009) and Breyfogle et al. (2010): (1) *analyzing curriculum materials from learners' perspectives,* (2) *doing the task as learners,* (3) *mapping learning trajectories,* and (4)

Figure 16.2

The Instructional Pyramid

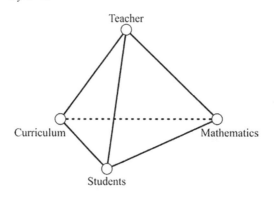

Source: Adapted from Dingman et al. (2021, p. 270).

reflecting and revising plans based on experiences in teaching and learning. As we coded, we ran into three problems. First, the curricular reasoning aspect of *analyzing curriculum materials from learners' perspectives* was too large. This curricular reasoning aspect seemed to be used any time a teacher reasoned about curriculum or students, which happened in many of their decisions. Second, we had a group of teachers' decisions that did not fit into any of the four curricular reasoning aspects. Third, we did not find any instances of teachers' reasoning by *doing the task as learners*. We began to wonder if this curricular reasoning aspect was part of analyzing curriculum materials from learners' perspectives. Before we could continue coding, we returned to the literature to determine ways in which we might separate *analyzing curriculum materials from learners' perspectives* to better define these for our research team. We also needed to determine other ways teachers reason when making mathematical decisions as they plan, implement, and reflect on lessons.

To address the first problem, we separated *analyzing curriculum materials from a learners' perspective* into two aspects: *Analyzing Curriculum Materials* and *Viewing Mathematics from the Learner's Perspective*. Our hope was that this change would help us discriminate among different teachers' reasoning in this larger category.

To address the second problem, we returned to the literature and built on Thompson's (2013, 2016) work with mathematical meanings. Thompson states that "teachers' mathematical meanings constitute their images of the mathematics they teach and intend that students have" (Thompson, 2016, p. 437). This seemed to be a key part of teacher reasoning that was not captured previously. Therefore, we created a new curricular reasoning aspect, *Considering Mathematical Meanings*.

To address the third problem, we determined that teachers doing the task as learners was an action that teachers do as they plan their lessons, but was not a way in which teachers were reasoning as they made decisions. Although many of the teachers discussed the tasks they selected or used, they were often reasoning about the learner perspective or the mathematics that they planned for students to learn from the task.

At this point in our research, we had identified five curricular reasoning aspects that described teachers' reasoning for their mathematics decisions. As we began to make sense of the coded data, we returned to the research literature to build on others' work. Olive et al. (2009), Rezat (2006), Rezat and Sträber (2012), and Tall (1986) had all advocated for a fourth vertex to be added to the Instructional Triangle to create a tetrahedron model, yet none had discussed the meaning of the edges in the model or the connection of the different vertices.

With our work on expanding the Instructional Pyramid to separate the content vertex into two elements—mathematics and curriculum—we began to wonder if we could overlay teachers' curricular reasoning on the Instructional Pyramid. Would the curricular reasoning aspects explain the connection between the different elements? This started our process of creating different models to demonstrate the different connections among the four elements in the Instructional Pyramid to the curricular reasoning aspects. Figures 16.3a and 16.3b illustrate just two versions of the model of the Instructional Pyramid with the curricular reasoning aspects overlaid. Our point in sharing the figures is not to have the reader understand all features of the figures, but to see the iterative process of curriculum research. These models demonstrate how the research team made sense of the Instructional Pyramid and the curricular reasoning aspects. However, another important part of this iterative process was to receive feedback from others in the field who are familiar with curriculum research, which had us continue to make sense of parts of the model that did not seem to work or make sense to others.

Our first version of the model (Figure 16.3a) was too complex and we attempted to describe too much in the model. We removed unnecessary parts that distracted from the four elements (*Teacher, Student, Mathematics,* and *Curriculum*) related to teachers' curricular reasoning aspects (i.e., edges) that teachers use as they make their curricular decisions. Although the iteration in Figure 16.3b is less complex, some of the curricular reasoning aspects are on multiple edges; in one case, multiple curricular

Figure 16.3a

First Model of Instructional Pyramid with Curricular Reasoning Overlayed

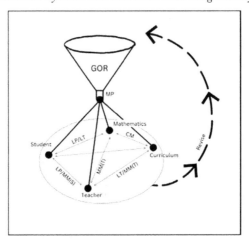

Figure 16.3b

Another Iteration of the Instructional Pyramid with Curricular Reasoning Overlayed

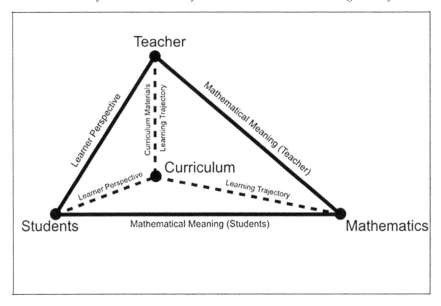

reasoning aspects are on one edge. For example, *Learner Perspective* is on the edge that connects the elements *Students* and *Teacher* as well as *Students* and *Curriculum*. Likewise, the edge between *Curriculum* and *Teacher* had two curricular reasoning aspects. As we continued discussions within our research team and shared our results with others in the field, we concluded that each edge should have a distinct curricular reasoning aspect if we wanted the model to provide teachers and educational leaders anything of importance. We returned to our coding and separated some curricular reasoning aspects to identify the specifics of teachers' reasoning between the different elements.

Figure 16.4 is our current model of the Instructional Pyramid with curricular reasoning overlayed. In this model, only one curricular reasoning aspect is on each edge of the Pyramid. Although the learner perspective is on two edges, we developed subcodes *Anticipating and Assessing* (A/A) and *Intentionality of the Task* (IT) to distinguish among the connections of *Students* and *Teacher* and *Students* and *Curriculum*. Similarly, mathematical meaning is on two edges, but the subcodes *Teacher's Mathematical Meaning* (TM) and *Student Mathematical Meaning* (SM) distinguish among the connections of *Teacher* and *Mathematics* and *Students* and *Mathematics*.

Figure 16.4

Another Iteration and Current Instructional Pyramid with Curricular Reasoning Overlayed

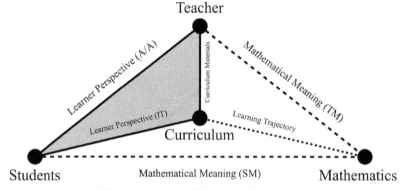

Source: Adapted from Dingman et al. (2021, p. 280).

RECOMMENDATIONS FOR CONDUCTING CURRICULUM RESEARCH

For the past 20 years, our team has conducted curriculum research to investigate the importance of curriculum for teaching and learning mathematics. We continue to learn and modify our own knowledge base about mathematics curriculum as we conduct more research, but also as we build on and extend others' work in the field. We now share four recommendations to assist other researchers who are on their journey of conducting research on mathematics curricula.

First, *be intentional about methodological decisions as you plan curriculum research studies*. In curriculum research, we have found that narrowing in on a specific topic within the curriculum (i.e., geometric transformations) and putting teachers and students in non-routine situations (i.e., using a new curriculum) creates transparent data for researchers to study. Intentional methodological choices will bound the data and keep the research focused on the research questions of interest.

Second, *expect a messy journey as you analyze the data*. Although published research does not depict this journey, it is necessary to move the field forward in curriculum research. Published manuscripts have gone through multiple tiers of editing and the final results are multiple iterations of researchers making sense of their data. The messiness of the journey is generally not included in published work because the field is interested in the findings rather than the messiness of the process.

Third, *expect iterations of consulting with other researchers, reading more literature, and analyzing data*. Researchers need to continually return to the research literature, even after starting a study. As researchers begin analyzing their data, they may have blinders that prevent them from seeing the data in other ways. It is important that researchers step back from their data to see a bigger picture, but also receive input from others who are not on the project. Our advisory board members pushed our research team to think in different ways about the data. Often, our advisory board meetings were the impetus for changing course, broadening our view, and building on others' work that was aligned with our research.

Fourth, *expect to be flexible during the data analysis*. Without listening to others and gaining their insight, some of the ideas that we found in our research would not have happened. If we had stayed with the framework of the Instructional Triangle, our data would not have fit and we may not have built the Instructional Pyramid model, allowing teachers and researchers to assess their own reasoning about curricular decisions. We had to be flexible and return to the drawing board multiple times. This was time consuming and tedious, but this flexibility allowed us to build on a large body of research and connect ideas in the field that had not been previously connected.

ACKNOWLEDGEMENT

This material is based upon work supported by the U.S. National Science Foundation under Grant Nos. 1561542, 1561554, 1561569, 1561617, 2201164, and 2201169. Any opinions, findings, and conclusions or recommendations expressed in this chapter are those of the authors and do not necessarily reflect the views of the National Science Foundation.

REFERENCES

Benson, L., Capuzzi-Feuerstein, C., Fletcher, M., Jakucyn, N., Klein, R., Marino, G., Miller, M. J., Powell, N. N., & Usiskin, Z. (2009). *The University of Chicago Mathematics Project: Geometry*. Wright Group/McGraw-Hill.

Boaler, J., & Brodie, K. (2004). The importance, nature and impact of teacher questions. In D. E. McDougall & J. A. Ross (Eds.), *Proceedings of the twenty-sixth annual meeting of the North American Chapter of the International Group for the Psychology of Mathematics Education* (pp. 774–782). Ontario Institute for Studies in Education.

Breyfogle, M. L., Roth McDuffie, A., & Wohlhuter, K. A. (2010). Developing curricular reasoning for grades pre-K–12 mathematics instruction. In B. Reys, R. Reys, & R. Rubenstein (Eds.), *Mathematics curriculum: Issues, trends, and future directions* (pp. 307–320). National Council of Teachers of Mathematics.

Brown, M. W. (2009). The teacher-tool relationship: Theorizing the design and use of curriculum materials. In J. T. Remillard, B. A. Herbel-Eisenmann, & G. M. Lloyd (Eds.), *Mathematics teachers at work: Connecting curriculum materials and classroom instruction* (pp. 17–36). Routledge.

Cai, J., & Howson, A. G. (2013). Toward an international mathematics curriculum. In M. A. Clements, A. Bishop, C. Keitel, J. Kilpatrick, & K. S. F. Leung (Eds.), *Third international handbook of mathematics education research* (pp. 949–978). Springer.

Cohen, D. K., & Ball, D. L. (1999). *Instruction, capacity, and improvement.* Consortium for Policy Research in Education Research Reports. https://www.cpre.org/sites/default/files/researchreport/783_rr43.pdf

Cohen, D., Raudensbush, S., & Ball, D. L. (2003). Resources, instruction, and research. *Educational Evaluation and Policy Analysis, 25*(2), 119–142. https://doi.org/10.3102/01623737025002119

Davenport, L. (2000). *Elementary mathematics curricula as a tool for mathematics education reform: Challenges of implementation and implications for professional development.* Center for the Development of Teaching (CDT) Paper Series. Education Development Center.

Dingman, S. W., Teuscher, D., Newton, J. A., & Kasmer, L. (2013). Common mathematics standards in the United States: A comparison of K–8 state and Common Core standards. *The Elementary School Journal, 113*(4), 541–564. https://doi.org/10.1086/669939

Dingman, S., Teuscher, D., Olson, T. A., & Kasmer, L. A. (2021). Conceptualizing curricular reasoning: A framework for examining mathematics teachers' curricular decisions. *Investigations in Mathematics Learning, 13*(4), 267–286. https://doi.org/10.1080/19477503.2021.1981742

Heck, D. J., Weiss, I. R., Pasley, J. D., Fulkerson, W. O., Smith, A. A., & Thomas, S. M. (2011). *A priority research agenda for understanding the influence of the common core state standards for mathematics* [Technical report]. Horizon Research.

Leatham, K. R. (Ed.). (2019). *Designing, conducting, and publishing quality research in mathematics education.* Springer. https://doi.org/10.1007/978-3-030-23505-5

Lloyd, G. M., Cai, J., & Tarr, J. E. (2017). Issues in curriculum studies: Evidence-based insights and future directions. In J. Cai (Ed.), *Compendium for research in mathematics education* (pp. 824–852). National Council of Teachers of Mathematics.

National Governors Association Center for Best Practices & Council of Chief State School Officers. (2010). *Common Core State Standards for Mathematics.* Retrieved September 9, 2023, from https://www.thecorestandards.org/Math/

Olive, J., Makar, K., Hoyos, V., Kor, L. K., Kosheleva, O., & Sträber, R. (2009). Mathematical knowledge and practices resulting from access to digital technologies. In C. Hoyles & J. B. Lagrange (Eds.), *Mathematics education and technology - Rethinking the terrain* (pp. 133–177). New ICMI Study Series, Vol 13. Springer. https://doi.org/10.1007/978-1-4419-0146-0_8

Remillard, J. T. (2005). Examining key concepts in research on teachers' use of mathematics curricula. *Review of Educational Research, 75*(2), 211–246. https://doi.org/10.3102/00346543075002211

Rezat, S. (2006). A model of textbook use. In J. Novotná, H. Moraová, M. Krátká, & N. Stehlíková (Eds.), *Proceedings of the 30th Conference of the International Group for the Psychology of Mathematics Education* (pp. 409–416). Charles University in Prague.

Rezat, S., & Sträber, R. (2012). From the didactical triangle to the socio-didactical tetrahedron: Artifacts as fundamental constituents of the didactical situation. *ZDM Mathematics Education, 44*, 641–651. https://doi.org/10.1007/s11858-012-0448-4

Roth McDuffie, A., & Mather, M. (2009). Middle school mathematics teachers' use of curricular reasoning in a collaborative professional development project. In J. T. Remillard, B. A. Herbel-Eisenmann, & G. M. Lloyd (Eds.), *Mathematics teachers at work: Connecting curriculum materials and classroom instruction* (pp. 302–320). Routledge.

Schoenfeld, A. H. (2011). *How we think: A theory of goal-oriented decision making and its educational applications*. Routledge.

Stein, M. K., Remillard, J., & Smith, M. S. (2007). How curriculum influences student learning. In F. K. Lester, Jr. (Ed.), *Second handbook of research on mathematics teaching and learning* (Vol. 1, pp. 319–369). National Council of Teachers of Mathematics.

Stigler, J. W., & Hiebert, J. (2004). Improving mathematics teaching. *Educational Leadership: Journal of the Department of Supervision and Curriculum Development, 61*(5), 12–17.

Tall, D. (1986). Using the computer as an environment for building and testing mathematical concepts: A tribute to Richard Skemp. *Papers in honour of Richard Skemp*, 21–36, University of Warwick.

Tarr, J. E., Chavez, O., Reys, R. E., & Reys, B. J. (2006). From the written to the enacted curricula: The intermediary role of middle school mathematics teachers in shaping students' opportunity to learn. *School Science and Mathematics, 106*(4), 191–201. https://doi.org/10.1111/j.1949-8594.2006.tb18075.x

Teuscher, D., Kasmer, L., Olson, T., & Dingman, S. (2016, July 24-31). *Isometries in new U.S. middle grades textbooks: How are isometries and congruence related?* [Paper presentation]. The Thirteenth International Congress on Mathematical Education, Hamburg, Germany.

Teuscher, D., Tran, D., & Reys, B. J. (2015). Common Core State Standards in the middle grades: What's new in the geometry domain and how can teachers support student learning? *School Science and Mathematics, 115*(1), 4–13. https://doi.org/10.1111/ssm.12096

Thompson, P. W. (2013). In the absence of meaning. In K. R. Leatham (Ed.), *Vital directions for research in mathematics education* (pp. 57–93). Springer. https://doi.org/10.1007/978-1-4614-6977-3_4

Thompson, P. W. (2016). Researching mathematical meanings for teaching. In L. English & D. Kirshner (Eds.), *Handbook of international research in mathematics education* (pp. 435–461). Taylor & Francis.

Tran, D., Reys, B. J., Teuscher, D., Dingman, S., & Kasmer, L. (2016). Analysis of curriculum standards: An important research area. *Journal for Research in Mathematics Education, 47*(2), 118–133. https://doi.org/10.5951/jresematheduc.47.2.0118

ENDNOTES

1. In the U.S., students in Grade 8 are generally 13–14 years old.
2. During this video clip, Mulan sings a song titled *Reflections* and there are multiple instances of her seeing her reflection in the water of a lake or a mirror on the wall.

CHAPTER 17

LESSONS LEARNED FROM STUDYING THE TEACHER INTENDED CURRICULUM

The Value of Capturing and Analyzing Attention

Kelsey Quaisley
Oregon State University

Lorraine M. Males
University of Nebraska-Lincoln

Our research focuses on the teacher intended curriculum with the goal of understanding how teachers use curriculum materials to plan lessons. In this chapter, we discuss how we overcame the challenge of studying the teacher intended curriculum by exploring teachers' noticing of the College Preparatory Mathematics (CPM) instructional materials for high-school mathematics and Go Math instructional materials for middle-school mathematics. Without being able to physically see what teachers see, our main challenges included: (1) knowing what teachers are looking at, especially when they are silent or quickly scrolling on a computer; (2) knowing how long, how often, and in what order teachers are looking at various aspects of materials; (3) knowing what teachers are NOT attending to; and (4) understanding what teachers are talking about during planning. Visualizations, together with eye-tracking video and eye-tracking metrics, allowed us to overcome these challenges and better understand teachers' attention. Thus, we were supported in understanding teachers' interpretations and responses to piece together a comprehensive story of teachers' transformation of the intended curriculum to the teacher intended curriculum. The overarching lesson for readers is that being able to physically see what teachers are seeing matters.

Keywords: curricular noticing; lesson planning; prospective secondary teachers; secondary mathematics teachers; teacher intended curriculum

Our research focuses on the *teacher intended curriculum* (Remillard & Heck, 2014; Valverde et al., 2002), also known as the teacher's plan, with the goal of understanding how teachers use curriculum materials to plan lessons. In this chapter, we discuss how we overcame the challenge of studying the teacher intended curriculum by exploring teachers' noticing of curriculum materials. We highlight the importance of and difficulties in capturing and analyzing teachers' attention to curriculum materials, and describe how we have used technology to mitigate these challenges.

CURRICULUM TRANSFORMATIONS AND THEORETICAL FRAMING OF THE RESEARCH

Before students have access to mathematics content, the intended curriculum goes through a series of transformations (Stein et al., 2007). These transformations are influenced by several factors, such as teachers' beliefs and knowledge, orientations towards curriculum materials, professional identities, understandings of students, access and support, and expectations in their teaching context (Remillard & Heck, 2014; Stein et al., 2007). It is critically important to understand the initial transformation between intended (i.e., the written curriculum as intended by the authors) and the teacher intended curriculum (i.e., the teacher's plan), as this transformation has implications for all other transformations (e.g., from teacher intended to enacted curriculum, and from enacted to learned curriculum). Moreover, understanding these transformations has the potential to increase students' opportunities to learn with the curriculum materials.

To frame our work, we draw from two complementary theoretical perspectives. First, we adopt the perspective that teachers' interactions with curriculum materials are participatory, meaning that there is a collaborative and dynamic interchange between teachers and curriculum materials (Remillard, 2005). Second, we adopt the view that teachers generate documents through a process of documentational genesis, in which features of the resources influence teachers' planning and the knowledge they develop (instrumentation), and teachers' knowledge then influences their decisions regarding different resources (instrumentalization) (Gueudet & Trouche, 2009). The result of documentational genesis is a document, which Gueudet and Trouche (2009) describe as schemes of utilization of a set of resources for the same class of situations across a variety of contexts. In other words, this process produces a set of documents consisting of a teacher's combined resources and plans of action.

To operationalize interactions teachers have with curriculum materials, we use the Curricular Noticing Framework (Dietiker et al., 2018), which is informed by research on professional noticing of children's mathematical thinking (Jacobs et al., 2010). Curricular noticing enables "teachers to recognize, make sense of, and strategically employ opportunities available within their curriculum materials" (Dietiker et al., 2018, p. 524). Curricular noticing is comprised of three interrelated skills (Males et al., 2015). *Curricular Attending* is looking at, reading, and recognizing aspects of curricular materials. *Curricular Interpreting* is making sense of that to which the teacher attended. *Curricular Responding* is making curricular decisions based on the interpretation (e.g., generating a lesson plan, a visualization, or enactment). Figure 17.1 illustrates the connections between these theoretical and analytical frameworks.

Figure 17.1

Theoretical Framework Used in the Project

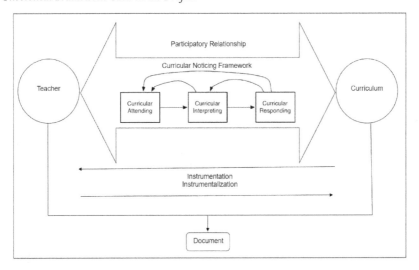

Because the transformation between intended curriculum and teacher intended curriculum influences other transformations, we hypothesized that curricular attention (what teachers look at or do not look at in the intended curriculum) may influence their interpretations (i.e., sense making) and eventual responses to the intended curriculum (i.e., what they actually use or do not use and how). In the next section, we describe how we chose to examine the teacher intended curriculum using curricular noticing—predominately, curricular attending—to gain an empirical, nuanced insight into the relationship between the teacher intended curriculum and the intended curriculum. We wanted to know: *What are teachers looking at in the curriculum materials? In what ways does that influence what they intend to do in the classroom?*

OUR INITIAL ATTEMPT AT CAPTURING CURRICULAR ATTENDING

When the second author (Males) initially used curricular noticing to investigate the interactions between teachers and materials, her research team observed prospective secondary teachers' (PST) and practicing teachers' planning sessions. These sessions were typically 30–120 minutes in duration, and consisted of a teacher sitting at a table with a printed copy of a lesson from a textbook along with a printed copy of the entire teacher edition volume/textbook from which the lesson came; a pair of researchers observed and took field notes from across the other side of the table. A video camera faced the participant, such that the researcher could only actually see the instructional materials in front of the participant, as shown in Figure 17.2. Each participant was asked to talk aloud while they planned their lesson, use highlighters and pens to mark up the instructional materials, and write their lesson plan on blank sheets of paper. From the researcher's perspective, it was difficult to read the curriculum materials and see what the teacher was seeing. Each planning session was transcribed for analysis.

Figure 17.2

Screen Capture of Video Recording from Initial Attempts at Capturing Attention

Source: From Dietiker, Kysh, Sallee, & Hoey. (2014). *Core Connections Algebra* (Dietiker & Kassarjian, Eds. 2nd ed. CPM Educational Program. Reprinted with permission.)[1]

Lessons Learned From Studying the Teacher Intended Curriculum

To analyze the participant's attention, the research team assigned the *Attend* code to portions of the transcript where teachers looked at, read, or focused on any portion of the written curriculum materials. They applied this code when teachers highlighted, used a writing utensil or their finger to follow along in the text, or when they read aloud a portion of the text. Some interesting findings were revealed when researchers examined the coded transcripts, marked-up lesson planning materials, and video recordings, such as what participants were likely *not* attending to in the materials based on portions of the lesson that were not marked-up. However, it was challenging to obtain meaningful understandings of participants' attention to the materials and how their attention might have been connected to their interpretations and responses. Many questions about participants' curricular noticing were difficult to address, such as the following:

- What portions of the materials did participants attend to for the longest amount of time or with the most frequency?
- What was the trajectory of the participant's attention that might have influenced them to attend to x?
- Did the participant not attend to written portions of the lesson that are not highlighted or verbally referenced in the transcript?
- To what extent did the participants attend to the unhighlighted visual representations, such as graphs, during their planning, and how might that have influenced their interpretations or later attention?

The research team's attempts to address these questions often seemed too inferential to draw meaningful conclusions about how teachers were interacting with the intended curriculum. The crux of the problem was that the team could not know, with certainty, exactly what the participant was looking at (or not) throughout the duration of the participant's planning session.

OUR CURRENT ATTEMPT AT CAPTURING CURRICULAR ATTENDING

To capture curricular attending, we previously relied solely on transcripts and video recordings focused on participants' instructional materials or their faces. Our current observations are much the same as before, except the participant now wears a pair of eye-tracking glasses that records video of their gaze, while a back-up camera records the participant's face and table in front of them.[2] Now when we analyze data, we find ourselves immediately turning to the eye-tracking video—watching it alongside the transcript—to make sense of what a participant was talking about.

Consider the following portion of a transcript that is a record of what occurred roughly 20 minutes into a naturalistic lesson planning session with a PST, Dan,[3] who talked aloud as they planned their lesson with electronic materials on their laptop. This transcript generates many questions about what exactly Dan was doing during the planning session.[4] See Table 17.1 for the portion of transcript and questions it generates.

Table 17.1

A Portion of Dan's Planning Session

Time	Transcript	Questions this Transcript Generates
21:14	Let's see how the book models it.	
21:31	Shade the grids to make three equal groups.	
21:48	**That's an interesting activity.**	What is an interesting activity?
21:54	Five, six. How are you supposed to divide them without like colors or something? Let's see. Use the decimal grids. I guess maybe they'd been using these or have used these in the past. How many are in each group? Six divided by three is equal to. Okay. **That really wouldn't be helpful for them. It's not the procedure that's more conceptual.**	What wouldn't be helpful? What is Dan referring to when he says that's more conceptual?
22:35	When using models to divide decimals, when might you use a grid? Tens instead of hundreds. Okay. Okay. I see there's 39. **Oh, okay. I see. That makes sense.**	What does Dan see? What makes sense to him?
22:52	Dividing using long division with whole numbers. Place the decimal point equivalent above. Okay. So it's kinda got like a nice little step by step there. I like that. So I guess maybe in the lesson we should start with dividing by whole numbers, kind of how the book models it there. You have two examples. **So I bet, I think they've seen this in fifth grade, but I'm not sure.**	What does he think they've seen exactly?
23:25	I think that's what [cooperating teacher name] so this might just be a big old review even for this part with decimals, bringing the decimal place up. But there are probably students that haven't seen this, so. Determined by the placement of the decimal divided. **That might be a good thing to put.**	What is a good thing for Dan to put?

(Table continued on next page)

Table 17.1 (Continued)

A Portion of Dan's Planning Session

Time	Transcript	Questions this Transcript Generates
24:03	Well let's see **if we can copy this.**	What is Dan copying? Where are they putting it?
24:18	Then maybe I'll give them like work with your table groups to see. Um, **so this would be, this is more of an explorer.**	What is more of an "explorer"?
24:40	So we have dividing decimals by whole numbers. I guess that kind of just tells them **maybe I should wait a slide to give them that.**	Wait a slide to give them what exactly?
24:47	And so maybe I give them a problem and see how they might approach it. And then I can come up with some questions. See yeah, **we'll do it.**	What does Dan mean by do it—do what exactly?
24:57	Perfect. Look at that. **I can just snip that,** the snipping tool is very nice. Um, if I can save as ha, we'll just save it to the desktop. This isn't my computer, so, okay. Well now to do and, oh maybe it's the new button.	What is Dan snipping with the snipping tool?
25:33	So yeah, **I think I give them one of these** and then we can work through it in groups. Um, yeah. And **I can scaffold them a little bit more with this.** Okay. Yeah.	Give them one of what? Scaffold them more with what?
26:03	**Let's make this bigger.** Oh gross.	Make what bigger?

A researcher reading through this transcript might think, "I will just follow along with the intended curriculum and their lesson slides to see what parts they were looking at." For some parts of the lesson, this follow-along strategy might be successful. For instance, "Use the decimal grids" and "3 equal groups" literally appear in the intended curriculum, so a researcher might feel reasonably certain that Dan was primarily looking at the division problem in Part A of the curriculum (6.39 divided by 3) around the times in the transcript when they said those phrases. And later when Dan said, "6 divided by 3 is equal to," a researcher might become more confident that Dan was considering the division problem in Part A throughout this section of the transcript.

We asked ourselves, "How can we be certain that Dan did not attend to the division problem in Part B (6.39 divided by 2.13)?" From watching the eye-tracking video, we know that there were often silent pauses in which Dan was looking around but not saying anything. The transcript alone does not capture all the sections of the intended curriculum that Dan scrolled through while constructing slides for students to see during the lesson, nor could a recording in which the camera was focused on the intended curriculum help us understand what Dan was attending to when completely silent. This attention is also not well captured with a video positioned to face the participant, because it is impossible to precisely determine the participant's gaze, as shown in Figure 17.3. Hence, a researcher can only know exactly what Dan was attending to (and not) if the researcher can see exactly what Dan was looking at.

Figure 17.3

Screen Capture of Video Recording Focused on Dan During Planning Session

Although Dan often read sentences or phrases directly from the intended curriculum, in most cases it was difficult to determine exactly what Dan was attending to, despite having the intended curriculum and slides to follow along. For instance, in the transcript, Dan mentions "That might be a good thing to put" and "Let's see if we can copy this." Later, Dan copies something with the snipping tool.[5] Is the portion of the curriculum in the first reference to be copied the same portion referenced in connection with the snipping tool? Unless the participant verbalizes every action with painstaking clarity, it is difficult, if not impossible, to know. Moreover, following along with the intended curriculum is difficult because of needing to simultaneously follow along with the evolving documents that PSTs generate during the planning session. For instance, when we watched the eye-tracking video, we learned that Dan was taking a screenshot of the

"Dividing Decimals by Whole Numbers" section title and the paragraph below it from the *Go Math* textbook, which was inserted into Slide 6 of their slideshow.

Eye-tracking video supported us in answering our questions regarding what Dan said was "a good thing to put" on their slide and what they were going to see "if they could copy." If a researcher just examined the final version of Dan's slideshow and read the transcript, however, they would not realize that the copied and pasted section of the textbook is no longer on Slide 6. In fact, it is no longer on the slideshow at all! Only from watching the video did we learn what Dan added to the slides and then deleted moments later. In the process of creating the slideshow, Dan edited wording, positioned activities on slides, re-ordered slides, added icons from other slides, renamed things, deleted things, and so on. None of that is captured in the finished version of a slideshow. Thus, any temptation to believe that what Dan was talking about must correspond to something Dan incorporated in the final version of the slideshow, word-processing document, or any other document is quickly dissolved.

As we have hinted before, another reason following along with the transcript is difficult is the multiplicity of documents Dan was attending to and generating. In the span of 3 minutes, Dan switched from attending to the slideshow, to Google Drive and email, to the intended curriculum, back to the slideshow, and then used the Windows search bar to create a Word document. But a researcher would not know this without watching Dan's gaze. Dan also split the screen and worked on the Word document and slideshow side-by-side—which adds an additional challenge for following along with the genesis of Dan's ideas and inspiration. Planning their lesson almost entirely electronically, Dan's attention constantly flickered among evolving sets of documents. So even if a researcher had copies of everything that Dan had during their planning session as they read through the transcript, they would need to do a considerable amount of detective work only to end up guessing at the answers to their questions. Ensuring that video cameras are positioned like a second pair of eyes and participants use eye-tracking glasses mitigate this guesswork.

CAPTURING ATTENTION USING HEAT MAPS AND GAZE PLOTS

Our nuanced capturing of attention matters because it increases the quality of our analysis. Not only are we able to address many of our questions about a participant's attention through their visual gaze, but we can further explore interpretations and responses by capturing attention more accurately. We use a combination of eye-tracking metrics (e.g., duration fixated on areas of interest, number of visits to areas of interest) to determine how much total time teachers devote to attending to the intended curriculum when planning, as well as to how much time they devote to attending to specific features or content (down to specific problems or activities). Additionally, given the diverse features and formats included in curriculum materials, these metrics can provide insight into the influence of features and format of the intended curriculum on the development of the teacher intended curriculum (e.g., Males & Setniker, 2019).

Visualizations, such as heat maps and gaze plots, provide insight into the concentration and sequence of attention during planning, respectively. When considered alongside the visual and verbal data from eye-tracking video, these visualizations support us in understanding how aspects of the intended curriculum might present points of conflict or struggle for teachers while planning, and how teachers' previous attention to the intended curriculum supports them in their later interpretations and responses to aspects of the intended curriculum.

Heat Maps

We use heat maps to examine the extent to which teachers attend to features of the intended curriculum. Those features teachers attend to the least indicate potential missed opportunities; those features teachers attend to the most suggest places they might have struggled, therefore warranting further investigation by researchers. Figure 17.4 is a heat map compiled from all four PSTs who participated in a staged planning interview.[6] It indicates that they struggled with interpreting the teacher versions of the CPM Educational Program student materials. The darker areas indicate increased attention to the suggested answer for Problem 2-11 and *Figure #* in the table. One PST, Wren, initially noted,

> I'm assuming that this answer [pointed to the suggested answer in bold] ... they ask me to write an equation at the top that represents the table below. But then they give me the equation?

Another PST, Cody, struggled to understand what the intended curriculum meant by *Figure #* and *# of tiles*, thinking that by tiles the authors meant a grid.

Gaze Plots

We use gaze plots to examine features of the intended curriculum that influence teachers' interpretations and responses. A gaze plot is a visualization that illustrates:

Figure 17.4

A Heat Map of Four PSTs' Attention to the Teacher Version of a CPM Student Page

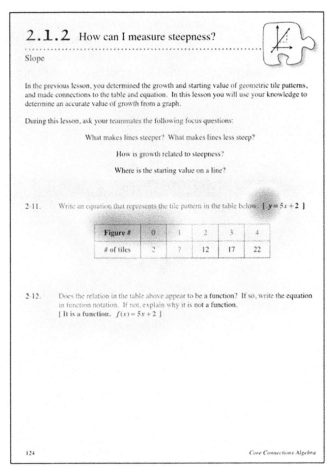

Source: From Dietiker, Kysh, Sallee, & Hoey. (2014). *Core Connections Algebra* (Dietiker & Kassarjian, Eds. 2nd ed. CPM Educational Program. Reprinted with permission.)

- What a participant looks at on the pages of the curriculum materials, as indicated by the location of the circles;
- The order in which participants looked at the location, as indicated by the numbers in the circles, with smaller numbers indicating earlier looks; and
- The time spent looking at locations, as indicated by the size of the circle, with longer durations represented by larger circles.

A gaze plot can be generated by the Tobii Pro Labs software (Tobii Technology, Inc., n.d.) for any portion of a video for one or a collection of participants.

Consider Figure 17.5, which presents the gaze plot of a practicing teacher, Paula. Recall that the numbers within the dots indicate the order in which Paula gazed at that portion of the text, the size of the dots correspond to the fixation duration or the amount of time that the teacher was looking at that portion of the text, and larger dots indicate longer fixation durations.

Figure 17.5

Gaze Plot Generated from a Portion of Paula's Planning Session

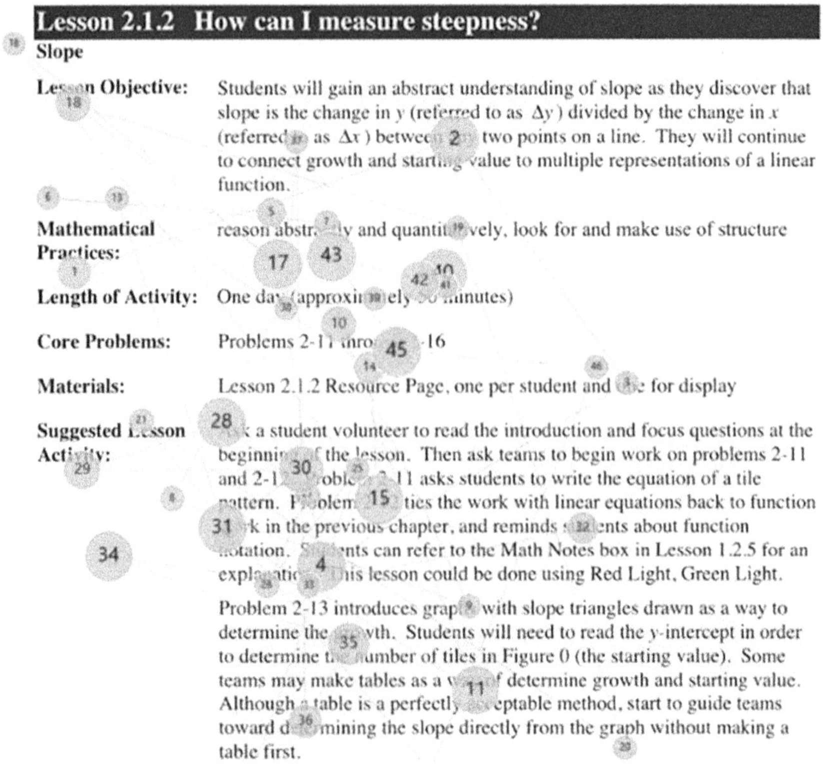

Lesson 2.1.2 How can I measure steepness?
Slope

Lesson Objective:	Students will gain an abstract understanding of slope as they discover that slope is the change in y (referred to as Δy) divided by the change in x (referred to as Δx) between two points on a line. They will continue to connect growth and starting value to multiple representations of a linear function.
Mathematical Practices:	reason abstractly and quantitively, look for and make use of structure
Length of Activity:	One day (approximately 50 minutes)
Core Problems:	Problems 2-11 through 2-16
Materials:	Lesson 2.1.2 Resource Page, one per student and one for display
Suggested Lesson Activity:	Ask a student volunteer to read the introduction and focus questions at the beginning of the lesson. Then ask teams to begin work on problems 2-11 and 2-12. Problem 2-11 asks students to write the equation of a tile pattern. Problem 15 ties the work with linear equations back to function work in the previous chapter, and reminds students about function notation. Students can refer to the Math Notes box in Lesson 1.2.5 for an explanation. This lesson could be done using Red Light, Green Light.

Problem 2-13 introduces graphs with slope triangles drawn as a way to determine the growth. Students will need to read the y-intercept in order to determine the number of tiles in Figure 0 (the starting value). Some teams may make tables as a way to determine growth and starting value. Although a table is a perfectly acceptable method, start to guide teams toward determining the slope directly from the graph without making a table first. |

Source: From Dietiker, Kysh, Sallee, & Hoey. (2014). *Core Connections Algebra* (Dietiker & Kassarjian, Eds. 2nd ed. CPM Educational Program. Reprinted with permission.)

Lessons Learned From Studying the Teacher Intended Curriculum   379

From the gaze plot, we see that Paula moved from a quick glance at Mathematical Practices (indicated by circle 1) to a longer look at the Lesson Objective (circle 2) followed by an even shorter gaze at the Materials (circle 3), and so on. An aspect of the gaze we found interesting relates to areas which Paula returned to multiple times, as indicated by the numbers within the circles being a greater distance apart. That is, Paula attended to a particular portion of the text long after her previous attention in that area. For example, Paula attended to the Lesson Objective three times (circles 2, 18, 37) and she did this with decreasing fixations (i.e., attention times). The third time she attended to the Lesson Objective was after attending to Suggested Lesson Activity 1 (circle 36), in which she read the sentence, "Although a table is a perfectly acceptable method, start to guide teams toward determining the slope directly from the graph without making a table first." Paula then turned to the student materials and began to work on Problem 2–13. See Figure 17.6 for Paula's work on the student materials.

Figure 17.6

Paula's Writing on a Portion of the Intended Curriculum

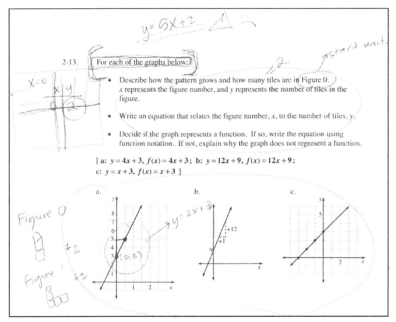

Source: From Dietiker, Kysh, Sallee, & Hoey. (2014). *Core Connections Algebra* (Dietiker & Kassarjian, Eds. 2nd ed. CPM Educational Program. Reprinted with permission.)

Consider how Paula's sequence of attention, particularly around the Lesson Objective, influenced her later attention, interpretation, and responses to the materials. Paula began interpreting the student materials by making sense of the complexity of the problem, which is captured in Figure 17.7. She first interpreted Problem 2–13 as asking for a description of the pattern from the previous question, so she created a table for the pattern using the function $5x + 2$, saying this is what her students would likely do. When reading the second bullet, she said the language confused her, and therefore she went back and read Problem 2–12. After returning to Problem 2–13 and reading the third bullet, she recognized that Problem 2–13 is asking the same thing as Problem 2–12. Furthermore, Problem 2–13 is asking her to complete each of the bullets for the graph. As she realized this, she drew brackets around "For each of the graphs below:" At this point she attended to Problem 2–13a and began to make sense of the graph, articulating what she expected her students to say: "Hopefully, they would see that point is (0, 3)." She then drew figure 0 with 3 tiles, and figure 1 with 5 tiles to the side, saying, "so they see that it is growing by 2 each time. Hopefully, that should jump out at them." After doing this, she turned to the second bullet and wondered,

> I'm trying to think about how I can get them to go from that [produces the circle on Problem 2-13a] to [draws the arrow, followed by $2x + 3$] $2x$ plus 3, because I don't really want to bring up slope-intercept form here because I haven't even used the word slope yet.

After about 20 seconds, she decided that she would have students go back to Problem 2–11 and ask them how the 5 and the 2 in the growth pattern affect the table, which prompted her to ask, "What was the goal of this lesson?" This question brought her back to look at the Lesson Objective followed by reading past what the Suggested Lesson Activity indicates about Problems 2-13 and 2-14. She decided that students might begin to consider how the line is growing by looking at the horizontal and vertical change, relating this to the $5x$ in the previous problem.

This small portion of our interview illustrates the center portion of the theoretical framework (from Figure 17.1). Paula's interpretations led her to attend to different aspects of the materials, which then allowed her to make progress in her interpretations and later in her decisions to respond. The gaze plot, along with the visual and verbal data from the eye-tracking video, suggests the influence of the Lesson Objective on Paula's eventual decision to emphasize horizontal and vertical change as a means of working towards understanding slope. Furthermore, Paula's continued revisiting of the Lesson Objective indicates that her connecting the goals of the lesson to students' tasks was not a linear process, but a cyclical one, which might

Figure 17.7

Screen Capture of Eye-Tracking Video During Paula's Planning

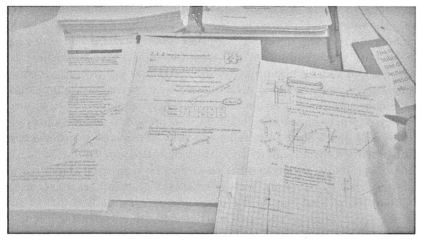

Source: From Dietiker, Kysh, Sallee, & Hoey. (2014). *Core Connections Algebra* (Dietiker & Kassarjian, Eds. 2nd ed. CPM Educational Program. Reprinted with permission.)

not have occurred or might have happened differently without her initial attention to the Lesson Objective.

Metrics and visualizations from eye-tracking video, along with the video itself, have greatly enhanced our analyses and understanding of the teacher intended curriculum. However, we believe that researchers can conduct meaningful studies of the teacher intended curriculum without expensive eye-tracking technology.[7] Although metrics and visualizations are powerful tools of inquiry, what has been invaluable to our analyses has been the ability to see what participants are doing from the perspective of the participants. One cost-effective means of examining teachers' attention to the intended curriculum or other resources they use during planning involves participants using wearable cameras, such as on a hat or headband, similar to that used in other classroom-based research on teacher noticing (e.g., Sherin et al., 2011). Although these cameras do not capture gaze data and are not as precise as eye-tracking glasses, they can provide researchers with insights into possible areas teachers are attending to (or not) for further inquiry.

CONCLUSION: ATTENTION MATTERS

In this chapter, we presented challenges in understanding the teacher intended curriculum and how we have overcome these challenges using eye-tracking technology. Our main challenges to capturing teachers' attention to the intended curriculum and the documents they generate during planning included: (1) knowing what teachers are looking at, especially when they are silent or quickly scrolling through electronic materials or using the internet; (2) knowing how long, how often, and in what order teachers are looking at various aspects of materials; (3) knowing what teachers are *not* attending to; and (4) understanding what teachers are talking about during planning. We discussed how eye-tracking metrics, visualizations, and the video itself allowed us to overcome each of these challenges, and thus better understand teachers' attention. This, in turn, supported us in understanding teachers' interpretations and responses and piecing together a more comprehensive story of teachers' transformation of the intended curriculum to the teacher intended curriculum.

Our overarching lesson to other researchers is that being able to physically see what teachers are seeing matters. Capturing teachers' visual attention has yielded empirical evidence demonstrating how the teacher intended curriculum is developed and how it is influenced by the intended curriculum (e.g., Males & Setniker, 2019; Quigley et al., 2022; Setniker, Males et al., 2019; Setniker, Quigley, & Males, 2019). As a result, we have grown in our understanding of the importance of focusing on the teacher intended curriculum, not only for theoretical development, but also for enhancing teacher educators' practice. We argue for more research on the teacher intended curriculum and encourage other researchers and teacher educators to take on this challenge.

ACKNOWLEDGMENTS

This research was supported by the CPM Educational Program (96531) and the U.S. National Science Foundation (DRL-1651836; DUE-1439867). We are incredibly grateful to the teachers who participated in our study.

REFERENCES

Burger, E. B., Larson, M. R., Dixon, J. K., Leinwand, S. J., Kanold, T. D., & Sandoval-Martinez, M. E. (2014). *Go Math Middle School Grade 6*. Houghton Mifflin Harcourt.

Dietiker, L., Kysh, J., Sallee, T., & Hoey, B. (2014). *Core Connections Algebra* (L. Dietiker & M. Kassarjian, Eds. 2nd ed.). CPM Educational Program.

Dietiker, L., Males, L. M., Amador, J., & Earnest, D. (2018). Curricular noticing: A framework to describe teachers' interactions with curriculum materials. *Journal for Research in Mathematics Education, 49*, 521–532. https://doi.org/10.5951/jresematheduc.49.5.0521

Gueudet, G., & Trouche, L. (2009). Towards new documentation systems for mathematics teachers? *Educational Studies in Mathematics, 71*, 199–218. https://doi.org/10.1007/s10649-008-9159-8

Jacobs, V. R., Lamb, L. L. C., & Philipp, R. A. (2010). Professional noticing of children's mathematical thinking. *Journal for Research in Mathematics Education, 41*(2), 169–202. http://www.jstor.org/stable/20720130

Males, L. M., Earnest, D., Dietiker, L., & Amador, J. (2015). Examining K–12 prospective teachers' curricular noticing. In T. G. Bartell, K. N. Bieda, R. T. Putman, K. Bradfield, & H. Dominguez (Eds.), *Proceedings of the 37th annual meeting of the North American Chapter of the International Group for the Psychology of Mathematics Education* (pp. 88–95). Michigan State University.

Males, L. M., & Setniker, A. (2019). Planning with curriculum materials: Interactions between prospective secondary mathematics teachers' attention, interpretations and responses. *International Journal of Educational Research, 93*, 153–167. https://doi.org/10.1016/j.ijer.2018.09.016

Quigley, K., Block, S., Zhou, Z., & Males, L. M. (2022, February 10-12). *The role of instructional resources in prospective secondary teachers' planning* [Conference presentation]. Twenty-sixth Annual Conference of the Association of Mathematics Teacher Educators, Henderson, NV.

Remillard, J. T. (2005). Examining key concepts in research on teachers' use of mathematics curricula. *Review of Educational Research, 75*, 211–246. https://doi.org/10.3102/00346543075002211

Remillard, J. T., & Heck, D. J. (2014). Conceptualizing the curriculum enactment process in mathematics education. *ZDM Mathematics Education, 46*(5), 705–718. https://doi.org/10.1007/s11858-014-0600-4

Setniker, A., Males, L. M., Quigley, K., & Block, S. (2019). The influence of high school teachers' curricular noticing on planning. In S. Otten, A. G. Candela, Z. de Araujo, C. Haines, & C. Munter (Eds.), *Proceedings of the forty-first annual meeting of the North American Chapter of the International Group for the Psychology of Mathematics Education* (pp. 120–124). University of Missouri.

Setniker, A., Quigley, K., & Males, L. M. (2019, February 7–9). *An eye tracking analysis of preservice teachers' attention to curriculum materials and implications for teacher education* [Conference presentation]. Twenty-third Annual Conference of the Association of Mathematics Teacher Educators, Orlando, FL.

Sherin, M. G., Russ, R. S., & Colestock, A. A. (2011). Accessing mathematics teachers' in-the-moment noticing. In M. G. Sherin, V. R. Jacobs, & R. A. Philipp (Eds.), *Mathematics teacher noticing: Seeing through teachers' eyes* (pp. 79–94). Routledge.

Stein, M. K., Remillard, J., & Smith, M. S. (2007). How curriculum influences student learning. In F. K. Lester, Jr. (Ed.), *Second handbook of research on mathematics teaching and learning* (pp. 319–369). National Council of Teachers of Mathematics.

Tobii Technology, Inc. (n.d.). Tobii Pro Labs [Computer Software]. Falls Church, VA.

Valverde, G. A., Bianchi, L. J., Wolfe, R. G., Schmidt, W. H., & Houang, R. T. (2002). *According to the book: Using TIMSS to investigate the translation of policy into practice through the world of textbooks*. Kluwer.

ENDNOTES

1. *Algebra Core Connections* from the CPM Educational Program is one of the two texts used in the study.
2. Our current observations include both the staged planning sessions mentioned previously, in which participants plan a hypothetical lesson using a set of curriculum materials chosen by the researchers, as well as naturalistic planning sessions, in which the participants plan real lessons they intend to enact with their current students.
3. Dan is a pseudonym.
4. The lesson under consideration was entitled Dividing Decimals from Grade 6 of the *Go Math Middle School* curriculum (Burger et al., 2014). The portion under consideration consisted of two pages: one on "Modeling Decimal Division," which included two problems with decimal grids provided for each (Problem A was $6.39 \div 3$; Problem B was $6.39 \div 2.13$), and the other on "Dividing Decimals by Whole Numbers."
5. The *snipping tool* is a screen capture tool that allows a participant to copy and paste portions of the text onto their slides.
6. In staged planning interviews, participants plan a hypothetical lesson using a set of curriculum materials chosen by the researchers.
7. At the time of purchase, the eye-tracking technology cost about $20,000 USD.

CHAPTER 18

OPPORTUNITIES AND CHALLENGES IN SURVEYING AND INTERVIEWING ELEMENTARY TEACHERS ABOUT THEIR MATHEMATICS CURRICULAR USE

Bima Sapkota
University of Texas Rio Grande Valley

Kristin Doherty
Michigan State University

Mona Baniahmadi
University of Texas, Austin

Nan Jiang
University of Arizona

Laurel Hendrickson
University of Arizona

Doris Fulwider
Purdue University

Amy M. Olson
Duquesne University

Marcy B. Wood
University of Arizona

Jill A. Newton
Purdue University

Many teachers use multiple curricular resources to teach mathematics, which differs from previous decades when teachers often taught from one curricular resource. To better understand how elementary teachers make decisions about which types and aspects of multiple mathematics curricular materials to use and how they create coherence among them, we administered a national survey to 524 teachers and conducted interviews with teachers from three different locations. While conducting this research, we experienced opportunities and challenges arising from conducting research during a pandemic, collecting data from teachers across the country and in our own local communities, and learning from our own findings. This chapter elaborates the lessons learned from the opportunities and challenges of each. It concludes with implications for conducting future research examining the complexities of elementary teachers' mathematics curriculum use in this understudied curricular landscape.

Keywords: elementary mathematics teachers; mathematics curricular use; survey methodology

INTRODUCTION

In this chapter, we share lessons learned from our research on U.S. elementary teachers' mathematics curricular use. We begin by introducing our study and situating it within the research literature, including our own past research on mathematics curriculum. Next, we elaborate on the lessons learned, including our experiences related to opportunities and challenges with an evolving study design, working with a data and analytics company, learning from our survey, and working with teachers. Crucial to our study's development and understanding was an ongoing need to be flexible in a dynamic curricular landscape and changing world.

Origin of the Study

The work related to this study began in 2012 when a group of mathematics education researchers applied for a small grant offered by the Center for the Study of Mathematics Curriculum (CSMC)[1] to investigate mathematics teacher educators' practices in the context of the *Common Core State Standards for Mathematics* (*CCSSM*) in the United States (National

Governors Association Center for Best Practices & Council of Chief State School Officers, 2010). Our research trajectory took a long and winding road involving nine faculty members, six graduate students, and two undergraduate students, leading to the current study of elementary teachers' use of mathematics curricular materials in the context of the COVID-19 pandemic and ubiquitous online resources. This journey was filled with successes and setbacks, including many accepted proposals and publications along with just as many unaccepted submissions and unfunded grant proposals.

We developed as researchers throughout this research journey. What was constant was the focus on the use of curriculum. However, we changed our focus from mathematics teacher educators to K–12 teachers and from *CCSSM* to curricular materials. Our metaphorical framing also shifted, from thinking about our work as capturing portraits of teachers' curriculum work to investigating how teachers were navigating the new frontier of online curriculum to exploring what tools teachers might need to support their curricular decisions. In the remainder of this chapter, we discuss challenges in our research, what we did to resolve them, and the lessons learned in addressing those challenges. We hope our story has insights to help readers along their own research journey.

The Current Study

Our current study arose from multiple conversations about how elementary teachers are making curricular decisions in a context that has undergone significant changes tied to *CCSSM* and to the increasing availability of materials on the internet. Teachers have unprecedented access to, and sometimes mandates for, the use of online curricular resources, including assessment tools, skill-focused platforms, and lesson banks (e.g., Teachers Pay Teachers, Pinterest, IXL, Prodigy) (Silver, 2022). Although teachers are using online resources (e.g., Gewertz, 2014; Monahan, 2015; Ross, 2015), how teachers interact with a variety of curriculum materials, including how they make connections across materials to provide students a coherent mathematical learning experience, is still underexplored.

As teachers have new opportunities to use online resources and lessons produced by other curriculum developers, including other teachers, they may face new challenges in creating a mathematically coherent experience for their students. Yet, this aspect of teaching has been relatively unsupported. Teachers have had to navigate the complex relationship between their experiences, instructional context, and perceived curricular design rationale to support student learning (Choppin, 2011; Keiser & Lambdin, 1996; Remillard & Bryans, 2004).

In our study, we sought to understand how teachers enact coherent instruction using multiple curriculum materials. In some contexts, teachers are provided a single set of curriculum materials and instructed to use those materials either strictly or flexibly (i.e., with curricular autonomy). The greatest challenge for teachers in terms of coherence in these contexts is to *understand* the underlying curriculum structure and storyline of the materials and to make instructional and curricular decisions that maintain or enhance the coherence of this storyline. In doing so, they often need to consider how mathematical ideas are developed across and within grade levels (Cai et al., 2014). In contrast, in contexts in which teachers are gathering and curating resources from a variety of sources, the central challenge is to *develop* a storyline or structure that will create coherence across these resources and inform both instructional and curricular decisions. We are researching how teachers respond in both situations.

LESSONS LEARNED

We experienced many opportunities and challenges in developing and conducting the study. Many lessons were tied to the unfolding COVID-19 pandemic. However, from the research process, we also learned more about administering national surveys, adjusting to emerging findings, and working with teachers. In particular, we learned that teachers had unique challenges using online resources during the pandemic, and these challenges have persisted since the pandemic. Thus, we included questions related to how their use of curricular resources was influenced by the pandemic.

Responding to Emerging Situations: Impact of COVID-19

Like many people conducting research at the start of the COVID-19 pandemic, we were forced to rethink our study in light of pandemic-related changes to schooling. This led to lessons about redesigning a study to make the most of changing contexts.

The overarching research question of our initial study design was "How do teachers make decisions related to curricular coherence across a range of curricular use patterns?" We also posed three sub-questions:

1. What are the patterns of teacher curricular resource use that exist across a range of curricular contexts?

2. Given differing patterns of teacher curricular resource use, what kinds of decisions about curricular coherence do teachers make and why?
3. How do curriculum toolkits co-developed with teachers support teachers in making decisions related to curriculum coherence?

Although we were guided by these questions, modifications were made to be responsive to the pandemic and other emerging factors influencing the curricular landscape in classrooms. Most notably, we shifted from the third sub-question to focus instead upon teachers' changing curricular use due to the COVID-19 pandemic. As a result, we included questions in our interview protocol to ask (a) which curricular materials teachers select, (b) to what extent they feel they have control over those resources, (c) how they establish coherence among curricular resources they select, and (d) how they collaborate with their colleagues (both inside and outside of their schools) to navigate the complexity of curricular materials. For example, we asked, "Describe all of the math curricular materials you might use with your students in a typical week. How do you use each material (e.g., whole class instruction, differentiation, assessment)? How often do you use each material?"

As initially planned, we began our inquiry using survey methods with both selected-option and constructed-response items to capture which curricular resources teachers used, how they used those resources, and what contextual influences impacted their use. Shortly after we finalized and piloted the survey and just as we were ready to launch the survey, the COVID-19 pandemic arrived, closing schools and radically changing instruction.

We put the survey on hold and waited for the pandemic to subside. As weeks passed, we realized the pandemic was not going to be a short-term event and we needed to rethink our plans. We reflected upon how the pandemic was impacting schools and teachers and decided a survey in spring 2020 was an inappropriate ask of teachers' time and energy, so we moved the survey to fall 2020. Also, we realized we were not going to be able to answer our original questions. Rather, the change in teaching and learning presented a new series of important questions. So, we developed a new plan, consulted with the program officer of our funding agency, and began revising our survey.

We shifted our focus to teachers' curricular experiences before and during remote instruction. This involved modification of our survey questions to ask about teachers' curricular context before the pandemic, how that context shifted with remote teaching, and the supports and challenges teachers faced in making the transition. For example, we originally planned to ask teachers to list each of the curricular resources they used.

We changed this single question to a series of questions in which we asked teachers to report the curriculum materials they used before the pandemic (September 2019 to March 2020) and during the pandemic (March 2020 to June 2020). If they changed resource materials, we asked them to explain why they stopped using one resource and started using another.[2]

We had to do more than change our survey; we also needed to remove one major study outcome. Our initial intention was to use the survey data to describe the U.S. curricular landscape broadly, and then to work with small groups of teachers to produce toolkits to (a) help teachers build a coherent curriculum for their students, (b) document teacher needs and if/how the tools could be designed to support teachers' curricular decisions, and (c) develop in-depth case studies of teachers' curriculum contexts. We planned to use the survey data to identify local teachers who could help co-develop and refine toolkits.

However, COVID-19 continued to impact our ability to recruit teachers even as we moved into the interview phase in fall 2021. Because many teachers were still teaching remotely or in limited in-person capacity, we could not enter teachers' work sites. Moreover, it was difficult to invite teachers to participate in a series of interviews while they returned to in-person teaching, especially when that work looked different from the work prior to COVID-19 or they were overwhelmed by professional needs in response to teacher shortages (e.g., loss of preparation time, larger class sizes). We needed to move away from developing a toolkit and instead focus on elaborating some of the surprising survey findings.

For example, our interview protocol provided teachers with the opportunity to explain how their use of online curricular resources changed during the pandemic. We developed a three-stage interview process, with two individual interviews with the same teachers, followed by focus-group interviews with teachers from the same school. We used several shorter individual interviews to respect teachers' time and levels of fatigue. Our results showed that teachers began to use more online resources during the COVID-19 pandemic. So, we framed interview questions to ask about the unique challenges associated with teachers' use of online and Teachers Pay Teachers resources. Indeed, in one of our studies, we have been investigating how teachers use online resources, including Teachers Pay Teachers and Pinterest.

In hindsight, our survey and interview changes made sense and resulted in a productive study. However, as we were rethinking our study, we were quite worried about changes. Although we knew the situation of the pandemic was unprecedented, we still felt responsible for following our research proposal and were initially reluctant to redesign the survey and shift our research questions. We decided to change the focus of the study after reaching out to other mathematics education researchers, to teachers, and to our program officer. *The major lessons we learned were the importance of pivoting and having multiple resources available to help make wise decisions in an uncertain time.*

Working With a Data and Analytics Company

Early rounds of feedback on our proposal included the suggestion to partner with a data and analytics company. We needed outside help for two reasons. First, we were challenged by the task of recruiting and collecting data from teachers across the nation, as well as attempting to recruit teachers who worked in varied school settings and with diverse students. Second, none of our university settings had the infrastructure to pay hundreds of teachers for taking the survey. The data and analytics company was well prepared to do each of these.

We were amazed at the number of teachers who needed to be contacted to obtain an adequate number of responses. People at the data and analytics company advised us that the unique stressors of the COVID-19 pandemic meant that only half a percent of those teachers who were recruited would agree to participate. In other words, to secure 500 completed surveys of teachers in particular grade levels and teachers serving in diverse school settings, we would need to send 100,000 emails. The scale of recruitment required for our survey was well beyond what we expected and exceeded our contact list of teachers. Ultimately, with help from people at the data and analytics company, we were able to achieve a sample size of more than 500 teachers, which included teachers in most states. We simply could not have done this without their help.

By consulting with experts at a data and analytics company, we learned there was a trade-off between the kinds of questions we asked and the likelihood of completed responses. We originally had many open-ended questions for which we hoped responses would be more than a word or two. The company's analytics consultants suggested that we shorten and streamline these questions because respondents would be more likely to complete the survey if they could answer by selecting from a few options rather than by typing multiple sentences. In addition, the consultants informed us that many teachers would be using their mobile devices, making it easier to answer selected-response items. Although we wanted more elaborated answers from teachers, it was more reasonable to ask survey questions requiring briefer responses. These changes to our survey meant that the follow-up interviews we had planned became necessary to develop deep understandings of curriculum decisions.

We felt fortunate that we worked with people at the data and analytics company, as they had essential expertise that resulted in an informative survey. We would recommend that others who want to conduct national surveys consider tapping the expertise of companies who do such work on a regular basis.

Connecting Survey and Interview Methods

Our survey produced findings that pushed our interview questions in unanticipated directions. We were surprised to learn that teachers were using as many as 10 different curricular resources to teach mathematics (Baniahmadi et al., 2021). We were also surprised that popular supplemental materials, such as Teachers Pay Teachers (48%) and Brainpop (32%), were referenced by more teachers than the most popular standard curricular materials such as *EngageNY* (21%) and *enVision* (25%) (Baniahmadi et al., 2021). Because many teachers were using multiple materials to teach mathematics, including supplemental materials, we altered our interview materials to ask more about these other materials and how teachers were making connections across them. We developed interview questions that focused on (a) how teachers selected, evaluated, adapted, and created curricular materials to teach mathematics, (b) what adaptations teachers made, including how and why, (c) how teachers created coherence across multiple curriculum materials, and (d) how curricular use varied across school contexts.[3] In order to learn from earlier findings and develop new areas of exploration, we learned to be flexible and adaptive at successive stages of the study.

Recruiting Teachers

Consultants at the data and analytics company assisted with our largest challenge of recruiting hundreds of teachers nationwide to participate in our survey. We relied on other methods to recruit teachers for the survey pilot, interview pilot, and full interviews. For these, we used convenience sampling (Teddlie & Yu, 2007) with teachers from local schools. Convenience sampling is a non-probability sampling technique in which we used our relationships with the local teachers and geographical convenience to recruit teachers for our study. Although a critique of convenience sampling might be that the population is not necessarily random, we were worried about finding teachers who would participate in lengthy interviews with researchers whom they did not know. This concern was compounded by the fact that our call for interviews began in fall 2021 as teachers were returning from the first full year of pandemic teaching and learning, suggesting that teachers would already be busier and more exhausted than before. This emphasizes the importance of respecting and compensating teachers' time and forming relationships with teachers and schools. Because the interviews required a smaller sample size and were more time-intensive, we offered each teacher $500 to participate in the interviews, compared to survey compensation of $5. Ultimately, convenience sampling allowed us to find teachers who agreed to participate as they already had a relationship with at least one member of our research team. We were cautious about which schools and teachers we asked to participate in the pilot because we did not want to exhaust local connections and limit access to community partners for full data collection. As we worked with a company for survey deployment and data collection, we were comfortable working with teachers with whom we had established relationships for pilots and interviews. Additionally, tapping existing relationships contributed to research that both teachers and researchers saw and experienced as building knowledge together.

SUGGESTIONS FOR FUTURE RESEARCH ON MATHEMATICS CURRICULUM

Here, we offer suggestions for conducting future research on mathematics curriculum, building from the lessons we learned with an evolving study design, teacher and school recruitment, and the ongoing need for flexibility. Given the complexity of understanding how and why teachers select and use various curricular materials to teach mathematics and create coherence among them, the implications we offer do not necessarily address this at the individual teacher level. Rather, multiple methods and studies will be required before the field has a better understanding of teachers' mathematics curricular use in an ever-changing landscape.

With regard to study design, we suggest using surveys and interviews together, as they can inform one another and collectively work toward a more thorough understanding of curricular use. In particular, surveys can be used to understand curricular use nationwide and then interviews can be leveraged to understand similarities and differences by context. Based on our experience, we suggest partnering with a data and analytics company to help recruit and disseminate surveys and provide compensation on a large scale. Such usage can reach more teachers than researchers would typically be able to achieve alone. With interviews, we suggest providing teachers with the questions in advance so they have time to unpack their curricular use and decision-making. We used this approach to allow teachers time to brainstorm responses, which we found provided rich understandings of their contexts. To understand more teachers' experiences with curriculum use, the field could benefit from researchers posting their full surveys, interview questions, and interview protocols so that others can use, modify, and build upon them.

Future research on mathematics curriculum might also benefit from conducting case studies with teachers in their classrooms to understand the lesson planning decisions they are making in-the-moment in contrast to our study which asked teachers to recall what they did in the past or their ideas for the future. Case studies could further illuminate how teachers are thinking about and creating a coherent mathematics curriculum for their students through the use of multiple materials. With regard to recruitment, the pandemic pointed out that online meeting platforms, such as Zoom, can be leveraged for reaching more teachers regardless of location. Importantly, this study highlights the need for future studies to be responsive to contexts, participants, or other issues, regardless of the initial plan, but still with a focus on the overarching research goals of understanding the changing curricular landscape that teachers need to address. We imagine the curricular landscape will continuously evolve, particularly as more curriculum materials become available online and more schools increasingly turn toward online learning tools.

CONCLUSION

We learned many lessons as we surveyed and interviewed teachers about their use of multiple mathematics curricular materials in a complex and ever-changing curricular landscape. In particular, we experienced opportunities and challenges through our evolving study design. We hope the lessons we learned can support future research in understanding how and why teachers use various curricular materials to teach mathematics and how they create coherence among them for their students. Given the expanding accessibility of curricular materials available to teachers and the increasing turn to using online learning materials in schools, understanding usage in a constantly evolving landscape will require varied studies using a range of research methods. In particular, we want to emphasize how important it will be to be responsive to the research context, as the curricular landscape may likely evolve as it is being studied.

ACKNOWLEDGMENT

Our work is supported by the U.S. National Science Foundation (NSF), Project Grant number 1908165. The statements and opinions reported here are those of the authors and do not reflect opinions of the NSF.

REFERENCES

Baniahmadi, M., Olson, A., Wood, M. B., Doherty, K., Newton, J., Sapkota, B., & Drake, C. (2021). Curriculum resources and decision-making: Findings from a national survey of elementary mathematics teachers. *Association of Mathematics Teacher Educators (AMTE) Connections Journal, 31*(2). https://amte.net/connections/2021/12/curriculum-resources-and-decision-making-findings-national-survey-elementary

Cai, J., Ding, M., & Wang, T. (2014). How do exemplary Chinese and US mathematics teachers view instructional coherence? *Educational Studies in Mathematics, 85*(2), 265–280. https://doi.org/10.1007/s10649-013-9513-3

Choppin, J. (2011). Learned adaptations: Teachers' understanding and use of curriculum resources. *Journal of Mathematics Teacher Education, 14*(5), 331–353. https://doi.org/10.1007/s10857-011-9170-3

Gewertz, C. (2014). Teachers a key source of Common-Core curricula, study finds. *Education Week.* https://www.edweek.org/teaching-learning/teachers-a-key-source-of-common-core-curricula-study-finds/2014/10

Keiser, J. M., & Lambdin, D. V. (1996). The clock is ticking: Time constraint issues in mathematics teaching reform. *The Journal of Educational Research, 90*(1), 23–31. https://doi.org/10.1080/00220671.1996.9944440

Monahan, R. (2015). *How Common Core is killing the textbook.* http://hechingerreport.org/how-common-core-is-killing-the-textbook/

National Governors Association Center for Best Practices & Council of Chief State School Officers [NGA & CCSSO]. (2010). *Common core state standards for mathematics.* https://www.thecorestandards.org/Math/

Remillard, J. T., & Bryans, M. B. (2004). Teachers' orientations toward mathematics curriculum materials: Implications for teacher learning. *Journal for Research in Mathematics Education, 35*(5), 352–388. https://doi.org/10.2307/30034820

Ross, T. (2015). The death of textbooks? *The Atlantic.* http://www.theatlantic.com/education/archive/2015/03/the-death-of-textbooks/387055/

Silver, D. (2022). A theoretical framework for studying teachers' curriculum supplementation. *Review of Educational Research, 92*(3), 455–489. https://doi.org/10.3102/00346543211063930

Teddlie, C., & Yu, F. (2007). Mixed methods sampling: A typology with examples. *Journal of Mixed Methods Research, 1*(1), 77–100. https://doi.org/10.1177/1558689806292430

ENDNOTES

1. CSMC was a research center funded by the U.S. National Science Foundation. It was housed at the University of Missouri and was a collaborative effort among the University of Missouri, Michigan State University, Western Michigan University, and Horizon Research.
2. For a copy of the final survey questions, contact the first author, Bima Sapkota, at bima.sapkota@utrgv.edu.

3. For a copy of the final interview questions, contact the first author, Bima Sapkota, at bima.sapkota@utrgv.edu

CHAPTER 19

LESSONS LEARNED FROM RESEARCHING ONLINE MATHEMATICS CURRICULUM SUPPLEMENTATION

Lara K. Dick
Bucknell University

Amanda G. Sawyer
James Madison University

Since 2016, our research team has investigated the growing phenomenon of teachers' online mathematics curriculum supplementation from virtual resource pools. Using Silver's (2022) Teacher Curriculum Supplementation Framework to situate our work, we describe lessons learned about data collection, data analysis, and publication when researching the supplement's source, teachers' reasons for and use patterns associated with online mathematics curriculum supplementation, and individual curriculum supplements. When collecting data, researchers must be creative with purchasing strategies, purposeful use of filters, and contacting teachers. When conducting data analysis, researchers should define a unit of analysis early in the investigation, be prepared to collapse data in justifiable manners, and be clear about limitations regarding the shifting online landscape to describe results accurately. When publishing research, researchers should carefully choose frameworks to situate their work within established curriculum research and be prepared to face challenges regarding copyright and individual's opinions of online curriculum supplementation.

Keywords: mathematics curriculum resources; online curriculum supplementation; teacherpreneurs; virtual resource pools

Prior to the rise of the internet, most schools' adopted curriculum for elementary-school mathematics included textbooks and associated materials with the understanding that this was "all" teachers needed to teach mathematics successfully (Browne & Haylock, 2004; Remillard, 2005). However, over the last decade, many teachers turned to various online virtual resource pools to supplement their mathematics curriculum. Virtual resource pools are websites that house curriculum resources for download (Torphy et al., 2020). Teachers find curriculum materials on social media sites like Pinterest, Instagram, and the arguably popular virtual resource pool TeachersPayTeachers (TpT). In spring 2023, TpT reported more than 7 million users and more than 7 million available curriculum resources, with over 1 billion resource downloads; TpT claims 85% of K–12 teachers in the U.S. "use TpT to save time, engage students, and learn from each other" (TeachersPayTeachers, 2023, para. 1).

Since 2016, our research team has investigated this growing phenomenon, predominantly focused on elementary mathematics curriculum supplementation resources found on TpT. Teacher supplementation can be described as any change teachers make to curriculum resources provided by officials at their schools or districts. One finding of a recent review of research is that "teacher curriculum supplementation is massive in scope" (Silver, 2022, p. 6). A contributing factor to the rise in popularity is the impact of social media as spaces where teachers both search for (e.g., Carpenter et al., 2020; Carpenter & Krutka, 2014) and also promote their self-created resources (e.g., Shelton & Archambault, 2019; Torphy et al., 2020). Thus, research within the space of online curriculum supplementation is imperative for the field to have a full understanding of curriculum usage in the classroom.

HOW WE DECIDED TO RESEARCH ONLINE MATHEMATICS CURRICULUM SUPPLEMENTATION

The impetus for studying online mathematics curriculum supplementation was a 2013 dissertation pilot study investigating implementation of reform-based teaching practices, such as the use of varied and numerous mathematical experiences that encourage students to understand and value mathematical thinking (National Council of Teachers of Mathematics [NCTM], 1989). The pilot study investigated how a practicing teacher implemented these ideas in her classroom; when asked how she selected resources for her first-grade students, she responded that she "Googled it" (Sawyer, 2017). This prompted the question of where teachers go online to search for and obtain supplemental mathematics curriculum resources and led to a study of preservice teachers' lesson planning. From this investigation, Sawyer, Dredger, and colleagues (2020) realized the role of teacher curation of curriculum supplements and identified various online resource pools that preservice teachers were using. In discussing the results of this work with Sawyer, Dick shared an anecdote regarding a teacher's request to the local Parent Teacher Student Association (PTSA) for a TpT gift card (personal communication, February 21, 2018). This request surprised both the principal and the PTSA, but highlighted the importance of studying curricular supplements found on virtual resource pools. Through further discussions, we decided to systematically study elementary and middle-school mathematics online curriculum supplementation. Over the past six years, we have learned a lot about how to research this emerging and ever-changing space of online curriculum supplementation.

CURRENT FRAMING OF OUR WORK

Before considering lessons learned from doing mathematics curriculum research in the space of online curriculum supplementation, we situate our work within the Teacher Curriculum Supplementation Framework (TCSF) (Silver, 2022). The framework includes four dimensions to guide research: (1) the teacher's reasons for supplementation; (2) the supplement's source; (3) the teacher's supplemental use pattern; and (4) features of the supplement itself. When we began this work in 2016, we surveyed 601 teachers across the U.S. to gather information about their reasons for supplementation and their supplemental use patterns. We asked where and how often they searched for elementary mathematics curriculum resources. In response, teachers described their reasons for supplementation and shared examples of curriculum supplements they obtained via searches on various virtual resource pools (Dick & Sawyer, 2022; Shapiro et al., 2019). From this early research, TpT was identified as the most popular virtual resource pool. So, we turned our attention to studying the curriculum supplements available across grades K–5,[1] studying different content domains[2] (Shapiro et al., 2019), and the levels of cognitive demand of the supplements (Dick et al., 2019; Dick et al., 2020/2021, Shapiro et al., 2021).

As we continued our research in this space, we studied the supplement's source by considering the teachers who create the curriculum supplements. Shelton and Archambault (2019) use the term *teacherpreneurs* to describe these teacher creators of curriculum supplements who use social media platforms like TpT to create, sell, and distribute curriculum resources to others. We gathered information about elementary mathematics teacherpreneurs' educational backgrounds, their teaching experience, and how they market their curriculum materials (Sawyer, Dick, & Sutherland, 2020) and are in the process of doing the same for middle-school mathematics teacherpreneurs.

Thus, we have experience researching all four aspects identified in the TCSF (Silver, 2022). We have learned lessons about data collection, data analysis, and publishing. We highlight the lessons learned across the four dimensions of TCSF to illustrate the scope of our research and sample findings, the difficulties we encountered, and how we have worked to overcome challenges.

DATA COLLECTION

Gaining access to data associated with curriculum that is part of virtual resource pools is complicated. For this work, data include the curriculum supplement itself, information regarding teachers who use curriculum supplements, and the teacherpreneurs who create the supplements. We learned how to approach data collection of all three, with some needing workarounds more than others.

Curriculum Supplements

Virtual resource pools like TpT contain both free and for-sale curriculum supplements. Funding is required to study the breadth of materials available on these sites, because some purchases of supplements are required. To conduct this research, both authors received internal funding from their universities, but TpT did not have a process for purchase orders nor for working with tax exempt institutions, such as universities, at the time we began our work. For one author, it took months to purchase the supplements identified for study. Discussions were necessary with both the university and with staff at TpT about the purposes of this research. TpT required assurance that the TpT user account was associated with a faculty member who would not distribute the downloaded curriculum supplements to teachers to be used in classrooms outside the university. Eventually, TpT created a tax waiver document to enable the university purchase. However, we then had difficulty with the setup of the TpT platform as a user on TpT could only place 300 supplements in their "shopping cart" at one time. Thus, we created two separate accounts (one holding 300 and the second holding the next 200) to purchase the needed curriculum supplements. Through these difficulties, we learned: *It is best to contact the virtual resource pool companies prior to purchasing supplements to discuss unique needs that a researcher may have.*

After purchasing, we learned to download all purchased supplements as soon as possible. Initially, we thought that a purchased supplement would remain in our TpT account indefinitely. Unfortunately, we discovered that supplements can be removed from TpT. We were planning to download supplements as we were analyzing them, but when we went to download, some links were dead (e.g., 31 of the 500 middle-school supplements, including 22 that we had purchased). We do not know if the dead links are a function of the TpT website or if a teacherpreneur chose to remove their supplement from circulation. In either case, this is an important consideration in researching online curriculum supplements: *It is best to move curriculum supplements outside of the platform as quickly as possible to ensure the full dataset is maintained.*

Another difficulty with obtaining curriculum supplements was hidden algorithms on the various platforms, which determine such things as the "top" supplements. For example, we learned from surveying teacherpreneurs (Sawyer, Dick, & Sutherland, 2020) that they can pay to be featured on TpT's platform. Thus, when we obtained TpT's "top" (i.e., first appearing) supplements, we were unsure as to how TpT defined them as "top." One workaround was to use filters more purposefully. In our most recent work, when obtaining middle-school mathematics supplements, we ran searches with filters for different mathematics content domains to bypass teacherpreneur supplements that may dominate one content strand. However, even with this change, all 500 middle-school mathematics supplements we obtained were ranked with top stars. Overall, it is important to remember: *TpT and other virtual resource pools do not share their search algorithms so researchers must carefully consider this potential limitation when accessing the supplements.*

Teachers' Reasons for Supplementation and Teachers' Supplemental Use Patterns

In our first survey of teachers who use online curriculum supplements for their elementary mathematics classrooms (Dick & Sawyer, 2022; Shapiro et al., 2019), we sought to understand teachers' use patterns and reasons for supplementation. Survey methodology has some limitations that eventually led us to conclude that interviewing is preferable to gather data from teachers. Specifically, a reviewer of a submitted manuscript noted that, because we distributed the survey via online platforms and e-mail, we most likely obtained teachers who are already online and may supplement their curriculum with online supplements more than others. In addition, we learned the importance of articulating the reasoning behind the phrasing of survey questions. For example, in our paper delving into teachers' reported reasons for choosing curriculum materials to supplement in their classrooms, we asked how they adapted "based on student needs," but the responses suggested the question was not specific enough and instead may have guided teachers' responses away from sharing adaptations focused on student learning. It was important to clarify our reasoning as we did in Dick and Sawyer (2022), where we explained that we were interested in how adaptations were made to elementary mathematics supplements specifically to enhance student learning.

Supplements' Source: Teacherpreneur Data

In 2019 when we collected data on the teacherpreneurs who created the "top" supplements on the TpT platform, we developed both a survey and an interview protocol (Sawyer, Dick, & Sutherland, 2020). TpT has an "Ask a Question" feature under each teacherpreneur's profile; we posted information about our study, invited the teacherpreneurs to complete the survey and to be interviewed. Before we contacted the full sample of top teacherpreneurs, officials at TpT blocked our account and asked us to stop all communications. We immediately contacted TpT's customer support line to explain our research and share our Institutional Review Board documents. We were told,

> TpT does not share contact information or other personal information about our customers with third parties, without our customers' permission. This is the case even if the need for that information is for non-commercial purposes. We appreciate your understanding of TpT's need to protect our customers' privacy. (TpT, personal communication, July 9, 2019)

Despite this setback, we were hopeful that things would be resolved because TpT asked to have a phone conversation with an author and the head of TpT's communication office. TpT wanted more information on the research and how we were analyzing our data. Although the conversation seemed amicable and finished with a discussion of a possible future collaboration, no further interaction occurred and our ability to contact teacherpreneurs through the TpT platform was never reinstated. As a result of this difficulty, we were only able to report results from the teacherpreneurs whom we contacted before our account was blocked; our dataset was reduced but we did complete the study (Sawyer et al., 2020).

To deal with this issue, we gathered teacherpreneur information from their TpT profiles and we attempted to contact them via other avenues. In their profiles, teacherpreneurs describe themselves, and sometimes include links to other social media sites at which they share their curriculum supplements (e.g., Facebook, Twitter, Instagram, blogs). We first tried contacting teacherpreneurs through Facebook messenger, but one of our personal Facebook accounts became restricted for a limited time because Facebook administrators also saw such contact as soliciting their users. We continue to have difficulty contacting teacherpreneurs to learn about their process of creating and sharing curriculum supplements. We have learned: *It is best to communicate with teacherpreneurs through their personal e-mails and blogs*. Sometimes this information is provided on their TpT user profiles, but if not listed, we learned that doing a reverse image search[3] of the teacherpreneur's TpT icon is the best way to find other means of contact. This is a tedious data collection process; however, once we find e-mail addresses or direct contact information via blogs, many teacherpreneurs have indicated interest in this research and are willing to discuss their philosophy around creating their curriculum supplements.

DATA ANALYSIS

After gaining access to data associated with curriculum supplements found on virtual resource pools, we had to choose and develop methods of analyzing all three types of data: the curriculum supplement itself, data from teachers who use curriculum supplements, and information about the teacherpreneurs who create the supplements.

Curriculum Supplements

There are many considerations when analyzing curriculum supplements from virtual resource pools. To gain an understanding of the different types of elementary mathematics curriculum supplements, we developed classification methods to capture the scope of items available. We identify characteristics of interest for each supplement, such as the intended gradeband, the content domain of the supplement, and the type of picture(s) (Brändström, 2005) included on the supplement. Although identifying these characteristics was straightforward, when attempting to analyze the results, we ran into problems because the size of our dataset would not always allow for statistical analyses to be run with integrity. For example, in considering the distribution of supplements across elementary grade-bands for the content domains, there were too few geometry supplements present in some of the grade-bands to run a chi-square analysis. We learned: *We must be prepared to collapse the data in justifiable manners*. For this example, we collapsed the grade bands into K–2 and 3–5 to permit statistical analysis across the elementary content domains (Shapiro et al., 2021).

In addition to classifying the curriculum supplements to understand the breadth of what is available, our main goal for analysis was to capture the quality of the mathematics tasks present within the individual supplements. Because supplements can range from posters, task cards, worksheets, units, and so on, we quickly recognized the need to define a unit of analysis. We chose Doyle's (1983) definition of academic tasks to guide this aspect of analysis (Sawyer et al., 2019) and used Smith and Stein's (1998) Task Analysis Guide (TAG) as a framework for data analysis (i.e., Shapiro et al., 2021). TAG was developed for individual problems rather than an entire worksheet or set of tasks. So, when analyzing a supplement that has different types of tasks (e.g., a worksheet with a fill-in-the blank definition section followed by naked number problems, and finally a story problem in which students are asked to draw a picture), we keep the unit of analysis as the individual tasks. We then determine the minimum, mode, and maximum level of cognitive demand that comprises the entire supplement. After this process is completed, chi-squared (Pearson, 1992) statistical analyses are run looking at minimum, mode, and maximum levels of cognitive demand for various features of the supplements (Dick et al., 2019; Dick et al., 2023; Shapiro et al., 2021).

Over the past few years, we have found many informative relationships regarding the TAG levels of supplements. For the elementary mathematics supplements, overall, we found that free supplements had a statistically significant ($p = .008$) lower level of minimum cognitive demand than supplements that require payment (we studied those costing less than $5). In contrast, for-sale supplements were statistically significantly ($p < .001$) more likely to reach the highest level of maximum cognitive demand. Taken together, these findings suggest that supplements considered *Doing Mathematics* are underrepresented (i.e., they do not account for 25% of the top supplements), considerably more free supplements include at least one lower level *Memorization* task, and for-sale supplements include at least one *Doing Math* task (2.3% of free versus 10.8% of for-sale) (Dick et al., 2023). When comparing TAG levels based on the content domain of the supplements, we found that geometry supplements had a greater than expected minimum level of *Memorization*; numbers and operations in base ten supplements had a greater than expected minimum level of *Procedures with Connections* ($p < .001$) (Dick et al., 2019). There are many differences between supplements within different content domains.

In our most recent work studying middle-school mathematics supplements, we continued to compare supplements that cost money and those that are free and looked at differences in TAG levels among content domains; but we were also interested in whether there were differences between TAG and the type of curricular supplement. Each of the 500 downloaded middle-school supplements was categorized by type. After collapsing categories through open coding, we identified nine supplement types in order of prevalence: worksheet, task cards/station, notes/organizer, game, match/sort, coloring activity, assessment, project, and other. In comparing their maximum level of cognitive demand, we found games and notes/organizer supplements were more likely to reach a lower level of cognitive demand, and project supplements were likely to reach a higher level of demand (Jockimo, 2023). This newer supplement classification based on type is ripe for further study.

In contrast, when we studied teacher-provided curriculum supplements that teachers had implemented in their classrooms, we looked for a different way to sort the supplements. We researched the structure of various virtual resource pools. This led to an analysis of whether the resource pools required any peer-review or other vetting before teacherpreneurs post their curriculum supplements. We described this work using the terms *sanctioned* versus *unsanctioned* resource pools (Shapiro et al., 2021). This emergent classification scheme led us to compare the cognitive demand associated with these teacher-provided curriculum supplements. Of the supplements the teachers provided, 85% were from unsanctioned resource pools, such as TpT or Pinterest; the other 15% were from sanctioned pools. Results of the chi-squared analysis showed a statistically significant difference between the TAG levels and whether the supplement came from a sanctioned or unsanctioned resource pool ($p < .001$). Specifically, the mode for supplements from the unsanctioned pools was more likely to reflect tasks with a lower level of cognitive demand and the mode for those from the sanctioned pools was more likely to reflect tasks with a higher level of cognitive demand (Shapiro et al., 2021). Thus, despite the popularity of virtual resource pools, it appears the cognitive demand of supplements found on unsanctioned pools is significantly lower than the cognitive demand of supplements found on sanctioned pools. However, more research is needed in this area, because this particular study was from 2018. A replication study would be informative.

Overall, *it is imperative to run these types of statistical comparisons for all three classifications of the cognitive demand of a curriculum supplement (min, mode, and max demand) and to consider what the results tell about the range of curriculum supplements available when considering cost, type of supplement, breadth of elementary grade bands, coverage of content domains, visual appeal, and type of resource pool in which the supplement is found.*

Teachers' Reasons for Supplementation and Teachers' Supplemental Use Patterns

The data we collected regarding teachers' use patterns and overall reasons for using elementary mathematics curriculum supplements found on virtual resource pools is from surveys; we have not yet surveyed middle-school teachers. Some of the data analysis is straightforward with clear methods available to study responses to questions, such as "how often do you supplement," with pre-provided response choices. For example, when asked, "How often do you use [free online activities or paid online activities] in your elementary mathematics instruction?," 314 teachers (63%) indicated that they used free supplements at least half of the time, and 198 teachers (40%) indicated that they used for-sale supplements at least half of the time. Only nine teachers (2%) indicated that they never use free supplements, and 94 teachers (19%) reported never purchasing mathematics supplements (Shapiro et al., 2019). In addition to reporting percentages, we have run statistical comparisons using various teacher characteristics. For example, using a chi-squared analysis, we found no significant relationship between a teacher's years of experience and their reported use of online supplements ($p = 0.941$) (Shapiro et al., 2019).

Data analysis of open-ended questions from surveys was more complicated and required a grounded-theory approach to coding responses. For example, when examining how teachers adapted curriculum supplements to use in their classrooms, we created a code book for different teacher-specified changes made to the supplements. We learned to be clear and concise when describing the open coding process that we used to collapse categories and determine themes (see Table 19.1). Through this process, we determined four themes, listed here in order of prominence: (1) adapted for different learners; (2) adapted for classroom implementation; (3) adapted for mathematics content; and (4) adapted for visual appeal. We found 99% of teachers identified some form of adaptation of their supplements, with 64% of teacher responses including descriptions regarding meeting the needs of different learners, such as English language learners or special education students (Dick & Sawyer, 2022). This result indicates a need for future work obtaining exact teacher adaptations to compare the downloaded curriculum supplement to the teachers' adaptations.

Supplement Source: Teacherpreneurs

To analyze information on teacherpreneurs, we learned that it is difficult to classify, and therefore to systematically analyze, various aspects of the information teacherpreneurs enter about themselves, both on their TpT profile pages as well as on other social media sites. For example, teacherpreneurs self-report their years of teaching experience, grade(s) taught, and educational backgrounds. We included an unknown category for when it was unclear whether a teacherpreneur included the current year in their teaching experience, or in fact, if they are currently teaching. Additionally,

Table 19.1

Example of Adaptation Category with Keywords and Definition

Adaptation Category	Open Coding Keywords	Description
Adapted for Different Learners	English Language Learners Needs Special Education Level of Difficulty: Easy/Hard/Difficult/Levels/Rigor/Advance/Struggle Differentiation Scaffold	The teachers adapted the activities to meet different learner needs.

Adapted for Classroom Implementation	Manipulatives
	Directions
	Time
	Structure:
	   Independent/Group/Partner/One-on-One
	   Centers
	Technology
	   Video/PowerPoint/iPad/Smart Board/Computer/Tech
		The teachers adapted the activities by adding elements or taking elements away to alter the implementation of the activity in their classroom.

Adapted for Mathematics Content	Standards Common Core/District/State Numbers Content Alignment Curriculum	The teachers adapted the activities by adding elements or taking elements away to alter the mathematical content.
Adapted for Visual Appeal	Format Font Design Size Picture/Image Visual	The teachers adapted the visual aesthetics of the activity.

because some teachers provide ranges of years of experience, we could not use quantitative data and had to switch to categories; however, it proved difficult to collapse categories into ranges that provided enough data in each range to run chi-squared analysis reliably and in a manner that made sense (Shapiro et al., 2019). Another issue was in looking at teacherpreneur educational background. Some reported degrees without details (e.g., BA, MAT) and others included details (e.g., BA in English, MAT in elementary education). We hoped to analyze the role of education in the types and level of cognitive demand of curriculum supplements the teacherpreneurs created, but learned this was not possible due to vast differences in the nature of the data provided. Overall, because there was no way to know if the teacherpreneur self-reported data was current, we learned that any analysis we attempted must clearly be presented and discussed alongside these limitations (i.e., Sawyer, Dick, & Sutherland, 2020; Shapiro et al., 2019).

We also learned *the importance of consistency in survey questions to provide a clear comparison across datasets*. For example, when we chose to compare the perceived importance of teachers' selection of supplements versus teacherpreneurs' creation of supplements, we used similar survey phrasing. When investigating elementary teachers' selection of online mathematics supplements as discussed in Shapiro et al. (2019), the survey we used included the item, "Please rank the importance of the following criteria you use when selecting elementary mathematics activities online." To help understand teacherpreneurs' creation of online mathematics supplements as reported in Sawyer, Dick, and Sutherland (2020), the survey we used included the item, "Please rank the importance of the following criteria you use when creating elementary mathematics activities." Because we used identical wording when surveying teachers who use curriculum supplements as well as teacherpreneurs who create supplements, we could directly compare what teachers report looking for in mathematics curriculum supplements versus what teacherpreneurs believed teachers look for when searching for supplements. We discovered discrepancies between the two groups (Sawyer, Dick, & Sutherland, 2020). For example, findings showed that teacherpreneurs believe teachers want free supplements aligned to standards but that are also visually appealing. Although teachers did rank alignment to standards as important, teachers did not rank visual appeal as important; in fact, they ranked it as one of the least important elements of the resources (Sawyer, Dick, & Sutherland, 2020). This and other similar findings contribute to our understanding of nuances related to teacherpreneurs who create online curriculum supplements and the teachers who use them, and show there is much to investigate further.

PUBLICATIONS

Lessons we learned about publications on research in the area of online curriculum supplementation are similar regardless of whether the research is on the curriculum supplements themselves, on teachers' reasons for supplementation and their supplemental use patterns, or on the teacherpreneurs. Thus, in this section, we share questions we learned to ask regarding what we can publish, where to publish, and perhaps most importantly, how to frame this work.

What Can We Publish?

After collecting and analyzing TpT curriculum supplements, we ran into copyright issues because the curriculum supplements under analysis are rightfully owned by the teacherpreneurs themselves. We cannot show images from specific supplements without written permission from the teacherpreneurs; as previously discussed, we had difficulty contacting teacherpreneurs. Our practitioner paper (Shapiro et al., 2021) was delayed in publication as we awaited permission to publish images of various TpT supplements. Because we had been restricted from contacting TpT teacherpreneurs when we reached out to them through the "Ask a Question" feature on the TpT website, we provided the journal editor with multiple options of TpT supplements that would work for our publication and the editor directly reached out to the teacherpreneurs and obtained permissions. Although we have occasionally been successful in receiving permission ourselves, we developed a publication workaround by creating our own examples of tasks found on various curriculum supplements that mirror the types and styles found on the downloaded supplement. We made the decision to take this approach with the paper by Dick and colleagues (2023) to reduce the number of necessary permissions from seven to three. Unfortunately, these types of publication decisions are sometimes necessary.

Where to Publish?

Research regarding online curriculum supplements that are created by teacherpreneurs lives in a fuzzy space. Some educational colleagues and reviewers have called our research "pop education"; others consider it analogous to research on textbook curriculum. We learned that individual opinions of this research space influence its dispersion and recognition in the field, especially in considering which research areas are willing to accept a "fit" into their space. Is this research *curriculum research*? Is this research on educational technology? What about mathematics education? We see our research at an intersection of these spaces, but primarily as mathematics curriculum research. However, editors and reviewers have not always agreed.

When submitting to curriculum journals, we were told to send our work to technology journals. When sending to technology journals, we were told our work is too focused on mathematics and we should send our work to mathematics education journals. When sending to mathematics education journals, we were told we are not investigating mathematics, we are studying technology or looking at curriculum. So, what have we learned? We learned to keep trying to find an appropriate publication venue and eventually someone will decide our work "fits" within their area. Thus, we have published in technology spaces (i.e., Sawyer et al., 2019; Shapiro et al., 2019), mathematics education spaces (i.e., Dick et al., 2019; Dick et al., 2023; Shapiro et al., 2021), and more general journals of education (Sawyer, Dick, & Sutherland, 2020; Sawyer, Dredger et al., 2020). More recently, we have found success with curriculum spaces within the field of mathematics education. We have given paper presentations under curriculum strands at the International Congress of Mathematical Education (Dick et al., 2020/1) and the Psychology of Mathematics Education–North America (Dick & Sawyer, 2022), and plan to continue presenting in these spaces. In general, we have learned: *The best strategy to support timely publications is to contact editors before formally submitting to determine editor beliefs about fit and overall interest in the topic.*

Because finding publishers can be difficult and the publication process is lengthy, the data described in manuscripts can quickly become out-of-date. Virtual resource pools are continually changing, and thus, it is hard to keep current and show consistency across research projects. For example, in 2018 when we downloaded free elementary TpT supplements, all were categorized as having four stars, the highest rating at that time, but TpT had changed its rating systems to five-stars before the paper was published (Sawyer, Dick, & Sutherland, 2020). Therefore, we included a disclaimer in the paper indicating that we collected supplements with the highest TpT rating possible at that time. In our most recent work with middle-school mathematics curriculum supplements, we documented the number of ratings for the supplements as a means to capture the popularity of a supplement. However, preliminary results indicate no difference between the mean number of ratings for a supplement and the cognitive demand of the supplement (Jockimo, 2023).

A similar issue occurred when we investigated Pinterest in Sawyer and colleagues (2019). Previously, the platform calculated the number of "pins" or number of times a person on their website collected and stored that image. Teachers who use supplements in their classrooms indicated they used the number of pins to determine quality when choosing supplements (Sawyer & Myers, 2018), thus we were looking to investigate this frequency as a measure of teacher perceived quality. However, in the midst of publishing this work Pinterest removed the counting of "pins" from the platform, making that research less relevant and impossible to compare across datasets. *The online environment is ever-evolving, and it is important to keep an eye on changes made on virtual resource pools and to adjust as needed to ensure timely publication.*

How to Frame This Work?

One early publication hurdle we faced was how to frame this work. We had to determine which frameworks applied and how to discuss curriculum supplementation within established theories. In our early work looking at teachers' reasons for supplementation, we built upon Coiro and colleagues' (2014) New Literacies Theory (NLT). NLT explains that individuals living in the digital age typically have different dispositions towards collecting, analyzing, and adapting curricular resources versus individuals who did not have the internet as a tool in their formative years. NLT was constructed to describe how these literacies have changed over time. Sawyer, Dredger and colleagues (2020) used NLT to develop the *critical curation theory* to help explain why we need to conduct research on teachers' use of virtual resource pools, specifically the ways teachers approach their selection of curriculum supplements (Shapiro et al., 2019). However, many reviewers questioned the application of NLT to online curriculum research because of its focus on digital literacy without a direct connection to curriculum.

In our work studying the supplements themselves (Sawyer et al., 2019), we adopted a modified version of the Mathematics Task Framework (Stein & Smith, 1998) as conceptualized by Wilhelm (2014). This allowed us to situate our study of the cognitive demand of the curriculum supplements within both the TAG levels of demand (Stein & Smith, 1998) and how others like Wilhem (2014) used TAG in practice to study curriculum. Although this well-respected framework is appropriate for this particular aspect of our work, reviewers pushed back, saying it was not sufficient to describe nuances related to online curriculum supplementation.

Our work is in a new area of curriculum research, so developing frameworks for it takes time. Silver's (2022) TCFS framework that we used as the framing for this chapter is based on many of our published works and draws upon literature we have read over the past six years. Silver's publication provides the field of online curriculum supplementation with a common language, and the overarching TCFS framework is one that the field can use moving forward. As a direct result of Silver's publication, we reached out to him and others Silver cited. We have had multiple virtual meetings discussing how our work overlaps and ways to continue moving forward to support teachers as they navigate the world of online curriculum supplementation; we also discussed publishing research in this fuzzy space. We learned: *An important aspect of publishing in newer curriculum research areas is developing relationships with others invested in doing the same types of work*. We learn much from others and our work is better from their input.

SUMMARY AND CONCLUSION

In summary, when collecting data regarding online curriculum supplements found on virtual resource pools that are not free, we recommend researchers contact the virtual resource pool companies prior to trying to obtain supplements to discuss any unique needs that a researcher may have regarding access to both the curriculum supplements and also to the teacherpreneurs who created them. To ensure the most accurate data are collected, use online filters purposefully, and after obtaining supplements, move them out of the online platforms to a more secure storage site as soon as possible. If collecting data on teacherpreneurs, be careful contacting them via social media platforms; instead, use reverse image searches of teacherpreneurs' logos to find direct personal contacts.

We recommend researchers approach data analysis with an eye towards flexibility due to the quickly changing landscape surrounding online curriculum resource supplementation. Be prepared to collapse data in justifiable manners and clearly define units of analysis as early in the investigation as possible. Also, researchers need to use varied statistical analyses to provide valid comparisons of data and be prepared to describe data analysis along with limitations when dealing with outdated or shifting information within the dataset.

When publishing research in the curriculum supplementation space, be aware that getting copyright permission to publish examples of supplements can be difficult. When permission is not granted, we recommend creating your own examples of various curriculum supplements that can be open-sourced; we recommend creating them in the spirt of common themes arising from the research, ensuring they are different and do not compromise any teacherpreneur's intellectual property rights. Also, understand that individual reviewers' or editors' opinions of the field of online curriculum supplementation influence dissemination, thus we recommend contacting editors ahead of submission to gauge fit and overall interest. Finally, and most importantly, be sure to select frameworks to situate your work in ways that are clear and serve to move the field of online mathematics curriculum supplementation forward.

We have learned a lot about researching online mathematics curriculum supplementation. Research can be difficult because of data access issues, the need to consider and/or create data analysis procedures, and various questions regarding publication. Despite these challenges, research in this area of mathematics curriculum supplementation is important. Teachers all over the world use virtual resource pools such as TpT to supplement their official mathematics curriculum (Shapiro et al., 2019). Still, little is known about either the supplements themselves or the teacherpreneurs who create them. Our work thus far has been focused on elementary mathematics curriculum and we are just beginning to research middle-school mathematics curriculum supplementation. To our knowledge, no one has yet done research regarding online high-school mathematics curriculum supplementation. We believe the lessons we have learned thus far directly apply to others interested in working within the online curriculum supplementation field of research and we hope others will decide to join us in this important and ever-evolving work.

ACKNOWLEDGMENT

This work was supported by the James Madison University Program of Grants for Faculty Assistance, the Martha B. and Guy M. Jones Educational Endowment, and the Bucknell Institute for Public Policy Research Grant. Any opinions, findings, and conclusions or recommendations expressed herein are those of the researchers and do not necessarily reflect the views of the supporting institutions.

REFERENCES

Brändström, A. (2005). *Differentiated tasks in mathematics textbooks: An analysis of the levels of difficulty* [Licentiate dissertation, Luleå University of Technology]. Digitala Vetenskapliga Arkivet. https://www.diva-portal.org/smash/record.jsf?pid=diva2%3A991116&dswid=-5206

Browne, A., & Haylock, D. (2004). *Professional issues for primary teachers*. Paul Chapman.

Carpenter, J. P., & Krutka, D. G. (2014). How and why educators use Twitter: A survey of the field. *Journal of Research on Technology in Education*, *46*(4), 414–434.

Carpenter, J. P., Morrison, S. A., Craft, M., & Lee, M. (2020). How and why are educators using Instagram? *Teaching and Teacher Education*, *96*, 103–149.

Coiro, J., Knobel, M., Lankshear, C., & Leu, D. J. (Eds.). (2014). *Handbook of research on new literacies*. Routledge.

Dick, L. K., & Sawyer, A. G. (2022). Investigating elementary mathematics teachers' adaptation of activities found on virtual resource pools. In A. E. Lischka, E. B. Dyer, R. S. Jones, & J. N. Lovett (Eds.), *Proceedings of the 44th annual meeting of the North American Chapter of the International Group for the Psychology of Mathematics Education* (pp. 123–131). Middle Tennessee State University.

Dick, L., Sawyer, A. G., & MacNeille, M. (2020/1). *Identifying the quality of teacher created curriculum shared via the Teachers' Pay Teachers online platform* [Conference presentation]. Paper presented at the 14th annual meeting of the International Congress on Mathematical Education, Shanghai, China and Online.

Dick, L. K., Sawyer, A., MacNeille, M., Shapiro, E., & Wismer, T. (2023). Exploring grades 3–5 mathematics activities found online. *Mathematics Teacher: Learning and Teaching PK–12, 116*(3), 174–183. https://doi.org/10.5951/MTLT.2021.0330

Dick, L., Sawyer, A. G., Shapiro, E., & Wismer, T. (2019). Influencing the influencers: Discussion of elementary mathematics online resources. *Association of Mathematics Teacher Educators Tech Talk Blog*. https://amte.net/tech-talk/2019/09/influencing-influencers

Doyle, W. (1983). Academic work. *Review of Educational Research, 53*(2), 159–199. https://doi.org/10.3102/00346543053002159

Jockimo, L. A. (2023, April 24). *Exploring middle school mathematics resources available on TeachersPayTeachers*. [Google Slides]. Mathematics Department, Bucknell University.

National Council of Teachers of Mathematics [NCTM]. (1989). *Principles and standards for school mathematics*.

National Governors Association Center for Best Practices & Council of Chief State School Officers [NGA & CCSSO]. (2010). *Common core state standards for school mathematics*. https://www.thecorestandards.org/Math/

Pearson, K. (1992). On the criterion that a given system of deviations from the probable in the case of a correlated system of variables is such that it can be reasonably supposed to have arisen from random sampling. In S. Kotz & N. L. Johnson (Eds.), *Breakthroughs in statistics* (pp. 11–28). Springer. https://doi.org/10.1007/978-1-4612-4380-9_2

Remillard, J. T. (2005). Examining key concepts in research on teachers' use of mathematics curricula. *Review of Educational Research, 75*(2), 211–246.

Sawyer, A. G. (2017). Factors influencing elementary mathematics teachers' beliefs in reform-based teaching. *The Mathematics Educator, 26*(2), 26–53.

Sawyer, A., Dick, L., Shapiro, E., & Wismer, T. (2019). The top 500 mathematics pins: An analysis of elementary mathematics activities on Pinterest. *Journal of Technology and Teacher Education, 27*(2), 235–263.

Sawyer, A. G., Dick, L. K., & Sutherland, P. (2020). Online mathematics teacherpreneurs developers on Teachers Pay Teachers: Who are they and why are they popular? *Education Sciences, 10*(9), 248. https://doi.org/10.3390/educsci10090248

Sawyer, A. G., Dredger, K., Myers, J., Barnes, S., Wilson, D. R., Sullivan, J., & Sawyer, D. (2020). Developing teachers as critical curators: Investigating elementary preservice teachers' inspirations for lesson planning. *Journal of Teacher Education, 9*(1), 1–19. https://doi.org/10.1177/0022487119879894

Sawyer, A. G., & Myers, J. (2018). Seeking comfort: How and why preservice teachers use internet resources for lesson planning. *Journal of Early Childhood Teacher Education*, *39*(1), 16–31.
https://doi.org/10.1080/10901027.2017.1387625

Shapiro, E., Dick, L., Sawyer, A. G., & Wismer, T. (2021). Critical consumption of online resources. *Texas Mathematics Educator*, *67*(1), 6–11.

Shapiro, E., Sawyer, A. G., Dick, L., & Wismer, T. (2019). Just what online resources are elementary mathematics teachers using? *Contemporary Issues in Teacher Education*, *19*(4), 670–686. https://citejournal.org/volume-19/issue-4-19/mathematics/just-what-online-resources-are-elementary-mathematics-teachers-using/

Shelton, C. C., & Archambault, L. M. (2019). Who are online teacherpreneurs and what do they do? A survey of content creators on TeachersPayTeachers.com. *Journal of Research on Technology in Education*, *51*(4), 398–414.

Silver, D. (2022). A theoretical framework for studying teachers' curriculum supplementation. *Review of Educational Research*, *92*(3), 455–489.
https://doi.org/10.3102/00346543211063930

Smith, M. S., & Stein, M. K. (1998). Selecting and creating mathematical tasks: From research to practice. *Mathematics Teaching in the Middle School*, *3*(5), 344–350.

Stein, M. K., & Smith, M. S. (1998). Mathematical tasks as a framework for reflection: From research to practice. *Mathematics Teaching in the Middle School*, *3*(4), 268–275.

TeachersPayTeachers. (2023). *Teachers Pay Teachers: About us.*
https://www.teacherspayteachers.com/About-Us

Torphy, K., Hu, S., Liu, Y., & Chen, Z. (2020). Teachers turning to teachers: Teacherpreneurial behaviors in social media. *American Journal of Education*, *127*(1), 49–76.

Wilhelm, A. G. (2014). Mathematics teachers' enactment of cognitively demanding tasks: Investigating links to teachers' knowledge and conceptions. *Journal for Research in Mathematics Education*, *45*(5), 636–674.

ENDNOTES

1. In the U.S., students in these grades would range in age from 4–11 years.
2. We used the content domains of the *Common Core State Standards for Mathematics* (National Governors Association Center for Best Practices & Council of Chief State School Officers, 2010).
3. A reverse image search is conducted by searching for a publicly available image's location on the internet using an uploaded or linked image.

SECTION V

RESEARCH ON THE ENACTED CURRICULUM

CHAPTER 20

FROM PARTICIPANTS TO PARTNERS

Lessons Learned From Enacted Curriculum Research in Early Numeracy Contexts

Arielle Orsini
Concordia University, Canada

Julie Houle
Lester B. Pearson School Board, Canada

Helena P. Osana, Alison Tellos, and Anne Lafay
Concordia University, Canada

The Early Numeracy Partnership (ENP) was established in 2018 in Eastern Canada to enhance the enacted numeracy curriculum in kindergarten classrooms and support children's numeracy development. In initial workshops on children's numeracy, teachers requested more practical tools for classroom instruction. Their request inspired us to create the Numeracy Kit for Kindergarten 5-year-olds (NyKK-5), *a resource for use in real-time as teachers interact with their students on core numeracy skills. This chapter describes three lessons we learned about designing, implementing, and testing the* NyKK-5 *in partnership with kindergarten teachers: (1) researchers must uncover barriers that impede teachers' ability to meet the objectives of the official curriculum; (2) successful partnerships are reciprocal when teachers and researchers prioritize common goals; and (3) teacher buy-in is essential to enrich the enacted mathematics curriculum successfully. We discuss implications of these three lessons for future curriculum research in early numeracy.*

Keywords: early numeracy; enacted curriculum; pedagogical tool; professional development; researcher-practitioner partnership

A strong numeracy foundation is critical for young children's mathematical development (Jordan et al., 2010; Mazzocco & Thompson, 2005; National Mathematics Advisory Panel [NMAP], 2008). When students are exposed to core numeracy principles, such as those related to counting and number knowledge, they acquire foundational concepts that support their later mathematical success (Ginsburg, Lee, & Boyd, 2008; Watts et al., 2014). For a variety of reasons, by the time some children enter kindergarten, their numeracy skills lag behind those of their peers (Starkey et al., 2004). Perhaps not surprisingly, children who fail to acquire these number concepts early continue to fall behind their peers in mathematical development as they progress through school (Duncan et al., 2007; Geary et al., 2013; Jordan et al., 2007; Jordan et al., 2009), with gaps widening over time (Aunola et al., 2004; Morgan et al., 2011). It thus becomes crucial for teachers and other practitioners to interact with young children about numeracy ideas, especially in kindergarten.

For the past six years, we have collaborated with kindergarten teachers in a number of schools in Canada to foster and sustain conversations in their classrooms about key numeracy concepts that have been shown to be predictive (in a correlational sense) of students' later mathematics achievement. Using a design-based research model (Cobb et al., 2003; The Design-Based Research Collective, 2003), we partnered with teachers and other educational stakeholders to provide professional development (PD) that would inform teachers' intended and enacted numeracy curriculum in kindergarten (Copur-Gencturk et al., 2019; Garet et al., 2001). Our objective is to design and create classroom-tested materials for teachers to use so that students acquire the conceptual foundation required for their mathematical success (Pepin et al., 2013; Remillard, 2005). Our overarching hypothesis is that teachers' enacted curriculum can be enhanced by augmenting their knowledge about the development of children's mathematics and by providing concrete resources for application in the classroom.[1]

This chapter describes the lessons we learned about designing and implementing numeracy materials for kindergarten teachers as we collaborated with a long-standing partner school district and its teachers. We use our research collaboration, the Early Numeracy Partnership (ENP), as a case study to illustrate the lessons. We begin by situating our research in current conceptualizations of the intended and enacted mathematics curriculum in early childhood settings. We then describe the historical trajectory of the ENP (Phases 0, 1, 2), with a focus on the increasingly iterative nature of our collaborations with teachers over the years. The chapter then moves to descriptions of the three lessons we learned:

1. Understanding barriers is critical for effective enactment of a numeracy curriculum in kindergarten.
2. Building trust with participating teachers enables them to adopt the role of partner for numeracy material design and appropriation.
3. Acquiring teacher buy-in for the adoption of curriculum resources allows teachers to become active agents of their own curriculum and classroom practice (Borko et al., 2010).

We conclude with future directions for curriculum research in early numeracy.

SITUATING THE CURRENT RESEARCH

In this section, we begin by describing the theoretical framework for our research and the educational context that brought the early numeracy needs of the partner district to our attention. We then outline the evolution of the Early Numeracy Partnership (2018–2022), which provides a backdrop for the lessons we learned about designing and implementing numeracy materials in kindergarten.

Theoretical Framework

Remillard and Heck's (2014) conceptualization of the mathematics curriculum provided a useful framework in which to situate our research on teachers' intended and enacted numeracy curriculum and the factors that influence it. The authors presented two primary components of the mathematics curriculum—the official curriculum and the operational curriculum. The official curriculum, often set by governing bodies, consists of statements that specify broad expectations for student learning. Often, but not universally, the official curriculum describes: (a) the aims and objectives for learning in school mathematics; (b) the contents of government-level assessments to track students' progress in acquiring those objectives; and (c) a designated curriculum, which provides instructional guidance and materials for teachers to achieve the objectives.

Meanwhile, the operational curriculum reflects what actually takes place in the classroom and consists of the reciprocal relationship among teachers' intended (i.e., planned) curriculum, their enacted curriculum, and students' learning outcomes. The intended curriculum, characterized by Remillard and Heck (2014) as the plans educators create for what and how they will teach, is largely guided by the designated curriculum. In turn, the intended curriculum directly influences the enacted curriculum, that is, the day-to-day classroom interactions that take place between teachers and students around specific tasks and lessons. Students' responses and teachers' interpretations of student thinking may subsequently inform teachers on how to modify their intended curriculum to meet their students' needs (Even & Gottlib, 2011; Even & Tirosh, 2002). Therefore, influenced by the targeted mathematics, teachers' pedagogical moves, the classroom tools and resources that are used, and teacher-student interactions, the enacted curriculum can deviate significantly from what is intended (Choppin et al., 2022; Remillard, Harris, & Agodini, 2014; Remillard et al., 2019; Roth McDuffie et al., 2018; Snyder et al., 1992), requiring teachers to adapt their practice in an ongoing, iterative fashion. In the context of this framework, our research objectives were to (a) work alongside partner teachers to create classroom-tested resources for the enacted numeracy curriculum in kindergarten, (b) test classroom applications of the resources on students' numeracy development, (c) refine the resources in an iterative process with partner teachers, and (d) use the findings to deepen our theoretical understanding of the enacted curriculum and the factors that influence it.

Current Research Context

Because of variations between official curricula across provinces in Canada, coupled with significant individual differences at the teacher and student levels that determine what actually occurs in classrooms, the ways in which mathematics curriculum takes hold in a given setting is highly variable and contingent on multiple actors and influences in the systems in which teachers work (Remillard & Heck, 2014). In our specific context, the overarching objective of the official preschool program is to foster in 4- and 5-year-old children the interest and curiosity about the world around them. To this end, teachers attend to student development in a number of areas, including social, physical-motor, and cognitive development. They are encouraged to foster thinking and reasoning skills in play-based contexts, which have as objectives, among others, to support the learning of new concepts related to subjects such as mathematics, science, and the arts. Numeracy skills, including reciting the rote counting sequence, recognizing small collections, using number words to enumerate sets, and manipulating concrete objects to perform simple arithmetic, are listed as behaviors that the teachers may or may not observe in classroom settings designed to foster exploration and play.

From our conversations with teachers, we learned that their intended curriculum was to achieve the objectives as outlined in the official preschool curriculum set by the provincial government. At the same time, several teachers have expressed that they did not know "what to do" to ensure that the curricular goals, as enumerated in the government's curriculum document, were met. It became clear that the teachers could rely on no official designated curriculum to structure their intended or enacted curriculum. In other words, the official curriculum provided no "instructional specificity" (Remillard & Heck, 2014, p. 710) toward addressing curricular aims and objectives, either in the form of textbooks, classroom materials, or other resources. What the teachers enacted may or may not have resulted in what they had intended, nor were the teachers equipped to make such judgments. Thereby, they received no useful or systematic feedback for the extension and refinement of their practices. Further, with their perceptions of the official curriculum as lacking instructional specificity and concrete applications, teachers had few useful signposts to determine whether the learning objectives in the official curriculum were being met. Because numeracy is not formally evaluated in the preschool program, teachers also had no clues about assessment content or outcomes to guide their practice.

The Early Numeracy Partnership

Although we had been working in the school district for several years, the Early Numeracy Partnership (ENP) was formalized in 2018 through a grant from the Canadian government with the objective of supporting kindergarten teachers in disadvantaged urban areas to engage in mathematical conversations with their students around key numeracy concepts. The ENP evolved over several research phases: Phase 0 in 2018–2019; Phase 1 in 2019–2021; and Phase 2 in 2021–2022.

Phase 0 (2018–2019)

In Phase 0, we provided three day-long workshops to five kindergarten teachers from three public elementary schools on the core numeracy skills that have been shown to predict later mathematics achievement, including subitizing; counting and enumeration skills; number knowledge, including transcoding and magnitude comparisons; simple arithmetic, explored through story problems and non-verbal calculation; and part-whole concepts (see Baroody et al., 2006; Jordan et al., 2010; Passolunghi & Lanfranchi, 2012; Purpura & Lonigan, 2013). Undergraduate and graduate research assistants visited classrooms twice a month over the course of the year to provide feedback to teachers about their students' thinking and to document any challenges teachers were experiencing enacting a curriculum that targeted key numeracy skills. Although a few activities were shared with the teachers during the workshops to illustrate how to support key numeracy skills, the teachers were encouraged to incorporate conversations around specific numeracy skills in their already established classroom routines. We compared the numeracy growth from the beginning to the end of the school year for students whose teachers received the PD (five teachers; 39 children) to the numeracy growth of those who were in one of two business-as-usual control groups (seven teachers; 63 children). The results revealed a positive correlation between the PD and children's numeracy development in key areas, including the one-to-one principle, cardinality, and transcoding (Osana et al., 2023; Sinda et al., 2021).

Phase 1 (2019–2021)

Overall, teachers appreciated the clarification and specific descriptions of the numeracy skills presented in the Phase 0 workshops because they provided a clearly delineated, and in some cases extended, description of the objectives listed in the official curriculum. With a better understanding of the end goal in terms of student outcomes, they were eager to try the suggested activities. A pedagogical mathematics consultant in the district was inspired to create a resource that teachers could use in real time to foster and maintain classroom conversations around key numeracy skills. The consultant recruited an interested kindergarten teacher to assist in the design of the resource and to validate its usability in real classroom contexts. Over the next two years, which constituted Phase 1 of the partnership, the consultant and teacher became active members of our research team to create the *Numeracy Kit for Kindergarten 5-year-olds* (*NyKK-5*).

During the first year of Phase 1, we translated targeted numeracy skills into operational suggestions for classroom use, selected accompanying manipulatives, and created the teacher manual and a collection of formative assessment documents. Our work culminated in a prototype for teacher use. We recruited four kindergarten teachers in the school district to assess all aspects of the kit, including its value in planning classroom activities, its usability during instruction, and its sensitivity to assess students' numeracy development formatively. After six weeks of using the kit in the classroom with their students, the teachers reported in two debriefing sessions on what worked and what needed improvement. We adjusted the prototype based on their feedback, resulting in a greatly improved original design.

Our conceptualization of the *NyKK-5* is that it acts as a guide and resource for the enacted numeracy curriculum in kindergarten. First, it targets six core numeracy skills (i.e., immediate quantity recognition, counting, number knowledge, non-verbal calculation, story problems, and combinations part-whole) which have been shown to be predictive of later mathematical success (Jordan et al., 2007; Jordan et al., 2009). Second, the kit consists of a collection of activity cards associated with each core numeracy skill for teachers to reference either as they plan their lessons or as a just-in-time resource during instruction (see Figure 20.1). Manipulatives and other materials (e.g., posters) provided in the kit accompany many of the activity cards. In this way, the cards serve to guide the teachers' pedagogical moves (Pepin et al., 2013; Remillard & Heck, 2014). The kit also contains activity cards to support the enactment of whole-class discussions about numeracy called "NyKK talks" (inspired by Parrish, 2011). The cards are not presented in a prescribed sequence, but rather as suggested ways to generate classroom interactions and conversations between teachers and students around numeracy. Other resources include a teacher training handbook and the formative assessment document, a tool teachers can use to document individual students' progress on each skill (Heritage, 2007).

Figure 20.1

Sample Activity Cards from the NyKK-5 Kit

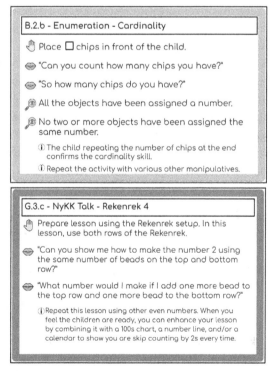

Note: A Rekenrek is a device that supports number-sense skills and representations and is composed of 20 beads; 10 are red and 10 are white, displayed on two rows.

Phase 2 (2021–2022)

Phase 2 of the ENP (2021–2022), supported by a second grant from the federal government, assessed the effectiveness of the revised toolkit for improving kindergarteners' numeracy development over the school year. Eight participating teachers were assigned to one of three PD conditions provided at the start of the academic year: (a) *NyKK-5*, where teachers received PD on children's early numeracy development and supplementary training on the *NyKK-5* itself ($n = 4$); (b) theory-only, where teachers received the same theory on children's numeracy as the *NyKK-5* group, but were not given the *NyKK-5* kit ($n = 2$); and (c) active control, where teachers received PD on children's broader cognitive development and received a box of assorted LEGO® bricks to address these domain-general skills in the classroom ($n = 2$). Using the *Purdue Early Numeracy Screener* (Purpura, 2021), a validated measure of young children's numeracy, research assistants collected student data at the beginning and at the end of the school year.

Between the months of February and April 2022, two research assistants visited classrooms of all participating teachers once a month, for a total of three visits, to observe their enacted curriculum based on the PD they received. Teachers in all three conditions also submitted weekly logs on specific practices that we had emphasized in each condition. Preliminary results showed that the students in both the *NyKK-5* and theory-only groups improved significantly from the beginning to the end of the year relative to the control group. The observational data and the teacher logs should provide a more complete account of the findings to identify why learning occurred (Cobb et al., 2003).

LESSONS LEARNED ABOUT ENACTED CURRICULUM RESEARCH IN KINDERGARTEN NUMERACY

We now turn attention to the specifics around three lessons learned from the project. The first lesson is about understanding the barriers the teachers faced when enacting a kindergarten numeracy curriculum. The second lesson highlights the need to build trust with our teacher collaborators to maximize the likelihood that the numeracy materials are adopted into their classroom practice. The third lesson focuses on the importance of buy-in so that teacher collaborators become active agents in the intended and enacted curriculum.

Lesson 1: Understanding Barriers

The first lesson we learned is that there were substantial barriers to the teachers' enactment of the official curriculum, which in turn impacted the numeracy practices that occurred in their classrooms. Shusterman et al. (2019) argued that researchers who attempt to make change in authentic educational settings must work to understand the "end-user context" (p. 17) and, we argue, to reveal the needs of the actors in that context rather than testing direct translations from the laboratory (Burkhardt & Schoenfeld, 2003; Collins et al., 2004; Shah et al., 2015). Schoenfeld (1999) succinctly summarizes this issue as "How do we make sense of the ways in which people use knowledge in circumstances different from the circumstances in which that knowledge was developed?" (p. 7).

Teachers in all three phases of the project clamored for tools and resources to guide their practice, including linguistic and cognitive tools. Although knowledgeable about their craft and fiercely dedicated to their students' success, most teachers felt unprepared to teach early numeracy and described feeling particularly stuck when attempting to help children who struggled. This sentiment was articulated by one teacher in Phase 0 during a workshop discussion about general principles for numeracy instruction, such as the importance of classroom discussion and the use of concrete objects. With considerable frustration, the teacher asked for directives on what to do and requested specific activities and materials that she could bring to the classroom in the short term to engage students in numeracy conversations. Several other teachers felt unprepared to design their own curriculum, even after having acquired considerable knowledge on children's numeracy development in Phase 0. Indeed, numerous teachers cited the lack of time to design and test their own activities that would target the core numeracy skills featured in the PD.

Although our current evidence is anecdotal, we surmise that the lack of instructional specificity in the official curriculum played a big role in teachers' frustrations. At the beginning of Phase 0, with no designated curriculum or consequential assessments of children's early numeracy as reference points, teachers were eager to learn about classroom activities that would support the alignment of their intended and enacted curricula. The teachers' shared experiences and frustrations made us aware of their needs and that context-specific solutions were needed. The *NyKK-5* kit addressed several of the challenges. For example, the activities target key numeracy skills while being versatile and easy to use. They require relatively little teacher planning, which frees time that can be allotted for instruction and intervention. Critically, the kit allows for a large amount of professional flexibility: It can be used one-on-one, in small groups, or with the whole class. Teachers use it not as a prescribed program with a set sequence of activities, but rather as a collection of concrete examples that could be used as springboards for the development of their own activities. This allows the teachers the flexibility to target specific skills with particular students according to their needs.

At the end of Phase 2, however, evidence of another barrier emerged. After all the student data were collected (before and after the PD), the partner teachers shared that they did not use any of the formative assessment documents that accompanied the kit. They cited a lack of time as well as the unwieldy nature of the format and organization of the tool. The teachers provided suggestions for restructuring the document to be more functional for teaching, and we took their feedback to create new versions for different instructional purposes. For instance, one version was designed to obtain a general sense of children's understanding of specific skills, and another sought to document the individual progress of specific children.

Learning about teachers' experiences with the formative assessment document is another example of the importance of accounting for the needs of practitioners in their own contexts, and it now features more prominently in our current research with our partner teachers. We are entertaining other potential explanations for the apparent lack of utility of the assessment document, such as teachers' beliefs about the nature and value of formative assessment, and their skill in assessing their students' knowledge on an ongoing basis during instruction (Bennett, 2011; Yan et al., 2021). It is also possible that teachers engaged in effective formative assessment using means other than the formative assessment document included in the kit, which we will investigate in our current research with kindergarten teachers.

Lesson 2: Building Trust: Moving from Participant to Partner

The second lesson we learned is that teachers must come to understand that successful partnerships must be truly reciprocal, namely that researchers and teachers can and should learn from each other. As outlined in Remillard's (2005) review, because their prior teaching experiences and beliefs about education play a large role in the curriculum they enact, teachers are active agents in their own practice and central decision makers about what occurs in their classrooms. As such, the adoption of outside resources or initiatives results in a variety of instantiations, interpretations, and modifications by the teacher, intentional or not (Geudet & Trouche, 2012; Remillard & Heck, 2014). We thus argue that the enacted curriculum, and research on characterizing and shaping it, should be conducted in tandem with teachers, who ultimately determine what happens in their classrooms and what their students learn (Donovan et al., 2013; Penuel et al., 2011; Shah et al., 2015). In fact, Penuel et al. (2007) reported that for teachers to adopt a new resource and remain committed to change, they must play a role in the development of the resource in a context that values their perspectives, concerns, and expertise.

In the initial stages of our research in early numeracy, we viewed the work with teachers as predominantly uni-directional—that is, not only did we consider ourselves as the "suppliers" of information about children's numeracy development, but we were not sufficiently attentive to the type of craft knowledge that was at our disposal when interacting with kindergarten teachers (Putnam & Borko, 2000; Ruthven & Goodchild, 2008). In Phase 0, our partner teachers contributed actively in workshop discussions and openly shared their experiences and challenges teaching numeracy to young children. We became aware of the importance of listening to their concerns and being open to customizing the PD to their needs.

This revelation was not sufficient for a truly bidirectional partnership to emerge. We noticed that many teachers appeared to view themselves primarily as learners rather than as contributors to joint knowledge construction (Shah et al., 2015; Smith et al., 2013). We entertained a variety of reasons for the teachers' perceptions of their role, the first of which was that they had little experience serving as true contributors to research initiatives. Typically, teachers enter PD spaces with the expectation of being taught, and in some cases, being told what to do. Thus, while reframing our own orientation to meet the teachers' needs, we worked to shift teachers' mindsets from participants to partners. This awareness has been at the forefront of our collaborations with partner teachers since that time, and actively working on the shift remains a key component of every subsequent iteration of our research.

We used four principal strategies to develop a sense of reciprocity in our work with teachers. First, we invited educators with considerable classroom experience to join our research team. A mathematics consultant and a kindergarten teacher, who both participated in Phase 0, were open to changing their intended curriculum. They brought their knowledge and experience to the design of the *NyKK-5*, contributing to the usability of the resource and the potential for it to address the challenges expressed by teachers throughout the collaboration. The educators' perspectives not only kept the work of the research team grounded in the authentic contexts in which teachers worked, but were also invaluable in interpreting teachers' feedback as we modified and refined the *NyKK-5* through subsequent phases of the research.

Second, during PD and debriefing sessions, we offered teachers clear explanations of the objectives of the research, stating that our goals could only be achieved with their help in testing the resource in classrooms and receiving honest and constructive feedback. In Phases 1 and 2, we additionally emphasized the iterative nature of the research and that our goal was to continue the refinement process in cyclical work with other teachers. Third, we asked questions about their experiences using the *NyKK-5* with their students for purposes of modifying the resource, which had the secondary effect of valuing the teachers' perceptions, opinions, and experiences. Our questions sought information about how they used the resource, which components were more and less useful for their instructional purposes, and how their students responded to various activities. Finally, it was critical for us to treat our partners with respect and to recognize the expert-level professional knowledge that is necessary to teach mathematics effectively (Ball et al., 2008; Shusterman et al., 2019). The teachers reacted positively when we acknowledged the challenges inherent in reforming the enacted curriculum (Suurtamm & Graves, 2011). Recently, one of our partner teachers expressed her gratitude for our recognition of the complexity of teachers' work: We had expressed that teaching mathematics is a highly complex and challenging activity and not the application of easy and quick instructional strategies. We contend that recognizing their demanding workload and expertise was needed to set the stage to build a trusting relationship.

Lesson 3: Teacher Buy-In

Working with our partner teachers made us aware of the importance of buy-in for the successful enactment of a mathematics curriculum. For an enacted curriculum resource to have lasting impact on teachers' practice, it is critical teachers not only implement the resource but also adopt it long term. Research has consistently shown the difficulty in getting teachers to internalize the core goals of curriculum reform (Boesen et al., 2014; Charalambous & Philippou, 2010; Heaton, 2000; Suurtamm & Graves, 2011). According to Gregoire's (2003) cognitive-affective model for conceptual change (CAMCC), practitioners' motivations for adopting curriculum reform depend in part on whether teachers view the reform as a challenge or a threat. If viewed as a challenge, teachers who ultimately accept the values of the reform will internalize and adopt a new curriculum. If viewed as a threat, teachers will engage in avoidance behaviors and either reject the reform efforts or only superficially implement a new curriculum into their classrooms. According to Gregoire (2003), adopting a challenge attitude requires a combination of high self-efficacy as well as coping resources, such as time, support, and subject-matter knowledge. In this section of the chapter, we use the CAMCC model to present the lessons we learned as we worked toward maximizing teacher uptake.

In our research, we realized that teachers needed to be convinced of the importance of early numeracy, the usefulness of using children's thinking as a springboard for instruction, and the value of formative assessment in moving children's mathematical thinking forward. We adopted an approach used by Shusterman et al. (2019) that involved sharing research findings about the development of early numeracy with teachers. Shusterman and colleagues (2019) found that clear explanations of research stimulated the teachers in ways that one-shot workshops typically do not (Guskey & Yoon, 2009). Therefore, in our first meetings with the teachers in all three phases of the ENP, we presented a study by Jordan et al. (2007) that revealed the predictive power of core numeracy principles, such as cardinality, number magnitude, and part-whole concepts, on children's mathematics achievement at the end of first grade. We also read and discussed excerpts from Ginsburg and Ertle (2008) about the importance of incorporating targeted and intentional interactions with young children in play contexts. Some teachers welcomed the message that the role of the preschool educator in the future academic success of children is essential. Others, in contrast, experienced uneasiness with the feeling that they may have let their students down with respect to early numeracy. Regardless of the response, stressing the importance of numeracy in the early years and increasing teachers' knowledge about effective learning environments for the development of children's numeracy provided teachers with coping mechanisms for changing their instructional practice.

As discussed in Lesson 1, identifying the teachers' challenges in enacting the official curriculum and reducing barriers for teachers' adoption of the *NyKK-5* also resulted in increased buy-in. Designing the *NyKK-5* as a collection of manageable activities that teachers could carry around to use on an as-needed basis (Shusterman et al., 2019), for example, helped reduce time constraints. Teachers were in charge of when and how to implement the *NyKK-5* resource and were given the freedom to modify and extend the activities to meet their own needs and those of their students. Inspired by the activity cards, some teachers developed their own numeracy games for their classrooms. Another teacher described how the activity cards in the *NyKK-5* provided the inspiration for new classroom routines. The flexibility of the resource increased teachers' ability to implement it within their classrooms, providing concrete evidence of uptake.

Further, we surmise that the observations made by the partner teachers when they used the *NyKK-5* in their classrooms were particularly motivating and contributed to uptake. One teacher in Phase 1, for instance, mentioned that she used several activities in the *NyKK-5* and the formative assessment document with two students in separate one-on-one sessions. Her a priori assessments of these two students diverged. From working with them over the course of the academic year, the teacher concluded that one had stronger numeracy skills than the other. Her use of the kit revealed the opposite pattern, however. The student who she assumed was strong could not successfully enumerate small collections of items. Conversely, the perceived struggling student was able to complete the same task successfully. The teacher professed, "I thought I knew; turns out I didn't know." She humbly admitted that for years she had relied on her intuition to ascertain whether her students were progressing adequately; after using the resource, she came to realize the disadvantages of using intuition for such purposes. Although this experience led the teacher to feel insecure about her ability to assess students' knowledge accurately, she embraced it as a challenge. Even though we have no data to explain the nature of her response, we presume that her extensive teaching experience and self-efficacy beliefs (Gregoire, 2003) can account for her adopting the resource. Her new knowledge of children's numeracy allowed for appropriate interpretations of the data yielded by the formative assessment documents. Moreover, she shared the experience with us during a debriefing session, allowing us to provide assurances and support (Gregoire, 2003). Together, her ability to cope with the experience mitigated a threat response and motivated more targeted attention to her students' numeracy development (Gregoire, 2003).

DISCUSSION AND CONCLUSION

The objectives of our research on the enacted numeracy curriculum in kindergarten have evolved over the last several years. Our initial goals were to test whether providing kindergarten teachers with knowledge about early numeracy and children's thinking would predict growth in their students' counting, number knowledge, arithmetic, and part-whole concepts. Our research team provided a year-long PD initiative to kindergarten teachers on ways to engage young children in conversations about numeracy in play-based classroom environments. Although the teachers were keen to learn about children's numeracy development, we received repeated requests for additional concrete activities for the classroom. This prompted the development of the *NyKK-5* toolkit, a resource for teachers to enact a numeracy curriculum to prepare all kindergarteners for mathematics in first grade and beyond.

Following a design-based research methodology, we spent subsequent years designing, informally and formally validating, and refining the *NyKK-5* in various kindergarten classrooms. In this chapter, we described the increasingly iterative nature of this work with a specific eye to the challenges we encountered in the process. In particular, we focused the discussion on three lessons learned that provide important implications for curriculum researchers in mathematics education. For those scholars whose focus is to merge theoretical and practical considerations in their work (Schoenfeld, 1999), the implications ultimately suggest strategies for high-quality teacher PD. Below, we present implications for curriculum research in the context of key characteristics of effective teacher PD as identified in the literature (i.e., Borko et al., 2010; Koellner et al., 2007).

We first identified significant barriers to the teachers' intent to enact the official curriculum. In our case, the main barrier was the lack of a designated curriculum in the official provincial preschool program. The partner teachers made it clear that they were eager, yet poorly equipped, to implement specific, concrete numeracy activities in their classrooms. When the official curriculum is wanting in terms of specificity and suggestions for practice, teachers need considerable support to enact a curriculum that will facilitate the development of children's numeracy skills. In a sense, then, the resource we created in collaboration with the teachers served as a stopgap designated curriculum that supported high-leverage pedagogical practices (Ball & Forzani, 2009) immediately relevant to their own contexts (Borko et al., 2010).

Second, we discovered that the need for reciprocal engagement between researchers and practitioners in design-based research was not obvious for teachers. We thus encouraged them to work in true partnership with us to make the materials as usable as possible in the classroom, which was entirely contingent on their experience and expertise. We worked to establish open communication: We shared the objectives of the research and our intentions for the *NyKK-5* resource, and we discussed our need for their honest feedback. The implication of this lesson for curriculum scholars is *understanding that reciprocity between teachers and researchers extends beyond a quid pro quo arrangement. It begins with a recognition of the degree to which all participants can learn from each other through collaboration built on respect and trust*. We learned that this recognition is not a given, nor is it obvious. Reciprocity hinges on convincing teachers that the objective of maximizing children's mathematical potential is shared by all participants in the partnership. We argue that true reciprocity lays the groundwork for the development of "professional learning communities" that are sustainable and impactful in the long term (Borko et al., 2010, p. 551).

The third lesson we learned was that teacher buy-in is critical for researchers who work with teachers to enhance the enacted curriculum. We found that providing research-based evidence of the predictive power of early numeracy development in children's mathematical success was effective for teacher buy-in because it clarified the reason we were asking teachers to reform their curriculum (Shusterman et al., 2019). Further, although the information acted as an initial stressor for some teachers, it served to promote deeper exploration and processing of the PD's content

(Gregoire, 2003). We additionally mitigated the stress of adopting the resource by increasing its versatility and maximizing its functionality with respect to teacher planning, implementation, and assessment of student learning. Keeping these objectives in mind, we relied heavily on teacher feedback and advice, involving them as active participants and keeping the collaboration grounded in inquiry, reflection, and exploration (Borko et al., 2010).

The lessons we learned in our work with teachers in the ENP are informative for curriculum researchers. Researchers should seek to understand the needs of the teachers, the nature of the available resources, and the requirements of the official curriculum, along with any gaps that may be impacting teachers' intended curriculum. In short, curriculum researchers should be aware of the particularities of the context in which the collaboration takes place (Stein et al., 1999). The development of a true partnership— that is, shifting the teacher's role from participant to partner—requires effort, extensive exchanges of information, and mutual respect and trust. Relatedly, buy-in is critical for teachers to adopt a resource that requires considerable adjustments to their knowledge and practice and is sustainable over the long term. The marriage between research and practice, where both components are equally valued, forms the base for effective curriculum resource design and, we suggest, is responsible for the success of the *NyKK-5*.

ACKNOWLEDGMENT

This research was supported by two grants to the third author (Helena P. Osana) from the Social Sciences and Humanities Research Council of Canada, 892-2018-1038 and 892-2021-0057.

REFERENCES

Aunola, K., Leskinen, E., Lerkkanen, M.-K., & Nurmi, J.-E. (2004). Developmental dynamics of math performance from preschool to grade 2. *Journal of Educational Psychology*, *96*(4), 699–713. https://doi.org/10.1037/0022-0663.96.4.699

Ball, D. L., & Forzani, F. M. (2009). The work of teaching and the challenge for teacher education. *Journal of Teacher Education*, *60*(5), 497–511. https://doi.org/10.1177/0022487109348479

Ball, D. L., Thames, M. H., & Phelps, G. (2008). Content knowledge for teaching: What makes it special? *Journal of Teacher Education*, *59*(5), 389–407. https://doi.org/10.1177/0022487108324554

Baroody, A. J., Lai, M.-l., & Mix, K. S. (2006). The development of young children's early number and operation sense and its implications for early childhood education. In B. Spodek & O. N. Saracho (Eds.), *Handbook of research on the education of young children* (2nd ed., pp. 187–221). Lawrence Erlbaum Associates.

Bennett, R. E. (2011). Formative assessment: A critical review. *Assessment in Education: Principles, Policy & Practice, 18*(1), 5–25. https://doi.org/10.1080/0969594X.2010.513678

Boesen, J., Helenius, O., Bergqvist, E., Bergqvist, T., Lithner, J., Palm, T., & Palmberg, B. (2014). Developing mathematical competence: From the intended to the enacted curriculum. *Journal of Mathematical Behavior, 33*, 72–87. https://doi.org/10.1016/j.jmathb.2013.10.001

Borko, H., Jacobs, J., & Koellner, K. (2010). Contemporary approaches to teacher professional development. In P. Peterson, E. Baker, & B. McGaw (Eds.), *International encyclopedia of education* (3rd ed., pp. 548–556). Elsevier. https://doi.org/10.1016/B978-0-08-044894-7.00654-0

Burkhardt, H., & Schoenfeld, A. H. (2003). Improving educational research: Toward a more useful, more influential, and better-funded enterprise. *Educational Researcher, 32*(9), 3–14. https://doi.org/10.3102/0013189X032009003

Charalambous, C. Y., & Philippou, G. N. (2010). Teachers' concerns and efficacy beliefs about implementing a mathematics curriculum reform: Integrating two lines of inquiry. *Educational Studies in Mathematics, 75*(1), 1–21. https://doi.org/10.1007/s10649-010-9238-5

Choppin, J., Roth McDuffie, A., Drake, C., & Davis, J. (2022). The role of instructional materials in the relationship between the official curriculum and the enacted curriculum. *Mathematical Thinking and Learning, 24*(2), 123–148. https://doi.org/10.1080/10986065.2020.1855376

Cobb, P., Confrey, J., diSessa, A., Lehrer, R., & Schauble, L. (2003). Design experiments in educational research. *Educational Researcher, 32*(1), 9–13. https://doi.org/10.3102/0013189X032001009

Collins, A., Joseph, D., & Bielaczyc, K. (2004). Design research: Theoretical and methodological issues. *Journal of the Learning Sciences, 13*(1), 15–42. https://doi.org/10.1207/s15327809jls1301_2

Copur-Gencturk, Y., Plowman, D., & Bai, H. (2019). Mathematics teachers' learning: Identifying key learning opportunities linked to teachers' knowledge growth. *American Educational Research Journal, 56*(5), 1590–1628. https://doi.org/10.3102/0002831218820033

Donovan, M. S., Snow, C., & Daro, P. (2013). The SERP approach to problem-solving research, development, and implementation. *Teachers College Record, 115*(14), 400–425. https://doi.org/10.1177/016146811311501411

Duncan, G. J., Dowsett, C. J., Claessens, A., Magnuson, K., Huston, A. C., Klebanov, P., Pagani, L. S., Feinstein, L., Engel, M., Brooks-Gunn, J., Sexton, H., Duckworth, K., & Japel, C. (2007). School readiness and later achievement. *Developmental Psychology, 43*(6), 1428–1446. https://doi.org/10.1037/0012-1649.43.6.1428

Even, R., & Gottlib, O. (2011). Responding to students: Enabling a significant role for students in the class discourse. In Y. Li & G. Kaiser (Eds.), *Expertise in mathematics instruction: An international perspective* (pp. 109–130). Springer. https://doi.org/10.1007/978-1-4419-7707-6

Even, R., & Tirosh, D. (2002). Teacher knowledge and understanding of students' mathematical learning. In L. English (Ed.), *Handbook of international research in mathematics education* (pp. 219–240). Lawrence Erlbaum.

Garet, M. S., Porter, A. C., Desimone, L., Birman, B. F., & Yoon, K. S. (2001). What makes professional development effective? Results from a national sample of teachers. *American Educational Research Journal, 38*(4), 915–945. https://doi.org/10.3102/00028312038004915

Geary, D. C., Hoard, M. K., Nugent, L., & Bailey, D. H. (2013). Adolescents' functional numeracy is predicted by their school entry number system knowledge. *PLoS ONE, 8*(1), e54651. https://doi.org/10.1371/journal.pone.0054651

Geudet, G., & Trouche, L. (2012). Teachers' work with resources: Documentational geneses and professional geneses. In G. Gueudet, B. Pepin, & L. Trouche (Eds.), *From text to 'lived' resources: Mathematics curriculum materials and teacher development* (pp. 23–42). Springer. https://doi.org/10.1007/978-94-007-1966-8

Ginsburg, H. P., & Ertle, B. (2008). Knowing the mathematics in early childhood mathematics. In O. N. Saracho & B. Spodek (Eds.), *Contemporary perspectives on mathematics in early childhood education* (pp. 45–66). Information Age.

Ginsburg, H. P., Lee, J. S., & Boyd, J. S. (2008). Mathematics education for young children: What it is and how to promote it. *Social Policy Report, 22*(1), 1–24. https://doi.org/10.1002/j.2379-3988.2008.tb00054.x

Gregoire, M. (2003). Is it a challenge or a threat? A dual-process model of teachers' cognition and appraisal processes during conceptual change. *Educational Psychology Review, 15*(2), 147–179. https://doi.org/10.1023/A:1023477131081

Guskey, T. R., & Yoon, K. S. (2009). What works in professional development? *Phi Delta Kappan, 90*(7), 495–500. https://doi.org/10.1177/003172170909000709

Heaton, R. M. (2000). *Teaching mathematics to the new standard: Relearning the dance.* Teachers College Press.

Heritage, M. (2007). Formative assessment: What do teachers need to know and do? *Phi Delta Kappan, 89*(2), 140–145. https://doi.org/10.1177/003172170708900210

Jordan, N. C., Glutting, J., & Ramineni, C. (2010). The importance of number sense to mathematics achievement in first and third grades. *Learning and Individual Differences, 20*(2), 82–88. https://doi.org/10.1016/j.lindif.2009.07.004

Jordan, N. C., Kaplan, D., Locuniak, M. N., & Ramineni, C. (2007). Predicting first-grade math achievement from developmental number sense trajectories. *Learning Disabilities Research & Practice, 22*(1), 36–46. https://doi.org/10.1111/j.1540-5826.2007.00229.x

Jordan, N. C., Kaplan, D., Ramineni, C., & Locuniak, M. N. (2009). Early math matters: Kindergarten number competence and later mathematics outcomes. *Developmental Psychology, 45*(3), 850–867. https://doi.org/10.1037/a0014939

Koellner, K., Jacobs, J., Borko, H., Schneider, C., Pittman, M. E., Eiteljorg, E., Bunning, K., & Frykholm, J. (2007). The problem-solving cycle: A model to support the development of teachers' professional knowledge. *Mathematical Thinking and Learning, 9*(3), 273–303. https://doi.org/10.1080/10986060701360944

Mazzocco, M. M. M., & Thompson, R. E. (2005). Kindergarten predictors of math learning disability. *Learning Disabilities Research & Practice*, *20*(3), 142–155. https://doi.org/10.1111/j.1540-5826.2005.00129.x

Morgan, P. L., Farkas, G., & Wu, Q. (2011). Kindergarten children's growth trajectories in reading and mathematics: Who falls increasingly behind? *Journal of Learning Disabilities*, *44*(5), 472–488. https://doi.org/10.1177/0022219411414010

National Mathematics Advisory Panel. (2008). *Foundations for success: The final report of the National Mathematics Advisory Panel*. U.S. Department of Education.

Osana, H. P., Orsini, A., MacCaul, R., Sindayigaya, Q., Provost-Larocque, K., & Lafay, A. (2023). Explorer la relation entre le développement professionnel des enseignants et les habiletés numériques des enfants de maternelle [Exploring the relationship between teachers' professional development and kindergarteners' numeracy skills]. *Revue Internationale de Communication et Socialisation* [*International Journal of Communication and Socialization*], *10*(1), 84–105.

Parrish, S. D. (2011). Number talks build numerical reasoning. *Teaching Children Mathematics*, *18*(3), 198–206. https://doi.org/10.5951/teacchilmath.18.3.0198

Passolunghi, M. C., & Lanfranchi, S. (2012). Domain-specific and domain-general precursors of mathematical achievement: A longitudinal study from kindergarten to first grade. *British Journal of Educational Psychology*, *82*(1), 42–63. https://doi.org/10.1111/j.2044-8279.2011.02039.x

Penuel, W. R., Fishman, B. J., Haugan Cheng, B., & Sabelli, N. (2011). Organizing research and development at the intersection of learning, implementation, and design. *Educational Researcher*, *40*(7), 331–337. https://doi.org/10.3102/0013189X11421826

Penuel, W. R., Roschelle, J., & Shechtman, N. (2007). Designing formative assessment software with teachers: An analysis of the co-design process. *Research and Practice in Technology Enhanced Learning*, *2*(1), 51–74. https://doi.org/10.1142/S1793206807000300

Pepin, B., Gueudet, G., & Trouche, L. (2013). Re-sourcing teachers' work and interactions: A collective perspective on resources, their use and transformation. *ZDM Mathematics Education*, *45*(7), 929–943. https://doi.org/10.1007/s11858-013-0534-2

Purpura, D. J. (2021). *The Preschool Early Numeracy Screener* [Measurement instrument]. PRO-ED. https://www.proedinc.com/Products/14630/pens-preschool-early-numeracy-screener.aspx

Purpura, D. J., & Lonigan, C. J. (2013). Informal numeracy skills: The structure and relations among numbering, relations, and arithmetic operations in preschool. *American Educational Research Journal*, *50*(1), 178–209. https://doi.org/10.3102/0002831212465332

Putnam, R. T., & Borko, H. (2000). What do new views of knowledge and thinking have to say about research on teacher learning? *Educational Researcher*, *29*(1), 4–15. https://doi.org/10.3102/0013189X029001004

Remillard, J. T. (2005). Examining key concepts in research on teachers' use of mathematics curricula. *Review of Educational Research*, *75*(2), 211–246. https://doi.org/10.3102/00346543075002211

Remillard, J. T., Harris, B., & Agodini, R. (2014). The influence of curriculum material design on opportunities for student learning. *ZDM Mathematics Education*, *46*(5), 735–749. https://doi.org/10.1007/s11858-014-0585-z

Remillard, J. T., & Heck, D. J. (2014). Conceptualizing the curriculum enactment process in mathematics education. *ZDM Mathematics Education*, *46*(5), 705–718. https://doi.org/10.1007/s11858-014-0600-4

Remillard, J. T., Reinke, L. T., & Kapoor, R. (2019). What is the point? Examining how curriculum materials articulate mathematical goals and how teachers steer instruction. *International Journal of Educational Research*, *93*, 101–117. https://doi.org/10.1016/j.ijer.2018.09.010

Roth McDuffie, A., Choppin, J., Drake, C., Davis, J. D., & Brown, J. (2018). Middle school teachers' differing perceptions and use of curriculum materials and the common core. *Journal of Mathematics Teacher Education*, *21*(6), 545–577. https://doi.org/10.1007/s10857-017-9368-0

Ruthven, K., & Goodchild, S. (2008). Linking researching with teaching: Towards synergy of scholarly and craft knowledge. In L. D. English & D. Kirshner (Eds.), *Handbook of international research in mathematics education* (2nd ed., pp. 575–602). Routledge.

Schoenfeld, A. H. (1999). Looking toward the 21st century: Challenges of educational theory and practice. *Educational Researcher*, *28*(7), 4–14. https://doi.org/10.3102/0013189X028007004

Shah, J. K., Ensminger, D. C., & Thier, K. (2015). The time for design-based research is right and right now. *Mid-Western Educational Researcher*, *27*(2), 152–171.

Shusterman, A., May, N., Melvin, S., Kumar, S., Blumenstock, S., Toomey, M., & Lewis, S. (2019). *Working in the research-to-practice gap: Case studies, core principles, and a call to action*. PsyArXiv. https://doi.org/10.31234/osf.io/qhxbn

Sinda, Q., MacCaul, R., Osana, H. P., Lafay, A., & Orsini, A. (2021, May 3–7). *Soutenir les activités numériques des enseignants de maternelle dans leur classe* [Supporting kindergarten teachers' numeracy activities in their classrooms] [Paper presentation]. Association canadienne-française pour l'avancement des sciences [French-Canadian Association for the Advancement of Science], Sherbrooke, QC, Canada.

Smith, G. J., Schmidt, M. M., Edelen-Smith, P. J., & Cook, B. G. (2013). Pasteur's quadrant as the bridge linking rigor with relevance. *Exceptional Children*, *79*(3), 147–161. https://doi.org/10.1177/001440291307900202

Snyder, J., Bolin, F., & Zumwalt, K. (1992). Curriculum implementation. In W. P. Jackson (Ed.), *Handbook of research on curriculum* (pp. 402–435). Macmillan.

Starkey, P., Klein, A., & Wakeley, A. (2004). Enhancing young children's mathematical knowledge through a pre-kindergarten mathematics intervention. *Early Childhood Research Quarterly*, *19*(1), 99–120. https://doi.org/10.1016/j.ecresq.2004.01.002

Stein, M. K., Smith, M. S., & Silver, E. (1999). The development of professional developers: Learning to assist teachers in new settings in new ways. *Harvard Educational Review*, *69*(3), 237–270. https://doi.org/10.17763/haer.69.3.h2267130727v6878

Suurtamm, C., & Graves, B. (2011). Developing mathematics classroom practices: The role of coherence and collaboration. *Journal of Education Research*, *5*(3/4), 335–359.

The Design-Based Research Collective. (2003). Design-based research: An emerging paradigm for educational inquiry. *Educational Researcher*, *32*(1), 5–8. https://doi.org/10.3102/0013189X032001005

Watts, T. W., Duncan, G. J., Siegler, R. S., & Davis-Kean, P. E. (2014). What's past is prologue: Relations between early mathematics knowledge and high school achievement. *Educational Researcher*, *43*(7), 352–360. https://doi.org/10.3102/0013189X14553660

Yan, Z., King, R. B., & Haw, J. Y. (2021). Formative assessment, growth mindset, and achievement: Examining their relations in the East and the West. *Assessment in Education: Principles, Policy & Practice*, *28*(5-6), 676–702. https://doi.org/10.1080/0969594X.2021.1988510

ENDNOTE

1. At this time, we have preliminary data to support our approach to PD, with plans to scale up the research to test the robustness of the effect, to extend the generalizability of the findings, and to test the conditions under which the materials are useful for student learning.

CHAPTER 21

ANALYZING ENACTED CURRICULUM

Seven Lessons Learned From Collecting, Coding, and Analyzing Large Sets of Qualitative Data

Kristy Litster
Valdosta State University

> The focus of this chapter is on seven lessons learned while collecting and analyzing qualitative evidence of student-enacted curriculum in an elementary mathematics classroom. It provides examples of strategies or practices that caused problems when collecting or analyzing the data as well as examples of strategies and practices that were productive. The seven lessons are: (1) Prizes can help collect consent forms; (2) Choose the right equipment; (3) Organize the data systems; (4) Transcribe some of the data yourself; (5) Rubrics increase coding reliability; (6) Numbers can be categorical data; and (7) Visual models are not always intuitive.
>
> **Keywords:** enacted curriculum; models; qualitative data

Evidence of student enacted curriculum, or what students do when engaging with the intended classroom curriculum, often comes from qualitative data sources. This chapter focuses on seven lessons learned when collecting and analyzing 1,234 pages of written evidence and 140 hours of verbal evidence of U.S. Grade 5 (ages 9–11) students' enacted curriculum when engaging with real-world mathematics task sets in small group settings (Litster, 2019).

LESSON 1: PRIZES CAN HELP COLLECT CONSENT FORMS

The first lesson learned is: *A small prize can facilitate collecting consent forms quickly and consistently*. This was found in a study where all students ($n = 101$) brought back a signed permission form and only four parents removed their child from the study (one per class). Each of the four non-participants provided justification that they would be out of town on one of the data collection dates. Based on previous experience having difficulty collecting forms from students, I wanted to try something different to increase the number of forms returned. First, I handed the permission forms to students myself, rather than having a teacher send them home with homework; second, I brought a box of inexpensive toy packs for prizes.

When introducing the form, I explained the study and let students know that if they brought back their form (signed yes or no), they would get to choose a prize from the box—first come, first served. I then left the prize box and some extra forms with each teacher. Teachers reported that over half of the students brought back the forms the next day. Once the other students saw their peers playing with their toys, they were motivated to return their forms, and most of the forms came back the following day. Within three days, all consent forms were returned and prizes collected. In summary, I learned that $27 spent on a variety of small toys was well worth the cost. Allowing students to get a prize for either "yes" or "no" may also have contributed to the high return rate.

LESSON 2: CHOOSE THE RIGHT EQUIPMENT

The second lesson learned is: *Choose the right equipment to collect the qualitative data. Choosing the right equipment to collect visual and verbal data can make or break a study*. Enacted curriculum takes place within a classroom where several students are often engaging in tasks or discourse at the same time. This can make it difficult to code if it is not clear what students are saying or doing.

In Litster and Wright (2019), I used cameras on tripods for different views and class conversations. However, I noticed that although the cameras showed an overview of the whole class from different angles; it was difficult to see what individual students or groups were doing. Additionally, the sound from the tripod camera picked up the teacher instructions or some students when only a few were talking, but provided garbled sound when all students were talking and working at the same time. I needed a strategy to collect data when every student was talking. The right equipment to complete this job was one GoPro camera per group, attaching the

GoPro to a child using a chest harness (Figure 21.1). The GoPro cameras had better sound and visuals than the tripod camera.

Figure 21.1

Attached GoPro on a Child's Chest Harness to Gather Verbal and Visual Evidence

The GoPro camera picked up clear conversations within about a two-foot radius and slightly softer recordings up to a three-foot radius of the device. That is, the camera picked up only conversations between students within the given group, and not conversations at neighboring tables. I collected data for this study in multiple classrooms and schools, each with unique furniture and classroom organization. In two of these classrooms, the arrangement of desks to form small groups occasionally placed a student at the edge of the three-foot radius. In these cases, the use of a back-up digital recorder, placed at the side of the group opposite the camera, was needed to fill in an occasional gap in the recordings. These recordings were used to collect verbal evidence of the student enacted curriculum.

Although the GoPro cameras were all purchased from the same source, I learned the quality of cameras and their batteries varied. Batteries generally provided 45–75 minutes of continuous life. The study was anticipated to take about 60 minutes, but generally took about 75 minutes due to transition times between tasks and instructions. Having backup batteries and ways to charge them was essential to guarantee continuous video coverage for each group (Figure 21.2). Charging the batteries using a docking station recharged multiple batteries to full power in under an hour. Another issue that arose was that two of the ten GoPro Cameras would overheat, causing a slight discomfort for the students and a decrease in battery life for the camera. Using GoPro brand batteries reduced this problem, as their battery casings are more heat resistant and their battery life was longer than generic batteries.

Figure 21.2

Camera, Backup Batteries, Backup Recorder, and Multi-Battery Charger

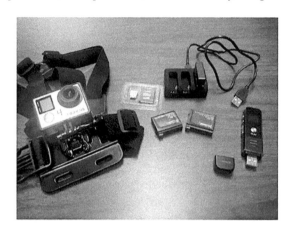

Buying multiple sets of digital equipment, such as the 10 sets used in this project, can seem like an expensive choice. I used funds from a grant to purchase the cameras, batteries, digital recorders, and charging stations for the project, and I have continued to use the materials for subsequent projects. Researchers may want to seek grant funding or talk with their institution's technology department about materials that are available to borrow to reduce costs associated with purchasing digital equipment. When purchasing materials, look online at multiple sources. For example, the Hero 4 was listed for one price on one website, but was $250 less expensive through another vendor. Also, I was able to use online sources to compare and read reviews of multiple styles of digital recorders to choose ones that would work best for my study. If using a grant or department purchase order, you may need to select vendors based on requirements of the funding agency. In some projects, I purchased more expensive equipment because the grant would only allow me to purchase from select companies.

The view from a GoPro camera provided a nice visual of student actions within the group. Because the camera was situated at chest level, the camera view would follow the body movements of the student to face people in the group who were speaking or writing related to the task. Unfortunately, it did not always pick up exactly what was being written by someone other than the camera holder. Using a variety of pen colors solved this dilemma. Each student in the group chose a different pen color and signed their name with their color. Any time they wrote on scratch paper or the assignment, they would use their chosen color. In this way, written actions captured on camera could be associated with the appropriate student based on the color of the written evidence of student enacted curriculum.

In summary, I learned that when collecting student enacted curriculum data with multiple groups of students working at the same time in a noisy environment; it is best to choose recording devices that catch the localized noise for each group. Back-up devices, batteries, and chargers will help avoid gaps in data collection. The use of colors, numbers, or symbols can help align visual evidence in the video with student actions.

LESSON 3: ORGANIZE THE DATA SYSTEM

The third lesson learned is: *Set up an organizational system to label, identify, and backup data sources before collecting any data.* This is especially important when working with students you do not know.

In one pilot study, I knew the students in the classroom. This meant I could easily match the video recordings, audio recordings, written work, and survey data when transcribing and coding data. However, the larger subsequent study took place in two schools and four classrooms where I was not familiar with the students. I premade folders for each group (numbered 1 to 9) with matching numbers on a camera, digital recorder, and written worksheets/scratch paper. The folders were in four colors (one set per class). Students wrote their name on the first page in the folder in their color pen so I could link their survey data with their written and verbal evidence of enacted curriculum. I downloaded the audio and video files in the three hours between the first and second class and placed them in their folders, Labeled 1–9; I then backed up the data on a second secure online drive before deleting the original data so I could use the devices for the second class.

After I collected data from the second class and went to download the files, I realized I could not save these new files as folders 1-9 because these numbers were already used. I wanted to be able to distinguish between classes and groups within classes for my coding and to quickly identify the associated folders (already labeled 1–9). I settled on re-naming the first files as class, then group. So, group 1 in the first class became Group 11 and group 1 in the second class became Group 21. This numbering system changed once again when I decided to transcribe one of the videos to make sure the sound quality was adequate. At this point, I realized I also wanted to distinguish different students in each group without using pseudonyms for all 97 students in the study. I added a third digit representing student in the group (1st digit = class, 2nd digit = group, 3rd digit = student). So, three students in the first group in the first class would be labeled 111, 112, and 113, respectively. I then linked this number to their survey data for easy triangulation of data and wrote the numbers on associated written evidence (group number for group work, individual number for individual work).

Before collecting data from the last two classrooms, I had the teachers pre-group their students and send me the names of students in each group. I was able to pre-assign their groups and numbers so I could just call out student names to form a group and collect the materials needed, which reduced transition times. Figure 21.3 shows examples of the alignment between files, surveys, and coding using this number system for Class 2, Group 2 (group 220), which contained three female students (221, 222, 223).

Figure 21.3

Alignment Between Data Labels in Files, Surveys and Coding

	A	B	C	D	E	F	G	H	I
1	ID	Gender	Low SES	Resource	Ethnicity	MathAttitude	Fractions	Decimals	
29	221	Female	No	No	White	20.5	30	19	
30	222	Female	Yes	No	White	28.5	34.5	23	
31	223	Female	No	No	White	38	38	33.5	

	A	B	C	D	E	F	G	H
	Group	Set	Intervention	Task	Intended DOK	Time Stamp	Student	Student Response
37	220	Diary	Collaborative	D2	2	10:50	221	No, just adding
38	220	Diary	Collaborative	D2	2			[all adding decimals silently]
39	220	Diary	Collaborative	D2	2	12:05	223	Ok, I got that one. 15, 49
40	220	Diary	Collaborative	D2	2		222	Seriously? how could that be 15 if this is 14? BEcause this doesn'
41	220	Diary	Collaborative	D2	2		223	But when you add all of these thingies together, or would we mult
42	220	Diary	Collaborative	D2	2	12:50	221	You would add. Like you would do this plus this plus this. blah. bl

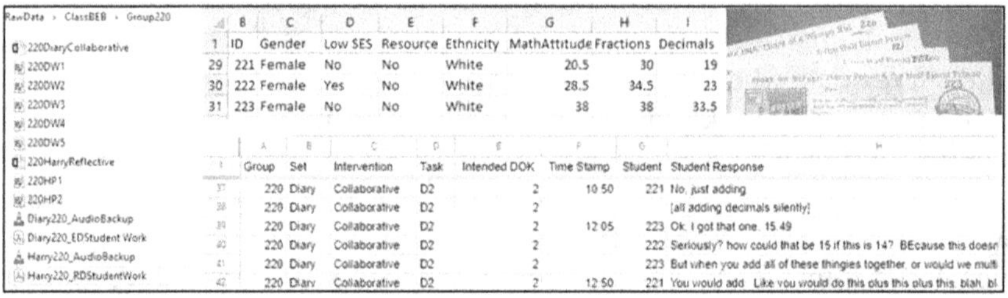

The upper right-hand corner of Figure 21.3 shows how physical student work was labeled with the pre-assigned student numbers linked to survey data (top, center of figure) collected at the same time as the signed permission forms. The bottom right of the figure shows how group and student numbers were also used in the transcribing (group, set, intervention, task, timestamp, student, student response) and coding (intended DOK [Depth of Knowledge]; Webb, 1999) of the data. On the left is a screenshot of the digital folder that contained my data spreadsheets for the two intervention comparisons (e.g., 220Diary Collaborative), the GoPro segmented video files for each intervention (e.g., 220DW1-5), the audio back-up files from the digital recorders (e.g., Diary220AudioBackup), and scanned copies of physical student work (e.g., Diary 220EDStudentWork).

I created the digital folders to organize data before collecting or downloading any data. This was helpful, as I only had one hour to download the GoPro video files between the collection of data for the last two classes. Alternatively, if there are back-to-back observations, it may be useful to purchase additional memory cards and swap them out for downloads at a later time. Having an external hard drive allowed me to access the data I needed offline, and the secure online folder allowed me to access data from a variety of locations. Using a secure online folder also helped with collaborative needs relating to double coding data and using multiple people to transcribe the data. However, unless a researcher's institution already provides a secure online folder, it can become expensive to obtain such access, especially for large audio or video files. If using an institutional secure online folder, check accessibility for coders or collaborators at other institutions. My institution did not allow access for some of my transcribers and coders, so I temporarily paid for an external online folder during transcribing and coding. I learned that it was better to keep a second physical back-up of data and just upload the files that were needed for that coding/transcribing cycle; removing those files after downloading the transcriptions and codes kept the data size within my limits. Once my secure online folders were empty, I could cancel the secure folder. Additionally, it is important to make a second physical back-up when changing institutions to be able to take data with you as permitted by research guidelines.

Before starting to collect data, decide on the software used to code the data, as this may affect how to save or upload data. Various software can be purchased to help with qualitative and quantitative coding. Some have great features to support coding, such as linking features between video, transcriptions, and codes in MAXQDA (www.maxqda.com; see example in Di Stefano et al., 2017), organizational coding features in NVivo (lumivero.com/products/nvivo; see example in MacDonald et al., 2020), links in GoReact between the video and the codes (get.goreact.com), or the structural equation modeling features of the MPlus software (www.statmodel.com; see example in Litster et al., 2020). However, if on a budget, sometimes free software is the best option.

When working with qualitative data, the pivot tables in Excel/Google Sheets were a great resource to identify common themes between classes, groups, individuals, and variables by isolating verbal responses related to key variables or demographics. For example, in one analysis, I looked specifically at the differences in timings of student discourse for students who typically struggle with mathematics. The left side of Figure 21.4 shows a sample pivot table that isolated verbal responses with collaborative discourse (row option) for students with this identifier (filter option). I was able to see a common theme of negativity. Adding rows *group* and *task* helped me later identify the specific conversations for a deeper analysis; changing the row option to a different intervention allowed me to compare differences between interventions (see Litster, 2020).

I found the use of pivot tables to identify frequencies, percentages, and basic statistics based on the parameters to be critical in "quantitizing" some of the data (Saldaña, 2015). The right side of Figure 21.4 shows a pivot frequency table comparing different strategy types (e.g., Importance, Order) at each intended DOK curriculum level (row) with coded enacted curriculum (column) displaying the results as percentages (value). There are

Figure 21.4

Using a Pivot Table for Thematic Coding and Creating Data Tables

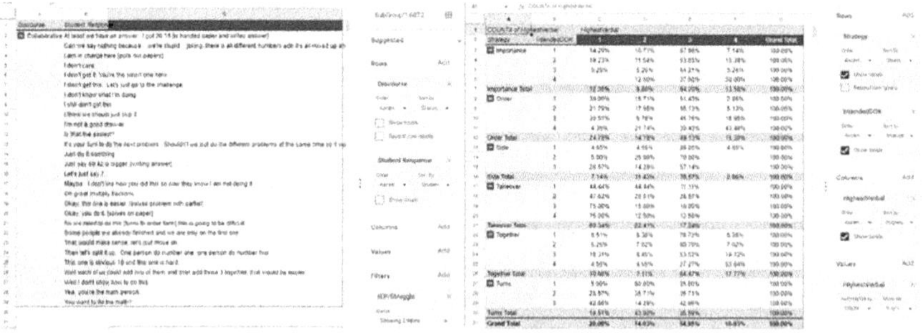

several tutorials online to explore different ways to create pivot tables and graphs (e.g., Excel Easy, n.d.).

Jamovi (www.jamovi.org) is free statistical software that was useful for my quantitative analyses. Although I was not able to do complex quantitative analyses using this software, it was perfect for the difference in proportions tests that are more common with qualitative data when analyzing enacted curriculum (Tashakkori & Teddlie, 2010).

In summary, I learned that it is important to use the same code for survey/demographic data, written data, video data, and audio data to triangulate data for each group. Paying extra for secure digital folders to share videos can help with short-term coding or transcribing for multiple coders, but is not recommended for long-term storage. Instead, use one new external hard drive to save primary data and a second to back-up and blind data. Finally, choose analysis software before collecting data as it may affect how to upload or transcribe the data, and do not overlook the benefits of free software.

LESSON 4: TRANSCRIBE SOME OF THE DATA YOURSELF

The fourth lesson learned is: *There are benefits to transcribing at least a portion of audio/video data yourself before hiring help for the remaining data.* As mentioned in Lesson 3, transcribing the data can help you organize how you want to label the participants (adults and children) in your videos. Although real names or pseudonyms might work for small groups of known children, numbers may be more feasible for large numbers of children. Transcribing some data yourself can also lead to benefits related to setting up the organizational scheme for transcribing the videos, conducting preliminary coding, identifying initial themes, as well as noting other possible interesting research questions to ask or case studies to explore.

When coding audio files, the transcriber focuses on what each person in the audio is saying, using their best judgement to distinguish the speaker and note an exact reproduction of the conversation sequence (McLellan et al., 2003). When transcribing video data, the actions that show physical enactment or dispositions relating to the curriculum are also important to create a full representation of what happened (Davidson, 2009). However, it can be difficult for new transcribers to distinguish between describing actions and interpreting actions. Consider the following scene in Figure 21.5.

Figure 21.5.

Transcriptions With Actions Versus Interpretations

Transcribed with Actions	Transcribed with Interpretations of Actions
261 I got [looks at 262's paper] yes, you finally got it right! 262 I got it right? [smiles at 261] 261 Yep, but add the decimal 262 [writes a dot between the 8 and 0 on paper and then on answer key write 588.0] Both [Read next problem silently and write 280 x 1.99 on scratch paper] 261 Oh wait that would be 0 [crosses out something on paper and writes over it]	261 I got [looks at 262's paper] yes, you finally got it right! 262 I got it right? [happy] 261 Yep, but add the decimal 262 [adds decimal, writes answer] Both [calculates next problem by multiplying 280 x 1.99] 261 Oh wait that would be 0 [fixes error]

In these short vignettes, the example on the left shows transcribed actions such as *smiles* or *crosses out*; the example on the right shows transcribed interpretations of these actions, such as *happy* or *fixes error*. By transcribing some of the data yourself, you can distinguish the level of detail needed for the transcribed actions. Is there a preference for the actions to be described or interpreted? What level of actions are needed (direction facing, amount of time engaged in different tasks, specific number or lines of text, hand movement, facial expressions, body language, etc.)? This is important when training transcribers, as the choices made for what to keep and what to ignore can impact how you ask and answer research questions (Davidson, 2009).

Transcribing also allows the researcher to decide what format is desired for the transcriptions. When I first started transcribing, I used Microsoft Word to detail the conversations and actions. However, because student responses were short, I was constantly working with formatting to keep track of the speaker. So, I switched to a spreadsheet view and kept track of the conversation using different lines (see Figure 21.3). In this way, I was able to keep track of the conversation and the speakers without constantly stopping the transcription to format responses. Also, when I started coding, the transcription was organized at the appropriate grain size. Other organizational aspects that I wanted within the transcriptions were timestamps and notes on the task, as students were working on multiple tasks during the video. If using specific technology for qualitative coding, consider the transcription requirements for the software and complete a sample analysis after the first few transcriptions to troubleshoot any issues with transcription formatting. This will save time by transcribing responses correctly the first time rather than reformatting all the data later.

A secondary benefit I discovered by transcribing a portion of my videos was that I was able to conduct preliminary coding and identify potential themes as I was transcribing. This helped me get a general idea of the expected overall outcomes. I was also able to note other possible research questions I may want to explore relating to the data.

In summary, coding a portion of the data allows you to set up the level of detail and organization needed from the transcriptions to answer the research question. Additionally, it can help identify preliminary results or themes as well as other potential research questions to consider.

LESSON 5: RUBRICS INCREASE CODING RELIABILITY

The fifth lesson learned is: *Using a detailed rubric to code data categorically or by magnitude can increase inter-rater reliability*. The rubric should be based on research for your curriculum area. For example, Figure 21.6 from Litster (2019) shows a sample rubric with sub-codes for different aspects of each category based on discourse quality and Depth of Knowledge (DOK) research (Bishop et al., 2016; Cobb & Yackel, 1996; Webb, 1999).

Providing examples and non-examples can help improve consistent coding for yourself as well as inter-rater reliability, as the examples and non-examples provide a context for the sub-codes. Consider the coding examples in Figure 21.7 that occurred during a training session, using the rubric in Figure 21.6.

In the initial training session for coders, the researcher reviewed the rubric and answered questions. Then, both the trainer and the coder coded a short conversation of a group working on a task. As seen in the codes from the two coders, there were some inconsistencies in the interpretations. For example, on line 3 (*Yep, but add the decimal*), coder 1 saw this as counterargument for accuracy (code 3c) whereas coder 2 saw this as identifying a component of a problem (code 1c). Using examples that show the difference between the categories and subcategories helps provide references. For example, providing examples for identifying a problem (*The*

Figure 21.6

Sample Detailed Rubric

Code	Category	Sub Code Descriptions
0	None	0a Student did not talk during the mathematical task 0b Responses were off-topic 0c Responses organized actions or responsibilities related to problem
1	Minimal DOK 1	1a Perform routine or one-step calculations 1b Recall facts or state answer 1c Identify problem or component of problem 1d Reiterate prior ideas, solutions, or indicate agreement
2	Considerable DOK 2	2a Share new solution strategy (similar efficiency/sophistication) 2b Share more efficient or accurate solution strategy 2c Share more sophisticated solution strategy
3	Substantive DOK 3	3a Generalizations regarding solutions or problems 3b Justification of a strategy/solution's accuracy or efficiency 3c Counterargument against a strategy/solution's accuracy or efficiency 3d Self-Assesses and adjusts strategy/solution for accuracy or efficiency
4	Extended DOK 4	4a Connect or apply solutions to multiple disciplines or contexts 4b Design a new mathematical model to inform the solution strategy 4c Analyze or synthesize information from multiple sources

Source: From Litster (2019).

Figure 21.7

Sample Coding

Coder1	Coder 2	Discourse		
1d	1d	261		I got [looks at 262's paper] yes, you finally got it right!
1d	1d	262		I got it right? [smiles at 261]
3c	1c	261		Yep, but add the decimal
3d	1a	262		[writes a dot between the 8 and 0 on paper.]
1b	0c	262		[On answer key writes 588.0]
1c	1a	Both		[Read next problem silently and write numbers on scratch paper]
3d	3d	261		Oh wait that would be 0 [crosses out on paper and writes over it]

number doesn't make sense) and providing a counterargument (*You don't need to multiply all the zeros because it is always 0. Just add them to the end.*) can help distinguish why *Add the decimal* is identifying a problem in their solution and not necessarily providing a full counterargument for accuracy. After using examples with the second coder, there was a higher rate of consistency on the same coding problems when analyzed using Krippendorff's (2013) Kalpha SPSS test.

In summary, it is important to train coders using detailed rubrics that include examples and/or non-examples to help increase inter-rater reliability. Be willing to modify examples as needed when questions arise.

LESSON 6: NUMBERS CAN BE CATEGORICAL DATA

The sixth lesson learned is: *Quantitized data may still be categorical, which can prohibit the use of parametric and non-parametric tests*. Parametric tests, such as *t*-tests, assume normal data distribution and variance, whereas non-parametric tests, such as Mann Whitney U tests, do not require assumptions of normality; however, both types require consistent intervals between measurements (Tashakkori & Teddlie, 2010). In Figure 21.6, the codes are numbered 0–4. In this type of magnitude coding, higher numerical magnitude represents a higher level of cognitive demand identified within the qualitative data. However, the difference between a 0 and 1 code is not the same difference as between the 1 and 2 or 3 and 4 codes. If the difference between two levels with magnitude coding cannot be assumed as the same difference between two other levels, parametric or non-parametric tests are inappropriate because the numerical data does not have consistent intervals. Rather, the data is categorical in nature.

There are a variety of quantitative analyses that can be used with categorical data. For example, difference in proportions tests, frequency tables, and chi-square tests of independence can work well with numerical data that are categorical in nature (Tashakkori & Teddlie, 2010). Missing data can also be an important category when looking at enacted curriculum. Missing data often arise when students skip required tasks or instructions through lack of time, lack of motivation, or oversight. Understanding when and why students do not enact the curriculum can provide important insights. For example, Figure 21.8 shows a side-by-side comparison of a frequency table that ignores percentages of missing data and a frequency table that includes missing data. Including the missing data in the analysis created an accurate picture of the overall enacted curriculum and also provided insight that the DOK 1 and 3 tasks were more likely to be skipped.

In summary, when qualitative data is quantitized using magnitude coding, the data is categorical in nature. Missing data is an important category to accurately represent and understand the enacted curriculum. Although categorical data cannot be evaluated using parametric or non-parametric tests, it is possible to use analyses that compare proportions and frequencies.

Figure 21.8

Missing Data Comparison

Ignores Missing Data						Includes missing data as "0"					
	Reflective Discourse						Reflective Discourse				
	Highest % Enacted Verbal CD						Highest % Enacted Verbal CD				
Task-Set 1—Harry / Intended DOK		1	2	3	4	Task-Set 1—Harry / Intended DOK		0	1	2	3	4
	1	35	10	51	5		1	20	28	8	41	3
	2	26	15	52	7		2	5	25	14	50	6
	3	28	5	53	15		3	37	17	3	33	10
	4	7	13	47	33		4	6	6	12	44	32

LESSON 7: VISUAL MODELS ARE NOT ALWAYS INTUITIVE

The seventh lesson learned is: *Visual models are not always intuitive*. Visual models are a great way to visualize relationships within student-enacted curriculum by combining data from multiple analyses. For example, heat maps differentiating quantitative results using hue and tint can reveal structural and hierarchical patterns to assist in quantitative comparisons; divergent color schemes could be used to contrast different elements or interventions (Kelleher & Wagener, 2011; Wilkinson & Friendly, 2009). For example, in Figure 21.8, the white boxes visually show less than 10% and the darker boxes visually show larger percentages. However, as the number of visual elements and data combinations increase, the intuitive nature of the visual model decreases. The tables in Figure 21.8 use organizational rows and columns to show different categories and shading to show the magnitude of the percentages. However, this is only one analyzed element. Adding in three more elements (Figure 21.9) allows for more comparisons within a single visual, but also decreases the ability of the reader to understand what is being represented intuitively.

In Figure 21.9, the added elements for two types of discourse and two sets of tasks now add two organizational layers of rows and columns. Litster (2019) shows how an additional visual element, specifically color, could be added to show which cells within each mini-table relate to levels of enactment that are *at*, *above*, or *below* the level of intended curriculum. With each new layer of data, it is important to provide clear labels for organizational features and keys for visual elements. Text preceding the visual needs clear explanations to orient the reader before they look at the

Figure 21.9

Increased Data Combinations

		Reflective Discourse					Exploratory Discourse				
		\multicolumn{10}{c}{Highest Enacted Verbal CD}
		0	1	2	3	4	0	1	2	3	4
Task-Set 1—Harry / Intended DOK	1	20	28	8	41	3	0	12	13	68	7
	2	5	25	14	50	6	1	4	7	81	6
	3	37	17	3	33	10	6	29	12	37	16
	4	6	6	12	44	32	6	6	0	23	35
Task-Set 2—Diary	1	34	10	12	41	3	2	11	28	58	0
	2	39	8	11	36	6	12	27	31	30	0
	3	45	11	4	33	7	20	30	6	39	5
	4	0	6	22	17	56	19	38	13	19	13

Key: Incomplete / CD > DOK / CD = DOK / CD < DOK

Saturation relative to cell %: 0% — 50% — 100%

representation. For example, typically before showing Figure 21.9, I might explain the labels and orientation of the visual, such as explaining *reflective discourse intervention* is on the left, *exploratory discourse* is on the right, *task-set 1* is represented in the top mini-tables, and *task-set 2* is represented in the bottom mini-tables. I might also explain how the *0 column* to the left in each mini-table represents the number of tasks not discussed (incomplete); the diagonal for each mini-table is the percentage enacted AT the intended levels (intended 2 = enacted 2), above the diagonal is enactment HIGHER than intended, and below the diagonal is enactment LOWER than intended; and relate the shading to the bottom left key for saturation. I would also explain how to read the mini-tables and tint by showing an example. For instance, the darkly shaded box with an 81 on the upper right mini-table means that 81% of the intended DOK2 tasks completed within Task-Set 1 by groups engaged in exploratory discourse were enacted at a verbal Cognitive Demand level 3 (higher than intended).

In summary, although visual representations can be a great way to show relationships in enacted curriculum, such as sequences or comparisons, clearly defined labels, answer keys, and textual introductions are important. These ensure that the reader understands the full picture from the results. Helping the reader understand how to read the model is just as important as the model itself.

FINAL REFLECTIONS

In conclusion, before you start to collect data for enacted curriculum research, it is important to consider preparations, such as recruiting participants, selecting appropriate equipment and setting up an organizational system. These preparations ensure that you are able to collect the data to answer the research question. Next, monitor the data while you are collecting it by downloading and storing data in secure locations, transcribing and coding at least part of the data to identify initial themes, and to facilitate training transcribers and coders. Using detailed rubrics and examples during this process will increase consistency and reliability. Finally, after the data are collected, ensure the use of appropriate tools and models to analyze the data and share results. If numerical data are categorical, analysis methods and visual representations that use proportions or frequencies can help visualize results. When using visual representations, use clear labels, keys, and introductions to help readers understand the models.

ACKNOWLEDGMENT

This work was supported by a Utah State University Dissertation Grant.

REFERENCES

Bishop, J. P., Hardison, H., & Przybyla-Kuchek, J. (2016). Profiles of responsiveness in middle grades mathematics classrooms. In M. B. Wood, E. E. Turner, M. Civil, & J. A. Eli (Eds.), *Proceedings of the 38th annual meeting of the North American Chapter of the International Group for the Psychology of Mathematics Education* (pp. 1173–1180). The University of Arizona.

Cobb, P., & Yackel, E. (1996). Constructivist, emergent, and sociocultural perspectives in the context of developmental research. *Educational Psychologist, 31*, 175–190. https://doi.org/10.1080/00461520.1996.9653265

Davidson, C. (2009). Transcription: Imperatives for qualitative research. *International Journal of Qualitative Methods, 8*(2), 35–52. https://doi.org/10.1177/160940690900800206

Di Stefano, M., Litster, K., & MacDonald, B. L. (2017). Mathematics intervention supporting Allen, an English Learner: A case study. *Education Sciences, 7*(2), 1–24. https://doi.org/10.3390/educsci7020057

Excel Easy. (n.d.) *Pivot tables.* https://www.excel-easy.com/data-analysis/pivot-tables.html

Kelleher, C., & Wagener, T. (2011). Ten guidelines for effective data visualization in scientific publications. *Environmental Modelling & Software, 26*(6), 822–827. https://doi.org/10.1016/j.envsoft.2010.12.006

Krippendorff, K. (2013). *Content analysis: An introduction to its methodology* (3rd ed.). SAGE.

Litster, K. (2019). *The relationship between small-group discourse and student-enacted levels of cognitive demand when engaging with mathematics tasks at different depth of knowledge levels* [Unpublished Doctoral dissertation, Utah State University]. https://digitalcommons.usu.edu/etd/7626/

Litster, K. (2020). Small-group discourse: Supporting the inclusion of elementary students that typically struggle with mathematics in cognitively demanding discourse. In A. I. Sacristán, J. C. Cortés-Zavala, & P. M. Ruiz-Arias (Eds.), *Mathematics education across cultures: Proceedings of the 42nd meeting of the North American Chapter of the International Group for the Psychology of Mathematics Education* (pp. 2009–2016). Cinvestav / AMIUTEM / PME-NA. https://doi.org/10.51272/pmena.42.2020

Litster, K., & Wright, B. J. (2019, Fall). Money makes sense: Understanding the standard division algorithm. *Utah Mathematics Teacher, 12*, 8–17.

Litster, K., Lommatsch, C. W., Moyer-Packenham, P. S., Novak, J., Ashby, M. J., Roxburgh, A., & Bullock, P. (2020). The role of gender on the associations among children's attitudes, mathematics knowledge, digital game use, perceptions of affordances, and achievement. *International Journal of Science and Mathematics Education, 19*(7), 1463–1483. https://doi.org/10.1007/s10763-020-10111-8

MacDonald, B. L., Hunt, J., Litster, K., Roxburgh, A., & Leitch, M. (2020). Diego's number understanding development through his subitizing and counting. *Investigations in Mathematics Learning, 12*(4), 275–288. https://doi.org/10.1080/19477503.2020.1824287

McLellan, E., MacQueen, K. M., & Neidig, J. L. (2003). Beyond the qualitative interview: Data preparation and transcription. *Field Methods, 15*(1), 63–84. https://doi.org/10.1177/1525822X02239573

Saldaña, J. (2015). *The coding manual for qualitative researchers*. SAGE.

Tashakkori, A., & Teddlie, C. (Eds.). (2010). *SAGE handbook of mixed methods in social & behavioral research* (2nd ed.). SAGE.

Webb, N. L. (1999). *Alignment of science and mathematics standards and assessments in four states* (Research Monograph No. 18). University of Wisconsin, Wisconsin Center for Education Research.

Wilkinson, L., & Friendly, M. (2009). The history of the cluster heat map. *The American Statistician, 63*(2), 179–184. https://doi.org/10.1198/tas.2009.0033

CHAPTER 22

EXAMINING CURRICULUM OPENINGS IN EARLY ALGEBRA

A Closer Look at Teachers' Interaction With Curricula

Despina Stylianou
City College of New York–CUNY

Ingrid Ristroph
The University of Texas at Austin

Boram Lee
The University of Texas at Austin

Eric Knuth
U.S. National Science Foundation

Ana C. Stephens
University of Wisconsin–Madison

Maria Blanton
TERC

Rena Stroud
Merrimack College

Angela Murphy Gardiner
TERC

This chapter focuses on the implemented or enacted curriculum, which consists of those aspects of the curriculum that occur in classrooms. We seek to extend and understand the notion of curriculum openings, *specifically in the context of early algebra. We look at instructional actions as afforded by the curriculum. We expand on these ideas and explore what we call* early algebra moments, *that is, opportunities that arise during classroom instruction that a teacher could potentially capitalize upon to address an important early algebra idea or practice. Finally, we explore relationships that might exist between teachers' practices around these curricular openings and students' early algebra learning. Lessons learned focus on the fidelity of teachers' implementation and the diversity in ways they approached the curriculum.*

Keywords: algebra moments; curriculum openings; early algebra

Over the past two decades, researchers have documented several critical instructional practices related to curriculum and its implementation that contribute to student learning in mathematics. These practices include selecting rich, cognitively demanding mathematical tasks, and maintaining students' engagement with these tasks at a high level (Stein et al., 2000), building on students' thinking (e.g., Cengiz et al., 2011; Fennema et al., 1996), and facilitating discussions that support students in connecting mathematical ideas within the curriculum (Smith & Sherin, 2020).

One instructional practice that has not received as much attention is teachers' interaction with *openings in the curriculum*. Remillard and Geist (2002) refer to "openings in the curriculum" as moments that afford teachers the opportunity to engage students more deeply in mathematics. These openings can be anticipated in advance, or they may arise spontaneously as students engage with the curriculum. In this chapter, we delineate and extend both anticipated and spontaneous curriculum openings in the context of an early algebra intervention. We refer to such openings as *early algebra moments*, defined as the opportunities that arise during the course of classroom instruction to address an important early algebra idea or practice. These moments can be anticipated moments (AMs) or spontaneous moments (SMs). AMs are moments that can be reasonably anticipated in advance of the lesson enactment based on examination of curricular materials; in contrast, SMs are moments that are not anticipated in advance but that arise in the course of instruction as curriculum openings. In this work, we sought to understand the ways in which teachers "took up" these early algebra moments to develop algebraic ideas in the context of a Grade 5 early algebra intervention. Finally, we explore any relationships that might exist between teachers' practices around these curricular openings and students' early algebra learning.

CONCEPTUALIZING EARLY ALGEBRA MOMENTS AND THEIR ENACTMENT

We value teachers as co-producers of knowledge, experts, and informants to research (Bromme, 2014; Kieran et al., 2012). The relationship between research and practice is bidirectional; research and practice communities have much to contribute to each other's work (Arbaugh et al., 2010). The professional knowledge of teachers can be characterized as a complex cognitive set of skills honed through formal training and years of practical experiences. This expertise shapes the ability to notice, interpret, and respond to students, often within a series of complex classroom events requiring immediate, improvisational teacher reactions (Borko & Livingston, 1989; Shulman, 1987).

It is from this perspective that we investigated teachers' implementation of an early algebra intervention. We examined the ways in which teachers engaged or not with anticipated early algebra moments as well as the ways in which spontaneous or unanticipated moments arose during their enactments of these lessons. The study of spontaneous moments, in particular, highlights the unexpected and often creative ways in which teachers implement lessons, and as such, serves to inform and refine research, curriculum, and professional development in early algebra—a domain of importance to students' future mathematics learning as well as a domain that presents instructional challenges for elementary school teachers (Hohensee, 2015).

Numerous researchers have documented critical moments within mathematics lessons that afford teachers opportunities to notice and act to further student reasoning and understanding (e.g., Fennema et al., 1996; Leatham et al., 2015; Stockero & van Zoest, 2013). Walkoe et al. (2022) refer to "moments of algebraic potential" that arise as extensions of the curriculum that teachers may or may not take up. Similarly, Stockero and van Zoest (2013) defined "pivotal teaching moments" (PTMs) as "instance(s) in a classroom lesson in which an interruption in the flow of the lesson provides the teacher an opportunity to modify instruction in order to extend or change the nature of students' mathematical understanding" (p. 127). Researchers have described such moments as "teachable moments" (Stockero & van Zoest, 2013), "mathematically significant pedagogical opportunities" (Leatham et al., 2015), "significant mathematical instances" (Davies & Walker, 2005), and "crucial mathematic hinge moments" (Thames & Ball, 2013). A commonality of these moments is that they are generally not anticipated in advance, and thus, require teachers to notice them in the moment.

Although studies around the occurrence of interesting algebraic moments have occurred in secondary algebra classrooms (Walkoe, 2015; Walkoe et al., 2022), less research has been done in the domain of early algebra. One exception is the work of Blanton and Kaput (2003, 2005), who documented the planned and spontaneous incorporation of algebraic thinking by teachers who took part in a professional development experience designed to develop their "algebra eyes and ears." We hope to add to the body of early algebra research by increasing our understanding of moments that afford opportunities to foster students' early algebraic thinking. Understanding these moments can inform the design of early algebra curricula that aim to highlight these moments. There is also a lack of empirical analysis of observations of multiple teachers teaching the same early algebra lesson. Our corpus of data provides a unique opportunity to compare and analyze common points of instruction for the ways in which teachers capitalized, or did not capitalize, on opportunities to engage students in algebraic thinking.

BACKGROUND OF THIS WORK

The study presented here is part of a broader series of studies on the teaching and learning of early algebra. These studies drew from Kaput's (2008) characterization of algebra to conceptualize early algebra in terms of four algebraic thinking practices–*generalizing, representing, justifying*, and *reasoning with mathematical structure and relationships*—as they occur in the context of the big algebraic ideas of *generalized arithmetic*; *equivalence, expressions, equations, and inequalities*; and *functional thinking* (Blanton et al., 2018). These big algebraic ideas and thinking practices framed an approach to early algebra instruction that included several core components: (1) an articulation of a research-informed curricular progression with associated learning goals; (2) an instructional sequence to address these goals; and (3) assessments to measure learning as students advance through the instructional sequence (see Fonger et al., 2018). The early algebra intervention implemented in these studies included 18 one-hour lessons per year over three years (Grades 3–5).[1] The lessons were designed to develop students' understanding and engagement with the aforementioned algebraic thinking practices and big algebraic ideas over time. (See also Blanton et al., this volume.)

METHODS

In this section, we discuss the data used for the study and how the data were collected and analyzed. As this work is a part of a broader study, we provide a summary of the overall design as well as the analyses that were specific to this particular chapter.

Data Corpus

The data corpus for the work presented here is drawn from a large-scale randomized study of the effectiveness of an early algebra intervention to increase Grades 3–5 students' algebra readiness for middle grades. The study involved approximately 3,200 students and 100 teachers at each grade level and occurred in 46 schools (23 experimental, 23 control) in three school districts (urban, suburban, and rural) in the Southeastern United States.

We focus in this chapter on data from the Grade 5 intervention. To understand the fidelity with which teachers implemented the intervention, a randomly selected subset of teachers was observed and videotaped twice—once in the fall and once in the spring—while teaching intervention lessons as part of their regular mathematics instruction. In particular, we consider two lessons, Lesson 3 and Lesson 12, for which we have a relatively large number of videotaped observations (17 and 16, respectively). Lesson 3 reviewed properties of multiplication addressed in previous lessons, including the Commutative Property of Multiplication, the Multiplicative Identity, and the Zero Property of Multiplication. Lesson 12 reviewed the use of area models to understand the Distributive Property. Lesson 3 was taught in the fall and Lesson 12 was taught in the subsequent spring.

Students completed a written assessment, designed as part of the broader study, consisting of 18 largely open-response early algebra items addressing the intervention's big algebraic ideas. The assessment was administered as a pretest at the beginning of Grade 3 and subsequently at the end of each academic year in Grades 3, 4, and 5.

Professional Development

All teachers who implemented the instructional intervention attended professional development consisting of a one-day workshop before the school year and subsequent half-day meetings once per month throughout the 3-year intervention. During these meetings, the intervention lessons were used to build teachers' content knowledge and pedagogical content knowledge around teaching early algebraic concepts and practices. Teachers were consistently asked to engage their students in the algebraic thinking practices of *generalizing, representing, justifying,* and *reasoning with relationships* as they occurred within the big algebraic ideas (e.g., *generalized arithmetic* for Lessons 3 and 12). Lesson tasks were structured to support this process; by implementing lessons as intended, teachers would at least engage in anticipated algebraic moments (AMs).

Data Coding and Analysis

Data coding began with the identification of *anticipated moments* (AMs) in Lessons 3 and 12, with an AM defined as an expected response to support students' understanding of an algebraic concept or foster an algebraic practice. For example, in Lesson 3 that focuses on the Commutative Property of Multiplication, an AM might occur when students are asked to consider if 3 × 5 is the same as 5 × 3 and why. The teacher may take advantage of the opportunities provided by such a task to deepen algebraic thinking by asking students to draw arrays to explore the Commutative Property, or to generate examples using different types of numbers (e.g., big and small, integers and decimals). Each lesson was reviewed by three team members, consisting of both senior researchers and graduate students, to identify those "openings" that had the potential to become early algebra moments. These pre-identified openings were coded as AMs. We identified 11 such AMs in Lesson 3 and 13 AMs in Lesson 12. (See Figure 22.1 for details on these AMs.)

After identifying the AMs in the written forms of Lessons 3 and 12, we investigated the implementation of these AMs in the teaching of the lessons. Team members viewed the videotaped lessons and characterized each identified AM as either taken up or not taken up by the teacher. Taking up a moment meant the teacher noticed the algebraic potential offered by the moment and leveraged the opportunity in a way that supported students' understanding of an algebraic concept or fostered an algebraic practice. Not taking up a moment meant that an opportunity to explore an algebraic idea present in the curriculum was missed during the actual teaching of the lesson; for example, a teacher may have skipped that part of the lesson.

We next identified spontaneous early algebra moments (SMs) as enacted in the 33 videotaped lessons; SMs are those unanticipated early algebra moments that arise during the enactment of a lesson. SMs often occur when students introduce unexpected, yet potentially fruitful, mathematical ideas that prompt teachers to take advantage of opportunities to advance students' algebraic thinking, even if this means deviating from the planned lesson. These unexpected turns call on teachers to address valuable moments by responding to students' thinking and extending the scope of the lesson while addressing big ideas in more depth. For example, a student may suggest the use of variables to represent the Commutative Property of Multiplication even though such a representation was not suggested in the curriculum materials. If the teacher chooses to take up this suggestion and invites the class to consider it further, the moment would be coded as a taken up SM. SMs that are taken up are close in nature to the "noticings" described by Van Es and Sherin (2008), as they describe the way in which teachers extend the curriculum script based on student strategies,

Figure 22.1

Anticipated Early Algebra Moments (AMs) Identified in Lessons 3 and 12

Lesson 3 AMs	Lesson 12 AMs
1. Review Multiplicative Identity as a generalized statement and as represented with variables or pictures	1. Represent a situation using an expression with variables
2. Review Zero Property of Multiplication as a generalized statement and as represented with variables or pictures	2. Create a table of values (input-output) for a function presented in words
3. Discuss the Commutative Property of Multiplication	3. Construct a graph to represent a given situation
4. Extend discussion of Commutative Property of Multiplication to number domains beyond whole numbers	4. Write a variable expression
5. Explicitly use the Commutative Property of Multiplication to solve the problem	5. Write the inequality $10h > 20$ to represent the comparison of two areas
6. Review Multiplicative Identity	6. Write the expression $10h + 20$ to represent the sum of two areas
7. Extend discussion of Multiplicative Identity to number domains beyond whole numbers	7. Write $10(h + 2)$ to represent the area of a rectangle
8. Review Zero Property of Multiplication	8. Write $10(h + 2) = 100$ to represent the area of the rectangle
9. Extend discussion of Zero Property of Multiplication to number domains beyond whole numbers	9. Write $10h + 20 = 100$ to represent the area of the rectangle using the Distributive Property
10. Generate general statements invoking properties such as "$a \times 1 \div a = 1$" or "$a \times (\frac{1}{a}) = 1$" using symbols or words	10. Discuss and compare two equivalent expressions (AM 8 and 9)
11. Justify the computation $23 \times \frac{1}{23} = 1$ using the Multiplicative Inverse	11. Solve an equation using "guess and test" and "unwinding" strategies
	12. Determine result of $7h - 3h$
	13. Highlight variable as fixed unknown versus varying quantity (e.g., h in $10h + 20 = 100$ versus h in $10h$)

act in ways that explore those strategies further, and capitalize on algebra understandings, even if that means deviating from their lesson plans. If the teacher ignores or fails to notice the potential of the student utterance, the moment would be coded as a SM not taken up. These SMs not taken up do not necessarily harm students or imply poor implementation of a lesson. A teacher may have a multitude of reasons for not taking up a particular SM. However, SMs represent important opportunities to extend an idea or to make connections that arise when students are potentially most receptive to such ideas and connections.

Codes for AMs and SMs taken up or not taken up were counted for each videotaped lesson observation. Initially, we examined the number of AMs and SMs taken up or not taken up using descriptive statistics. To explore the relationship between the extent to which teachers take up AMs and SMs and student learning of early algebra, we examined the relationship between student performance on the algebra assessments and the number of AMs/SMs taken up. As noted earlier, students were assessed at the beginning of Grade 3 and at the end of Grades 3, 4, and 5. To examine the relationship of interest, we created a predicted score based on prior achievement and a residual, which was the difference between the student's predicted and actual scores. In the Level 1 model, $GR5COM_{ij}$ is the average predicted posttest score in Grade 5 for student i and teacher j. $GR3PRECOM_{ij}$ is used to denote the actual Grade 3 pretest score for each student, $GR3POSTCOM_{ij}$ is used to denote the Grade 3 posttest score for each student, and $GR4COM_{ij}$ is used to denote the Grade 4 posttest score for each student. We used the following multilevel linear model to create these residuals:

Level-1 Model

$GR5COM_{ij} = \beta_{0j} + \beta_{1j}(GR3PRECOM_{ij}) + \beta_{2j}(GR3POSTCOM_{ij}) + \beta_{3j}(GR4COM_{ij}) + r_{ij}$

Level-2 Model

$\beta_{0j} = \gamma_{00} + u_{0j}$

$\beta_{1j} = \gamma_{10}$

$\beta_{2j} = \gamma_{20}$

$\beta_{3j} = \gamma_{30}$

The residual is the difference between the students' predicted and actual scores. We looked for correlations between the number of AMs and student growth, and the number of SMs and student growth.

RESULTS

We first present examples of AMs and SMs we observed in our viewing of the 33 videotaped lessons. We then present data on the frequency of AMs and SMs identified and taken up by teachers in each lesson. Because the ultimate goal is to investigate the impact of teachers' interactions with AMs and SMs on students' early algebra learning, we present relationships the data revealed between the frequency with which teachers took up AMs and SMs and students' growth on the algebra assessment.

Illustrative Examples of AMs and SMs

To illustrate the variety of ways in which teachers engaged with anticipated and spontaneous early algebra moments in the curriculum, we share some excerpts from lesson videos that are representative of teachers' responses to anticipated and spontaneous moments. These excerpts serve to highlight qualitative differences in the nature of teachers' instructional practices around these algebra moments.

Lesson 3 AM

We start with an example of an AM from Lesson 3 that was taken up by the teacher. In this excerpt, the teacher engaged students in a review of the Commutative Property of Multiplication. The teacher presented students with the claim, "If one multiplies two numbers in any order, they still get the same result."

> **Teacher:** So, what does this mean?
> **Student:** You can take like 2×6 and 6×2 and it's 12.
> **Teacher:** He is saying you can take 2×6 and 6×2 and your answer is going to be the same. That's true for those numbers. But is it true for *all* numbers? […] You've proven it for those two numbers. *But is it true for any two numbers?*
> **Student:** "a times b equals c"?
> **Teacher:** Let me write that on the board. What are a, b, and c?
> **Student:** Variables!
> **Teacher:** He said $a \times b = c$. But, if what he said is true, $a \times b = c$, what else could it be?

Student: $c = b \times a$ [Teacher writes on the board what she hears the student say.]

Student: "b times a equals c"; a, b, and c are variables so it's true for all numbers

Teacher: So, if a, b, and c are variables, they can stand for any number!

Student: Yes!

Teacher: I heard some people say that they recognize a property there.

Student: Commutative

Teacher: What is a Commutative Property?

Student: The order doesn't matter.

Teacher: Yes, the order doesn't matter, your answer would be the same. So, if we have these [points at equations above], we could write it this way [writes $a \times b = b \times a$].

This teacher "took up" the AM, an invitation presented in the written lesson plan to discuss the Commutative Property of Multiplication. Although there was no justification for why $b \times a = a \times b$ demonstrates that the property is true for all numbers, the goal of the lesson was for students to explore the property and "explain your thinking with words and pictures" about why the Commutative Property of Multiplication holds for different types of numbers. Hence, we believe that the teacher "took up" the AM.

Lesson 12 AM and SM

Lesson 12 focused on the Distributive Property, and students were asked to express the areas of the shaded, white, and overall rectangular figure given in Figure 22.2 using variable expressions, equations, and inequalities. The goal of this task was to encourage use and discussion of the Distributive Property and to build students' experiences writing variable expressions, equations, and inequalities. Students were asked to consider the various rectangular areas to write an inequality that relates the shaded rectangular area to the white rectangular area. The teacher materials emphasized the inequality $10 \cdot h > 10 \cdot 2$ as a targeted inequality for students to produce. The following excerpt is an example of teachers' enactment of this AM:

A student stands at the front of the classroom after writing "$10 \cdot h > 10 \cdot 2$" on the board.

Teacher: Alright, tell us what you wrote.

Student: Ten times h is greater than 10 times 2.

Figure 22.2

A Diagram Designed to Invoke Use and Discussion of the Distributive Property

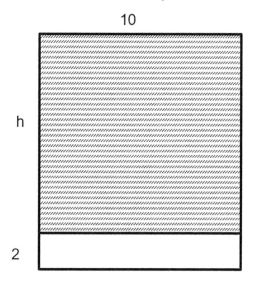

Teacher: [*Addressing the class*] Does everybody see how that is correct? If you look at your picture, you see that the area of the grey rectangle is larger than the white one. Does that make sense?

All 16 teachers who were observed teaching Lesson 12 addressed this AM. We also identified a number of SMs that arose as students engaged with this task. The following SM occurred when a teacher took up some of her students' claims that h could represent different values, and that this would result in different inequalities. The lesson as written did not suggest that teachers address this point; this observation was brought to the lesson by students.

Teacher: I hear some people ... some really good conversations. Leah, can you finish what you were saying about the areas of the white and the grey?

Student A: So, we said that the white part represents the 10 × 2... equals 20. And we think that one is greater because.... Well, you don't know what h represents.

Teacher: Absolutely. Can I get some agreement, disagreement? Maybe someone who agrees that the white (area) is greater? Someone who disagrees, okay? What do you say?

Student B: We agree because 10 times 2 is 20, and 10 times h is ... um, unknown.

Student C: Wait!... (*inaudible classroom discussion*)

Teacher: Alright, so 20 is greater than ... we don't know (*teacher writes "$20 > 10 \times h$" on the board*).

Student D: I kinda agree but I kinda disagree because we don't know what $10 \times h$ is ... we don't know what h is—maybe it's a 100, but maybe it's 15, so it could go both ways. We don't really know which one is bigger.

Student E: h can be any number. It's a variable.

Student F: But you don't have h right now!

Teacher: Okay. Student G, what are you thinking?

Student G: Because in the graph the h is bigger than the 2.

Teacher: In this specific picture? The h is bigger than the 2? [*gesturing to a projected figure*].... Based on the picture, we could make an assumption ... so then $20 < 10h$?

Student H: Well, you didn't say that—you didn't say "based on the picture."

Teacher: Right, right.... Well, what number could h be that would make it smaller than 20?

Students: [*overlapping talk*] One.... Zero!

Teacher: So, if it's zero, it [*gesturing to $10 \times h$ in "$20 > 10 \times h$"*] would be zero.

Student J: Also, one.

Teacher: Yep, I heard a couple of people say "one." If it's one, then it would be 10.

Student K: So, which is right?

Student L: We don't know yet.

Teacher: I think it's really cool—it depends on your argument.

In this excerpt, the teacher skillfully took up a student's claim that relied on the meaning of a variable as an unspecified quantity. The teacher revoiced the claim to the whole class and solicited the reasoning of others. A lively class-wide discussion ensued, with students engaging collaboratively as they made and justified other claims by reasoning algebraically about the values the variable could take on and what the consequences would be. The teacher noticed and made productive improvisational decisions to elicit student thinking about an important algebraic idea (i.e., that the value of h was unknown and that varying values of h would result in differing inequalities).

Although we observed teachers take up many SMs, a significant proportion of SMs were not taken up. For example, the teacher curriculum materials for Lesson 12 did not anticipate that students might express the need to assign a specific value for h to write an expression for the areas of the rectangles or that they might view the rectangular figure as a specific case, but this is in fact what many students did. Several teachers did not take up opportunities to engage students in discussion about their desire to assign a specific value to the variable h before an expression for the total area of the two rectangular areas could be determined. In one teacher's classroom, several students expressed a need to physically measure with a ruler or an improvised ruler (e.g., repeated measurement with the width of their finger or tick marks on a paper) to determine a numerical value for h. One student, using a pencil as a ruler, insisted that the length of h was half of 10 and thus $h = 5$. The teacher's response was "Okay, then plug that in for your h." The teacher did not address the generic nature of the rectangular figure nor what that meant in terms of variable representation.

Frequency of AMs and SMs in the Two Lessons

We now share descriptive results of the extent to which teachers took up the AMs and SMs in the lessons. We then proceed to examine possible relationships between AMs and SMs taken up by the teachers and students' performance on an early algebra assessment.

Lesson 3: Frequency of AM and SM

Lesson 3 contained 11 AMs (see Figure 22.1). So, across the 17 videotaped lessons observed, there were 187 possible AMs to be taken up. Of those 187 AMs, 108 AMs (58%) were taken up and the remaining 79 AMs (42%) were not. The AMs not taken up were most often the result of teachers not addressing those parts of the lesson (for example, a teacher pressed for time may have skipped parts of the lesson). Figure 22.3 shows the AMs taken and not taken up by each teacher.

Figure 22.3

AMs Taken and Not Taken Up in Lesson 3

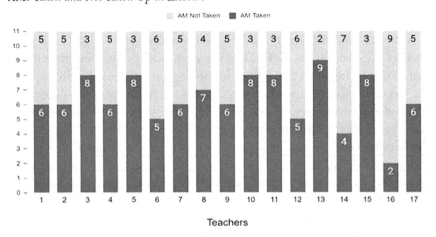

We identified 57 SMs across the 17 observations of Lesson 3. Fifty-one (81%) of these SMs were taken up by teachers and 6 (11%) were not. On average, there were 3 SMs taken up per observed lesson. Figure 22.4 offers a summary of SMs taken or not taken up across the 17 teachers.

Figure 22.4

SMs Taken and Not Taken Up in Lesson 3

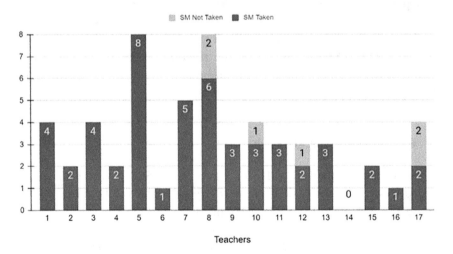

Examining Curriculum Openings in Early Algebra 471

Lesson 12: Frequency of AM and SM

Lesson 12 had 13 identified AMs; across the 16 teachers' enactments of the lesson, there were 208 AMs. Of those 208 AMs, 145 AMs (70%) were taken up and 63 AMs (30%) were not. Of the 13 possible anticipated moments for each lesson, the average number of anticipated moments taken by each teacher is about 9 and the average number of anticipated moments not taken up is about 4 (see Figure 22.5). As was the case in Lesson 3, the reason for teachers not taking up an AM was most often due to skipping that part of the lesson.

Figure 22.5

AMs Taken and Not Taken Up in Lesson 12

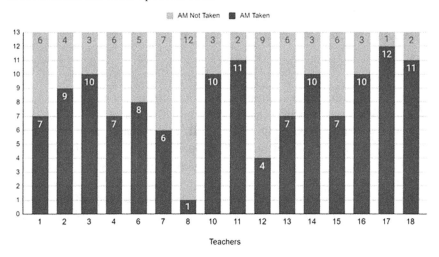

Note: Teachers 5 and 9 did not enact the lesson, so only 16 teachers are represented in the figure.)

We observed 56 SMs across the 16 observations of Lesson 12. Forty-nine (87.5%) of these SMs were taken up by teachers and 7 (12.5%) were not (see Figure 22.6). On average, there were again 3 SMs taken up per observed lesson.

Relationships Among AMs, SMs, and Student Performance

We modeled students' predicted scores for their post Grade 5 assessment. Then, for each student, we created a predicted score based on prior achievement and a "residual," which was the difference between the student's predicted and actual scores.

Figure 22.6

SMs Taken and Not Taken Up in Lesson 12

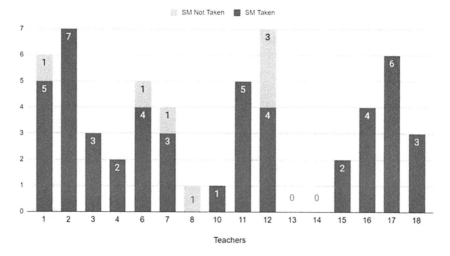

We saw a positive correlation in the relationship between the number of AMs taken up by teachers and student growth as measured by the early algebra assessments (see Figure 22.7). In other words, students whose teachers took advantage of more AMs during Lessons 3 and 12 experienced the most growth from Grade 3 pretest to posttest Grade 5 early algebra assessment.

We also saw a positive correlation in the relationship between the number of SMs taken up by teachers and student growth as measured by the early algebra assessments (see Figure 22.8). Once again, students whose teachers took advantage of more SMs during Lessons 3 and 12 experienced the most growth in their posttest early algebra assessment.

DISCUSSION

We began this chapter by asking whether it is possible to identify aspects of teachers' curriculum implementation that may be related with increase in student performance, specifically their capitalizing on opportunities offered by the curriculum to develop algebraic ideas. We presented a framework for considering the anticipated and spontaneous early algebra moments, or curricular "openings," that presented themselves in the context of an early algebra intervention. We provided examples of these moments and reported on the extent to which they were or were not taken up by teachers over the course of 33 lesson observations. We suspected that finding

Figure 22.7

Student Growth as It Relates to AMs Taken Up in Lessons 3 and 12

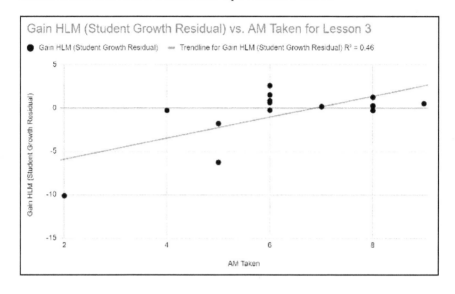

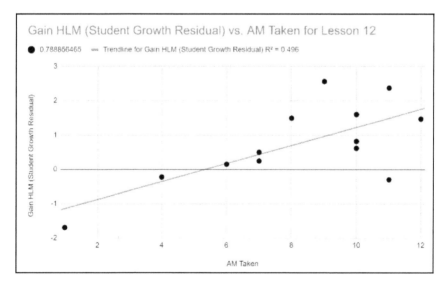

Figure 22.8

Student Growth as It Relates to SMs Taken Up in Lessons 3 and 12

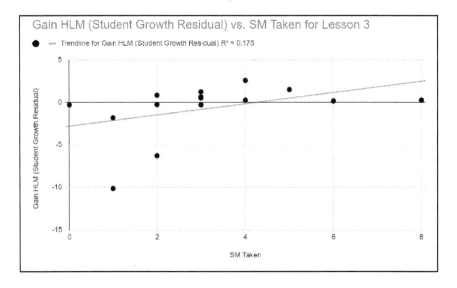

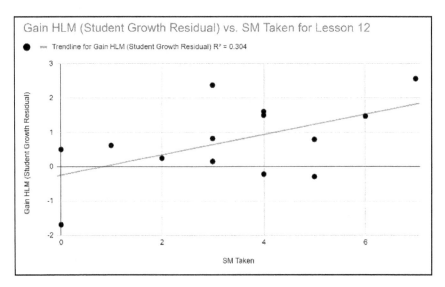

such instructional practices or actions have the potential to help advance our understanding regarding both curriculum and teaching. Here, we discuss the implications of our findings by attending to both curriculum and teaching as well as to the lessons we learned as researchers engaged in the investigation and promotion of early algebra teaching and learning.

Not surprisingly, the algebra moments that the research team anticipated would surface during the implementation of each lesson (AMs) were indeed addressed by the majority of teachers. Interestingly, those algebra moments that were not prescribed in the lesson nor anticipated by the research team, and that arose spontaneously during the implementation of each lesson (SMs), were also addressed by a number of teachers. Previous work suggests that teachers' responsiveness to these spontaneous moments is an important characteristic of "good" instruction. Teachers' abilities to take up openings in the curriculum and to identify important mathematical moments that arise outside of the curriculum have been identified as a potentially important aspect of teaching, and also important to student learning.

Our findings are both somewhat surprising and optimistic—in particular with respect to teachers' abilities to engage their students with early algebra ideas and practices—given the often negative characterizations of elementary school teachers' mathematics knowledge. In the 33 lessons we were able to observe, teachers' early algebra "eyes and ears" (Blanton & Kaput, 2003) were relatively well developed. Further research is needed to understand why some early algebra moments—both anticipated and spontaneous—were not taken up by teachers. In some cases, there may certainly be reasonable professional explanations supporting a teacher's decision to not take up a moment, including practical limitations such as time and classroom interruptions. In other cases, it may be that the moment was indeed missed, that is, a teacher did not notice an opportunity as it arose during instruction. Such cases speak to the need for professional development as well as for curricular attention to instructional opportunities and explanations of anticipated student thinking if teachers are expected to engage their students in early algebra in meaningful ways.

The positive relationship between the extent to which teachers took up AMs and SMs and student learning supports the importance of teachers taking advantage of curriculum openings to push their students' algebraic thinking forward. Simply put, teachers' increased implementation of the curriculum with fidelity to anticipated algebra moments, and with an eye toward unanticipated curricular opportunities to respond to students' thinking, was correlated with increased student learning. Additionally, increased teacher attention to students' reasoning, correct or incorrect (including misunderstandings and misconceptions), and invitations to discussions on these issues as they organically arose, were also correlated with improved student performance on these concepts.

CONCLUSIONS AND LESSONS LEARNED

In this chapter we explored the idea of early algebra moments, both those moments that arise as anticipated from the intended curriculum and those moments that arise spontaneously during instruction as the curriculum is implemented. The first lesson we learned is a hopeful one: *By examining the same two lessons implemented by 33 teachers, we observed the variety of ways in which teachers engage with these moments, highlighting the diversity and number of approaches*. As curriculum researchers, we were afforded the rare opportunity to view the same lesson implemented by several teachers, offering important insights that may not be as obvious when viewing lessons implemented by the same teacher. For example, we took notice of those aspects of the curriculum that teachers tend to notice and take up in their lessons as well as those ideas they may not take up. We noted how these ideas are implemented similarly or differently, and the ways in which they are implemented.

These results offer promise with respect to teachers' implementation of early algebra in the elementary classroom. *Teachers were willing and able to meet the challenging ideas of early algebra and implement intervention curriculum with fidelity* (see also Stylianou et al., 2019)—a lesson that offers optimism and confidence in teachers' skill and interest in the curriculum. We know already that decisions regarding curriculum implementation choices often come down to questions of values. Hence, we presume that teachers value the ideas of early algebra and they are willing to implement them to a high degree of fidelity.

This study did not measure teachers' mathematical content knowledge, so we cannot discuss whether prior mathematical knowledge was a factor in teachers' responsiveness to anticipated and spontaneous early algebra moments. The larger project in which teachers took part provided relatively extensive professional development workshops (see Stylianou et al., 2019). As we did not measure the effect of these workshops on teachers' knowledge, we cannot argue that professional development impacted teachers' skill in implementing the curriculum with fidelity. Hence, another lesson we learned: *When implementing new curriculum, it is important to measure both pre-existing mathematical knowledge and skill and the potential effects of professional development on teachers' content knowledge and professional knowledge*. Although we did not measure these effects, teachers' abilities to capitalize on and amplify anticipated and spontaneous early algebra moments give us reason to hypothesize for future studies that professional development experiences help teachers develop familiarity with the curriculum, and support teachers' efforts to develop their early algebra eyes and ears.

Another lesson learned is also a surprisingly simple and profound one: *Students will learn what they have had the opportunity to study*. Students in elementary grades can successfully engage with early algebra concepts when teachers give them the opportunity to do so. This finding and lesson reiterates earlier work that offers lessons on the importance and feasibility of teaching algebraic concepts to children in early grades (see, e.g., Carraher et al., 2016). A corollary to this finding is that students will not learn what they have not had the opportunity to study. In this study, we saw a correlation between the number of early anticipated and spontaneous algebra moments taken up by teachers and students' performance on the early algebra assessments and students' learning in general.

Finally, the results offer insights into understanding how the early algebra curricular materials support teachers to anticipate students' algebraic thinking and how teachers' spontaneous responses to early algebra moments can further students' algebraic thinking. By studying and spotlighting teachers' facilitation of these moments, we hope to amplify high quality teaching of early algebra. Additionally, understanding algebraic moments that were not taken up in classrooms could serve to hone professional development around noticing and attending to students' algebraic thinking.

ACKNOWLEDGMENTS

The research reported here was supported in part by the U.S. National Science Foundation (NSF) under Award DRL-1721192. Any opinions, findings, and conclusions or recommendations expressed in this material are those of the authors and do not necessarily reflect the views of the NSF.

We are grateful for the assistance of Jordan Burkland, Hangil Kim, and Bethany Miller who participated in earlier stages of the project, particularly in the coding of the videos.

REFERENCES

Arbaugh, F., Herbel-Eisenmann, B., Ramirez, N., Knuth, E., Kranendonk, H., & Reed Quander, J. (2010). *Linking research and practice: The NCTM research agenda conference report*. National Council of Teachers of Mathematics.

Blanton, M., Brizuela, B., Stephens, A., Knuth, E., Isler, I., Gardiner, A., Demers, L., Fonger, N., & Stylianou, D. (2018). Implementing a framework for early algebra. In C. Kieran (Ed.), *Teaching and learning algebraic thinking with 5–12 year-olds: The global evolution of an emerging field of research and practice* (pp. 27–49). Springer.

Blanton, M. L., & Kaput, J. J. (2003). Developing elementary teachers' algebra eyes and ears. *Teaching Children Mathematics, 10*(2), 70–77.

Blanton, M., & Kaput, J. J. (2005). Characterizing a classroom practice that promotes algebraic reasoning. *Journal for Research in Mathematics Education, 36*(5), 412–446.

Borko, H., & Livingston, C. (1989). Cognition and improvisation: Differences in mathematics instruction by expert and novice teachers. *American Educational Research Journal, 26*(4), 473–498.

Bromme, R. (2014). *Der lehrer als experte: Zur psychologie des professionellen wissens* (Vol. 7). [The teacher as an expert: On the psychology of professional knowledge]. Waxmann Verlag.

Carraher, D., Schliemann, A., Brizuela, B., & Earnest, D. (2016). Arithmetic and algebra in early mathematics education. In E. Silver & P. Kenney (Eds.), *Lessons learned from research: Helping all students understand important mathematics* (pp. 109–122). National Council of Teachers of Mathematics.

Cengiz, N., Kline, K., & Grant, T. J. (2011). Extending students' mathematical thinking during whole-group discussions. *Journal of Mathematics Teacher Education, 14*(5), 355–374.

Davies, N., & Walker, K. (2005). Learning to notice: One aspect of teachers' content knowledge in the numeracy classrooms. In P. Clarkson, A. Downton, D. Gronn, M. Horne, A. McDonough, R. Pierce, & A. Roche (Eds.), *Building connections: Theory, research and practice—Proceedings of the 28th Annual Conference of the Mathematics Education Research Group of Australasia* (pp. 273–280). Mathematics Education Research Group of Australasia [MERGA].

Fennema, E., Carpenter, T. P., Franke, M. L., Levi, L., Jacobs, V. R., & Epson, S. B. (1996). A longitudinal study of learning to use children's thinking in mathematics instruction. *Journal for Research in Mathematics Education, 27*(4), 403–434.

Fonger, N. L., Stephens, A., Blanton, M., Isler, I., Knuth, E., & Gardiner, A. M. (2018). Developing a learning progression for curriculum, instruction, and student learning: An example from early algebra research. *Cognition and Instruction, 36*(1), 30–55.

Hohensee, C. (2015). Preparing elementary pre-service teachers to teach early algebra. *Journal of Mathematics Teacher Education, 20*(3), 231–257. https://doi.org/10.1007/s10857-015-9324-9

Kaput, J. J. (2008). What is algebra? What is algebraic reasoning? In J. J. Kaput, D. W. Carraher & M. L. Blanton (Eds.), *Algebra in the early grades* (pp. 5–17). Lawrence Erlbaum/Taylor & Francis Group.

Kieran C., Krainer K., & Shaughnessy, J. M. (2012). Linking research to practice: Teachers as key stakeholders in mathematics education research. In M. Clements, A. Bishop, C. Keitel, J. Kilpatrick, & F. Leung (Eds), *Third international handbook of mathematics education* (pp. 361–392). Springer International Handbooks of Education, Vol 27. Springer.

Leatham, K., Peterson, B., Stockero, S., & van Zoest, L. (2015). Conceptualizing mathematically significant pedagogical opportunities to build on student thinking. *Journal for Research in Mathematics Education, 46*(1), 88–124.

Remillard, J. T., & Geist, P. (2002). Supporting teachers' professional learning by navigating openings in the curriculum. *Journal of Mathematics Teacher Education, 5,* 17–34.

Shulman, L. (1987). Knowledge and teaching: Foundations of the new reform. *Harvard Educational Review, 57*(1), 1–23.

Smith, M., & Sherin, M. (2020). *The five practices in practice.* National Council of Teachers of Mathematics.

Stein, M. K., Smith, M. S., Henningsen, M. A., & Silver, E. A. (2000). *Implementing standards-based mathematics instruction: A casebook for professional development.* Teachers College Press.

Stockero, S., & van Zoest, L. (2013). Characterizing pivotal teaching moments in beginning mathematics teachers' practice. *Journal of Mathematics Teacher Education, 16*(2), 125–147.

Stylianou, D., Stroud, R., Cassidy, M., Gardiner, A., Stephens, A., Knuth, E., & Demers, L. (2019). Putting early algebra in the hands of elementary school teachers: Examining fidelity of implementation and its relation to student performance. *Journal for the Study of Education and Development—Infancia y Aprendizaje* (Special Issue: Early Algebra), *42*(3), 523–569.

Thames, M. H., & Ball, D. L. (2013). Making progress in U.S. mathematics education: Lessons learned–past, present and future. In K. Leatham (Ed.), *Vital directions for mathematics education research* (pp. 15–44). Springer.

Van Es, E. A., & Sherin, M. G. (2008). Mathematics teachers' "learning to notice" in the context of video classroom interactions. *Journal of Technology and Teacher Education, 10*(4), 571–596.

Walkoe, J. (2015). Exploring teacher noticing of student algebraic thinking in a video club. *Journal of Mathematics Teacher Education, 18,* 523–550. https://doi.org/10.1007/s10857-014-9289-0

Walkoe, J., Walton, M., & Levin, M. (2022). Supporting teacher noticing of moments of algebraic potential. *Korean Journal of Education Research in Mathematics, 32*(3), 271–286.

ENDNOTE

1. In the United States, students in grades 3–5 are typically ages 8–11.

CHAPTER 23

LESSONS LEARNED IN STUDYING THE IMPACT OF TEACHING *MATHEMATICS IN CONTEXT* ON STUDENT ACHIEVEMENT

Mary C. Shafer
Northern Illinois University

In 1992, the U.S. National Science Foundation (NSF) funded projects to develop curricular materials reflecting the reform vision of school mathematics recommended by the National Council of Teachers of Mathematics in the Curriculum and Evaluation Standards for School Mathematics. *One project led to the development of* Mathematics in Context *for Grades 5–8. Later, NSF funded the* Longitudinal/Cross-Sectional Study of the Impact of Teaching Mathematics Using Mathematics in Context on Student Achievement. *In this chapter, I discuss lessons learned as we conducted the study in the reality of schools. Variations found in key variables draw attention to the need to examine effects of the culture in which student learning is situated, yet controlling potential sources of variation is difficult in classroom settings. When data collection and analysis methodologies are designed to take the variations that are encountered in school settings into consideration, research of high quality can be conducted.*

Keywords: enacted curriculum; mathematics; middle school; student achievement; summative evaluation

In 1992, the U.S. National Science Foundation (NSF) funded projects to develop curricular materials to reflect the reform vision of school mathematics espoused by the National Council of Teachers of Mathematics (NCTM) in the *Curriculum and Evaluation Standards for School Mathematics* (NCTM, 1989). One of the funded projects was awarded to the National Center for Research in Mathematical Sciences Education (NCRMSE) at the University of Wisconsin–Madison. With assistance from the Freudenthal Institute at the University of Utrecht in The Netherlands, a comprehensive mathematics curriculum for Grades 5–8, *Mathematics in Context* (*MiC*), was developed and field-tested prior to being published in 1997–1998 by Encyclopaedia Britannica. Later in the decade, NSF funded two grants to NCRMSE: (1) to conduct a four-year study of the impact of *MiC* on student mathematical performance, *The Longitudinal/Cross-Sectional Study of the Impact of Teaching Mathematics Using* Mathematics in Context *on Student Achievement*, and (2) to analyze the data gathered for the longitudinal and cross-sectional studies. In this chapter, I draw from the longitudinal and cross-sectional studies of *MiC* to discuss four aspects of lessons we learned from this research, regarding: conducting research in large urban settings; use of a structural research model; the relationship between the enacted curriculum and student performance; and the effects of treatment fidelity.

THE INTENDED CURRICULUM: *MATHEMATICS IN CONTEXT*

Mathematics in Context (*MiC*) was different from conventional textbooks prevalent in middle schools at that time. *MiC* included 40 curriculum units (10 units at each grade level[1]), associated teacher guides, and two ancillary units. In *MiC*, students are encouraged to deepen their understanding of algebra, geometry, probability, and statistics, in addition to number. The pedagogy underlying *MiC* was based on Dutch Realistic Mathematics Education (RME) (Gravemeijer, 1994), an approach to school mathematics deemed consistent with the NCTM *Standards*. The goals of *MiC* reflect an assumption that students need to participate in the mathematization of reality and can do that by exploring aspects of several mathematical domains. As a result, students gradually shift from creating "models of" problem situations to "models for" mathematical reasoning and problem solving (Gravemeijer, 1994). Through lessons that allow students to solve problems using a variety of strategies, teachers encourage students to discuss interpretations of problem situations, express their thinking, and react to different levels and qualities of solution strategies shared in their group. Students solve problems at different levels of abstraction, falling back to more concrete, less abstract strategies whenever they feel the need. Through exploration, reflection, and generalization, the curriculum supports students progressing from context-specific situations to more abstract mathematical reasoning. This process is called *progressive formalization*. This type of instruction is far removed from two-page lessons that emphasize memorization and independent seatwork often seen with traditional curricula.

The study was designed to investigate three questions:

1. What is the impact of the *MiC* instructional approach on student performance?
2. How is this impact different from that of traditional instruction on student performance?
3. What variables associated with classroom instruction account for variation in student performance?

LESSONS LEARNED ABOUT CONDUCTING RESEARCH IN LARGE URBAN SCHOOL DISTRICTS

Because of differences between *MiC* content and instructional approach and conventional mathematics curricula at the time, it was important to study the ways teachers enacted *MiC* in various school settings. NSF program officers asked us to work in districts that had begun reform initiatives. We wanted to see how *MiC* would work in urban, low-income school districts. We found that searching for and contracting with school districts was difficult work.

Participation of School Districts

During the mid-to-late 1990s, when systemic reform initiatives were taking place around the country, many school district administrators were unwilling to participate in a comparative research study. They were concerned they might be seen as providing some students with opportunities to learn mathematics in more stimulating and engaging ways while other students continued to study conventional curricula already in their schools. These administrators did not want to be perceived as rejecting reform efforts (Romberg & Shafer, 2008, p. 22). Nevertheless, key administrators in two districts were very interested in the study, and we met with the administrators at both sites to further explain *MiC*, the study, and the subcontracts. Administrators in these two districts, referred to as Districts 1 and 2, agreed to participate in comparative research between *MiC* and non-*MiC* (comparison) classes. The assignment of subjects was nonrandom. District administrators were asked to select schools that were representative of the district population, rather than selecting schools with extremely low- or high-achieving student populations. Principals of the selected schools chose the study teachers. Teachers were asked to select classes with average mathematical abilities (Romberg & Shafer, 2008, p. 23). *MiC* was available in its commercial form for the first time, and some *MiC* teachers had taught one to four field-test or pre-publication units prior to the study. Comparison teachers used curricula already available in the schools. Data would be gathered on all research variables to enable us to understand the enacted curriculum and its relationship to student performance.

Later, two other districts, referred to as Districts 3 and 4, agreed to participate in a modified research design in which only *MiC* was taught and data on classroom interaction was not gathered. Consequently, the number of *MiC* teachers grew in comparison to teachers using conventional curricula (Romberg & Shafer, 2008, p. 23). Also, in District 4, because of the substantial dispersion of incoming Grade 5 students among Grade 6 mathematics classes, only middle-school students (Grades 6–8) participated in the study.

Lessons Learned During Data Collection

The subcontracts we negotiated with administrators in each district contained a detailed scope of work for all study participants, which was discussed with district personnel prior to signing contracts. However, in Districts 1 and 2, administrators did not carry through with our requests for selecting schools that were representative of the district population. Administrators in all districts also struggled to keep intact the groups of students necessary for the longitudinal studies, to choose classes of average ability, to assign incoming sixth-grade students to classes of middle-school study teachers, and to provide standardized test scores for participating students (Romberg & Shafer, 2008, p. 163).

District Coordinators

In each district, an on-site coordinator was selected by district administrators to work with the research team. In District 1, the on-site coordinator was the district mathematics specialist. She worked directly with the superintendent and the director of the curriculum to select study schools and identify teachers for participation. In District 1, the same on-site coordinator was available during all the study years. In District 2, the study research plan was discussed with the directors of their government-funded grant programs, and they worked with district administrators to select study schools. However, a different on-site coordinator was selected each year, partially due to district restructuring. Therefore, each year the liaison in District 2 was unfamiliar with research expectations and problems in data collection that had arisen the prior year. In District 3, the on-site coordinator was the principal of the only middle school in the district. The coordinator maintained regular contact with the principal of the only Grades 3–5 school in the district and with the district superintendent. The on-site coordinator highly supported teachers in their previously made decision to use *MiC* as their curriculum by giving encouraging feedback, handling parental concerns, and providing monthly meetings for teachers to discuss *MiC* units and teaching *MiC*. Transition of study students from Grade 5 to Grade 6 was handled smoothly. In District 4, initial discussions about the study were with a university faculty member who worked with one of the school districts within a large urban school system. The on-site coordinator was the assistant principal for mathematics and science at the large middle school selected for the study. Therefore, fifth-grade students did not participate in this district. The coordinator supported teachers' changes in curriculum and instruction through personal monthly meetings with each teacher, during which they discussed reform and research-based ideas for mathematics teaching and learning (Romberg & Shafer, 2008, p. 54).

We kept in regular contact with the on-site coordinators to remind them of contractual agreements and to support teachers who were teaching *MiC* for their first full school year. In our conversations, we discussed immediate needs in our data collection, specifying dates for administering study assessments, and scheduling the spring scoring and summer institutes that we provided each year as incentives to each district for study teachers' professional development opportunities. On-site coordinators' recommendations about sending teaching logs and study assessments to and from teachers were invaluable.

Following Students Longitudinally

A major challenge was monitoring the performance of study students longitudinally. One research goal was to follow groups of students for three years from Grade 5 through Grade 7, from Grade 6 through Grade 8, and for two years from Grade 7 to Grade 8. Knowing that students would transfer from elementary schools to middle schools for Grades 6–8, we worked with on-site coordinators to explore the feeder patterns of elementary study students into middle schools. On-site coordinators then were asked to work with principals of middle schools to ensure that incoming study students would be placed in classes with the same Grade 6 mathematics teacher already in the study or who would begin participation in that year.

However, keeping study students together rarely occurred, and we were unable to follow a substantial number of students who began the study in Grade 5. Of the Grade 5 students who completed both study assessments in the first study year, 90% in District 1, 96% in District 2, and 43% in District 3 did not participate in the study for all three years (Romberg & Shafer, 2008, p. 25). Attrition also occurred because of parental choice of their child's school each year in District 1, or lack of parental approval for their child to continue in the study during their middle-school years in all districts.

On-Site Observers

A major success in working with Districts 1 and 2 was finding reliable observers. Administrators recommended retired mathematics teachers to gather classroom observation data and conduct principal and teacher interviews. Observers worked with graduate research team members to gain interrater reliability for the classroom observation instrument developed for the study (Shafer, Wagner, & Davis, 1997a), first through using videotaped lessons, then in mathematics classes in the observer's district. After interrater training, observers observed each study teacher once a month and compiled a report for each observation. These reports were sent to the research team for analysis. Observers also learned how to use the interview protocols designed by the research team to gather more inclusive information from the principals and teachers they interviewed (Shafer, Davis, & Wagner, 1997, 1998). The interviews provided details on beliefs about mathematics instruction, the school's vision of teaching and learning mathematics, and cultural and organizational situations that worked together to provide the learning environment in the school.

Our ability to make contracts with the observers and to develop interrater reliability for observations was crucial to gather details of classroom interaction, instructional materials used, and descriptions of lessons. Observers also asked teachers questions in brief pre- and post-observation interviews to gather information about the lesson from the teachers' perspectives. We were fortunate that the observers continued their work throughout the study. Because observation reports were based on a consistent set of criteria, the research team was able to identify changes in instruction and opportunity to learn over time.

Study Teachers

Over the study years, 37% of the sixth-grade teachers, 41% of the seventh-grade teachers, and 46% of the eighth-grade teachers completed the study as requested. We learned that teachers frequently changed within a school year, with some students having two or three mathematics teachers per year. Teachers took positions at a non-study school, were on family leave, or accepted an administrative position. Others decided to resign from participation in the study (Romberg & Shafer, 2008, p. 25). For example, in one middle school in District 2, standardized tests substantially influenced teachers' mathematics instruction; large amounts of time were devoted to test preparation, so *MiC* was rarely used. This became such a serious issue for teachers that the principal intervened on their behalf. The principal stated that district funding for their school was directly tied to student results on the traditional portion of the state testing program, and the principal perceived that students would not perform well if they studied *MiC*. The principal strongly encouraged the teachers to withdraw from the study, and they withdrew in the second semester of the first study year (Shafer, 2004, p. 53). Replacement teachers and their students in other schools were found for the study. However, data collection for students was limited to one semester of the school year in the initial school and in the replacement school. In the same district, the study was confounded when an eighth-grade teacher taught *MiC* in the fall semester but from a traditional algebra textbook in the second semester. These changes affected the comparability of groups of students in the research.

Data collection was also compromised when some teachers administered tests on days of scheduled classroom observations. Some did not cancel classroom observations when school schedules changed, or when they were not in class because of professional development or illness. Other teachers decided not to complete teaching logs that added to our understanding of instruction in their study classes, even though they were given a monthly stipend for each log sent to the research team (Shafer, 2004, p. 69).

Nevertheless, we learned many examples of good teaching practice. For example, in journal entries and interviews, teachers noted differences in student engagement during instruction because students had studied *MiC*. Teachers developed ways to allow students to do more mathematical thinking in class. For example, teachers noted how they integrated students' suggestions into instruction. When teaching geometry lessons on two-dimensional representations of three-dimensional objects, Ms. Piccolo[2] emphasized that "what we see is determined by our perspective" and described that "we gave examples and changed positions as we looked at models." Ms. Piccolo allowed students to introduce different strategies that provided access to problem solutions, noting:

> Using blocks [to model objects in a diagram], one student suggested to put the model on the diagram in the book and lift the book to eye level at each [camera position] to determine the correct matching of side views [to cor-

responding camera positions]. Some students lifted the book and others got on their knees in front of their desks. It worked! The majority of the students were successful after using this strategy. (District 2, Grade 5, *MiC*, Piccolo, Journal entry 9/16/1997; Shafer, 2004)

Although some observations were compromised, the depth of data collection enabled us to understand classroom interaction and led to identifying differences in methods of instruction and use of instructional materials among study teachers. The research team learned about teacher and student successes during classroom interaction. We were encouraged that *MiC* would be beneficial in diverse student populations as students learned mathematics in depth and with understanding.

Conditions Affecting Study Assessments

Study assessments were administered in May or June of each year. Many students completed study assessments in an array of less-than-ideal conditions. For example, in District 2, a Grade 5 teacher reviewed assessment instructions for such a long time that students did not have the opportunity to complete the assessments. In District 3, Grade 6 students completed study assessments after three weeks of practice for and completing standardized testing. In District 4, study assessments were taken between two sets of district testing. As these examples illustrate, the results of some assessments were not necessarily reflective of students' understanding and application of mathematics.

In District 2, teachers relayed that students seemed not to put forth the effort to develop reasoned responses that the research team had hoped for because study assessments did not affect their grades or promotion from one grade level to the next. Many students began to solve tasks, but did not complete them, so their responses did not earn full credit. A few students wrote notes to us in their assessment booklets; others made drawings that had nothing to do with the mathematics or the contexts used in a set of items. We learned that it is important for study teachers to encourage their students to think about the items and develop thoughtful responses. We worked with on-site coordinators and teachers to plan when they would administer the assessments and the appropriate conditions for administering them.

LESSONS LEARNED ABOUT THE RESEARCH MODELS

The *MiC* curriculum is intended to develop content from informal to formal, abstract mathematics over grade levels. Therefore, conducting a longitudinal study over years allowed the research team to track growth in student performance. Rather than try to control sources of variation, we used a structural research model to gather, interpret, and report variations in classroom interactions when teachers implemented *MiC* or conventional curricula in classrooms (Romberg & Shafer, 2008, p. 9). We developed a modified research model as the analysis progressed.

The Structural Research Model

The research model used to gather information for the study was an adaptation of a structural model for monitoring changes in school mathematics (see Figure 23.1). The model is composed of 14 variables in five categories: *Prior*; *Independent*; *Intervening*; *Outcome*; and *Consequent*. The theoretical interrelationships between the variables are represented by arrows in the model. The sequence of variables from *Prior* to *Consequent* was logical. In addition, the identification of potential variables in each category gave us a starting point for the development of research instruments for the study.

Figure 23.1

Model for the Monitoring of School Mathematics

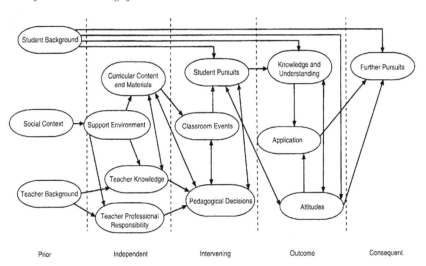

Source: *The Impact of Reform Instruction on Student Mathematics Achievement: An Example of a Summative Evaluation of a Standards-Based Curriculum*, Edition 1, by Thomas A. Romberg and Mary C. Shafer, ©2008, p. 11, by Routledge. Reproduced by permission of Taylor and Francis Group.

Quantifying the variables in the structural model was arduous and extensive. In the year prior to data collection, we developed the instruments we planned to use, and pilot-tested and scaled data gathered from local schools and schools in a neighboring state involved in field-testing *MiC*. We are pleased that we were able to create good measures for many variables. For each variable, data were gathered with attention to multiple sub-variables, and an index was created to illustrate differences for each variable and corresponding sub-variables. For analysis purposes, we generally found that measures of the variables and sub-variables were statistically collinear. However, intellectually, the sub-variables were different and made sense with respect to research literature (Romberg & Shafer, 2008, p. 166). Eventually, the information gathered for the sub-variables was distilled into a single variable, but we could go back to the sub-variable indices and written descriptions to detail aspects of a teacher's instructional approaches or to illustrate events that transpired in a classroom.

Even as we developed good measures for some variables, we were unable to design adequate measures for other variables for teachers and students. At a time when good measures of teachers' mathematical background were in beginning stages at other research centers, we worked to develop good questionnaires for this purpose. However, district administrators and teacher union representatives discouraged any attempts to measure teacher background knowledge. For students, we requested to use student standardized test score data as a measure of student background knowledge. Although the districts provided the scores, tests were different in each district and item data were not available. We felt that the best we could do was to use each student's percentile rank as a score, assuming that the norm-populations of the various tests were similar. Also, the *Student Attitude Inventory* (Shafer, Wagner, & Davis, 1997b) that we developed and situated in research literature proved to be of little use because initial scores were very positive, limiting the possibility of measuring growth. In addition, in studying the Consequent variable, our plan was to gather information on students' transition into high-school mathematics and the number and type of mathematics courses in which they intended to enroll in high school. However, our attempts to follow eighth-grade study students into high school, obtain parental permission, and ask students to complete this questionnaire were not productive. The few responses we received were not enough to make conclusions about the Consequent variable.

Simplified Research Model

In 1998–1999, when Romberg was a Fellow at the Center for Advanced Study in the Behavioral Sciences at Stanford, he met with the statistics working group headed by Lincoln Moses. After reviewing the indices and analysis plan for the study, the group recommended the creation of four composite variables using data gathered from all the variables in the structural research model. This simplified model describes the relationship between variations in classroom achievement (*CA*), aggregated by content strand or total performance, that can be attributed to variations in prior achievement (*PA*), method of instruction (*I*), opportunity to learn with understanding (*OTLu*), and school capacity (*SC*). This relationship can be expressed as $CA = PA + I + OTLu + SC$ (Romberg & Shafer, 2008, p. 13). The simplified research model proved to be important in the study. The composite indices, based on one or more sub-indices developed for each variable in the structural research model, were used to subdivide and describe treatment groups, and in regression analysis to determine the impact of the variables on student achievement.

Classroom Achievement

The creation of the composite classroom achievement scale (*CA*) via a subcontract with the Australian Council for Educational Research was critical as the primary outcome measure for the study. *CA* was developed using results from the two sets of assessments designed for the study: the External Assessment System (EAS) (Webb et al., 2005); and the Problem-Solving Assessment (PSA) (Dekker et al., 1997–1998). Each set included grade-specific assessments, one for each of Grades 5–8. Each of the four EAS assessments consisted of publicly released tasks from the 1992 National Assessment of Educational Progress (NAEP), 1996 NAEP, and the 1996 Third International Mathematics and Science Study (TIMSS).[3] Each assessment contained items evenly divided among four content strands: number, geometry, algebra, and statistics/probability. Seventy percent of the items were multiple-choice; the other items were constructed response. All items were scored using the same rubrics as in the NAEP or TIMSS. To examine growth over time, a selection of items of moderate difficulty was repeated on each assessment. These items were referred to as *anchor items* (Romberg & Shafer, 2008, pp. 56–57).

The PSA was designed under a subcontract by researchers at the Freudenthal Institute in The Netherlands to align with reform goals of problem solving, communication, reasoning, and connections. All assessment items were constructed-response problems that were presented in contexts, with multiple items associated with each context. Items were written with the intent that students' strategies would progress from informal models and calculations to more formal symbolic notations across grade levels. All responses were scored with partial credit scoring, and some answers were coded by solution strategy. The general directions for the PSA requested that students write out their solution strategies. Although the PSA was designed as part of a study evaluating the impact of *MiC*, the assessments were not curriculum specific. All students, regardless of the curriculum studied, should be able to solve the items successfully. PSA items were accessible in a variety of ways so students who relied on informal strategies as well as students who used more sophisticated strategies experienced success with the problems.

In calculating *CA* scores, the results of students who completed both the EAS and PSA at the end of each year of their participation were included in the analysis. Because students' responses to assessment items were essential in the development of the progress map for *CA*, missing scores were not statistically imputed (Romberg & Shafer, 2008, p. 81). Ratings for individual students were calculated and achievement bands were determined.

Six levels of achievement were derived from the typical set of knowledge, skills, and understandings students demonstrated as they engaged with the test items (Turner & O'Connor, 2005). The descriptions of the *CA* achievement bands were consistent with both the Program for International Student Assessment (PISA) definition of mathematization (Adams & Wu, 2002) and Romberg's view of mathematical literacy (Romberg, 2001; Romberg & Shafer, 2008, p. 60).

The *CA* progress maps provided pictorial representations of the variable of *mathematical competence*. Progress maps were useful to monitor changes in student achievement over time and to compare achievement for various groups, such as all students at a particular grade level in a particular year, or by district, curriculum taught, teacher/student groups (i.e., students in one or more classes in the same grade who have the same teacher during a school year), gender, and ethnic group (Romberg & Shafer, 2008, p. 60). Band descriptions for the study's progress map are shown in Figure 23.2.

Figure 23.2

Progress Map Band Descriptions for the Longitudinal/Cross-Sectional Studies of Mathematics in Context on Student Performance[4]

> *Band 6:* Students typically need to identify the key elements of the problem; show an extensive working solution; and provide a summative statement in response to the problem. Students are typically required to identify, compare, or combine elements of the problem, draw on assumptions based on real-world knowledge, and provide a fully justified conclusion supported by work, explanation, or reasoning.
>
> *Band 5:* A high level of mathematization is required in order to respond fully to items in this band, as students are typically required to translate contextualized real-world problems into mathematical terms, and then identify and use an appropriate mathematical strategy and use a range of tools to solve the problems. Students typically need to provide a correct answer accompanied by a complete explanation of the work needed to arrive at the solution that takes into account the key points identified in the problem.

(Figure continued on next page)

> *Band 4:* A moderate level of mathematization is required, as students are typically required to translate either a contextualized or a non-contextualized, generally non-routine problem into mathematical terms. For contextualized problems, the solutions tend to depend on the application of a formula or relationship (e.g., proportionality of corresponding side lengths) with which it is expected that the student is familiar. Non-contextualized items also tend to depend on the application of specific knowledge (e.g., recognize an algebraic expression).
>
> *Band 3:* Mathematization in this band is limited to the application of mathematical tools in order to solve predominantly routine problems and any contextualized problem-solving tasks that generally require only relatively simple computations for their solution.
>
> *Band 2:* Mathematization at this level is limited to the application of routine procedures to standard or familiar contextual representations of problems.
>
> *Band 1:* Partial credit items in this band tend to be of the type in which less credit is given when only some of the required criteria are met or when criteria are met, but an explanation is lacking.

Source: The Impact of Reform Instruction on Student Mathematics Achievement: An Example of a Summative Evaluation of a Standards-Based Curriculum, Edition 1, by Thomas A. Romberg and Mary C. Shafer, ©2008, pp. 59–60, by Routledge. Reproduced by permission of Taylor and Francis Group.

A progress map for the distribution of classroom achievement in *MiC* classrooms in District 1, Grade 5, in Year 1, by teacher/student group is shown in Figure 23.3.[5] Students in Teacher 31's class performed significantly higher than the students of any other teacher in this grouping, and Teacher 19's class performed significantly higher than students of the other four teachers. However, a restricted range of scores was apparent for Teachers 4, 19, 31, and 49. Such restriction suggests homogeneous grouping or tracking of students, which was confirmed in interviews. Teacher 14 taught three study classes, one of which was low in prior achievement, which may have accounted for the greater variation in performance for Teacher 14's students (Romberg & Shafer, 2008, p. 66).

Prior Achievement

Multiple instruments were used for describing students' prior achievement (*PA*). We administered the *Collis-Romberg Mathematical Problem-Solving Profiles* (Collis & Romberg, 1992) as a diagnostic test at the beginning and end of each student's participation. The *Student Attitude Inventory* created for the study was completed at the beginning of each student's participation and in the spring of each year. We also asked districts to provide their standardized test data for study students, which were gathered for the year preceding the study as a baseline and again in the spring of each study year.

Figure 23.3

Distribution of Classroom Achievement in MiC Classrooms in District 1, Grade 5 in 1997–1998, by Teacher

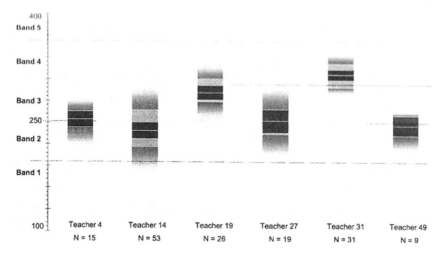

Source: The Impact of Reform Instruction on Student Mathematics Achievement: An Example of a Summative Evaluation of a Standards-Based Curriculum, Edition 1, by Thomas A. Romberg and Mary C. Shafer, ©2008, p. 66, by Routledge. Reproduced by permission of Taylor and Francis Group.

Standardized test data were used as *PA* in the simplified research model to examine initial differences among study classes. The standardized tests varied by district, so percentiles were used for consistency when examining various groups of student scores. The assumption was that classes of study students were comparable to the groups of students on which the norm-referenced scores were developed. We believed that comparing percentile scores, derived from comparison to norm-referenced groups, could be useful in comparing student groups over study years and might validate information resulting from the study's *CA* scores (Romberg & Shafer, 2008, pp. 60–61). *CA* results from the previous year were also used as measures of *PA* in Years 2 and 3.

Instruction

In this study, instruction was described through the three Intervening variables in the structural research model: *pedagogical decisions* (the teacher's decisions in defining the enacted curriculum); *classroom events* (interactions that promote learning); and *student pursuits* (the nature of students' engagement in learning). Using the underlying single dimension of teaching mathematics for understanding, the Instruction Total was composed of five major categories: *unit planning*; *lesson planning*; *mathematical interaction during instruction*; *classroom assessment practice*; and *student pursuits during instruction*. A total of 19 sub-variables from these categories were compiled to make the Instruction Total. Quantitative analyses were used to distill the results from 19 indices into one composite index *Instruction* (Shafer, 2005, pp. 28–34). Because the indices were composed of three to six levels, indices were weighted so each would have an equal emphasis in the analysis. The weighted sum was referred to as the Instruction Total. Using a correlation matrix, we investigated the strength of the correlations between the variables used in defining the composite variable and the Instruction Total. We found that 14 of the 19 variables correlated well with the Instruction Total. A principal component factor analysis revealed that the same 14 variables contributed to the two factors that accounted for a significant amount of variance among the variables. As a result, only the 14 variables were included in further analyses. The Instruction Total was then recalculated for each teacher. Cluster analysis led to the classification of teachers into six groups. For each group of teachers, common characteristics from the sub-variables of instruction were identified. Descriptions of each group of teachers were created by using the qualitative evidence gathered through observations, interviews, and teaching logs. These groupings of common characteristics are referred to as levels of instruction for the composite index I shown in Figure 23.4.

Figure 23.4

Levels of the Composite Index for Instruction

Level 6: Most reflective of teaching for understanding. Teachers emphasized conceptual understanding, and students' solutions, generalizations, and connections were thoroughly discussed. Teachers frequently posed questions focused on articulation of thinking, understanding mathematics, or reasonable solutions. Classroom assessment practice attended to problem solving and reasoning. Feedback that emphasized making sense of mathematics and solution strategies was provided by teachers and students.

Level 5: Reflective of teaching for understanding. Teachers emphasized conceptual understanding, and the mathematical work was shared by students and their teacher. Classroom assessment practices focused on student explanations of reasoning and procedural understanding. Feedback was consistent with Level 6.

Level 4: Attempted to teach mathematics for understanding. Teachers attempted to teach for understanding, but most lessons focused on procedural understanding. Students generally accepted and used procedures presented by their teachers, although they were encouraged to develop their own strategies. Teachers supplemented texts with additional exercises, mini-lessons, contexts, or review. Classroom assessment practice focused on student explanations regarding procedures. Teacher feedback was related to concepts and contexts, and student-student feedback consisted of answers and steps in procedures.

(Figure continued on next page)

> *Level 3: Limited attention to conceptual understanding.* Limited attention was given to teaching for understanding. Students generally used procedures presented by their teachers. Some attention was given to articulation of thinking and reasonable solutions, and teachers occasionally added a different context or review. Evidence from homework, classwork, and at times student explanations was used in classroom assessment practice. Teacher feedback was related to concepts, contexts, and procedures, and student-student feedback consisted of sharing answers.
>
> *Level 2: Focus on procedures.* Students were expected to accurately use procedures demonstrated by their teachers. Limited changes were made in response to student difficulties or misunderstandings. Evidence from homework and classwork was used in classroom assessment practice. Teacher feedback emphasized procedures, and student-student feedback was minimal.
>
> *Level 1: Underdeveloped lessons.* No formal lessons were presented. Procedures were demonstrated to individual students, and students depended on their teacher to do the mathematical work. Frequent confusion or misunderstandings were evident, but teacher feedback was inattentive to these, and student-student feedback was non-existent.

Source: *The Impact of Reform Instruction on Student Mathematics Achievement: An Example of a Summative Evaluation of a Standards-Based Curriculum*, Edition 1, by Thomas A. Romberg and Mary C. Shafer, ©2008, pp. 40–41, by Routledge. Reproduced by permission of Taylor and Francis Group.

Opportunity to Learn With Understanding

Opportunity to learn has been interpreted more broadly in this study as a student's opportunity to learn mathematics with understanding (*OTLu*) (Romberg & Shafer, 2008, pp. 49, 51). *OTLu* was described through the Independent variable *curricular content and materials* (the content and materials used in defining the enacted curriculum) and the Intervening variable *classroom events* (the interactions among teacher and students that promote learning mathematics with understanding). As conceived in this study, *OTLu* was composed of three overarching categories: *curricular content*; *modification of curricular materials*; and *teaching for understanding*. Teaching for understanding contained four subcategories: the *development of conceptual understanding*; the *nature of student conjectures about mathematical ideas*; the *nature of connections within mathematics*; and the *nature of connections between mathematics and students' life experiences*. As with the qualitative and quantitative methodologies used in defining the composite index *Instruction*, the subcategories for *OTLu* were distilled into four levels, as shown in Figure 23.5.

Treatment Fidelity

An extensive set of classroom observations, teacher logs, journal entries, and teacher questionnaires was used to triangulate teacher data and to

Figure 23.5

Levels of the Composite Index for Opportunity to Learn With Understanding

> *Level 4: High level of OTLu.* Attention was given to all content areas, content was taught in depth, and few modifications were made to curricular materials. Portions of lessons focused on conceptual understanding, student conjectures related to validity of particular statements, connections among mathematical ideas were clearly explained by the teacher, and connections between mathematics and students' life experiences were apparent in lesson.
>
> *Level 3: Moderate level of OTLu.* Although the content was taught in depth, only one or two content areas were taught during the school year. Supplementary activities were occasionally used. However, limited attention was given to conceptual understanding. Student conjectures were related to making connections between a new problem and problems previously seen. Connections among mathematical ideas were briefly mentioned, and connections between mathematics and students' life experiences were reasonably clear when explained by the teacher.
>
> *Level 2: Limited level of OTLu.* For teachers using *MiC*, few content strands were taught due to slow pacing, and for some *MiC* teachers, supplementary materials subsumed the curriculum. Teachers using conventional curricula taught vast content as disparate pieces of knowledge, laden with prescribed algorithms. Few modifications to curricular materials were made, and supplementary activities were occasionally used. All teachers provided limited attention to conceptual understanding, which was consistent with Level 3.
>
> *Level 1: Low level of OTLu.* Teachers taught vast content as disparate pieces of knowledge, laden with prescribed algorithms. *MiC* teachers added supplementary materials to the point that these materials subsumed *MiC*. Teachers using conventional curricula presented content in haphazard ways; no adherence to a textbook guideline was evident. Conceptual understanding was not promoted, and connections were not encouraged.

Source: The Impact of Reform Instruction on Student Mathematics Achievement: An Example of a Summative Evaluation of a Standards-Based Curriculum, Edition 1, by Thomas A. Romberg and Mary C. Shafer, ©2008, p. 52, by Routledge. Reproduced by permission of Taylor and Francis Group.

write descriptions of lessons that portrayed the enacted curriculum of study teachers in Districts 1 and 2 (Shafer, 2005). As shown in the *Instruction* and *OTLu* composite indices, some *MiC* teachers taught only a few *MiC* units using conventional pedagogy or included supplementary materials to the extent that the materials subsumed the *MiC* curriculum. Also, some teachers teaching conventional curricula used more reform-oriented pedagogy and engaged students in discussions and projects. The results from the *Instruction* and *OTLu* composite indices led us to distinguish three treatment groups among the 79 study teachers in Districts 1 and 2: 28 *MiC* teachers who taught the intended *MiC* curriculum, referred to as the *MiC* treatment group; 23 teachers who taught their intended conventional curriculum, referred to as the Conventional treatment group; and 28 *MiC* teachers who regularly used supplementary materials with *MiC* units or taught *MiC* with conventional pedagogy, referred to as the *MiC* (Conventional) treatment group. An additional group included three teachers in the Conventional treatment group who displayed reform characteristics. Due to very low student sample sizes for these teachers, this treatment group was not used in further analyses (Romberg & Shafer, 2008, pp. 93–94).

Variations in *treatment fidelity*, the extent to which the enacted curriculum aligns with the intended curriculum, reinforces the need to go beyond simple comparison of student achievement scores in studying the impact of a curriculum. Because variation in the enactment of curriculum occurs, describing treatment fidelity and taking into account the variation are critical (Romberg & Shafer, 2008, p. 44). In this study, the analysis of student performance accounted for this variation by examining CA student performance results in the three treatment groups. To understand differences in student performance results, we examined teacher ratings from the *Instruction* and *OTLu* composite indices and reviewed lesson descriptions that were written for teachers in the three treatment groups. These investigations allowed us to make sound conclusions about the impact of treatment fidelity on student achievement.

School Capacity

The composite variable School Capacity (SC) was specified from variables in the original research model: school context (Prior variable) and support environment (Independent variable). Newmann et al. (2000) summarized school capacity as "the collective power of the full school staff to improve student achievement for all students in the school" (p. 261). Through the work of the school's professional community, a vision of student learning, professional inquiry, and control over school policy and activities develops. School capacity is strong when all programs and initiatives are focused on the vision for student learning and professional inquiry. The principal nurtures school capacity through effective leadership that aligns school policy with the vision and sets the tone for professional development opportunities for faculty and staff. Quality technical resources, such as curricula, assessments, and equipment, contribute to obtaining the established goals.

Two major categories of school capacity were described by Newmann and Associates (1996): cultural conditions and structural conditions. In this study, we used subcategories of these conditions to examine school context and support environment from the structural research model. The four subcategories we used for cultural conditions were *shared vision for mathematics teaching and learning between principal and teachers; administrative support; school as a workplace;* and *support for innovation*; the three subcategories used for structural conditions were *collaboration among teachers*; *work structure*; and *influence of standardized testing*. An index for each of the seven subcategories was developed from principal and teacher interviews and two teacher questionnaires. These indices were distilled into one composite index for *SC*. The composite index *SC* contained five levels of teachers' perceptions of school capacity. Examples of teacher perceptions of School Capacity are provided for three levels of *School Capacity* in Figure 23.6.

WHAT WE LEARNED IN CONDUCTING THE PLANNED ANALYSIS

Over the three study years, we collected data to investigate relationships between the variables in the structural research model. The planned analysis included 17 separate studies. In the eight grade-level-by-year studies, we examined student classroom achievement (*CA*) results by grade level in one study year. For example, Figure 23.3 is a progress map for District 1 Grade 5 *MiC* teacher/student groups in 1997-1998. Cross-sectional studies included three cross-grade comparisons and three cross-year comparisons. In cross-grade comparisons, we examined student results at different grade levels by curriculum in each year in each district (e.g., in District 1, students in Grade 7 and students in Grade 8 who studied *MiC* in the third year). In cross-year comparisons, we investigated *CA* results at the same grade level by curriculum in the same district over time (e.g., in District 1 all Grade 6 students who studied *MiC* in the first year, and all Grade 6 students who studied *MiC* in the second year). Longitudinal studies involved studying the *CA* results of individual students who used a curriculum and participated in the study for two or three years (e.g., Cohort A included students who participated in the study in Grades 5, 6, and 7). Each type of study had associated assumptions; statistical testing; multiple representations of the data, such as tables, graphs, and progress maps; checks for differences between *CA* results and standardized test scores; and written results with links to instructional approaches, opportunity to learn with understanding, and school capacity. Overall, the student achievement results in all the studies imply that, regardless of other differences, when *MiC* is enacted as intended, students' achievement was higher than when *MiC* was taught with conventional pedagogy or was supplemented with other materials or when students studied conventional curricula (Romberg & Shafer, 2008, p. 123).

Considering the 17 studies, we were very ambitious. We just did not have enough resources to adequately gather data on all variables in several school districts over three years; we had to modify our plans. Initially, we planned to use structural equations to investigate the relationship of the variables in the structural model with the classroom as the unit of analysis

Figure 23.6

Levels of the Composite Index for School Capacity with Teacher Illustrations

---

Level 5: High level of school capacity. In District 2, eighth-grade MiC teacher Ms. Keeton felt that principal and teacher visions for mathematics teaching and learning were clearly defined and generally aligned. She felt that she received very strong administrative support in terms of clearly communicated expectations, support for selecting instructional materials, changes in instructional practice, and changes in policy. Ms. Keeton felt that she had a high influence in planning and teaching mathematics, an average level of influence over curriculum, and less influence over the content of professional development programs and discipline. She felt faculty and staff were committed to academic excellence and that teachers supported one another in their efforts to improve instruction. Ms. Keeton met regularly with other teachers on her team, and they discussed mathematics content, particular MiC units, instructional and assessment methods, and program evaluation.

Level 3: Average level of school capacity. In District 1, seventh-grade MiC teacher Mr. Bartlett felt that the teacher and principal visions in the school were aligned on some ideas but were incompatible on others. He felt that he received strong administrative support in terms of clearly communicated expectations, support for selecting instructional materials, and changes in instructional practice, but limited support for implementing changes in policy. Mr. Bartlett felt that he had a moderate level of influence over educational decisions regarding curriculum, discipline, the content of professional development programs, and mathematics planning and teaching, but limited influence over textbook selection. Mr. Bartlett felt that faculty and staff were committed to academic excellence, although he felt an average level of support from other teachers. Formal meetings with other mathematics teachers in the school were held infrequently, and teachers met informally only a few times to discuss mathematics curriculum, instruction, and assessment.

Level 1: Very low level of school capacity. In District 1, eighth-grade MiC teacher Ms. Reichers transferred from one middle school to another during the third year of the study. She had looked forward to working with the other 8th-grade MiC teacher at her school, but this collaboration did not take place. Ms. Reichers felt that the principal and teacher visions in the school were aligned on some ideas but were incompatible on others. She felt that she received weak administrative support in terms of clearly communicated expectations, support for selecting instructional materials, changes in instructional practice, and changes in policy. Ms. Reichers felt that she had an average level of influence in planning and teaching mathematics and limited influence over educational policies related to curriculum, content of professional development programs, and discipline. She felt that faculty and staff were not very committed to academic excellence, and that teachers only somewhat supported one another in their efforts to improve instruction.

Source: The Impact of Reform Instruction on Student Mathematics Achievement: An Example of a Summative Evaluation of a Standards-Based Curriculum, Edition 1, by Thomas A. Romberg and Mary C. Shafer, ©2008, pp. 45–46, by Routledge. Reproduced by permission of Taylor and Francis Group.

for the eight grade-level-by-year studies in classes using *MiC* in all four districts. Instead, because we had complete sets of data for only two districts, we chose to make descriptive comparisons using the *CA* scale and its content subscales as dependent variables. Considerable variation in student performance was apparent in every comparison we examined (Romberg & Shafer, 2008, p. 160).

To look for differences between classrooms in which *MiC* and conventional curricula were enacted in the same districts, a quasi-experimental study was planned. We learned that quasi-experimental studies are unsatisfactory because there is no way to control all the sources of variation in classrooms. As illustrated in this study, no matter how we examined the experimental units, they were not equivalent. This finding led us to identify three treatment groups by using the composite indices *Instruction* and *OTLu*. We believe that using a structural model and quantifying variables in the model is the best way to conduct studies of the enactment of curriculum in schools (Romberg & Shafer, 2008). The analyses of the grade-level-by-year and cross-sectional studies suggest that the prior achievement of students and teachers' enactment of *MiC* influenced *CA* among students at various grade levels and in different districts. When the enacted *MiC* aligned with the intended *MiC*, student achievement was significantly greater than when instruction and opportunity to learn with understanding were less aligned with the instructional approach of *MiC*. In longitudinal analyses, students were the units of analysis, and there was a small sample size. Yet the results from multiple student cohorts suggested that *MiC* had a considerable impact on *CA* over two or three years. The results also suggest that student achievement in the *MiC* Conventional group was similar to the achievement of students who used conventional curricula. The results imply that if teachers chose to use a reform curriculum like *MiC*, they should enact it as intended. Regardless of other differences, if *MiC* is enacted as intended, students' performance will be better than with a partial enactment of *MiC* and better than studying a conventional curriculum.[6]

CONCLUSION

The challenges we faced in conducting comparative longitudinal research in schools seemed overwhelming at times. Controlling potential sources of variation is difficult in classroom settings. However, as this study demonstrates, impact studies of high quality can be conducted in classrooms when data collection and analysis are designed to take into consideration the variations encountered in these settings (Romberg & Shafer, 2008, p. 164). These variations draw attention to the need to study the effects of the classroom and school settings in which student learning is situated, rather than relying solely on achievement data, when analyzing the impact of a standards-based curriculum.

The use of a structural research model, together with the development of composite indices based on the model, is an effective way to plan and conduct curricular research in school settings. This process led to our understanding of instruction that transpired in classrooms, students' opportunity to learn mathematics content in depth and with understanding, and the capacity of study schools to support mathematics teaching and learning. In this research, the *Instruction* and *OTLu* composite indices led to the description of three treatment groups that differed in the ways teachers enacted their mathematics curricula. Examining student performance results overall and by treatment groups helped us to understand the impact of each treatment. After controlling for the influence of students' prior achievement that favored the *MiC* treatment, most important to the impact of *MiC* is to enact the curriculum as it was intended. Regardless of other variables associated with instruction, the results imply that when *MiC* is enacted as intended, student achievement will be higher, especially after two or three years, than when students experience *MiC* with conventional pedagogy and supplemental materials or when they study a conventional curriculum.

ACKNOWLEDGMENT

The research reported in this article was supported by U.S. National Science Foundation grants REC-9553889 and REC-0087511 to the Wisconsin Center for Education Research, School of Education, University of Wisconsin–Madison, and to Northern Illinois University. This chapter draws on the monographs and technical reports from the research (http://micimpact.wceruw.org/) and the book that followed (Romberg & Shafer, 2008). Any opinions, findings, or conclusions are those of the author and do not necessarily reflect the views of the National Science Foundation, the University of Wisconsin–Madison, or Northern Illinois University.

REFERENCES

Adams, R. J., & Wu, M. L. (Eds.). (2002). *PISA 2000 technical report*. Organisation for Economic Co-Operation and Development Publications.

Collis, K., & Romberg, T. (1992). *Mathematical problem-solving profiles*. Australian Council for Educational Research.

Dekker, T., Querelle, N., van Reeuwijk, M., Wijers, M., Fejis, E., de Lange, J., Shafer, M. C., Davis, J., Wagner, L., & Webb, D. (1997–1998). *Problem solving assessment system*. University of Wisconsin–Madison.

Gravemeijer, K. (1994). Educational development and developmental research in mathematics education. *Journal for Research in Mathematics Education, 25*(5), 443–471.

National Center for Research in Mathematical Sciences Education & Freudenthal Institute (Eds.). (1997–98). *Mathematics in Context*. Encyclopaedia Britannica.

National Council of Teachers of Mathematics. (1989). *Curriculum and evaluation standards for school mathematics*.

Newmann, F. M., & Associates. (1996). *Authentic achievement: Restructuring schools for intellectual quality*. Jossey-Bass.

Newmann, F. M., King, M. B., & Youngs, P. (2000). Professional development that addresses school capacity: Lessons from urban elementary schools. *American Journal of Education, 108*(4), 259–299. https://doi.org/10.1086/444249

Romberg, T. A. (2001). *Designing middle-school mathematics materials using problems set in context to help students progress from informal to formal mathematical reasoning*. University of Wisconsin-Madison.

Romberg, T. A., & Shafer, M. C. (2008). *The impact of reform instruction on student mathematics achievement: An example of a summative evaluation of a standards-based curriculum*. Routledge, Taylor and Francis Group.

Shafer, M. C. (2004). Conduct of the study. In T. A. Romberg & M. C. Shafer (Eds.), *Purpose, plans, goals, and conduct of the study*. (Longitudinal/cross-sectional study of the impact of teaching mathematics using *Mathematics in Context* on student achievement: Monograph 1). University of Wisconsin–Madison.

Shafer, M. C. (2005). *Instruction, opportunity to learn with understanding, and school capacity*. (Longitudinal/cross-sectional study of the impact of teaching mathematics using *Mathematics in Context* on student achievement: Monograph 3). University of Wisconsin–Madison.

Shafer, M. C., Davis, J., & Wagner, L. R. (1997). *Principal interview: School context*. (*Mathematics in Context* longitudinal/cross-sectional study: Working Paper No. 12). University of Wisconsin–Madison.

Shafer, M. C., Davis, J., & Wagner, L. R. (1998). *Teacher interview: Instructional planning and classroom interaction*. (*Mathematics in Context* longitudinal/cross-sectional study: Working Paper No. 3). University of Wisconsin–Madison.

Shafer, M. C., Wagner, L. R., & Davis, J. (1997a). *Classroom observation scale*. (*Mathematics in Context* longitudinal/cross-sectional study: Working Paper No. 6). University of Wisconsin–Madison.

Shafer, M. C., Wagner, L. R., & Davis, J. (1997b). *Student attitude inventory*. (*Mathematics in Context* longitudinal/cross-sectional study: Working Paper No. 7). University of Wisconsin–Madison.

Turner, R., & O'Connor, G. (2005). The development of a single scale for mapping progress in mathematical competence. In T. A. Romberg, D. C. Webb, M. C. Shafer, & L. Folgert (Eds.), *Measures of student performance* (Longitudinal/cross-sectional study of the impact of teaching mathematics using *Mathematics in Context* on student achievement: Monograph 4) (pp. 27–66). University of Wisconsin–Madison.

Turner, R., O'Connor, G., & Romberg, T. A. (2004). *Scale for mapping progress in mathematical competence.* (*Mathematics in Context* longitudinal/cross-sectional study: Working Paper No. 49). University of Wisconsin–Madison.

Webb, D. C., Romberg, T. A., Shafer, M. C., & Wagner, L. (2005). Classroom achievement. In T. A. Romberg, D. C. Webb, M. C. Shafer, & L. Folgert (Eds.), *Measures of student performance* (Longitudinal/cross-sectional study of the impact of teaching mathematics using *Mathematics in Context* on student achievement: Monograph 4) (pp. 7–26). University of Wisconsin–Madison.

ENDNOTES

1. Middle schools in the U.S. typically contain Grades 6–8, with students ranging in age from 11 to 14. Elementary schools in the U.S. typically contain Grades K–5, with students ranging in age from 5 to 11.
2. Teacher and observer names are pseudonyms.
3. The NAEP Questions Tool is located online at https://www.nationsreportcard.gov/nqt/searchquestions. TIMSS: IEA's Third International Mathematics and Science Study—Mathematics Items Released Set for Population 2 (Seventh and Eighth Grades) is located online at https://timssandpirls.bc.edu/timss1995i/TIMSSPDF/BMItems.pdf.
4. For examples of assessment items that illustrate each band of the progress map developed for this study, see Turner et al. (2004).
5. In the progress map for each teacher, the black area indicates the 95% confidence interval. The white line within the 95% confidence interval represents the mean score for students in the teacher's class. The results of 50% of the students are shown in the light grey area, and the results of 90% of the students are shown in each map. When the confidence intervals do not align in any way, the means are considered to be statistically different.
6. Summary from *The Impact of Reform Instruction on Student Mathematics Achievement: An Example of a Summative Evaluation of a Standards-Based Curriculum,* Edition 1, by Thomas A. Romberg and Mary C. Shafer, ©2008, p. 160, by Routledge. Reproduced by permission of Taylor and Francis Group.

CHAPTER 24

LESSONS LEARNED FROM A MULTI-SITE STUDY OF AN ALGEBRA SUPPORT CURRICULUM

Deborah Spencer, June Mark, Mary Beth Piecham, Julie K. Zeringue, Kelsey Klein
Education Development Center (EDC)

Laura M. O'Dwyer
Boston College

In this chapter, we describe challenges that emerged in conducting research in the context of year-long algebra support courses for ninth graders underprepared for Algebra 1. Our research, funded by the U.S. National Science Foundation, is examining the implementation and impact of Transition to Algebra (TTA), *an innovative curriculum designed for algebra support courses. Our study uses a quasi-experimental design with propensity score analyses to examine the impact of* TTA *on students' algebra achievement and attitudes towards mathematics as well as patterns of implementation of* TTA *in treatment schools. We discuss three main challenges: (1) accounting for variation in how algebra support courses are designed and organized in study sites; (2) managing a study in which contextual factors contributed to program instability; and (3) identifying, adapting, and designing instruments appropriate and sensitive enough to measure student learning outcomes. Strategies for addressing these challenges and implications for research on curriculum interventions are discussed.*

Keywords: algebra; curriculum implementation; high school; mathematics education; methodology

In the United States, many students enter high school underprepared for Algebra 1, a foundational course for advanced mathematics and college and workforce readiness, and pivotal for staying on track to high school graduation (Achieve, 2008; National Mathematics Advisory Panel, 2008; Silver et al., 2008). Curriculum directors of school districts in the nation's largest cities have reported that more than half of students enter ninth grade a grade level or more behind in mathematics (Council of the Great City Schools, 2009). Over the past decade, national assessment results show mathematics scores for the lowest performing eighth-grade students (i.e., students at the 50th percentile and below) declined, even prior to accounting for additional learning loss associated with the pandemic (Stephens et al., 2022; U.S. Department of Education, 2022a, 2022b). Failing to pass Algebra 1 is the primary reason students cite for dropping out of high school (Schachter, 2013); in some districts, more students fail Algebra 1 more often than any other course (e.g., Ham & Walker, 1999; Helfand, 2006). Although these challenges must be addressed comprehensively throughout pre-K–12 mathematics education, there is a pressing and urgent need to help underprepared ninth-grade students succeed in Algebra 1, as the stakes are high.

High schools across the country utilize extended learning time as a strategy to support underprepared students, enrolling them in two mathematics courses in ninth grade—a standard Algebra 1 course, and a supplemental algebra support course that often takes the place of an elective in students' schedules. The use of an additional period for algebra support is widespread, but there is no standard scope and sequence for such a course. Moreover, few curriculum materials exist, and teachers of these additional periods are, in many schools, given little in the way of guidance, materials, or professional development. As a result, what schools offer during this additional mathematics period varies significantly in approach, quality, and coherence. Our prior study of district algebra supports indicates that materials used in such courses are assembled primarily by teachers themselves and often found through digital sources (Mark et al., 2012).

Transition to Algebra (*TTA*) is a coherent year-long curriculum designed for use in algebra support courses. *TTA*, developed at the Education Development Center (EDC) with funding from the U.S. National Science Foundation (NSF), focuses on building algebraic habits of mind, ways of thinking that mathematicians use in their work (Cuoco et al., 1996, 2010) and that are critical in supporting the transition from arithmetic to algebra. The curriculum's underlying design principle is that students underprepared for Algebra 1 can benefit from very specific help in building the logic of algebra by connecting arithmetic patterns and algebraic structure. As a group of researchers at EDC and Boston College, we (the authors) have been conducting a large-scale, multi-site study to examine the impact and implementation of *TTA* in high schools around the country. We have encountered several challenges in conducting this research on a supplemental mathematics course that is outside of the typical mathematics course sequence, and that is targeted for students who may not be at grade level. To our knowledge, there is little in the literature that addresses challenges of studying curriculum interventions in this context, and we hope that frank discussion of the challenges we have faced can serve future research in efforts to build knowledge in this area.

WHAT IS KNOWN ABOUT ALGEBRA SUPPORT COURSES

The research base on algebra support courses and guidance for districts and schools regarding how to implement them effectively is limited (Nomi & Allensworth, 2009; Wilson & Kelly, 2018). The most extensive studies on the use of additional mathematics instructional time have been done in the Chicago Public Schools (CPS), which in 2003 began requiring all entering ninth graders[1] scoring below national norms on an eighth-grade assessment to take two periods of algebra (Nomi, 2012; Nomi & Allensworth, 2009, 2013). In addition to the structures of extended learning time in Algebra 1, the CPS policy provided teachers with curricular resources, lesson plans, and professional development. Studies examining the effects of double-period algebra found the policy led to substantial improvements in students' algebra test scores. However, Algebra 1 failure rates were unchanged for students in the double-period courses and increased for higher-skilled, non-double-period algebra students. Researchers suggest the policy-induced changes in the composition of Algebra 1 classes contributed to changes in classroom peer ability levels, which can influence the rigor of instruction, the classroom learning climate, and teacher expectations, and consequently affect passing rates (Nomi, 2012; Nomi & Allensworth, 2009, 2013). Reasons for improved test scores among double-period algebra students included the strong curriculum and instructional supports provided to algebra support teachers, and increased use of interactive pedagogical practices (Nomi & Allensworth, 2013), leading researchers to conclude strategies such as double-period algebra are more successful when "deep supports for teaching" are included (Durwood et al., 2010, p. 7).

One study examining the long-term effects of double-period algebra in CPS found students who received two periods of algebra instruction were more likely than non-double period algebra students to earn more high school credits overall, graduate from high school, and enroll in college, with the largest impact for students with below-average reading skills (Cortes et al., 2015). Researchers cited the intervention's focus on building verbal and analytical skills in algebra contexts as reasons for the larger impact. In a similar study, the lowest performing students benefitted less from double-period algebra, suggesting a more intensive intervention may have been needed in addition to the double-period intervention (Cortes & Goodman, 2014). These findings underscore the importance of well-designed interventions that are targeted to students' different skill levels. Although informative, a limitation of the CPS double-period algebra policy studies is the lack of insight on the curriculum programs that were used in algebra support courses, specifically how consistently the programs were used across schools, the content that was covered in supplemental algebra courses, or how instruction was organized in the double-period courses (Nomi & Allensworth, 2009).

An earlier study in this area examined the effects of *Transitions to Advanced Mathematics*, a catch-up course for underprepared ninth-grade students that focuses on the development of foundational skills emphasized in middle school and more advanced reasoning skills (Balfanz et al., 2002). This course, which originated as part of a comprehensive school reform model, differed from a typical double-period algebra course structure because the extended learning time in Algebra 1 was offered in block semesters, with students taking the equivalent of a full-year catch-up course in the first semester followed by a full-year Algebra 1 course in the second semester (Balfanz et al., 2004). An experimental study found *Transition to Advanced Mathematics* students showed higher gains on an intermediate skills assessment at the end of the first semester, and their overall performance in Algebra 1 was significantly higher in graphing linear equations than, and was equal in other content areas to, the comparison group who received algebra support and Algebra 1 instruction for the full year. There was also evidence that students' attitudes towards mathematics in the treatment group improved as a result of teachers' improved instructional practices. Unfortunately, for both groups, the additional instructional time in Algebra 1 did not lead to significantly improved achievement scores at the end of the year, with the majority of students performing below the 50th national percentile on a standardized Algebra 1 assessment (Balfanz et al., 2012; Sweet, 2010).

The findings from these studies show promise for additional instructional time designed to help underprepared students develop the knowledge and skills they need to be successful in Algebra 1, and suggest that both a strong curriculum and strong instructional supports for teachers are critical. This is consistent with what is widely understood about mathematics curriculum-based interventions. A synthesis of research on STEM^2 interventions found that programs that included both new curriculum material and professional development to support teachers' implementation and pedagogical content knowledge had larger impacts compared to programs with only one of these components (Lynch et al., 2019).

The existing research also points to some unanswered questions. Given the mixed results on student learning outcomes, there remains a strong need for evidence-based insights on how to improve outcomes for underprepared students on a broad scale. Despite an emerging consensus that extra time must be paired with strong curricula and teacher supports to have positive effects, there is no consensus among researchers or practitioners about which curricular approaches may be effective. The need to evaluate the effectiveness of interventions designed for use in these contexts is critically important not only to the field, but strong implementation research is also critical for schools to know what is working in terms of the curricular content of those courses, and for whom. This is especially important for district administrators who are seeking evidence-based interventions to address student learning needs effectively.

TRANSITION TO ALGEBRA CURRICULUM

The typical emphasis of algebra support courses is on remediation, review, and test preparation (Mark et al., 2012). Materials used in such courses are often teacher developed or pulled from online sources (Mark et al., 2012), which can limit careful and coherent development of foundational algebraic ideas. In this section, we describe the algebra support curriculum developed at EDC and a study initiated to determine the impact of the curriculum on achievement and attitudes of ninth-grade students underprepared for Algebra 1.

Description of the Curriculum

Transition to Algebra (*TTA*) (Mark et al., 2014) represents a novel approach. It is one of few curricula designed for use in full-year algebra support courses; *TTA* can be used in a support course taught concurrently alongside any Algebra 1 curriculum, providing flexibility in implementation. It employs several distinctive features: daily *Mental Mathematics* activities develop fluency with algebraic properties of operations; *Puzzles* build logic and increase perseverance; *Explorations* provide opportunities for sustained reasoning; and *Student Dialogues* model mathematical discourse and develop academic language. The *TTA* materials include *Teaching Guides* and *Student Worktexts* that cover standard algebra topics through three pervasive mathematical themes:

- *Expecting mathematics to make sense:* extending the logic of arithmetic so operations behave as expected; deriving algebraic ideas and principles from arithmetic knowledge.
- *Using intuitive visual models*, e.g., using number lines to represent distance and location; using area models to organize multiplying, dividing, and factoring.
- *Building broad-use algebraic habits of mind and problem-solving stamina*, e.g., using abstraction from experience, logic, and structure; strategies to understand problems.

For example, in *Unit 1: Language of Algebra*, students see algebra as a language for expressing patterns and relationships. They build intuition and develop algebraic language through "number tricks" as they move from pictures to words, and finally to algebra notation (see Figure 24.1) to track the transformations of an unknown starting number.

Figure 24.1

A Number Trick Using Pictures, Words, and Algebraic Notation

Words	Maria	Pictures	Asher	Description of Pictures	Abbreviation
Think of a number.	7		3	bucket	b
Add 5.	12		8	bucket and 5 apples	$b + 5$
Multiply by 2.	24		16	2 buckets and 10 apples	$2b + 10$
Subtract 2.	22		14	2 buckets and 8 apples	$2b + 8$
Divide by 2.	11		7	bucket and 4 apples	$b + 4$
Subtract your original number.	4		4	4 apples	4

Source: *Transition to Algebra*, Unit 1. (Reprinted with permission.)

In *Unit 5: Logic of Algebra*, students formalize the commonsense logic they developed earlier to make sense of the rules for solving equations and systems of equations by performing only operations that preserve balance and substituting equivalent values as needed. Students use mobile puzzles (see Figure 24.2) to learn about equivalence and to visualize the common sense behind algebraic solving steps, helping them transition from concrete to abstract representations.

In other units, the number line is used as a tool for reasoning about integers, the relationships between integers including order and distance, and the operations of addition and subtraction, first with numbers and then generalized to variables. By "zooming in" on the number line, students experience that decimals and fractions on the number line continue to follow the same structures and logic as integers.

Figure 24.2

Mobile Puzzles

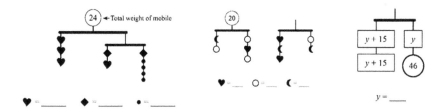

Source: Transition to Algebra, Unit 5. (Reprinted with permission.)

Students also use area models (see Figure 24.3) to build a common-sense foundation for multiplying algebraic expressions by examining multiplication in the context of area. They multiply integers and numerical expressions and then extend their understanding to multiplying algebraic expressions and identifying equivalent expressions. Area models are also used for factoring, which is seen as a form of "un-multiplying."

Figure 24.3

Area Models

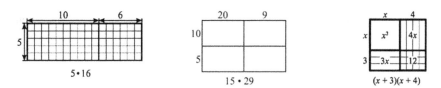

Source: Transition to Algebra. (Reprinted with permission.)

In development, *TTA* underwent extensive classroom testing and revision based on feedback from teachers, students, and mathematics leaders across the country. A 2011–2012 field test in 11 classrooms with 165 students at two high schools in Massachusetts investigated *TTA*'s impact on student competence in algebra. Findings provided evidence of score gains between the pretest and final *TTA* assessments; average gains were statistically significant overall and within each content area. However, the study was small and did not use an experimental design or comparison group. A quasi-experimental study was conducted in the same year using a regression discontinuity design with a sample of 183 students to gauge the impact of *TTA* on students' performance on a modified version of the

Massachusetts state mathematics assessment.[3] Results were inconclusive about the effects of *TTA* on student competence in algebra, due to a small sample size and limited variation in students' assignment scores. Methodological lessons about the limitations of regression discontinuity designs in educational impact evaluations were reported in the *American Journal of Evaluation* (Louie et al., 2016). Our results suggested that requirements of the design—including the large sample size needed and the necessity of maintaining adherence to a strict cutoff—were not a good fit for studying the efficacy of *TTA*.

The *Supporting Success in Algebra* Study

The *Supporting Success in Algebra* study built on prior work and aimed to determine the impact of *TTA* on students' algebra achievement and attitudes toward mathematics compared with outcomes for similar ninth graders receiving business-as-usual instruction in algebra support courses. EDC and Boston College collaborated on this effort, with EDC managing the project, leading recruitment, and supporting teachers and mathematics leaders in implementation, and Boston College leading instrument development, data collection, and analysis. The study was funded by the NSF in 2016.

One challenge we encountered was determining a research design that would yield useful information about the impact of *TTA* on students' learning. We knew that a regression discontinuity design would not work, even with a dramatic increase in sample size, because it was unlikely that schools would be willing to rely solely on a test score with a strict cutoff point to determine enrollment in the algebra support course (nor could we, in good conscience, recommend that they do so). In addition, the regression discontinuity design would compare students above the cutoff (who received both the additional instructional time in the algebra support course and *TTA*) to students below the cutoff (who received neither), which would not isolate the effects of *TTA* from the effects of the additional instructional time in mathematics.

We considered whether a randomized control trial (RCT) with delayed *TTA* implementation for the control group would be possible. However, RCTs are particularly difficult to carry out in curriculum effectiveness studies (Ginsburg & Smith, 2016). So, we sought feedback from administrators in several large districts across the country about the feasibility of the design and consulted with other researchers conducting similar studies. We found that administrators would be reluctant to participate in a study that asked them to comply with random assignment, even with a delayed treatment model. In many districts, individual high schools have significant autonomy in determining which curricula are used, and high-school teachers and administrators feel an urgency to implement interventions that address underprepared ninth-grade students. Randomizing would mean delaying treatment in circumstances where mathematics leaders were anxious to try new approaches with students for whom support was urgently needed. We ultimately decided on a quasi-experimental design, with propensity score analysis to reduce selection bias threats, and additional data collection to identify patterns of implementation and teacher adaptation of *TTA* through interviews and observations in a subset of treatment schools.

Participation in the study was limited to schools that offered algebra support courses in 2018–2019 and provided the equivalent instructional time of a full year-long course. Treatment schools (47 schools, with 95 participating teachers) committed to using *TTA* in 2018–2019; Houghton-Mifflin Harcourt donated all the *TTA* student and teacher materials, and teachers received in-person and online professional development. The following year, post-data collection, comparison schools (33 schools, 59 teachers) received the *TTA* teacher materials as well as professional development.

In designing this study, we recognized that our work with treatment and comparison group teachers and leaders across the country would give us the opportunity to learn more about the structure and design considerations of algebra support courses in a large number of high schools in varied contexts, outside the specifics of using the *TTA* materials. This area is under-explored in the field and relevant to understanding the context for curriculum implementation. Using teacher surveys, teacher and administrator interviews, and site visits, we explored how districts designed and structured algebra support courses, and the challenges of supporting the course and sustaining program improvement over time.

Overall, the following research questions guided the study.

1. What is the impact of *TTA* on the algebra achievement and attitudes toward mathematics of ninth-grade students underprepared for Algebra 1 compared with similar students receiving business-as-usual instruction?
2. In what ways do teachers use and adapt *TTA* for use in their classrooms? Do differences in curriculum use relate to student outcomes?
3. How do districts vary in their algebra support course models, including teacher supports?
4. What are the characteristics of the student population enrolled in algebra support courses?[4]

CHALLENGES IN STUDYING THE IMPACT OF *TRANSITION TO ALGEBRA* ON ALGEBRA ACHIEVEMENT AND ATTITUDES

The challenges we describe in the remainder of this chapter largely pertain to answering our first research question, addressing the impact of *TTA* on student outcomes. We share these challenges—namely the variation in how algebra support courses were designed and organized, the contextual factors that contributed to program stability or instability, and the instruments used to measure student learning—and highlight implications for future research.

Variation in Design and Organization of Algebra Support Courses

We encountered significant variability in how schools organized the algebra support course, in teacher staffing, student placement, and course credits offered, that had the potential to affect how curriculum was used in that setting. Given these differences in the context for algebra support, we developed profiles for each of our 31 participating districts, drawing on interviews conducted with district and school mathematics leaders and teachers, teacher surveys, and field notes compiled during site recruitment, data collection, and professional development. These profiles helped us to describe the variation across our sites and understand the contextual factors affecting implementation.

A key variation that affected the use of instructional materials was the differences in course goals across districts. Algebra support courses are different than "core" mathematics courses, such as Algebra 1, in that they typically do not have an official curriculum conveyed through content standards or even a district-prescribed scope and sequence that serves as the designated curriculum (Remillard & Heck, 2014). There are usually no district-adopted instructional materials for the course, and therefore the materials used in the course have not typically served to provide a default scope and sequence as textbooks often do in core courses.

Across our study sites, there was variation in how administrators and teachers perceived the purpose of the course, and those differences played out in teachers' use of instructional materials. In some sites, the algebra support course was seen as intended primarily to help students understand the content and skills addressed in the Algebra 1 course, with a greater focus on just-in-time support and instruction to help students access and succeed with the Algebra 1 coursework. As one mathematics leader described, "It's almost like an extra period for them to really solidify the content from Algebra I and then that way they're more successful in Algebra I and passing rates are higher."

In contrast, teachers and leaders in other districts believed that the course should primarily address foundational algebraic concepts and skills, mathematical content, and practices that students may not have had an opportunity to learn in earlier grades. They felt that the role of the course was to ensure that students understood core concepts and developed positive attitudes that enabled them to persist and succeed in Algebra 1. As described by one teacher:

> For me the goal is [to] actually have these students see that they are capable of doing the work in Algebra 1 ... so they can start to see ... the same concept presented in different ways. And I think that really helps to reinforce in their mind what that concept actually means.

In some sites, teachers using *TTA* were trying to balance both goals: building conceptual understanding of critical algebraic ideas; and directly supporting the grade-level content covered in the Algebra 1 course. Managing these potentially competing goals affected teachers' use of *TTA*. If the course was intended to directly support Algebra 1, the primary instructional materials for the algebra support course were often the Algebra 1 textbooks; because the course operated as an extension of Algebra 1, the scope and sequence of the Algebra 1 course drove the use of the *TTA* units. For example, one teacher described using *TTA* as a supplemental resource, so when the Algebra 1 class was in the chapter on slope and linear equations, the teacher went to the *TTA* unit to find additional materials on distance and slope so that algebra support students would have a better understanding of what slope means and how to calculate it. In this case, *TTA* was used to fill gaps in the Algebra 1 materials, and to support stronger student understanding and access to key algebraic ideas. In another district, competing demands on the algebra support block, such as test preparation and social emotional goals, led to the use of *TTA* for part of the instructional time, with teachers being asked to use other resources for the rest of the period.

In several districts, mathematics leaders worked with teachers to manage these multiple goals, carving out instructional time within a school week that balanced use of *TTA* with direct support for Algebra 1. One district suggested a 60/40 split of instructional time; another, 80/20. Several district mathematics leaders in our study worked with algebra support teachers to create a common pacing guide that outlined a district specific *TTA* scope and sequence that aligned to the Algebra 1 scope and sequence, using some units out of order and skipping other units. This approach created some unintended effects. For example, in one site, the district pacing guide dictated the use of Unit 10 (factoring polynomials using area models) but not Unit 4 (the unit in which area models are first introduced in *TTA*), so students ran into confusion; eventually the district adjusted their pacing guide. Because we often see this sort of adaptation with instructional materials use, we had expected that modifications would be made by teachers and administrators to meet their own needs, and we provided support for these modifications, working closely with teachers and administrators. However, we found that the modifications were greater than we had anticipated. Such rearrangement of units within the curriculum significantly affected the mathematical storyline and coherence intended in *TTA* and highlighted challenges in studying the use and impact of a carefully developed curriculum.

These adaptations created a challenge in measuring teacher use of *TTA*. Using our district profiles and mixed methods analysis with surveys on use and interviews, we did additional work to understand how the *TTA* materials were being used, whether *TTA* or the Algebra 1 textbook was the primary curriculum, and which units were being used in each site and in what sequence. We found a mix of teacher use patterns, including closely and regularly using *TTA*, selective use of the *TTA* units and lessons, and intense use (where *TTA* was used exclusively, but for shorter periods of time in the school year). Varying patterns in the use of lesson components were also evident. This investigation helped us determine how to adapt our measures of *TTA* use.

Our original intent had been to use an approach to measuring the extent of use of *TTA* by determining if teachers had enacted the material in each unit (Heck et al., 2012). Our unit surveys asked teachers to indicate if they taught each lesson using primarily *TTA* (*PTTA*), taught primarily from *TTA* with some supplementation (*STTA*), taught primarily from alternatives to *TTA* (*PALT*), or if the content was not taught. Prior studies (Tarr et al., 2012; Tarr et al., 2013) have used a calculation of the extent of textbook implementation (*ETI*) that weights those lessons that have been supplemented or taught primarily from another source. Following that model, the calculation of *ETI* for our study would have been:

$$ETI = \frac{PTTA + \frac{2}{3}STTA + \frac{1}{3}PALT}{\text{total number of lessons in } TTA}.$$

Our investigation into patterns of use also found that many treatment teachers reported "supplementing" in ways that they believed enhanced the implementation of *TTA*—adding exit tickets; adding an opening focus question to pique student interest; using a teacher-made game; or making direct connections to the students' Algebra 1 classes. This data made us consider whether it made sense, in the algebra support class context, to weight those lessons that were taught with some supplementation, as it was unclear that supplementation detracted from *TTA* use. Conversely, our investigation into patterns of use confirmed a clear distinction between teaching lessons primarily from *TTA*, and teaching lessons primarily from another source and using *TTA* as the supplement (for example, pulling only practice problems or puzzles to enhance Algebra 1 textbook lessons). Retaining the weighting for lessons taught primarily from other sources made sense. Our resulting modified *ETI* (denoted *mETI*) calculation was as follows:

$$mETI = \frac{\text{PTTA} + \text{STTA} + \frac{1}{3}\text{PALT}}{\text{total number of lessons in } TTA}.$$

We also added an additional measure to determine the extent teachers were using the intended elements within lessons and units. To identify elements that would make up our *TTA*-specific "component use" measure, we obtained input from *TTA* developers and reviewed the *TTA* materials, including recommendations shared with teachers in professional development. Some critical elements, such as the *Important Stuff* (activities addressing the core ideas) and *Mental Mathematics*, were present in every lesson. Other elements were present at the unit level, such as *Explorations* (longer, project-like activities) and *Unit Assessments*. Survey questions asked teachers to indicate the frequency in which they used these key lesson and unit components, on average, at three points in the school year. Additional data sources included teacher interviews, open-response questions from teacher surveys, and contextual information about the model for algebra support from district profiles. We developed a rubric associated with high, medium-high, medium-low, or low levels of component use. Component use was assessed independently by two researchers using the rubric. All conflicting ratings were resolved through discussion.

We found that *mETI* and component use operated independently. For example, one teacher with a high *mETI* (indicating use of a significant portion of the *TTA* units and lessons) exhibited a medium-low level of component use, using *Important Stuff* and *Additional Practice* problems but rarely other components. In contrast, a teacher who had a low *mETI* score reported using only a few *TTA* units for a portion of the year, but exhibited a high level of component use, frequently using key components when teaching a *TTA* unit, including *Important Stuff*, *Mental Mathematics*, and *Dialogues*. Both *mETI* and component use proved useful measures, targeting different aspects of *TTA* use and allowing us to capture variation more fully.

In summary, we learned several important lessons related to organization of algebra support courses:

- *Developing profiles of school districts and their use of TTA was essential to understanding the instructional goals for algebra support courses and how those goals influenced implementation, and ultimately student achievement.*
- *Using both quantitative and qualitative data sources in a mixed methods approach strengthened our understanding of the implementation context.*
- *Researchers need to determine measures to compare the level of implementation of a support curriculum aligned with both the developers' intentions and with an understanding of the implementation context, in order for such measures to effectively capture differences in use.*

Program Instability Due to Limited Funding and Competing Priorities

Algebra support courses are more susceptible to year-to-year budget fluctuations and changes in district or school priorities than core mathematics courses. Many schools in our study faced uncertainty about the number of sections of the algebra support course that could be offered in a given year, given budget constraints, and therefore delayed teacher assignments to the support course until close to the start of school. This late assignment made it difficult for teachers to plan for the coming school year, or to participate in professional development prior to teaching the course. In some schools, the uncertainty around budget issues led to other decisions about teacher staffing for the algebra support courses—for example, assigning all of the algebra support classes to one non-tenured teacher (rather than distributing them across several teachers in the mathematics department) so that course assignments could be adjusted more readily if funding was not available. In one participating district, half of the teachers originally assigned to teach the support courses were given pink slips[5] two weeks before the start of school (and one week before *TTA* professional development). Overall, this phenomenon had an impact on our ability to recruit schools to participate in the study and to ensure that treatment teachers could participate in professional development and other supports offered for *TTA* implementation.

Due to budget restrictions and staffing limitations, it was also common for schools in our study to have more students who were eligible for the algebra support course than they had seats available. Some schools selected students randomly from those eligible for the course to determine enrollment; in others, course placement was determined by counselors and teachers who considered which students were most likely to benefit from the course. In some cases, course schedules proved a deciding factor; if limited sections of the course were offered, and students had other complicated scheduling requirements, they might not be able to enroll. In some schools, students whom teachers felt strongly could benefit from the course, but were not enrolled in it at the beginning of the year, were then assigned to the course later in the year when there were open seats in a class.

Competing school initiatives also contributed to student mobility in and out of the course. The pressure to support students to pass Algebra 1 and state tests can create a sense of urgency in a school that our teachers and administrators often noted in interviews. It was common for schools to take a "kitchen sink approach," participating in multiple initiatives at one time—individual tutoring, online skills practice sessions, or programming to support socio-emotional learning. In addition, students identified for algebra support were often supported with other services (including supports for English language learners, learning difficulties, or other content areas such as English Language Arts) and enrollment in the algebra support course could be affected by conflicting scheduling. These factors contributed to fluctuations in course enrollment that were more common than is typical for core mathematics courses such as Algebra 1.

Of course, instability in enrollment affects teachers' ability to implement instructional activities and support students' progress over time. For the purposes of our study, it also created some havoc in data collection, both for tracking primary data (e.g., algebra assessments and assessments of students' attitudes towards mathematics) and matching primary data to secondary data. It proved helpful to include course enrollment data in our secondary data requests; dates of course entrance and exit helped us determine which students had received a sufficient dose of *TTA* to be included in analyses. Term and semester grades also proved helpful. Although we did not need interim grades for measuring student outcomes, they helped by providing triangulation on the question of whether a student was enrolled in the course in a given time period, especially when course enrollment dates were unreliable. Because enrollment fluctuated most in the fall semester, it was also helpful that we switched our administration of student assessments from a pretest design to one using students' eighth-grade state test scores as the baseline measure—thus avoiding the pitfall of missing pretest data for students who enrolled after the school year began.

Establishing program stability for an algebra support course can be quite challenging but is a critical pre-condition for supporting program improvement over time. We were fortunate to have some district mathematics leaders participating in our study who were particularly adept at creating this kind of stability—maintaining funding and staffing across program years, creating structures to support teachers' development of expertise with the algebra support course, developing a multilayered approach to student assignment, or making the course credit bearing to promote implementation and stability over time for both students and teachers. Future research could investigate more fully how effective mathematics leaders can leverage existing resources to support coherent and stable program development over time.

For other researchers investigating curriculum implementation and impact in a multi-site algebra support course context, *we recommend a plan to invest substantial resources in recruitment, particularly in efforts to understand both the school and district context for the course.* This is because algebra support courses varied across schools within a single district, as well as across districts. We were also helped by our plan to collect impact data in year two of our intervention, which allowed us to learn more about context than could be learned in the recruitment phase. We were also able to pilot our instruments and adapt our intervention design to be more flexible.

Instruments to Measure Student Learning

For federally funded studies, the What Works Clearinghouse standards (2022) are the recognized metric by which intervention studies, including studies of curriculum impact, are assessed. As part of meeting these standards, baseline equivalence between the treatment and control group on the outcome measure(s) should be assessed for the analytic sample. In addition, instruments deployed to collect outcome measures must have face validity, provide reliable scores, be sensitive enough to differentiate between students, be able to detect change or growth, and be appropriate for the student population. If measures do not meet these criteria, researchers may not be able to find a difference between groups (e.g., between *TTA* students and students receiving business-as-usual instruction) even if a difference exists, resulting in statistically null results even if the instructional materials (e.g., *TTA*) are practically impactful.

In our original design for this study, conceived in 2015, we intended to use two measures of algebra performance. First, we planned to use the *Partnership for Assessment of Readiness for College and Careers* (PARCC) (2016) consortia's ninth-grade algebra end-of-course exam to ensure comparability on state assessments across states and to conduct project activities efficiently. Second, we intended to use the CTB/McGraw-Hill *Acuity* series (CTB/McGraw-Hill LLC, 2007a), which includes an *Algebra Readiness Exam* (pretest) and an *Algebra Proficiency Exam* (posttest), to compare groups before the intervention (pretest), as covariates in subsequent analyses (pretest), and to estimate the impact of *TTA* on the algebra achievement of ninth-grade students underprepared for Algebra 1 compared with similar students receiving business-as-usual instruction (posttest). We experienced difficulties with this plan for both measures.

Existing Measures of Student Learning in Algebra Appropriate for Multi-Site Studies

State tests for accountability reporting are widely used to evaluate the impact of interventions on student outcomes, particularly in large-scale studies. However, for students who are likely to perform below grade level, these assessments may lack precision at the lower end of the performance distribution. Also, because they are designed to assess grade-level benchmarks, they are less sensitive to representing small changes or growth in student learning over time (Cook et al., 2015; Guryan et al., 2021). In addition, multi-state studies cannot leverage a common assessment across states. Nationwide testing consortia, such as PARCC and Smarter Balanced, were promising solutions to this challenge; however, the adoption of these consortia has not been sustained, and at the time of this study, only 16 states were using consortia-based assessment systems (Gewertz, 2019). During recruitment, we found use varied greatly even across those states (e.g., a consortia assessment was adapted in a particular state, the ninth-grade test was not being offered in a given year in another), and the context was ever-changing. Therefore, multi-state studies, including ours, are often left to use state-administered assessments and then perform statistical adjustments, such as z-score transformations, to place scores on the same scale for analysis (May et al., 2009).

In addition to the use of state tests, curriculum intervention study researchers often develop and administer their own assessments to estimate the effect of an intervention. The benefit of this approach is that the assessment can be tailored to assess the specific skills expected to be acquired and the question formats can mirror how students are exposed to the content. The downside, however, is that the assessment may be so tailored to the intervention that the scores generated do not support inferences for how students might perform on state tests or other types of tests used for accountability purposes. To avoid this pitfall, we sought a preexisting assessment with evidence of reliability and validity. After a thorough review of available assessments, all but one, the *Acuity* assessment series (CTB/McGraw-Hill LLC, 2007a), were deemed inappropriate given their purpose (e.g., state testing, progress monitoring) or format (e.g., computer-based administration, too long, large scoring burden). Our intention was to use the *Acuity Algebra Readiness Exam*, a 40-item multiple-choice assessment that aims to measure students' preparation for successful performance in algebra as a baseline measure, and the similar *Acuity Algebra Proficiency Exam* as an outcome measure. Of note, the exam booklet for the Readiness Exam was 20 pages and responses were to be recorded on Scantron answer sheets.

We administered the Readiness Exam in the fall of 2018, with teachers administering the exam during class time. No incentives were provided for completion beyond the encouragement of the teacher in the usual classroom context. Several challenges were encountered in using this exam. Most items on the exam were substantially harder than expected based on data provided in the technical report, ranging in item difficulty from 0.11 to 0.51 (CTB/McGraw-Hill LLC, 2007b). Relatedly, the distribution of scores centered around 10 out of 40 possible points with many item difficulties around 0.30, providing evidence that guessing was a predominate test-taking behavior. Additional test-taking behaviors observed were selecting the same response for all items (e.g., all Bs), creating patterns on the sheet, and returning blank sheets. In addition, the amount of missing data increased as the test progressed, indicating the items were too difficult and/or the assessment was too long to complete. Finally, teachers reported the exam looked too much like a standardized assessment and many of their students did not complete the exam in the allotted time or they gave additional class time to complete it beyond what was suggested. They also reported the following student test-taking behaviors: putting their head down/not starting the exam, doodling in the exam booklets after one or two minutes, and working on the exam for only part of the administration time.

In summary, we found the *Acuity Algebra Readiness Exam* was too difficult (high item difficulty and skewed score distribution) and too closely resembled a standardized assessment (unsuitable in format and length), which we believe is what was reflected in students' attitudes towards the assessment (low motivation and perseverance). We concluded that the test was unlikely to provide valid estimates of students' knowledge. So, we made the decision to develop our own assessment for administration at the end of the year. In addition, because the *Acuity Algebra Readiness Exam* did not perform as expected, we decided to use students' eighth grade assessment scores as the baseline measure rather than the *Acuity Algebra Readiness Exam*.

Developing a Measure of Student Learning in Algebra Appropriate for Multi-Site Studies

Our aim was to develop an end-of-year assessment that was at an appropriate difficulty level for our population, accessible in format and length, and that would provide valid and reliable estimates of students' proficiency with algebra content. We created an exam similar in format to a classroom assessment, broke the exam into three content-focused parts, and included more items targeted at the easier end of the difficulty range. These changes were made intentionally with the hope of increasing students' engagement. The three-part assessment was administered to over 3,700 students, with favorable results. Specifically, the range of item difficulties increased, item difficulties were approximately normally distributed, rates of missing item-level data decreased, assessment completion rates increased, and test-taking patterns and behaviors improved. Overall, the new assessment shows evidence of reliability and validity.

Implications of Assessment Development for Future Research and Instrument Development

Standardized measures of Algebra 1 content that are both psychometrically robust and easily administered in a multi-site study context are not widely available. This is particularly true when measures must be sensitive enough to differentiate among students who typically perform poorly in mathematics. Therefore, *assessment development efforts, like our own, are needed to conduct rigorous research in algebra support contexts; otherwise, researchers and evaluators cannot provide compelling evidence on the impact of promising interventions.* Although additional evidence should be collected, the psychometric improvements seen in the spring assessment may indicate some directions for future measurement development, particularly the need to make assessments accessible in format and appropriate in difficulty for the population being studied.

CONCLUSION

In this chapter, we discuss three main challenges that emerged in our study of *TTA* use, a curriculum designed for use in a supplemental course that is outside of the typical mathematics course sequence and designed to support students who may not be on grade level. These challenges included accounting for variation in how algebra support courses are designed and organized in our study sites; managing a study in which contextual factors contributed to program instability; and identifying, adapting, and designing instruments that are appropriate and sensitive enough to measure student learning outcomes.

This work has highlighted some strategies for responding to these challenges and implications for future research. Namely, in our work, we found that *using multiple data sources in a mixed methods approach strengthened our analyses, particularly in considering the question of measuring teachers' use of TTA.* By asking teachers parallel questions on the same constructs in our surveys and our interviews, we were able to compare qualitative and quantitative data, looking for patterns, variations, and contradictions in the data. This increased our understanding of the variations in school implementation contexts and strengthened the validity of our findings.

Further, our understanding of the factors that contribute to program instability highlights the need to investigate further how mathematics leaders support coherent program improvement, especially in intervention contexts. *Better understanding of the perspectives of district and school mathematics leaders, along with their goals and challenges, can serve to help refine the intervention design and identify additional supports needed for implementation.*

Our work with existing instruments to measure student learning of algebra points to the need for better assessments and measures appropriate for conducting rigorous research in algebra support contexts. We recommend building adequate time to pilot proposed measures with the intended student population, and adapting the measures as needed before implementation.

Our study was a multi-site study working with schools in over 30 districts. Working in one school, one district, or even fewer districts may have reduced the variability we found in implementation contexts. However, we still found variation across schools within districts, and working in fewer sites may limit the generalizability of any findings. Future researchers should carefully consider the tradeoffs in situating studies within a single district context versus multiple sites.

Although challenging, it is critically important to conduct research on interventions designed to support students performing below grade level in mathematics, and in algebra specifically. We are hopeful that our study contributes to a better understanding of the contexts for algebra support, and highlights measurement and implementation needs.

REFERENCES

Achieve, Inc. (2008, May). *The building blocks of success: Higher-level math for all students* [Policy Brief]. Achieve. http://www.achieve.org/BuildingBlocksofSuccess

Balfanz, R., Byrnes, V., & Legters, N. (2012). *A randomized trial of two approaches to increasing mathematics achievement for underprepared freshmen.* Johns Hopkins Center for Social Organization of Schools.

Balfanz, R., Legters, N., & Jordan, W. (2004). *Impact of the talent development ninth grade instructional interventions in reading and mathematics in high-poverty high schools.* CRESPAR/Johns Hopkins University. https://files.eric.ed.gov/fulltext/ED484524.pdf

Balfanz, R., McPartland, J., & Shaw, A. (2002, April). *Re-conceptualizing extra help for high school students in a high standards era.* U.S. Department of Education, Office of Vocational and Adult Education. https://files.eric.ed.gov/fulltext/EJ903936.pdf

Cook, P., Dodge, K., Farkas, G., Roland, G., Fryer, J., Guryan, J., Ludwig, J., Mayer, S., Pollack, H., & Steinberg, L. (2015). *Not too late: Improving academic outcomes for disadvantaged youth.* Working Paper 15-01. Institute for Policy Research, Northwestern University. https://harris.uchicago.edu/files/inline-files/not_too_late_IRP%20version%202015.pdf

Cortes, K., & Goodman, J. (2014). Ability tracking, instructional time, and better pedagogy: The effect of double-dose algebra on student achievement. *American Economic Review, 104*(5), 400–405. https://www.aeaweb.org/articles?id=10.1257/aer.104.5.400

Cortes, K., Goodman, J., & Nomi, T. (2015). Intensive math instruction and educational attainment: Long run impacts of double-dose algebra. *The Journal of Human Resources, 50*(1), 108–158. https://muse.jhu.edu/article/571926 N1

Council of the Great City Schools. (2009, Spring). Council of the Great City Schools high school reform survey, 2006–2007. *Urban Indicator.* https://eric.ed.gov/?id=ED505530

CTB/McGraw-Hill LLC. (2007a). *Acuity algebra.*

CTB/McGraw-Hill LLC. (2007b). *Acuity algebra readiness technical report.*

Cuoco, A., Goldenberg, E. P., & Mark, J. (1996). Habits of mind: An organizing principle for mathematics curricula. *Journal of Mathematical Behavior, 15*(4), 375–402. https://doi.org/10.1016/S0732-3123(96)90023-1

Cuoco, A., Goldenberg, E. P., & Mark, J. (2010). Organizing a curriculum around mathematical habits of mind. *Mathematics Teacher, 103*(9), 682–688. https://eric.ed.gov/?id=EJ886974

Durwood, C., Krone, E., & Mazzeo, C. (2010). *Are two algebra classes better than one? The effects of double-dose instruction in Chicago* [Policy Brief]. Consortium on Chicago School Research. https://eric.ed.gov/?id=ED512287

Gewertz, C. (2019, March 7). Only 16 states still share Common-Core tests, survey finds. *Education Week.* https://www.edweek.org/teaching-learning/only-16-states-still-share-common-core-tests-survey-finds/2019/03

Ginsburg, A., & Smith, M. S. (2016). *Do randomized controlled trials meet the "gold standard:" A study of the usefulness of RCTs in the What Works Clearinghouse.* American Enterprise Institute. https://www.aei.org/wp-content/uploads/2016/03/Do-randomized-controlled-trials-meet-the-gold-standard.pdf

Guryan, J., Ludwig, J., Bhatt, M. P., Cook, P. J., Davis, M. V. J., Dodge, K., Farkas, G., Freyer Jr., R. G., Mayer, S., Pollack, H., & Steinberg, L. (2021). *Not too late: Improving academic outcomes among adolescents.* (NBER Working Paper No. 28531). National Bureau of Economic Research. https://www.nber.org/system/files/working_papers/w28531/w28531.pdf

Ham, S., & Walker, E. (1999). *Getting to the right algebra: The Equity 2000 initiative in Milwaukee Public Schools.* MDRC. https://eric.ed.gov/?id=ED441896

Heck, D. J., Chval, K. B., Weiss, I. R., & Ziebarth, S. W. (2012). Developing measures of fidelity of implementation for mathematics curriculum materials enactment. In D. Heck, K. Chval, I. Weiss, & S. Zeibarth (Eds.), *Approaches to studying the enacted curriculum* (pp. 67–87). Information Age Publishing.

Helfand, D. (2006, January 30). A formula for failure in L.A. schools. *Los Angeles Times.* https://www.latimes.com/local/la-me-dropout30jan30-story.html

Louie, J., Rhoads, C., & Mark, J. (2016). Challenges to using the regression discontinuity design in educational evaluations: Lessons from the *Transition to Algebra* study. *American Journal of Evaluation, 37*(3), 381–407. http://dx.doi.org/10.1177/1098214015621787

Lynch, K., Hill, H. C., Gonzalez, K. E., & Pollard, C. (2019). Strengthening the research base that informs STEM instructional improvement efforts: A meta-analysis. *Educational Evaluation and Policy Analysis, 41*(3), 260–293. https://doi.org/10.3102/0162373719849044

Mark, J., Goldenberg, E. P., Fries, M., Kang, J. M., & Cordner, T. M. (2014). *Transition to algebra*. Heinemann.

Mark, J., Louie, J., & Fries, M. (2012, April 23–25). *Supporting students to succeed in algebra: District strategies and resources* [Conference presentation]. Annual meeting of the National Council of Supervisors of Mathematics, Philadelphia, PA. https://ttalgebra.edc.org/resources#presentations

May, H., Perez-Johnson, I., Haimson, J., Sattar, S., & Gleason, P. (2009). *Using state tests in education experiments: A discussion of the issues* (NCEE 2009-013). National Center for Education Evaluation and Regional Assistance, Institute of Education Sciences, U.S. Department of Education. https://files.eric.ed.gov/fulltext/ED511776.pdf

National Mathematics Advisory Panel. (2008). *Foundations for success: The final report of the National Mathematics Advisory Panel*. U.S. Department of Education. https://files.eric.ed.gov/fulltext/ED500486.pdf

Nomi, T. (2012). The unintended consequences of an algebra-for-all policy on high-skill students: Effects on instructional organization and students' academic outcomes. *Educational Evaluation and Policy Analysis, 34*(4), 489–505. https://doi.org/10.3102/0162373712453869

Nomi, T., & Allensworth, E. (2009). Double-dose algebra as an alternative strategy to remediation: Effects on students' academic outcomes. *Journal of Research on Educational Effectiveness, 2*(2), 111–148. https://doi.org/10.1080/19345740802676739

Nomi, T., & Allensworth, E. (2013). Sorting and supporting: Why double-dose algebra led to better test scores but more course failures. *American Educational Research Journal, 50*(4), 756–788. https://doi.org/10.3102/0002831212469997

Partnership for Assessment of Readiness for College and Careers [PARCC]. (2016). *PARCC accessibility features and accommodations manual: Guidance for districts and decision-making teams to ensure that PARCC summative assessments produce valid results for all students* (5th ed.). https://files.eric.ed.gov/fulltext/ED573573.pdf

Remillard. J. T., & Heck, D. (2014). Conceptualizing the curriculum enactment process in mathematics education. *ZDM Mathematics Education, 46*(5), 705–718. https://doi.org/10.1007/s11858-014-0600-4

Schachter, R. (2013). Solving our algebra problem: Getting all students through Algebra 1 to improve graduation rates. *District Administration, 49*(5) 43–46. https://eric.ed.gov/?id=EJ1013968

Silver, D., Saunders, M., & Zarate, E. (2008). *What factors predict high school graduation in the Los Angeles Unified School District*. California Dropout Research Project. https://www.issuelab.org/resources/11619/11619.pdf

Stephens, M., Erberber, E., Tsokodayi, Y., & Fonseca, F. (2022). *Changes between 2011 and 2019 in achievement gaps between high- and low-performing students in mathematics and science: International results from TIMSS* (NCES 2022-041). U.S. Department of Education. National Center for Education Statistics, Institute of Education Sciences.
https://nces.ed.gov/pubsearch/pubsinfo.asp?pubid=2022041

Sweet, T. (2010). *An analysis of a ninth grade mathematics intervention* [Technical report]. Carnegie Mellon University. https://kilthub.cmu.edu/articles/journal_contribution/An_Analysis_of_a_Ninth_Grade_Mathematics_Intervention/6586421/1

Tarr, J. E., Grouws, D. A., Chávez, Ó., & Soria, V. (2013). The effects of content organization and curriculum implementation on students' mathematics learning in second-year high school courses. *Journal for Research in Mathematics Education, 44*(4), 683–729. https://doi.org/10.5951/jresematheduc.44.4.0683

Tarr, J. E., McNaught, M. D., & Grouws, D. A. (2012). The development of multiple measures of curriculum implementation in secondary mathematics classrooms: Insights from a three-year curriculum evaluation study. In D. Heck, K. Chval, I. Weiss, & S. Zeibarth (Eds.), *Approaches to studying the enacted curriculum* (pp. 89–115). Information Age Publishing.

U.S. Department of Education. Institute of Education Sciences, National Center for Education Statistics, National Assessment of Educational Progress [NAEP]. (2022a). *NAEP Report Card: 2022 NAEP Mathematics Assessment. Highlighted results at grades 4 and 8 for the nation, states, and districts.*
https://www.nationsreportcard.gov/highlights/mathematics/2022/

U.S. Department of Education. Institute of Education Sciences, National Center for Education Statistics, National Assessment of Educational Progress [NAEP]. (2022b). *NAEP long-term trend assessment results: Reading and mathematics.*
https://www.nationsreportcard.gov/highlights/ltt/2022/

What Works Clearinghouse. (2022). *What Works Clearinghouse standards handbook, version 5.0*. U.S. Department of Education, Institute of Education Sciences, National Center for Education Evaluation and Regional Assistance.
https://ies.ed.gov/ncee/wwc/handbooks

Wilson, J., & Kelly, D. (2018). Supplemental supports for currently struggling students. In P. Cobb, K. Jackson, E. Henrick, T. M. Smith, & the MIST team (Eds.), *Systems for instructional improvement: Creating coherence from the classroom to the district office* (pp. 169–177). Harvard Education Press.
https://eric.ed.gov/?id=ED583100

ENDNOTES

1. In the U.S., ninth-grade students are generally 14–15 years of age.
2. STEM stands for Science, Technology, Engineering, and Mathematics.
3. The state mathematics assessment is part of the Massachusetts Comprehensive Assessment System (MCAS).

4. We are still finishing conducting our analyses of impact and will be reporting those findings elsewhere.
5. Pink slips are notices given to employees that their position is being eliminated and/or their job is being terminated.

CHAPTER 25

STUDYING THE ENACTED SECONDARY MATHEMATICS CURRICULUM

Lessons Learned From the University of Chicago School Mathematics Project

Denisse R. Thompson
University of South Florida

Sharon L. Senk
Michigan State University

The University of Chicago School Mathematics Project (UCSMP) has developed curricula for secondary schools and studied curricular effectiveness since 1983. In this chapter, we share lessons learned from studying the enactment of the curriculum in three major areas: content; instructional processes; and student achievement. In each area, we share lessons about methodological issues related to studying enactment that may be of interest to other curriculum researchers and illustrate those lessons with data from evaluations of the curriculum enactment. We also reflect on other aspects of curriculum enactment that might be investigated in the future.

Keywords: enacted mathematics curriculum; opportunity to learn; secondary mathematics; University of Chicago School Mathematics Project (UCSMP)

In the United States, textbooks are a prominent feature of most secondary mathematics classrooms, with 70% of teachers in college-preparatory courses reporting using a textbook for their most recent instructional unit

(Hayes, 2019). Curriculum resources, especially textbooks, influence education in three important ways:

1. They are instruments of change in the mathematical content that is taught and the objectives of teaching mathematics;
2. They are designed to influence instruction in mathematics, such as innovative classroom practices or pedagogical approaches; and
3. They are instruments to enhance students' knowledge and change students' beliefs and attitudes related to mathematics. (Rezat et al., 2021)

However, as Thompson and Senk (2010, 2014b) report, the intended curriculum of the textbook is not necessarily what teachers actually enact. Using data from textbook evaluation studies of seven different mathematics courses, Thompson and Senk (2010) report that the percent of textbook lessons taught by teachers varied widely, with at least one teacher of each course teaching less than 50% of the textbook lessons. Even for one chapter taught by all teachers in an algebra course, the opportunities for students to extend knowledge through assigned homework also varied, for both basic and applied questions. In a detailed study of geometry teachers teaching from the same textbook, Thompson and Senk (2014b) found large differences in the percent of lessons taught on the topic of congruence, the homework assigned from these lessons, the use of whole-class or small-group instruction, and the use of technology.

As Kilpatrick (2003) notes: "Two classrooms in which the same curriculum is supposedly being 'implemented' may look very different; the activities of teacher and students in each room may be quite dissimilar, with different learning opportunities available" (p. 473). And, as recommended by Remillard et al. (2009), understanding how teachers implement their curriculum is essential to making conclusions about its effectiveness.

Since its inception in 1983, the secondary component of the University of Chicago School Mathematics Project (UCSMP) has written textbooks for grades 7–12,[1] and most recently for grades 6–12, that have attempted to implement national recommendations for mathematics education about updating the mathematics content (e.g., recommendations in documents from the National Council of Teachers of Mathematics [NCTM], 1980, 1989, 2000), modifying methods of teaching, and improving students' achievement.[2] As part of its development process, researchers have investigated teachers' implementation and students' achievement when using UCSMP textbooks. Since the development of the first drafts of textbooks in the mid-1980s, UCSMP researchers have conducted field trials of its materials to inform revisions prior to commercial publication and to investigate what students learn from their use. Over time, more and varied data about implementation have been collected, as recommended by Confrey and Stohl (2004), to enhance insights about textbook implementation.

In this chapter, we build on the work of Thompson and Senk (2014a) regarding lessons learned from studying enactment of the UCSMP secondary textbooks. We first describe the context of the UCSMP evaluation research, then share lessons learned in three sections based on uses identified by Rezat et al. (2021): textbooks as determinants of mathematics content, as guides for instruction, and as potential opportunities for students' learning. In each section, we describe the instruments used to study enactment and illustrate lessons learned from analysis of the collected data. The final section addresses issues related to future research.

CONTEXT OF THE UCSMP EVALUATIONS

The overall design, methods, and types of instruments used were similar across the evaluation studies for different UCSMP courses conducted from 2005 to 2008. The methodology and results of each study can be found in the technical reports by Thompson (2018, 2019, 2020) and Thompson and Senk (2018), available on the UCSMP website.[3] Additional discussion of the instruments can be found in Thompson and Senk (2012).

Evaluation studies of UCSMP textbooks were done on the field-trial editions, not on the final commercial versions of those materials. Teachers using the UCSMP materials began the school year under less-than-ideal conditions. They did not have access to the entire textbook at the beginning of the school year; instead, they started with soft-bound volumes of only the first three to four chapters, with the remaining chapters delivered in parts during the school year. They had minimal Teacher Notes and only one sample chapter test for each chapter. Teachers received no professional development prior to the use of the materials. However, they were invited to the university twice during the academic year to discuss the materials; such meetings essentially provided informal professional development.

Each study had both a formative and a summative component. The formative component was designed to gather information about the use of UCSMP field-trial editions to aid the developers in revising them for commercial publication. Both quantitative and qualitative data were gathered about what aspects of the materials were used or not used, and what issues arose during classroom implementation. The summative component was designed to assess the effectiveness of the UCSMP materials. Thus, classroom implementation of the UCSMP materials was compared to implementation of comparable materials, and achievement of students in UCSMP classes was compared to that of students in comparable classes, using a quasi-experimental matched-pair pretest-posttest design (Campbell & Stanley, 1963).

Schools were solicited from across the United States, in different types of communities (rural, small town, suburban, urban), and in both public and private settings. Because the evaluations were conducted over an entire school year, we recruited schools with relatively stable populations whose administrators would commit to keeping classes intact at the end of semesters. Participating schools typically involved pairs of classes matched on the basis of student achievement on one or two pretests, with one class in each pair using the UCSMP field-trial materials and the other class using the textbook already in place at the school. Because schools have their own ethos toward learning, we generally sought matched classes in the same schools.

A variety of written data were collected from the teachers and students in both the UCSMP and comparison classes: pre- and post-teacher questionnaires; chapter-by-chapter evaluation forms; post-student survey; and pre- and post-student assessments. Observations were made in all participating classes to identify any contextual factors that might have influenced implementation but that would not have been easily identified from the written instruments; teacher interviews were conducted in connection with these visits.

In this chapter, data used to illustrate lessons learned come from evaluation studies of the final four UCSMP secondary courses: *Geometry (Geo)* (Benson et al., 2006); *Advanced Algebra (AA)* (Flanders et al., 2006); *Functions, Statistics, and Trigonometry (FST)* (McConnell et al., 2007); and *Precalculus and Discrete Mathematics (PDM)* (Peressini et al., 2007). Specifically, data are from 79 UCSMP classes taught by 38 teachers in 36 schools in 14 states.

As Kilpatrick (2003) and Thompson and Senk (2014a) argue, the class is the appropriate unit of analysis for investigating students' achievement when studying curricular effectiveness. Teachers typically teach classes of students and make assignments for those classes; they do not teach individual students. Classes have their own culture, which is influenced by teachers' instructional practices as well as the mix of students. This has been an important *premise* underlying all UCSMP evaluation studies since the inception of the project:

> *Analyses need to be shared at the class level. That is, for student data, the class and not the student needs to be the unit of analysis. Each matched pair of classes is a ministudy; looking at patterns across the mini-studies provides information about overall effectiveness of a textbook.*

LESSONS LEARNED ABOUT STUDYING MATHEMATICS CONTENT

Each UCSMP secondary course includes some topics typically included within other textbooks used in the early 2000s at that level, as well as some unique topics or approaches. To study the implemented content, data primarily came from chapter evaluation forms completed for each chapter a teacher taught. Figure 25.1 contains a sample of the type of data collected.

Figure 25.1

Data Sources on Individual Chapter Evaluation Forms

- An indication if the lesson was taught and a rating of its quality on a scale from *disaster* through *excellent*
- The number of days spent on each lesson taught
- The questions assigned for homework for each lesson taught, with a rating on the quality of the questions using the same scale as for the lessons
- Comments about the overall sequence and level of difficulty of the chapter
- Comments about specific lessons or features (e.g., the use of a specific technology or specific instructional feature)

From these data, three types of analyses were conducted: the number and percent of textbook lessons taught; the lesson-by-lesson coverage (i.e., which specific lessons were taught); and the coverage of lessons with content unique to the course with rationales about why the lesson was or was not taught. Each analysis provides insights into the implemented content of interest for both the developers and other curriculum researchers. Taken together, they enable the construction of detailed profiles of the content taught within a class, providing comparisons in content coverage between classes within and across schools. They also provide valuable information to the textbook developers about what may be feasible for inclusion in the development of innovative materials (see Usiskin, this volume).

Lessons Taught

Even when teachers are all using the *same* textbook, there is often far more variability in the percent of lessons taught than might be expected. Figure 25.2 illustrates the percent of lessons taught by teachers in each of the four UCSMP high-school courses. The minimum percent of lessons taught was 33% in *FST* and the maximum was 92% in *Geometry*. The median percent of lessons taught was somewhat comparable across courses, ranging from 56% in *PDM* to 67% in *Geometry*. In *FST*, some classes studied twice as many lessons as other classes.

Figure 25.2

Percent of Lessons Taught by Teachers in Four UCSMP High-School Courses

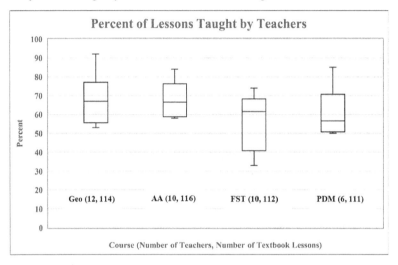

The percent of lessons taught in the textbook provides a global perspective of content taught. Teachers can teach comparable percents of lessons yet teach quite different content. Analyzing lesson-by-lesson coverage at the individual teacher level allows researchers to provide nuanced comparisons of content taught across teachers.

Figure 25.3 illustrates lesson-by-lesson coverage among the *AA* teachers, using a display similar to those of Tarr et al. (2006) in their study of curriculum enactment. Such analyses were completed in Excel to compare and contrast implemented content.

We wondered about potential causes for such variability in lessons taught. On the beginning-of-the year questionnaire, teachers indicated the amount of class time available for instruction each day of the week, their number of years teaching the particular course, and previous experience with any UCSMP courses or the specific UCSMP course in the study. During the interviews conducted in conjunction with classroom visits, teachers had an opportunity to discuss how the year was going. From these data sources, we hypothesized possible factors for differences in lessons taught.

One possible factor is the amount of class time available and how it is allocated. Across classes in the four studies, instructional time varied from

Figure 25.3

Lesson Coverage Among Teachers Using AA

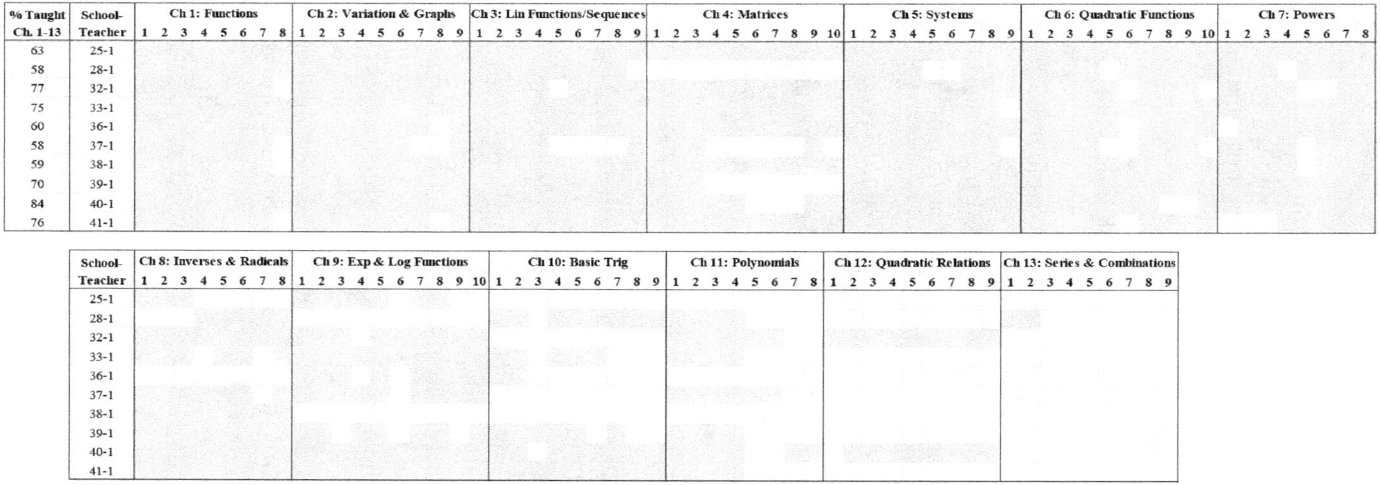

% Taught Ch. 1-13	School-Teacher	Ch 1: Functions 1 2 3 4 5 6 7 8	Ch 2: Variation & Graphs 1 2 3 4 5 6 7 8 9	Ch 3: Lin Functions/Sequences 1 2 3 4 5 6 7 8 9	Ch 4: Matrices 1 2 3 4 5 6 7 8 9 10	Ch 5: Systems 1 2 3 4 5 6 7 8 9	Ch 6: Quadratic Functions 1 2 3 4 5 6 7 8 9 10	Ch 7: Powers 1 2 3 4 5 6 7 8
63	25-1							
58	28-1							
77	32-1							
75	33-1							
60	36-1							
58	37-1							
59	38-1							
70	39-1							
84	40-1							
76	41-1							

School-Teacher	Ch 8: Inverses & Radicals 1 2 3 4 5 6 7 8	Ch 9: Exp & Log Functions 1 2 3 4 5 6 7 8 9	Ch 10: Basic Trig 1 2 3 4 5 6 7 8 9	Ch 11: Polynomials 1 2 3 4 5 6 7 8	Ch 12: Quadratic Relations 1 2 3 4 5 6 7 8 9	Ch 13: Series & Combinations 1 2 3 4 5 6 7 8 9
25-1						
28-1						
32-1						
33-1						
36-1						
37-1						
38-1						
39-1						
40-1						
41-1						

Note. School-Teacher indicates school number-teacher number, so 25-2 is Teacher 2 at School 25; shaded cell indicates lesson was taught.

200 minutes/week in several classes in the *AA* and *FST* samples to 400 minutes/week in one class in the *FST* sample.[4] Among each course, some classes met every day, with class periods ranging from 41 to 80 minutes per day. However, in each course, there were also some classes that did not meet every day; rather, they had extended time (i.e., more than 55–60 minutes of instruction) on some days of the week and no instruction on other days. For instance, in the *FST* study, classes in one school met 55 minutes per day in a four-day rotation, meeting on three of every four days. Although teachers could sometimes do more than one lesson during an extended class period, this was not always possible; likewise, when teachers had short time periods of less than 50 minutes, they were often pressed to complete a lesson in a single day.

A second factor influencing lessons taught was the teacher's prior UCSMP experience. Teachers who had previously taught from a UCSMP textbook typically reported teaching more lessons than those without that experience, perhaps because they were better able to predict lessons likely to cause difficulties for students. For instance, among teachers with prior UCSMP *AA* experience, the percent of lessons taught ranged from 70 to 84%; among those without that prior experience, the percent ranged from 58 to 63%. Specifically, the teachers at Schools 33 and 41 both had prior experience with the UCSMP course and taught comparable percents of the text (75% and 76%), even though they had differences in weekly instructional time (235 and 300 minutes). In contrast, teachers at Schools 28, 36, and 38 had no prior UCSMP experience; they had among the least amount of weekly instructional time (200 to 225 minutes) and taught among the least percent of lessons (58%, 60%, and 59%, respectively).

Figure 25.3 enables the reader to identify those lessons taught by the majority of teachers in the study. Most lessons in the first eight chapters of *AA* were taught, with the exception of lessons in Chapter 4 on *Matrices*; the teacher at School 28 omitted this chapter based on district content expectations. At least 80% of the lessons in the first eight chapters were taught by 80% of the teachers. However, among the last five chapters, only Chapter 9 on *Exponential and Logarithmic Functions* had lessons taught by the majority of the teachers. Only three teachers taught more than 50% of the lessons within the last five chapters.

Such data about lesson coverage were used by authors to revise the textbook for commercial publication, with lessons frequently skipped providing ideas on what to revise in the final editing. Analysis of lessons taught in each textbook identifies a canonical set of topics for a course across various teachers, schools, and districts. The data suggest that content crucial to a subsequent course likely needs to be included within at least the first two-thirds of the textbook. Likewise, developers of a subsequent course should not assume that students have had an opportunity to learn content in the final third of the textbook of the previous course.

New Mathematical Content

In addition to standard content taught in other textbooks for similar courses, each UCSMP textbook contains content unique to the UCSMP curriculum. Table 25.1 gives the number and sample topics of such lessons. Lessons with unique content ranged from 12% (*FST*) to 16% (*PDM*) of the total number of lessons.

Table 25.1

Details About Lessons With Unique Content for Four UCSMP Secondary Textbooks

Textbook	No. of Lessons With Unique Content	Total No. of Lessons in Textbook	Sample Lessons With Unique Content
Geometry	17	114	• Size transformations • Transformations with music
Advanced Algebra	18	116	• Matrices for transformations • Fitting models to data
FST	14	112	• Logistic models • Designing simulations
PDM	18	111	• Complex numbers & transformations • Algorithms for sorting lists

When a chapter contained a lesson with unique content, the chapter evaluation form contained questions asking about that lesson, such as the following:

> Did you teach the lesson on *Transformations with Music*? If so, how did students respond to the lesson? Discuss the provided lesson plan. How did you use the accompanying website application? (*Geometry*)

Overall, the number of lessons with unique content taught by field-test teachers varied considerably across each course. For instance, in *Geometry*, a lesson on *Size Transformations* was taught by 11 of the 12 teachers, with one teacher noting, "It was a great way to apply parallel lines to something else besides the coordinate plane ... it gave students another reason to use the vocabulary of transformations that they learnt in the lesson on rotations earlier" (*Geo*, 34-1). In contrast, only half taught the lesson connecting

Transformations and Music, with the lesson providing an opportunity for co-teaching with the music teacher or for music-oriented students to take the lead in class.

In *AA*, lessons with unique content that were in chapters with standard content, such as *Fitting Models to Data* in the chapter on variation and graphs, were generally taught by at least 80% of teachers. However, lessons with unique content, such as *Matrices for Transformations* in the chapter on matrices, were often omitted. Some teachers did not realize the importance of transformations for later course content; others skipped most of that chapter because matrices were not part of their school's curriculum for this course.

Another factor influencing the teaching of unique lessons was prior UCSMP experience and/or content familiarity. Four of the five *AA* teachers with previous UCSMP experience taught at least 13 (72%) of the unique lessons. In contrast, of the five *AA* teachers with no prior UCSMP experience, only two taught at least 13 of the unique lessons.

Validating Content Taught Through End-of-Course Assessments

The lessons taught and not taught in a classroom are one measure of the mathematics students have a chance to learn. Assessments provide another measure because teachers tend to assess only content they have taught and expect students to have learned. Thus, teachers' perspectives on the appropriateness of the content of end-of-course assessments can serve as a check against teachers' responses to other measures of content implementation.

Because no content-specific standardized tests exist for courses beyond first-year algebra, UCSMP personnel had to create their own course-specific assessments. Some may wonder if such assessments are biased in favor of UCSMP. So, it is essential to collect data to document the appropriateness of an assessment for both UCSMP students and peers in the matched classes.

Based on the work of Schmidt et al. (1993), we developed an Opportunity-to-Learn (OTL) Questionnaire for each posttest assessment. For *every* posttest item, teachers responded to the question: "During this school year, did you teach or review the content needed for your students to answer the item correctly?" Thompson and Senk (2017) note that such an OTL question for each item "allows for nuances that only a teacher might know, such as if a test item addresses an aspect of a topic not taught in a particular class or if an application item is likely to be unfamiliar" (p. 4). For instance, some teachers respond *Yes* to an OTL question for a skill-based item requiring the quadratic formula, but *No* to an OTL question when the quadratic formula is needed for an application of projectiles.

Information from the OTL question provides evidence about the extent to which a test is fair to all students in each class in a study. It also provides additional information about the implemented content obtained from other sources. Consider *Advanced Algebra*. On a multiple-choice posttest, across 10 teachers using the UCSMP textbook, the OTL ranged from 69% to 100%, with a median of 86%. On a second multiple-choice test, OTL ranged from 50% to 100%, with a median of 72.5%. On the free-response test, OTL ranged from 62% to 100%, with a median of 85%. This variability in OTL on the assessments among teachers using the same textbook mirrors the variability in the lessons taught within the textbook. Even more variability is often observed when considering content implementation among teachers using different textbooks, and this variability can and should help interpret achievement differences, as we discuss in a later section.

Summary

The previous discussions suggest the following lessons about studying enacted content:

- *Analyzing data about mathematics content from multiple perspectives, such as percent of lessons taught, lesson-by-lesson coverage, and rationales for teaching specific lessons with unique content, provides information useful to curriculum developers for improving the quality of their materials.*
- *Analyzing mathematics content taught from multiple perspectives enables researchers to understand what mathematics topics teachers consider important or what might be the actual implemented curriculum across various teachers.*
- *Analyses of implementation data by teachers allow developers and researchers to understand why particular topics are taught or not taught.*
- *Information on the teacher's perspective on opportunity to learn the content of items on end-of-course assessments allows researchers to compare teachers' reports on content taught with their expectations that students should have mastered that content and is a means of validating reports about content taught.*

LESSONS LEARNED ABOUT STUDYING INSTRUCTION

Understanding curriculum enactment involves more than identifying *what* mathematics students have a chance to learn. *How* teachers engage students with the mathematics content is also important. Some design features embedded within a textbook have the potential to influence instructional practices by facilitating how teachers engage students with content during a lesson or via homework. Design features are one means by which developers send messages to teachers about the intersection of content and instruction. In this section, we describe how we investigated instructional practices, and we share what we learned from these data collection methods.

Curriculum-Specific Design Features

UCSMP textbooks include several curriculum-specific features that provide opportunities for teachers to use active learning without having to develop them on their own. Some, such as *Guided Examples*, *Quiz Yourself*, and *Activities*, are labeled as such and are intended to be used within the normal flow of a lesson. Other features, such as an emphasis on reading the text and the use of technology, are not explicitly labeled because they underlie the philosophy of the texts and overall expectations that developers think teachers should have for students.

Guided Examples, Quiz Yourself, and Activities

In addition to typical worked examples found in many textbooks, UCSMP textbooks include *Guided Examples* (*GE*) to scaffold solutions by providing blanks for students to complete as they work through the solution; in essence, *GE* provide students with practice on how they might start the solution to a problem.[5] *Quiz Yourself* (*QY*) questions offer opportunities for teachers to check students' understanding of the lesson content through short questions in the text margins whose answers are provided at the end of the lesson; teachers might have students complete these as part of a lesson discussion. *Activities* appearing within or between lessons encourage students to work on an extended task or an exploration or use technology to make conjectures; these might be done individually or collaboratively with partners or in small groups. Although *Activities* had been included in the second editions of the UCSMP materials, the *Guided Examples* and *Quiz Yourself* features were new to the third edition.

Figure 25.4 shows some of the data sources and questions used to determine if and how the teachers used these design features.

Analyses of use of these features was primarily for formative evaluation purposes because the comparison textbooks in the matched classes typically did not have these features. Rather than attempt to quantify usage of

Figure 25.4

Data Sources and Questions Related to Activities, Guided Examples, and Quiz Yourself Features

---

Individual Chapter Evaluation Forms

- Did you use the Activity in [specific lesson]? Why or why not?
- How are you using the Quiz Yourself (QY) throughout each chapter? How are students responding to the QY?
- How are you using the Guided Examples? What reactions do students have?

Teacher Interviews (conducted in conjunction with classroom visits)

- How have you used the activities embedded in the curriculum materials? (Probe for student reactions, frequency of use, reasons for using or not using, etc.)
- How have you used the Guided Examples? (Probe for students' reactions, frequency of use, reasons for using or not using, etc.)

---

the *Activities*, it was more useful to the developers to record teachers' individual responses about their usage and the rationales for their decisions, such as the following.

> *Activities* ... are more memorable than just handing the students the information. It is easy for me to refer to "the table they created." (AA, 36-1)

> I love the *Activities*, and ... discovery learning ... when they learn it on their own, they're going to learn it so much better than if I teach it. (FST, 25-2)

Some *Activities* were perceived as too long for use during class, particularly in classes with short time periods, or required too much preparation. Such comments helped the developers determine which *Activities* to keep or what modifications to make during revisions before commercial publication.

Responses to interview questions and comments on the individual chapter evaluation forms suggest that many teachers implemented these features as intended and used them to inform their instruction. For instance, one *Geometry* teacher used *Activities* and *Guided Examples* to identify where a unit was headed and the underlying logic of the unit, as well as providing guidance for designing additional examples for use in class.

The *Quiz Yourself* and *Guided Examples* were often used together to assess students' understanding of the lesson, as the following comment demonstrates:

> For every lesson, we do all the *QY* and the *Guided Examples*. [We do these to reinforce what] they have learned in the lesson. We do it together and right

after they can get feedback, we go over the lesson.... I think they [*QY* and *GE*] are very good because they guide the students.... From the very beginning, I emphasized that that's part of the lesson and they are expected to do the *QY* and *Guided Examples*. (*PDM*, 50-1)

Although many teachers and students liked the *Guided Examples*, others reported that their students did not like them as they seemed like "pre-homework." Regardless of how teachers rated these design features, qualitative feedback from their use gave insight to the developers about which to keep as is, modify, or omit. They also helped researchers measure the extent to which teachers implemented the instructional practices expected within the UCSMP program.

Reading

Each UCSMP secondary textbook is written with the expectation that students would read the lesson and potentially do some learning on their own before instruction by the teacher. This feature was incorporated because of a belief that students need to read to learn mathematics and learn mathematics through reading to prepare for future mathematics study. Also, students should read as it is often not feasible to address all aspects of a lesson during class. So, it was important to know the extent to which teachers and students expected to read and discuss the reading. Figure 25.5 reports the data collected on this issue.

Table 25.2 illustrates how data from the end-of-year questionnaire can be summarized quantitatively. Except for *PDM*, teachers generally expected students to read at least *2–3 times per week*. Also, at least half of the teachers in each course except *PDM* reported discussing the reading in class.

The quantitative analyses allow comparison of this instructional practice by UCSMP teachers with teachers using other textbooks. For instance, the results in Table 25.2 contrast sharply with those of Hayes (2019) who found that only 15% of high school teachers of comparable courses reported that they had students read from the textbook at least once a week. Likewise, quantitative analyses of responses to these same questions from students provide a means of assessing the extent to which teachers' reported expectations about reading are internalized and implemented by students. Thompson (2022) found that students using the UCSMP curriculum were more likely to report reading and discussing the reading and to value the importance of reading to learn mathematics than students in non-UCSMP comparison classes.

Figure 25.5

Data Sources and Questions Relative to Reading Mathematics

End-of-Year Teacher Questionnaire

- How often did you expect students to read their mathematics textbook? (*almost every day* to *almost never*)
- How often did these things happen during this mathematics class? (*daily* to *never*)
  o Teacher read aloud in class.
  o Students read silently in class.
  o Students read aloud in class.
  o Students discussed the reading in class.

Supplementary End-of-Year Teacher Questionnaire

- To what extent is the reading level appropriate?
- To what extent does the structure of the text promote the learning of mathematics through reading of the text?

Teacher Interviews (conducted in conjunction with classroom visits)

- What are your expectations for students to read the text? (Depending on the response, probe for how they handle reading, the reading level, the difficulty of reading, any other issues related to reading, etc.)

Student End-of-Course Survey

- Same questions about reading as on end-of-year teacher questionnaire
- Individual student reporting about how often they actually read their textbook

---

Table 25.2

Percent of Teachers Reporting Expecting Students to Read and Discuss the Reading

Course (number of teachers)	Expecting Students to Read the Textbook		Expecting Students to Discuss Reading in Class	
	Daily	2–3 times/week	Daily	Frequently
Geometry ($n = 12$)	75	17	8	42
Advanced Algebra ($n = 10$)	60	10	0	60
FST ($n = 10$)	70	20	20	60
PDM ($n = 6$)	33	33	0	17

Analyses of qualitative data from the interviews and open-ended questions also influenced potential revisions. The concerns teachers raised about the reading level of a lesson were used by developers to edit lessons for the commercial version. Comments from teachers about how they integrated reading were used when writing Teacher Notes and providing support for subsequent users about how to bring reading into their mathematics instruction, an instructional practice that is often new for many teachers.

Some teachers expected little or no reading because they thought the reading level was too high for their student population. Others specifically chose to use a UCSMP textbook because of the reading expectation: "the only reason why we adopted UCSMP is because of the reading. We know it's a high-level reading content.... The use of bolding the vocabulary, boxing theorems and definitions, and emphasizing the vocabulary helps students learn to read math" (*Geo*, 29-1). Still others expected students to read but realized that "I as a teacher could do a better job modeling that [reading mathematics text] for them.... It's something as a teacher that I know I want to keep working on" (*AA*, 39-1).

Those who engaged students in reading often had students pre-read a lesson prior to discussion in class and noted benefits to those who took the assignment seriously. Comments from two teachers illustrate this idea:

> I wanted them to [engage in] ... prereading looking for boldfaced words, looking for some of the major concepts and definitions, and highlighting them ... just so that they could get those key words planted [in their heads].... Read through it not being too concerned if something didn't make sense, and ... go back a third time and really try to understand the examples. (*FST*, 48-1)

> My expectation is that they read the sections.... Because the ones that read through [the sections] pick up the fine details ..., and the ones that don't, I think they are getting the basic idea but not grasping the details. (*PDM*, 47-1)

Comments about reading, together with other data about content and instruction, help provide profiles of teachers' practices that can help understand achievement differences among students in relation to the implementation of a textbook as intended by the developers.

New Technology

Since 1983, UCSMP authors have written lessons with the assumption that students had access to appropriate technology. In the most recent iteration of materials, students were expected to have a graphing calculator in each of the last four courses for use in class and for homework. In *Geometry*, the calculator should also have had a dynamic geometry drawing tool. In the other three courses, the graphing calculator should have included a computer algebra system (CAS), with statistical tools also required in *FST*. So that access to technology would not be an issue during evaluation studies, UCSMP directors arranged for loaner calculators to schools in sufficient quantities that they could be loaned to students for use in class and at home. It was up to teachers whether they would actually loan the calculators to students.

As with other instructional features, data about technology use were collected from several sources, with some data collected from teachers and some from students to assess the alignment of teachers' and students' views on technology. See Figure 25.6.

Most teachers of *Geo*, *AA*, and *PDM* loaned the relevant calculators to their students; but when the calculators were perceived as not user-friendly, teachers did not loan them to students. CAS was unfamiliar to most teachers, and the time needed to learn the CAS-capable technology posed challenges for them. One *AA* teacher (36-1) noted: "I had never used an 89 [TI-89 graphing calculator] before.... So that first chapter was difficult because problems would come up and I would go and ask people from the department 'how do you do this on the 89?'" Another *AA* teacher, who had used an earlier edition of *AA* and who loaned the calculators to students, was not happy with the introduction of CAS at this level. Although she considered CAS acceptable for precalculus students, she thought CAS was too powerful for students in advanced algebra; nevertheless, she regularly used a graphing calculator because "it allows us to accomplish much more than we would be able to do without it" (*AA*, 32-1). Other teachers sometimes limited the use of CAS calculators because of concerns that students would not be able to use CAS in subsequent courses.

In contrast, only half of the *FST* teachers loaned the CAS-capable calculators to their students, who typically had access to graphing calculators. In addition to citing the steep learning curve for CAS, many *FST* teachers reported that CAS was not sufficiently used throughout the textbook. That is, if teachers are going to invest time in learning new technology themselves and in helping students learn the technology, the technology needs to be truly integral to the course.[6]

Teachers often noted that the presence of technology changed the nature of problems they used and the connections they could make: "[previously] they were just looking at the very simplest applications of what we were doing. But now I feel like we can apply things a little bit more in-depth since they have the calculators" (*AA*, 36-1). Others were concerned about the balance between what students should do with technology and without:

Figure 25.6

Data Sources and Questions Relative to Use of Calculator Technology

---

Individual Chapter Evaluation Forms

- Did you as the teacher use calculators in this chapter? If so, how were they used?
- Did students use calculators in this chapter? If so, how?
- What comments or suggestions do you have about the incorporation of calculator technology in this chapter?
- [on Chapter 1 only] Did you check out the loaner calculators to students?
- [when a specific technology feature was introduced] How helpful was this technology feature in this chapter?

Teacher Interviews (conducted in conjunction with classroom visits)

- In what ways are you using the available technology?
- In a broad sense, how has the presence of calculator technology influenced how you have approached the course? (Probe for both content and instructional strategies.)
- How have your students responded/reacted to the technology integration?

End-of-Year Teacher Questionnaire

- About how often did students use calculator technology during this mathematics class? (*almost every day* to *almost never*)
- For what did your students use calculator technology in this mathematics class? (e.g., *doing computations, graphing equations, making tables, analyzing data*)

"because the calculator is so powerful, it's making me re-evaluate … how I am going to teach this. What am I going to make them know and what is okay for the calculator to do?" (*AA*, 41-1).

When asked about potential benefits of the technology to students' learning, teachers noted that students could spend more time analyzing situations, had more confidence when using the technology, or were more motivated to learn when using technology. But teachers also expressed concerns that students relied too much on the technology, sometimes to the detriment of understanding what they were doing.

Types of Questions Assigned

In UCSMP textbooks, four types of questions appear in every lesson.[7] *Covering the Ideas* (*CTI*) questions focus on basic aspects of lesson content;

Applying the Mathematics (*ATM*) questions require students to engage with the content at a higher level, connect ideas across concepts, and typically involve application problems; *Review* (*R*) questions provide opportunities to work with content over several lessons and chapters to develop mastery. The recommendation in each UCSMP textbook is for teachers to assign all *CTI*, *ATM*, and *Review* questions in each lesson. So, the extent to which teachers assign these different levels of problems gives insight into their expectations for students in relation to higher-order thinking.

One analysis identified those questions assigned by at least two-thirds of the teachers teaching a particular lesson. That data, together with ratings of the questions (*disaster* to *excellent*) enabled developers to make reasoned decisions about editing question sets prior to commercial publication.

Determining the percent of each of these types of questions assigned by teachers again highlighted considerable variability across teachers and classes, with comparable patterns across courses. Teachers generally assigned a larger percent of the basic *Covering the Ideas* questions, a somewhat lower percent of the higher cognitive level *Applying the Mathematics* questions, and often only a small percent of the *Review* questions.

Understanding differences in assigned questions across teachers provides a nuanced view of teachers' expectations for students' level of thinking and adherence to developers' intentions for opportunities to practice. Table 25.3 provides a detailed look at the assignment of these types of questions among the *AA* teachers. Six of the 10 *AA* teachers assigned at least 80% of the most basic questions (*CTI*) but only three assigned at least 80% of the applying questions (*ATM*). Four of the ten assigned fewer than half of the *Review* questions. The relatively small percent of *Review* questions assigned by teachers raises concerns about mastery of content, given that UCSMP textbooks are designed with the use of continual review as one means of developing mastery. Teachers who skipped lessons would thus need to skip any review questions from those lessons when such review questions appeared in later lessons taught.

Results in this table can be compared across teachers in various ways. Although teachers at Schools 28, 36, and 38, who had no prior UCSMP experience, taught comparable percents of lessons, the teacher at School 38 assigned a lower percent of the *Applying* problems, that is, the higher-order thinking problems, than the other two; likewise, the teacher at School 38 assigned very few *Review* problems, perhaps impacting students' mastery of the content. These differences in the nature of problems assigned can be used to help understand any potential differences in student achievement.

Table 25.3

Percent of Question Types Assigned by AA Teachers in Lessons Taught

Type of Question	School-Teacher									
	25-1	28-1	32-1	33-1	36-1	37-1	38-1	39-1	40-1	41-1
CTI	90	66	68	87	64	69	82	88	84	88
ATM	91	82	61	71	55	69	47	71	80	60
Review	87	59	45	73	39	29	12	51	64	59
Overall	90	68	61	79	56	59	56	75	78	73

Note: Overall is based on the total number of questions assigned in all lessons taught.

During the interviews, teachers were queried about how they made decisions about assigning homework. Some teachers attempted to follow the developers' expectations that students would typically complete most questions in each lesson. One teacher remarked:

> I'm a big homework assigner in math classes "cause you can't learn math without doing math. ... those people who do the homework get the better grade and you know when people say, 'oh how come I scored so poorly on the test,' I say let me see your review problems and they say, 'well I didn't do them....'" There's a direct correlation. (*Geo*, 09-1)

Other teachers reported making decisions based on identifying those questions that focused on essential ideas within a lesson. Still others selected questions based on students' potential difficulties with the problems, particularly if they were concerned that students would struggle too much or not attempt problems that seemed too hard. For example:

> I make sure I do all the *Covering the Reading*, and that is where a lot of the drill is, and then I choose some of the applications; the math, probably the math problems that I just like, the problems that I think will be on the test, problems that I find interesting. Then, I bring in the *Review* problems, because we've got to get some spiraling going. (*FST*, 47-1)

> I try to think about it ... my top kids could do these few problems and my bottom kids could for sure do these problems and then put some in the middle as well. (*PDM*, 39-1)

Classroom Organization

Teachers generally teach *classes* of students, not students individually, so it is useful to consider how classrooms are structured. A sampling of the types of data collected related to these issues is shown in Figure 25.7.

Figure 25.7

Data Sources and Questions about Classroom Organization and Lesson Time Allocation

---

End-of-Year Teacher Questionnaire

- About what percent of class time each week did you devote to instruction in the following arrangements? (*whole class, small cooperative groups, individual seatwork, other*)
- About what percent of a typical lesson is devoted to the following activities? (*warm-up exercises, review of homework, introduction of new content, management issues, other*)
- Think about your mathematics class this past year. About how often did you do each of the following in your mathematics instruction? (*almost never* to *almost all*)
   - Introduce content through formal presentations
   - Pose open-ended questions
   - Have students listen and take notes during presentations by the teacher
   - Engage the whole class in discussions
   - Have students work in small groups
- Think about your mathematics class this past year. How important to you in your teaching were each of the following? (*of little importance* to *of highest importance*)
   - Help students learn mathematical concepts
   - Help students learn how to solve problems
   - Help students learn to explain ideas in mathematics effectively

Classroom Visits

- Notation of classroom furniture arrangement

Teacher Interviews (conducted in conjunction with classroom visits)

- Describe your typical classroom structure in terms of how students work. (Probe for teacher directed lesson, students working in small groups, students working on activities)

Data concerning the distribution of responses by teachers to the question about use of various instructional strategies is shown in Figure 25.8. It shows considerable variability across teachers *within* a course, but remarkable consistency in the pattern of variability *across* courses. UCSMP teachers were more likely to use whole-class instruction than small group or individual seatwork, results that are consistent with data reported in the national survey by Hayes (2019).

Figure 25.8

Boxplots of Percent of Weekly Time Spent in Various Instructional Arrangements

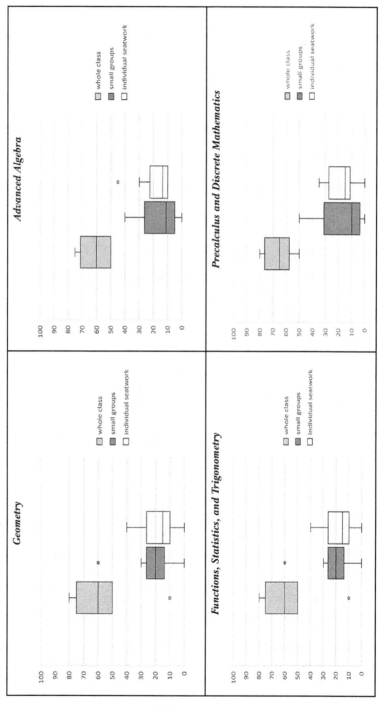

In addition to deciding on instructional arrangements for the class, a teacher also has to decide the nature of the instructional activities in which to have students engage. Table 25.4 reports responses for the percent of a lesson spent reviewing homework or introducing new content. As with organizational structure, teachers in all UCSMP courses are relatively consistent in how they allocate time across several activities. In general, teachers spent about a fourth of their time reviewing homework and about half their time engaged with new content. Taken together, data about classroom organization and data about time allocated in a typical lesson suggest there appears to be an unstated *norm* to how a mathematics lesson in a UCSMP secondary class occurs.

Table 25.4

Teachers' Reported Percent of Typical Lesson Spent on Various Activities

Activity	Geo (n = 12)		AA (n = 10)		FST (n = 10)		PDM (n = 6)	
	mean	s.d.	mean	s.d.	mean	s.d.	mean	s.d.
Review of Homework	24.8	9.6	27.0	10.1	27.0	7.5	27.5	21.2
Intro of New Content	59.0	16.7	49.0	5.7	55.4	14.8	49.2	20.4
Other	15.0	11.5	23.0	9.5	17.6	12.8	23.3	20.7

Note: Other includes warm up exercises, attendance, and classroom management.

In retrospect, our survey and interview questions about organizational structure and lesson time allocation are very broad and may have been unclear. In particular, the phrase *introduction of new content* may have seemed to some like a formal lecture. However, the responses to related questions about frequency of certain practices (e.g., formal presentations, posing open-ended questions), together with interview responses, suggest that lessons were often more active than formal lectures. For instance, one teacher said: "I'd say there is a lot of teacher focused work ... but with a high interaction with the students, where they feel comfortable asking questions and butting in and having me clarify different points" (*FST*, 48-1).

Individual teacher perspectives or school philosophy often accounted for classroom structure. One *AA* teacher (37-1) noted: "I'm doing more teacher directed, and I think that may be my comfort level." In contrast, one *FST* teacher (43-1), who taught at a Montessori high school, reported spending more than 50% of each week with students working in small cooperative groups, which was very much aligned with the school's philosophy. How the use of *Guided Examples* and *Activities*, discussing the reading, or using technology played into the allocation of lesson time in the limited options on the questionnaire provided is not clear.

Summary

The discussion about instruction in these sections leads to the following lessons.

- *Both qualitative and quantitative data about classroom instruction provide useful information to curriculum developers and researchers. In particular, qualitative data that explicates rationales for particular instructional approaches and how they are implemented provide feedback to developers for curricular revisions and potential teacher support.*
- *Looking at data by individual teachers allows curriculum researchers to identify instructional differences that might account for differences in student achievement.*
- *Students' perspectives on instructional practices provide a means to validate teachers' self-reported responses about instruction. Student data provides insight into the extent to which students have internalized the instructional messages sent by their teachers.*
- *Interviewing teachers provides opportunities for detailed discussions about issues related to the consistent use of technology in teaching advanced courses, such as supports to learn how to use the technology, beliefs about the appropriateness of technology use in that course, and perceptions about integration of technology into the materials to warrant the time spent learning it.*
- *Analyzing data across courses, as in Tables 25.2 and 25.4, and Figures 25.2 and 25.8, allows researchers to investigate trends across courses and compare results from our studies to those from other researchers. This suggests the importance of utilizing consistent measures across multiple studies.*

LESSONS LEARNED ABOUT STUDYING STUDENTS' ACHIEVEMENT

The main purpose of teaching mathematics is to help students learn mathematics. In most schools, students' learning is assessed on unit tests and end-of-year assessments.[8] As indicated earlier, we believe that students' achievement should be analyzed by class rather than by individual student. In this section, we describe how teachers' reported Opportunity-to-Learn measures may be incorporated into reports of student achievement on end-of-year assessments and the lessons learned from such analyses.

In the section on mathematics content, we discussed the use of teachers' reports of opportunity to learn the content on end-of-course assessments to validate content enacted in a classroom. Often, researchers report achievement results for all items on an assessment, regardless of whether students have had an opportunity to learn or review needed content. One way to provide more nuanced information about achievement is to analyze scores based only on those items for which teachers have reported "Yes" to the OTL question mentioned previously.

We typically construct two subtests of items based on the OTL responses (Thompson & Senk, 2001, 2012). For each teacher, or pair of teachers in a school with matched classes, we report results on a *Fair Subtest* consisting of only those items for which the individual teacher or both teachers in the pair reported "Yes" to students having a chance to learn the content needed for the item. The *Fair Subtest* is different for each pair of teachers or school, but it provides a perspective on what students learn when given the opportunity. To provide a comparison of achievement across *all* classes participating in a study, we also construct a *Conservative Subtest* consisting of only those items for which all teachers in the study responded "Yes" to the OTL question. This is often a small set of items, and sometimes mostly skill items, because it is highly influenced by the teacher who responds "Yes" to the least number of items. But such a test provides a level playing field across classes regardless of textbook used, especially for the purposes of achievement comparisons across classes.

Achievement on the entire test, the *Fair Subtests*, and the *Conservative Subtest* often provide different pictures of student achievement. So, controlling the opportunity to learn the content can influence conclusions drawn about student achievement, and is particularly important when comparing achievement for students using different textbooks. For example, the third edition study of *AA* involved 10 pairs of classes comparing students using the UCSMP curriculum to a non-UCSMP comparison curriculum (Thompson, 2019). On one multiple-choice posttest, there were five significant differences in means when using all 35 items, all in favor of the UCSMP classes; a matched-pairs t-test found the overall difference in achievement to be significant. However, the OTL percent reported by the teachers ranged from 57% to 86% of the assessed items. The school-level *Fair Subtests* ranged from 18 to 27 of the 35 items depending on the pair; once again, there were five significant differences among class pairs with all favoring UCSMP. Across the 10 teachers in the five schools, there were just 10 of the 35 items for which all teachers reported "Yes" on the OTL form. On this *Conservative Subtest* there were three significant differences in pairs in favor of UCSMP and one difference in favor of the non-UCSMP curriculum; however, a matched-pairs t-test found no overall significant difference in achievement among students using the two curricula. Thus, quite different conclusions about textbook effectiveness can be made depending on which test is used. We typically report results for all three tests, as well as individual item results by class, with an indication of whether the teacher had reported that students had a chance to learn the content of the item.

The detailed analysis of instructional practices reported earlier led us to explore statistical models of relations between instructional practices and achievement. Regression analyses on posttest achievement showed that pretest knowledge and the percent of lessons taught were significant positive predictors of achievement on all posttests for *FST* and *PDM* (Thompson, 2018, 2020), but this was not the case for *AA* and *Geo* (Thompson, 2019; Thompson & Senk 2018). In addition, the percent of questions assigned was a positive predictor of posttest achievement for only some posttests for some courses.

Senk et al. (2018) investigated instruction of UCSMP *Geometry* teachers in which the pretest achievement of students in the classes was comparable, but the posttest achievement was different. Teachers of classes that had higher than expected end-of-course achievement seemed to have enacted the curriculum more aligned with the developers' intentions than other teachers with respect to engagement with proof, applied problems, and some aspects of reading. But it was difficult to draw conclusions about how engagement with the *Activities*, with writing about mathematics, or with the use of concrete materials influenced achievement. Senk et al. attribute some of this difficulty to an insufficient depth of questions on the instruments about these instructional practices.

The discussion about studying students' achievement leads to the following lessons:

- *When curriculum researchers develop their own assessments, often because course specific standardized measures are unavailable, opportunity to learn is one way to document that assessments are fair across schools and different curricula.*
- *Reporting students' achievement in ways that control for opportunity to learn the content needed for the item enables researchers to make unbiased comparisons of curricular effectiveness.*
- *To develop statistical models that relate students' achievement to instructional practice, curriculum researchers need to develop robust measures of instructional practice that the curriculum developers believe influence achievement.*

Studying the Enacted Secondary Mathematics Curriculum 555

THOUGHTS FOR FUTURE RESEARCH

From 2005 to 2008, we collected data from year-long evaluation studies of seven different courses, each with 20–30 UCSMP and comparison teachers. If we were conducting similar studies in the future, it would be worthwhile to investigate how technology might help in data collection and analysis. Although we provided chapter evaluation forms to teachers electronically, and many completed their forms via computer as Microsoft Word documents, we had to enter that data into spreadsheets and other templates for analysis. Having teachers input data via technology so that it is incorporated directly into an underlying template for analysis would make analysis more efficient. The continual evolution of technology is likely to provide many options for use in the future and to enable researchers to scale up the size of their studies.[9]

As we indicated earlier, our research has always been conducted on field-trial versions of the textbooks. As a result, we collected some data specifically for the purposes of determining what revisions might be needed prior to commercial publication. If research were conducted on the commercial versions, some questions we asked may still be important, but others might be less important. Given teachers' many responsibilities, researchers need to be sensitive to their requests for teachers to complete detailed questionnaires. Perhaps stratified sampling of questions across teachers would enable a broader range of questions without taxing teachers; the challenge will be to ask enough teachers to respond to particular questions to be able to draw inferences about practice.

Most of the data we collected on classroom implementation was self-reported by teachers and students on questionnaires. Although we attempted to triangulate data by asking related questions in interviews and through observations of one to two days per class, if resources permit, more extended observations in a sample of classes would be worthwhile to validate self-reports. But, we do believe that if questions are asked multiple times in different ways, and if students also report on instructional practices as appropriate, then researchers can have faith in the results when consistent patterns of responses are obtained.

The large amount of data collected provides many opportunities for future research. The directors of the evaluation were focused on analyzing the data to inform the textbook developers as they were making revisions prior to commercial publication. But much of the data could be analyzed in more detail. For instance, doctoral students or other researchers might undertake focused study on some specific aspect of the data. Hauser (2015) and Karadeniz (2015) investigated detailed aspects of technology use in the last two courses; Yu (2015) investigated how types of homework questions might influence achievement.

Across all seven evaluation studies, the same types of instruments were used, with comparable questions asked on the instruments across the courses. Now that analyses of all seven evaluation studies are complete, it is possible to investigate how the mathematics curriculum is enacted across all UCSMP secondary courses and what differences might exist based on the course level or student population. Tables 25.2 and 25.4 and Figures 25.2 and 28.8, comparing and contrasting enactment by teachers of *Geometry*, *Advanced Algebra*, *FST,* and *PDM*, are a first step in identifying researchable questions. If researchers studying enactment of varied curriculum or textbooks begin to use comparable instruments or at least a set of common items, it may be easier to make comparisons of enactment across the different curricula.

Finally, we believe it would be useful to hypothesize potential aspects of enactment that might be predictors of achievement upfront and ensure that sufficiently detailed data are collected to enable those aspects to be included in statistical models. For example, use of reading and use of appropriate technology are underlying design features of all UCSMP secondary curricula. In our regression models (e.g., Senk et al., 2013, 2018), we developed indices to document implementation of these features, together with data on percent of lessons taught or percent of questions assigned. But these were broad measures. More specific questions are likely needed to highlight differences between teachers and determine how these differences influenced student achievement. We also considered how item response theory rather than raw scores could be used in such models (Senk et al., 2018). The modeling of students' achievement we have done is a start, but clearly, we could do more in another iteration of such studies.

REFERENCES

Benson, J., Capuzzi, C., Fletcher, M., Klein, R. J., Marino, G., Miller, M. J., & Usiskin, Z. (2006). *Geometry* (Third Edition, Field-Trial Version). University of Chicago School Mathematics Project.

Campbell, D. T., & Stanley, J. C. (1963). *Experimental and quasi-experimental design for research*. Rand McNally.

Confrey, J., & Stohl, V. (2004). *On evaluating curricular effectiveness: Judging the quality of K-12 mathematics evaluations*. National Academies Press.

Flanders, J., Karafiol, P. J., Lassak, M., McMullin, L., Sech, J. B., Weisman, N., & Usiskin, Z. (2006). *Advanced algebra* (Third Edition, Field-Trial Version). University of Chicago School Mathematics Project.

Hauser, L. A. (2015). *Precalculus students' achievement when learning functions: Influences of opportunity to learn and technology from a University of Chicago School Mathematics Project Study* [Unpublished Doctoral dissertation, University of South Florida].

Hayes, M. L. (2019). *2018 NSSME+: Status of high school mathematics*. Horizon Research.

Karadeniz, I. (2015). *UCSMP teachers' perspectives when using graphing calculators in advanced mathematics* [Unpublished Doctoral dissertation, University of South Florida].

Karadeniz, I., & Thompson, D. R. (2018). Precalculus teachers' perspectives on using graphing calculators: An example from one curriculum. *International Journal of Mathematical Education in Science and Technology, 49*(1), 1–14. https://doi.org/10.1080/0020739X.2017.1334968

Kilpatrick, J. (2003). What works? In S. L. Senk & D. R. Thompson (Eds.), *Standards-based school mathematics curricula: What are they? What do students learn?* (pp. 471–488). Lawrence Erlbaum.

McConnell, J. W., Brouwer, S., Brown, S. A., Ives, M., Karafiol, P. J., McCullagh, R., & Usiskin, Z. (2007). *Functions, statistics, and trigonometry* (Third Edition, Field-Trial Edition). University of Chicago School Mathematics Project.

National Council of Teachers of Mathematics [NCTM]. (1980). *An agenda for action: Recommendations for school mathematics of the 1980s*.

National Council of Teachers of Mathematics [NCTM]. (1989). *Curriculum and evaluation standards for school mathematics*.

National Council of Teachers of Mathematics [NCTM]. (2000). *Principles and standards for school mathematics*.

National Governors Association Center for Best Practices & Council of Chief State School Officers [NGA & CCSSO]. (2010). *Common core state standards for mathematics*. https://www.thecorestandards.org/Math/

Peressini, A. L., Canfield, W. E., DcCraene, P. D., Rockstroh, M. A., Viktora, S. S., Wiltjer, M. H., & Usiskin, Z. (2007). *Precalculus and discrete mathematics* (Third Edition, Field-Trial Edition). University of Chicago School Mathematics Project.

Remillard, J. T., Herbel-Eisenmann, B. A., & Lloyd, G. M. (Eds.). (2009). *Mathematics teachers at work: Connecting curriculum materials and classroom instruction*. Routledge.

Rezat, S., Fan, L., & Pepin, B. (2021). Mathematics textbooks and curriculum resources as instruments for change. *ZDM Mathematics Education, 52*, 1189–1206. https://doi.org/10.1007/s11858-021-01309-3

Schmidt, W. H., Wolfe, R. G., & Kifer, E. (1993). The identification and description of student growth in mathematics achievement. In L. Burstein (Ed.), *The IEA study of mathematics III: Student growth and classroom processes* (pp. 59–99). Pergamon Press.

Senk, S. L., Thompson, D. R., Chen, Y.-H., & Voogt, K. J. (2018). Exploring models of secondary geometry achievement. In P. Herbst, U. H. Cheah, K. Jones, & P. Richard (Eds.), *International perspectives on the teaching and learning of geometry in secondary schools: Contributions to the 13th ICME Congress* (pp. 265–282). Springer.

Senk, S. L., Thompson, D. R., & Wernet, J. (2013). Curriculum and achievement in Algebra 2: Influences of textbooks and teachers on students' learning about functions. In Y. Li & G. Lappan (Eds.), *Mathematics curriculum in school education* (pp. 515–541). Springer.

Tarr, J. E., Chávez, Ó., Reys, R. E., & Reys, B. J. (2006). From the written to the enacted curricula: The intermediary role of middle school mathematics teachers in shaping students' opportunity to learn. *School Science and Mathematics, 106,* 191–201.

Thompson, D. R. (2018). *An evaluation of the third edition of UCSMP Precalculus and Discrete Mathematics* [Technical report]. University of Chicago School Mathematics Project.

Thompson, D. R. (2019). *An evaluation of the third edition of UCSMP Advanced Algebra* [Technical report]. University of Chicago School Mathematics Project.

Thompson, D. R. (2020). *An evaluation of the third edition of UCSMP Functions, Statistics, and Trigonometry* [Technical report]. University of Chicago School Mathematics Project.

Thompson, D. R. (2022). What role might the textbook play in integrating reading into mathematics instruction? *International Journal of Science and Mathematics Education, 20,* 141–162. https:/doi.org/10.1007/s10763-022-10268-4

Thompson, D. R., & Senk, S. L. (2001). The effects of curriculum on achievement in second-year algebra: The example of the University of Chicago School Mathematics Project. *Journal for Research in Mathematics Education, 32,* 58–84. https://doi.org/10.2307/749621

Thompson, D. R., & Senk, S. L. (2010). Myths about curriculum implementation. In B. Reys, R. Reys, & R. Rubenstein (Eds.), *Mathematics curriculum: Issues, trends, and future directions* (pp. 249–263). National Council of Teachers of Mathematics.

Thompson, D. R., & Senk, S. L. (2012). Instruments used by the University of Chicago School Mathematics Project to study the enacted curriculum. In D. J. Heck, K. B. Chval, I. R. Weiss, & S. W. Ziebarth (Eds.), *Approaches to studying the enacted mathematics curriculum* (pp. 19–46). Information Age Publishing.

Thompson, D. R., & Senk, S. L. (2014a). Lessons learned from three decades of textbook research. In K. Jones, C. Bokhove, G. Howson, & F. Fan (Eds.), *Proceedings of the International Conference on Mathematics Textbook Research and Development (ICMT-2014)* (pp. 51–58). University of Southampton.

Thompson, D. R., & Senk, S. L. (2014b). The same geometry textbook does not mean the same classroom enactment. *ZDM Mathematics Education, 46*(5), 781–795. https://doi.org/10.1007/s11858-014-0622-y

Thompson, D. R., & Senk, S. L. (2017). Examining content validity of tests using teachers' reported opportunity to learn. *Investigations in Mathematics Learning, 9*(3), 148–155. https://doi.org/10.1080/19477503.2017.1310572

Thompson, D. R., & Senk, S. L. (2018). *An evaluation of the third edition of UCSMP Geometry* [Technical report]. University of Chicago School Mathematics Project.

Yu, Y. (2015). *The influence of types of homework on opportunity to learn and students' mathematics achievement: Examples from the University of Chicago School Mathematics Project* [Unpublished Doctoral dissertation, University of South Florida].

Studying the Enacted Secondary Mathematics Curriculum 559

ENDNOTES

1. In the United States, students in grades 7–12 generally range in age from 12–19 years.
2. See the UCSMP website (https://ucsmp.uchicago.edu/secondary/curriculum/) for descriptions of each book and sample lessons from the first three books written to supplement content related to the Common Core State Standards for Mathematics (National Governors Association Center for Best Practices & Council of Chief State School Officers, 2010).
3. https://ucsmp.uchicago.edu/secondary/research_reports/downloadable_technical_reports/
4. In *Geometry*, one school completed the entire course during only the second semester, with 490 minutes of weekly instruction. The data in the paragraph are for classes that met for the entire school year.
5. Samples of these features can be found on the UCSMP website (https://ucsmp.uchicago.edu/secondary/ucsmp-ccss/ccss-lesson-list/) in the additional lessons written for *Pre-Transition Mathematics, Transition Mathematics*, and *Algebra* to address content in the middle-grades *Common Core State Standards for Mathematics* that are not included in these courses. The structure of these features is similar in the last four secondary courses.
6. Detailed perspectives from the *FST* teachers about their use of such technology can be found in Karadeniz and Thompson (2018).
7. Each lesson also includes one or more *Exploration* questions, which teachers may or may not assign depending on the question. We do not discuss these questions in this chapter.
8. In the UCSMP evaluation studies, results on end-of-year assessments did not count toward students' course grade.
9. Some possibilities for such technology are available at https://www.brookings.edu/essay/digital-tools-for-real-time-data-collection-in-education/.

SECTION VI

PERSONAL REFLECTIONS ON CURRICULUM DEVELOPMENT AND RESEARCH

CHAPTER 26

LESSONS LEARNED FROM THE DEVELOPMENT OF INNOVATIVE MATHEMATICS MATERIALS

Zalman Usiskin
University of Chicago

Discussed here are lessons learned from the development and testing of the secondary school mathematics materials by the University of Chicago School Mathematics Project (UCSMP), with particular emphasis on the use of technology, mathematical language, and newer content areas. Discussed also are inherent difficulties of testing innovations, such as using the traditional experimental-comparison research paradigm.

Keywords: curriculum; evaluation; innovation; technology; UCSMP

Developing innovative curriculum materials for a mass market involves a change from the *traditional paradigm* under which most widely-used materials are created. In the traditional paradigm, a curriculum framework is developed by experts from governmental or professional organizations. This framework is translated into textbooks and other materials that are analyzed to make certain they conform. Teachers are expected to implement these materials in a faithful manner and tests are developed whose items also are faithful to this official curriculum. And, to evaluate the effectiveness of the curriculum, students are tested and the results examined. Based on these results and on new developments in mathematics and education, the experts' curriculum is periodically re-examined and the process is cycled. This process is used in many countries. In Japan, for example, this strategy has been followed for many decades, with about a 10-year cycle (Yoshikawa, 2008).

A variant of the traditional paradigm is the *test-influenced paradigm*, in which curricula are designed based either on the strengths and weaknesses found in analyzing results of standardized or national tests, or on the prospectuses of what will appear on an important test. This paradigm is often found in the development of curricula to prepare students for college entrance tests, but it is also found in earlier grades where student performance on a test (or tests) may determine eligibility for certain courses or programs.

Whereas the traditional and test-influenced paradigms are top-down in the organizational structure of most nations, the *innovative paradigm* typically is bottom-up in that it begins with work in classrooms with materials that differ in their objectives from commonly-used existing mathematics materials. Innovation may occur at any of the sizes of curriculum, with problems, lessons, units, courses, or a multiple-year curriculum. This paradigm requires that materials be classroom-tested during the process of writing as well as after a finished product exists.

To have significant usage of innovative mathematics materials in the United States, any developer has to confront three long-standing beliefs held by a large percent of the public as well as by many researchers and policymakers:[1]

1. that the fundamental ideas of mathematics itself do not change over time;
2. that mathematics is learned basically through memorization of facts and algorithms; and
3. that, with regard to mathematics, a person either "has it" or does not, and—related to this inevitability—that there is a standard age to learn important mathematics ideas.

In the United States, changing these beliefs has proved to be enormously difficult and presents a fundamental barrier to the widespread use of innovative materials.

THE DEVELOPMENT AND TESTING OF UCSMP MATERIALS

The author directed the development of the materials of the University of Chicago School Mathematics Project (UCSMP) for grades 7–12 from its inception in 1983 through three editions.[2] The reader can access details of the thinking and design principles that went into these materials in Usiskin (2003, 2007), as well as in UCSMP newsletters.[3]

Among the innovations, that is, ideas that differed from those in the commonly used textbooks of the time, were the following:

a. incorporating real-world data and applications throughout to develop concepts as well as to apply and sustain them (Usiskin, 1989);
b. utilizing geometric transformations in the development of congruence, similarity, and symmetry in geometry, as well as in the study of graphs of functions;
c. requiring that students have course- and grade-level appropriate calculators at all times, both in class and for homework;
d. incorporating four dimensions (skills, properties, uses, and representations) into the discussion and questions for all concepts, and testing on all these dimensions (Usiskin, 2015);
e. designing a curriculum that would have a full first course in algebra in eighth grade for the preponderance of students in order to allow an extra year for the inclusion of substantial material on statistics and discrete mathematics into the standard college-bound curriculum (Usiskin, 2020);
f. encouraging schools to offer the same courses but at different years, thus enabling all students to take a course, such as first-year algebra, when they have the prerequisite knowledge; and
g. writing exposition in all lessons with the expectation that it be read and encouraging this expectation by asking students questions about the reading.

Innovations (a) and (b) countered belief (1), that the basic ideas of mathematics do not change. Experiences developing textbooks before UCSMP began (Coxford & Usiskin, 1971; Usiskin, 1969, 1975, 1979, 2014/1975; Usiskin & Coxford, 2014/1972) had led me to conclude that these changes in the standard curriculum were feasible. Innovations (c) and (d) specifically countered belief (2), that the important aspects of mathematics are skills and algorithms to be memorized and executed with paper and pencil. Innovations (e) and (f) regarding use of UCSMP texts[4] countered belief (3), that only some people "have it" in mathematics and that having it means having it by a certain age. This UCSMP belief arose from the fact that in other countries, including Japan and the Soviet Union (whose textbooks we translated[5]) and some European countries, much of the standard high-school algebra and geometry content in the U.S. was taught in grades 7 and 8. Innovation (g) found in the UCSMP texts might be considered more as instructional than curricular and is not of the same ilk as beliefs (a) through (f), as it countered beliefs about student behavior rather than about mathematics. Still, this belief, like all others, required teachers using the materials to modify their teaching.

Testing the curricular materials attempted to take these innovations into account. The UCSMP research was standard, following the belief that the most important outcomes to test are how well students perform in the standard curriculum. This meant that the performance of students using the UCSMP materials was compared to the performance of students in a comparison group of classrooms who were matched with prerequisite knowledge. Although this is a standard practice in curriculum research, it works against innovation. To be equitable to the two curricula, the same attention should be given to testing the comparison group on the innovative curriculum, but this is generally considered as being unfair to the comparison group.

In 1990, after seven years of work of the University of Chicago School Mathematics Project, and before the U.S. National Science Foundation had embarked on its large-scale funding of over a dozen other projects, I mused about some of the lessons we had learned (Usiskin, 1991, 1993). Regarding technology, we had found that new technology requires changes in the sequence of the school mathematics curriculum. Calculators (at that time very new in classrooms) were easy to implement, but computers (at that time in their infancy in classrooms) were not so easy. With respect to real-world applications (at that time in their infancy in classrooms), we found that it was possible to teach pure and applied mathematics simultaneously, and that rich problems enabled more students to get involved in the action rather than fewer. In implementing multiple years of the UCSMP materials, we had found that it was difficult for teachers to take advantage of students having studied from previous UCSMP texts unless at least 80% of a class had studied from the prior course. Otherwise, most teachers felt an obligation to review the content of the previous course for the entire class.

At that time I wrote, "We have an extraordinary amount of evidence that people want to change" and "Educators may be concerned about whether their students will be able to handle the [innovative] material, and they may be concerned about whether their teachers can handle the material, but these concerns are not enough to keep them from using our curriculum" (Usiskin, 1991, p. 10). We were operating from the innovative paradigm perspective in which we focused on those who were using or interested in using our materials, rather than on those who were using traditional materials that had been developed in the traditional way.

To obtain any significant usage in schools, any curriculum developer also has to face the additional reality that commercial publication of textbooks in the U.S. is a competitive business. Publishing companies have editorial, marketing, sales, and production staffs with their own beliefs and whose jobs depend on the success of their products. Furthermore, in many states—including the most populous ones—textbooks have to make an "adoption list" in order for schools to use state money for their purchase. These lists contain, in some cases, specific guidelines as to the content that must be (or, occasionally, must not be) in materials in order to be adopted.[6] These guidelines are drawn by committees within the states and often follow or adapt national reports. In all states, the ultimate views as to the materials used in schools are decided in local school districts, by teachers and/or school boards.

Consequently, all materials, not just those that are innovative, are subject to reactions from various parties, not just of mathematics teachers. For instance, there may be opinions heard by interested parties or by specially formed "focus groups" convened by publishers interested in knowing how their materials will fare before they put money into their commercial availability. Reactions to the materials may be influenced by the contents of national reports, by state committees, and the decisions of school boards, of administrators and parents, as well as the views of publishers' sales forces, editorial staffs, and management. Significant percentages of many of these constituencies have beliefs that are in line with the three long-standing beliefs mentioned earlier. These beliefs contribute to a skeptical view at best and provide little support for any significant changes in school mathematics curricula. The number and variety of these constituencies make it natural and somewhat imperative that objective criteria be used to compare curricula. Of all the objective criteria that might be used to compare materials in this competitive arena, *student performance* is the obvious preference.

TESTING AN INNOVATIVE CURRICULUM BEFORE PUBLICATION

The development process for each UCSMP text usually included a *pilot year* in which materials were taught by authors or people the authors knew well so that there was constant feedback that could be used to revise the lessons. During this year, teachers had tests and quizzes they could give in their classrooms but there were no comparable classes and no formal analysis of results, though, when appropriate, teachers would often report how well an idea was presented compared to times they had taught it previously. Teachers typically had nothing to compare with the innovative material. A revised version of the materials was taught in a *formative year*. In that year, classes in a small number of schools were taught by teachers close enough to the university to be brought in several times during the year for their comments on how things were going. Again, there was no formal comparative study. Finally, after a second revision, materials were taught in a *summative year* in which classes around the country were matched (if possible) with corresponding classes not using the materials, observed by a researcher, and students and teachers completed pretests, posttests, and various questionnaires. Detailed reports of many of the summative studies can be found on the UCSMP website.[7] Additionally, these studies are discussed in several published papers (Senk et al., 2014, 2018; Thompson & Senk, 2001, 2014) as well as in Thompson and Senk (this volume).

Regardless of the development paradigm, most commonly it is the traditional curriculum that is the "ideal" to which we compare the effectiveness of textbooks, describe the implemented curriculum, and measure student performance, with the ultimate criterion being how student performance when using the new materials fares in comparison to performance of students using materials adhering to the official curriculum. In an era where we have seen remarkable changes in what mathematics is important to teach and how mathematics is done, basing the effectiveness of a curriculum on traditional and out-of-date curricular objectives reinforces the idea that mathematics does not change. This research paradigm is damaging to significant progress in mathematics learning in schools because it stifles any major change in what and how we teach.

By the use of this research paradigm, three broad types of innovations that are and have been stifled are the use of technology, changes of definitions of important concepts, and new areas of study. Here are lessons we have learned about each type of innovation.

Use of Technology

A half-century ago, in the early 1970s, the first hand-held calculators appeared. The implications were immediately enormous. The ability to obtain answers to arithmetic calculations *solely* using paper and pencil became obsolete. At a minimum, the curriculum needed to take into account the existence of these calculators in everyone's everyday lives and the ubiquity of automatic calculation in business and commerce at all levels. For starters, the time spent on *how* to obtain sums, differences, products, quotients, and powers needed to be diverted to *when* these operations are appropriate. But, perhaps even more important, because the curriculum dealing with number and operation had been based on those calculations, a major rethinking of the scope, sequence, and timing of the curriculum needed to be undertaken.

To this day, this major rethinking has not taken place in the U.S. For instance, in the *Common Core State Standards for Mathematics* (National Governors Association Center for Best Practices & Council of Chief State School Officers, 2010), "Use appropriate tools strategically" is one of the eight standards for mathematical practice and calculators are one of the tools mentioned. Nevertheless, calculators are *not* mentioned even once in any standard or its description in grades K–8.

The situation is somewhat better with respect to graphing technology, which first appeared in hand-held form in the 1980s. In the United States, the annual standardized tests that appear in the primary and lower secondary grades do not appear in the upper secondary grades. Instead, there are tests to judge readiness in mathematics for college-level study, and the two most-taken of these tests, the ACTs and the SATs, have since the 1980s allowed four-function, scientific, and graphing calculators. As a consequence, graphing calculators have been in widespread use in classrooms for many years and curricular materials assume that students have them available. Furthermore, with the COVID pandemic and with many students having to access their classrooms using a computer, the use of computer graphing programs has increased.

Since 1990, the National Assessment of Educational Progress (NAEP) in the U.S. has dealt with the issue of testing in an environment where calculators exist by having "calculator" blocks and "calculator-free" blocks of items at all grades at which it tests, grades 4, 8, and 12.[8] In recent years, this practice has been made easier to apply because the NAEP assessment itself is done online, and the functionality of the calculator can be adapted both to the grade level and the particular question being asked. Teachers, too, often have calculator-allowed and calculator-free portions of tests they administer in their classrooms.

However, for developers of innovative curricula the situation is more problematic, because an important thing to learn is *whether* students use a calculator or not on a particular item. Forcing the decision on students by having items on which calculators are not allowed does not provide enough information for the developer.

A situation comparable to calculators in the primary and elementary grades exists with computer algebra systems (CAS)-capable calculators. These have been available since the 1990s and play much the same role in the learning of algebra as four-function and scientific calculators play in arithmetic. The existence of this technology begs for a re-examination of the algebra curriculum from the earliest encounters that students have with symbolic algebra. However, because testing often drives the curriculum, some large-scale assessments and some tests of new curricula continue to have calculator-free and calculator-allowed sections. This strategy supposedly gives appropriate attention to the power of today's technology but ignores the ubiquitous use of calculators and computers in the world outside the classroom.

It is obvious that utilizing calculators to replace arithmetic calculations, such as multi-digit multiplication and long division, and using CAS technology to deal with algebraic calculations, such as factoring and division of polynomials, would release a significant amount of instructional time. But just as significant would be the major changes required in the overall scope and sequence of curricular materials, in retraining teachers to teach the newer materials, and in the assessment of student performance.

In a study of 11 teachers using the UCSMP course *Functions, Statistics, and Trigonometry* during the 2007–2008 school year, Karadeniz and Thompson (2018) report that teachers had mixed views about the technology that was integrated into the materials. Several of the teachers found it overwhelming to learn the CAS technology while teaching a new curriculum, and they were concerned about external assessments that did not allow that technology. These views certainly played into the finding that graphing calculator technology was used more than CAS technology, and underscore the need for teachers to have experience with the advantages and issues dealing with CAS technology before we can expect its widespread use.

Changing Definitions

Definitions are important in mathematics, in regulations of all sorts, and in everyday conversations, not only because they identify the subjects of discussion but also because they provide instances of where ideas lie in relation to other ideas. Although some people seem to think that every term in mathematics has a preferred definition, alternate definitions are found in all areas. For example, with numbers, it is generally agreed that whole numbers are integers; integers are rational numbers; rational numbers are real numbers; real numbers are complex numbers. But where do *natural numbers* fit in this hierarchy? To some people *natural number* is a synonym for *positive integer*, but to others *natural number* refers to a non-negative integer, the difference being in the exclusion or inclusion of 0. This small difference does not significantly affect properties, but it has been the subject of many discussions.[9] Likewise, in algebra, the words *variable* and *function* have multiple interpretations. But it is in geometry where the implications of definitions have traditionally been given the most consideration.

Fundamental to the study of geometry are the concepts of *congruence* and *similarity*. Traditionally, these concepts have been introduced via the study of congruent triangles. Some texts also discuss congruent segments, angles, and circles, with a different definition for each type of figure, namely:

- Two *segments* are congruent if they are of the same length.
- Two *angles* are congruent if they have the same measure.
- Two *triangles* are congruent if and only if there is a correspondence between their vertices such that corresponding sides are congruent and corresponding angles are congruent.
- Two *circles* are congruent if and only if their radii are equal.[10]

However, when congruence is approached through geometric transformations, specifically, through the four types of distance-preserving transformations (isometries)—reflections, translations, rotations, and glide reflections—then figures of all types (polygons, curves, drawings, etc.) are included in a single definition, as follows:

> Two *figures* are congruent if and only if one is the image of the other under a distance-preserving transformation (isometry); or equivalently, under a composite of reflections, rotations, translations, and glide reflections; or, equivalently, under a composite of reflections. (*Geometry*, Benson et al., 2016; Usiskin, 2018)

It is certainly unfair to test students on transformations who have not studied them nor to speak of congruent quadrilaterals or congruent ellipses if students have not seen any definition of congruence that applies to them.

The same difficulty applies to the concept of similarity. Traditionally, similarity has been defined for specific types of figures. For example,

> Two *triangles* are similar if and only if there is a correspondence between their vertices such that corresponding sides are proportional and corresponding angles are congruent. (SMSG, 1961)

With transformations, as with congruence, there can be a single definition that applies to all figures:

> Two *figures* are similar if and only if one is the image of the other under a distance-multiplying transformation; or, equivalently, under a composite of a dilation (size transformation) and an isometry. (*Geometry*, Benson et al., 2016; Usiskin, 2018)

Perhaps the most discussed definition in geometry is the definition of *trapezoid*. In the study of properties of quadrilaterals, most textbooks in the U.S. define a trapezoid as a quadrilateral with *exactly* one pair of parallel sides, leading to the hierarchy in Figure 26.1, in which properties of figures at each level are also properties of all figures connected at lower levels (Usiskin et al., 2008). This is an *exclusive hierarchy*, because parallelograms are excluded from being trapezoids.

But an increasing number of textbooks and virtually all geometers use the definition that a trapezoid is a quadrilateral with *at least* one pair of parallel sides. This leads to an *inclusive hierarchy* in which all parallelograms are trapezoids. See Figure 26.2.

Figure 26.1

Traditional Hierarchy Showing Properties of Quadrilaterals

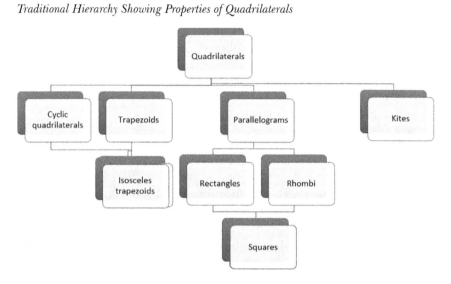

Figure 26.2

Inclusive Hierarchy of Quadrilaterals

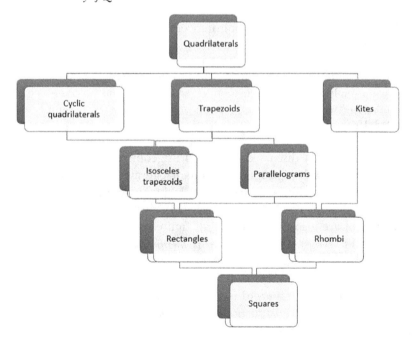

Now consider the following statement (Figure 26.3) that might be given to students to assess their understanding of the relationships among various types of quadrilaterals.

Figure 26.3

Assessing the Relationship Among Types of Quadrilaterals

> Which is a true statement about trapezoids?
> a. All trapezoids are parallelograms.
> b. Some but not all trapezoids are parallelograms.
> c. No trapezoids are parallelograms.

The majority of secondary school geometry teachers in the U.S. teach that the correct answer is choice C, because their textbooks define *trapezoid* as a quadrilateral with exactly one pair of parallel sides. But a minority of textbooks and virtually all geometers would state that the correct answer is choice B, using the definition that a trapezoid is a quadrilateral with at least one pair of parallel sides.[11]

Questions about kites give rise to an even more complex situation. Three non-equivalent definitions are in use as shown in Figure 26.4.

Figure 26.4

Three Non-Equivalent Definitions of Kite

- Definition 1: A *kite* is a quadrilateral with two distinct pairs of equal sides. Thus, all rhombuses are kites.
- Definition 2: A *kite* is a quadrilateral with two distinct pairs of equal sides of different lengths. Thus, no rhombuses are kites.
- Definition 3: A *kite* is a convex quadrilateral with two distinct pairs of equal sides of different lengths. Thus, no rhombuses are kites and also the diagonals must intersect.

Source: Based on Usiskin et al. (2008).

These three different definitions lead to three different hierarchies and different understandings of the relationships of the various quadrilaterals. Are rhombuses special kinds of kites? Is a square a special kind of kite? Must kites be convex? Students who have learned different definitions will answer the same question differently. Results on questions like these confound anyone studying the comparative performance of students, and so the topic may not even be tested. Teachers who are aware of the existence of different definitions of such terms as *trapezoid* and *kite* may find that their textbook's definition is not the one they would use. The quandary can lead to classroom discussions of the arbitrariness of definitions and the implications of using different definitions, thus using this inconsistency as an opportunity to teach an important broad lesson. But that discussion does not resolve the difficulties of testing on this topic.

Introducing New Areas of Study

In their initial development, each UCSMP course included material that was unconventional in the course it would replace, for example, exponential growth and decay in the first algebra course rather than the second course; transformations to develop congruence and similarity in geometry; matrices to represent transformations in the second algebra course, whereas earlier they were only used to solve linear systems; substantial amounts of statistics in *Functions, Statistics, and Trigonometry*; substantial amounts of discrete mathematics in *Precalculus and Discrete Mathematics*. Decades later, some of these ideas have entered mainstream curricula in the U.S., but when first introduced in the UCSMP curricula it would have been unfair to test students in the comparison classes on them.

Because the UCSMP curriculum added an extra year from the concentrated study of algebra to the concentrated study of precalculus mathematics, a few areas were not studied by UCSMP students at the same course level as students in comparison classes. One of these was polynomial long division, which was moved from its traditional place in the first algebra course to *Precalculus and Discrete Mathematics*. To test UCSMP first-year algebra students on such content would also be unfair.

New areas of study need not be large areas of courses covering months of work. Innovative units covering several weeks of work present the same challenges, particularly if a unit is centered around particular applications of mathematics rather than using applications to accompany particular topics from pure mathematics. For instance, how can there be a fair comparison between students who have studied a unit on mathematics and climate change, or mathematics and polling, with other students who have not had the opportunity?

Attempting to Mediate Differences in Opportunity to Learn

In their analyses of studies of UCSMP courses, Thompson and Senk (2001, 2014, 2017) dealt with the quandaries of varying opportunities to learn (OTL) mathematics content in several ways. One way is by having teachers of UCSMP students and teachers of the comparison students examine all test items given, and then comparing student performance only on those items taught by both UCSMP and comparison teachers in matched classes. Those items that both teachers in the matched classes identify as taught constitute what was called a *Fair Test*. Thompson and Senk also identified items taught by *all* teachers in the study and called this a *Conservative Test* because it often consisted of a small set of items. These are reasonable strategies for identifying and assessing on those items taught to both groups, but these strategies exclude those traditional topics that the comparison classes might have studied and the innovative topics that the UCSMP classes studied. (See also Thompson & Senk, this volume.)

As evident in the technical reports of the evaluation studies on the UCSMP website, in addition to reporting results for an entire test or some subset of test items, the researchers report results at the individual item level, together with the OTL responses from the teachers. This additional level of detail does provide evidence of how UCSMP students perform on content they have studied that the comparison students have not, and also provides some sense of how comparison students might perform on unexpected content. The reverse is also true because item-level results illustrate how UCSMP students exposed to a problem-solving curriculum might address content they have not studied or not studied in the same level of detail as the comparison students.

However, even a *Fair Test* may not give an adequate picture of the effectiveness of innovative curriculum materials. Thompson and Senk (2014) compared how 12 UCSMP *Geometry* third edition teachers dealt with the topic of congruence, a set of 43 lessons that were designed to take 48–58 days. The percent of lessons taught by the teachers studied ranged from 60% to 100%, with a median of 93%. All teachers taught the lessons on proofs involving congruence, a longstanding traditional topic. And, all taught the geometric transformations that were essential to the mathematical development of congruence. In contrast, content such as frieze patterns, tessellations, and connections between geometry and the real world, content that the UCSMP authors viewed as contributing to a deep understanding of the mathematics, were viewed by some teachers as enrichment rather than as required content and skipped. Teachers never gave the new content a chance to see what their students might learn. These results show how difficult it is for teachers to teach new content, especially if that content is not assessed at state or district levels.

THOUGHTS FOR FUTURE RESEARCH ON INNOVATIVE CURRICULA

The difficulty of substantially different opportunities to learn is intractable and logically impossible to overcome: if students have not been taught something, testing them on that thing is unfair, particularly if the idea is new enough that they would not have had an opportunity to learn the content in another place. How can we break the logjam in which substantially different curricula cannot be given a fair test and so are not given serious consideration? If we wish to study how innovative material fares in the classroom, we have to test and report findings in some way that produces results that can be replicated.

One course of action is to replace the classic experimental versus comparison model—used in comparing student performance on innovative curricula with those using traditional curricula—with a performance model tailored to the innovative curriculum, with tests given to those students alone. The question this research model treats would then be: How well do students learn the innovative material? This is a question that always needs to be considered, and one hopes for acceptable performance. However, even acceptable performance on the innovative material does not guarantee its wide use beyond the original developers and early adopters and users of the materials.

UCSMP research shows that unless there is some external incentive—such as an important test—to learn content that is not traditional, the pressure of time and the level of comfort teachers have with traditional material are significant factors causing teachers to skip the newer curricular material. It seems inevitable that for an innovative curriculum to gain universal acceptance and, once accepted for its ideas to be conveyed in the classroom, some top-down pressure is needed.

REFERENCES

Benson, J., Klein, R., Miller, M. J., Capuzzi-Feuerstein, C., Fletcher, M., Marino, G., Powell, N. N., Jakucyn, N., & Usiskin, Z. (2016). *Geometry* (3rd edition). UChicagoSolutions.

Coxford, A., & Usiskin, Z. (1971). *Geometry—A transformation approach*. Laidlaw.

Karadeniz, I., & Thompson, D. R. (2018). Precalculus teachers' perspectives on using graphing calculators: An example from one curriculum. *International Journal of Mathematical Education in Science and Technology, 49*(1), 1–14. https://doi.org/10.1080/0020739X.2017.1334968

Merseth, K. K. (1993). How old is the shepherd? An essay about mathematics education. *Phi Delta Kappan, 74*(7), 548–554.

National Governors Association Center for Best Practices & Council of Chief State School Officers. (2010). *Common core state standards for mathematics.* http://www.thecorestandards.org/Math

School Mathematics Study Group [SMSG]. (1961). *Geometry, Student's Text, Part I.* Yale University Press.

Senk, S. L., Thompson, D. R., Chen, Y.-H., & Voogt, K. J. (2018). Exploring models of secondary geometry achievement. In P. Herbst, U. H. Cheah, K. Jones, & P. Richard (Eds.), *International perspectives on the teaching and learning of geometry in secondary schools: Contributions to the 13th ICME Congress* (pp. 265–282). Springer.

Senk, S. L., Thompson, D. R., & Wernet, J. (2014). Curriculum and achievement in Algebra 2: Influences of textbooks and teachers on students' learning about functions. In Y. Li & G. Lappan (Eds.), *Mathematics curriculum in school education* (pp. 515–540). Springer.

Thompson, D. R., & Senk, S. L. (2001). The effects of curriculum on achievement in second-year algebra: The example of the University of Chicago School Mathematics Project. *Journal for Research in Mathematics Education, 32*, 58–84. https://doi.org/10.2307/749621

Thompson, D. R., & Senk, S. L. (2014). The same geometry textbook does not mean the same classroom enactment. *ZDM Mathematics Education, 46*(5), 781–795. https://doi.org/10.1007/s11858-014-0622-y

Thompson, D. R., & Senk, S. L. (2017). Examining content validity of tests using teachers' reported opportunity to learn. *Investigations in Mathematics Learning, 9*(3), 148–155. https://doi.org/10.1080/19477503.2017.1310572

Usiskin, Z. (1969). *The effects of teaching Euclidean geometry via transformations on student achievement and attitudes in tenth-grade geometry* (Publication No. AAT 7014670), University Microfilms.

Usiskin, Z. (1975). *Advanced algebra with transformations and applications*. Laidlaw.

Usiskin, Z. (1979). *Algebra through applications*. National Council of Teachers of Mathematics.

Usiskin, Z. (1989). The sequencing of applications and modelling in the University of Chicago School Mathematics Project (UCSMP) 7-12 curriculum. In W. Blum, J. S. Berry, R. Biehler, J. D. Huntley, G. Kaiser-Messmer, & L. Profke (Eds.), *Applications and modelling in learning and teaching mathematics* (pp. 176–181). Ellis Horwood.

Usiskin, Z. (1991). What we have learned from seven years of UCSMP. *UCSMP Newsletter No. 8*. Retrieved April 17, 2023, from https://ucsmp.uchicago.edu/resources/conferences/1990-11-02/

Usiskin, Z. (1993). Lessons learned from the University of Chicago School Mathematics Project. *Educational Leadership, 50*(8), 14–18.

Usiskin, Z. (2003). A personal history of the UCSMP secondary curriculum. In G. M. A. Stanic & J. Kilpatrick (Eds.), *A history of school mathematics* (pp. 673–736). National Council of Teachers of Mathematics.

Usiskin, Z. (2007). The case of the University of Chicago School Mathematics Project—Secondary Component. In C. R. Hirsch (Ed.), *Perspectives on the design and development of school mathematics curricula* (pp. 173–182). National Council of Teachers of Mathematics.

Usiskin, Z. (2014). Applications of groups and isomorphic groups to topics in the standard curriculum, grades 9–11. In B. J. Reys & R. E. Reys. (Eds), *We need another revolution: Five decades of mathematics curriculum papers by Zalman Usiskin* (pp. 226–248). National Council of Teachers of Mathematics. (Reprinted from "Applications of groups and isomorphic groups to topics in the standard curriculum, grades 9–11, Part I," 1975, *The Mathematics Teacher*, 68[2], 99–106.)

Usiskin, Z. (2015). What does it mean to understand some mathematics? In S. J. Cho (Ed.), *Selected regular lectures from the 12th International Congress on Mathematical Education* (pp. 821–842). Springer.

Usiskin, Z. (2018). Approaching Euclidean geometry through transformations. In Y. Li, W. J. Lewis, & J. J. Madden (Eds.), *Mathematics matters in education* (pp. 233–244). Springer.

Usiskin, Z. (2020). The timing of the first concentrated study of algebra in the past century in the United States. *Mathematics Teacher: Learning & Teaching PK–12, 113*(6), 524–526.

Usiskin, Z., & Coxford, A. F., Jr. (2014/1972). A transformation approach to tenth-grade geometry. In B. J. Reys & R. E. Reys (Eds.), *We need another revolution: Five decades of mathematics curriculum papers by Zalman Usiskin* (pp. 292–302). National Council of Teachers of Mathematics. (Reprinted from "A transformation approach to tenth-grade geometry," 1972, *The Mathematics Teacher, 65*[1], 21–30.)

Usiskin, Z., Griffin, J., Witonsky, D., & Willmore, E. (2008). *The classification of quadrilaterals: A study of definition*. Information Age Publishing.

Yoshikawa, S. (2008). Education ministry perspectives on mathematics curriculum in Japan. In Z. Usiskin & E. Willmore (Eds.), *Mathematics curriculum in Pacific Rim countries—China, Japan, Korea, and Singapore: Proceedings of a conference* (pp. 9–22). Information Age Publishing.

ENDNOTES

1. See Merseth (1993) for an elaboration of some of these beliefs.
2. When Usiskin became overall director of UCSMP in 1987, Sharon Senk became co-director with Usiskin of the secondary-level component of UCSMP and remained so through the development of the 2nd and 3rd editions of the materials.
3. Available at https://ucsmp.uchicago.edu/newsletters/.
4. From 1983–1997, UCSMP developed two editions of texts and supporting materials for each of grades 7–12 and from 2001–2008 a third edition that included materials for Grade 6. Details can be found at https://ucsmp.uchicago.edu/secondary/overview/.
5. See https://ucsmp.uchicago.edu/resources/translations/.
6. Adoption typically lasts for 5–7 years, during which time the state covers a school's purchase costs of adopted texts.
7. https://ucsmp.uchicago.edu/secondary/research_reports/

8. See https://nces.ed.gov/pubs92/web/92060.asp.
9. The reader is encouraged to go online and search "Is 0 a natural number?"
10. In the U.S., these definitions were popularized by the *Geometry* text of the School Mathematics Study Group (SMSG), the most-used of the "new math" projects of the 1960s (SMSG, 1961).
11. In the United States, the lack of agreement on a definition of *trapezoid* has in recent years extended to the elementary school, where two of the most used curricula take different sides of this issue.

CHAPTER 27

DISENTANGLING MATHEMATICS CURRICULUM FROM CONTEXTUAL ENMESHMENT

Lessons Learned Across Eight Studies

Thomas E. Ricks
Louisiana State University

Research on curriculum is inherently problematic because curriculum is enmeshed in related educational contexts. In this chapter I describe several lessons learned across eight mathematics curriculum research studies covering multiple countries with many colleagues. After a brief description of each study, I summarize overarching themes and lessons that cut across the studies, including theoretical and practical benefits of complex systems theory to curriculum research, benefits of juxtaposition, metaphorically casting wide nets to catch exotic fish, and the danger of bureaucratic overreach. I conclude with recommendations for future mathematics curriculum research.

Keywords: complexity theory; internationalization; juxtaposition; mathematics curriculum and reform recommendations; teacher preparation

INTRODUCTION

Mathematics curriculum research—an evolving sub-domain of mathematics education scholarship—still faces significant hurdles (Heck et al., 2012; Suurtamm et al., 2018; Thompson et al., 2018; Thompson & Usiskin,

2014; Tienken, 2008). The research described in this chapter has primarily focused on trying to disentangle curriculum from surrounding contextual enmeshment. Evidence of curriculum's entanglement is demonstrated by the proliferation of theoretical sub-categorizations attempting to merely *define* what it is that researchers are trying to investigate (Remillard & Heck, 2014; Remillard & Reinke, 2018). All curriculum is tightly bound by contexts like culture, language, beliefs about pedagogy/epistemology, scholarly paradigms, local pressures, and more global concerns such as government mandates (Hill et al., 2005; Kilpatrick et al., 2006). The work of disentanglement is only in its infancy (Li & Lappan, 2014; Rogers, 2017).

I begin this chapter by briefly describing eight curriculum research studies in which I have been involved. I use these to discuss specific lessons I have learned about curriculum research, categorized as overarching themes that cut across multiple studies.

DESCRIPTION OF EIGHT CURRICULUM RESEARCH STUDIES

I have learned many lessons from eight separate but related research studies involving mathematics curriculum research over my 20-year career. Each research study used Glaser's (1965) four-part constant comparative method—comparing, integrating, delimiting, and writing—to develop a theory related to mathematics curriculum for wide application in efforts toward mathematics education reform. I found the constant comparative methodology to be particularly helpful in juxtaposing educational issues because of its broad holistic approach, something recommended by a variety of educational researchers (Lather, 2004; Tierney, 2001).

Study One: Modified Lesson Study

In this study (Ricks, 2003, 2011), I investigated how collaborative and reflective environments (Dewey, 1933; Schön, 1983) provided by modified Japanese lesson study (Lewis, 2002) improved both intended and enacted public-school curriculum. The study focused on two four-person groups of secondary mathematics teacher candidates preparing to teach lessons addressing specific, local, public secondary school mathematics curricular goals. Part of the study involved probing teacher candidates' understanding of how the lesson-study mathematics topic—identified for the group by the cooperating teacher in whose classroom they would later teach their lesson-study-refined lesson—fit local curricular objectives. I used personal, qualitative interview questions such as: "Where does this *topic* fit into the [cooperating teacher's public school] curriculum?... Any curriculum that

you know of that would be helpful [to your lesson study planning process]? What is your experience using this curricula?" (Ricks, 2003). This study demonstrated modified lesson study's potential as a vehicle to achieve specific curricular goals in methods courses, something explored more in Study Seven.

Study Two: Summer Institute

In the second study, I investigated a multi-week, summer professional development institute for invited U.S. secondary mathematics teachers as they learned to teach reform-oriented, state-mandated curriculum. The official announcement inviting applicants to the institute stated: "Activities in the institute will focus on open-ended investigations ... and the integration of state and district curriculum requirements into instruction" (Nipper et al., 2011, p. 386). In particular, the study revealed developing tensions as the reform-oriented, enacted curriculum of the institute clashed against the anticipated curriculum the participating teachers would teach. The "teachers commented on how much they enjoyed ... conducting open-ended [mathematical] investigations, but ... said that they could never apply [such pedagogy] in their own classrooms. They saw a narrow curriculum, ... insufficient time, and their students' abilities ... as barriers to implementation" (Nipper et al., 2011, p. 384). This preliminary disentanglement of curriculum from anticipated pedagogy is insightful: Despite experiencing for themselves the benefits of reform-oriented curriculum's intended pedagogy in the institute, the teachers seemed unable to transform their own positive experiences into perceived, anticipatory action for their next-semester students. One participant's reflective lament captures well the general feeling of all participants: "The extensive curriculum ... required for students to learn in 180 days ... makes it nearly impossible to implement significant amount[s] of open-ended investigations ... we as teachers simply cannot take enough time to thoroughly investigate problems such as these" (Nipper et al., 2011, p. 385).

Study Three: The Mathematics Class as a Complex System

The third study (Ricks, 2007, 2016) involved an investigation of eight pre-algebra and algebra classes of three specially chosen teachers from a complex systems perspective to understand how expert mathematics teachers successfully enact reform-oriented curricula. The theoretical framework of complexity theory undergirded the study. This theory describes how various relationships between agents (in this case, the class members) expressing a spectrum of characteristics give rise to super-systemic phenomena lacking in the individual agents' behavior—the whole essentially becomes more than the sum of the parts. A key aspect of complexity theory examines how the system finds unpredictable, oscillatory equilibrium by continually balancing opposing tensions, for example, how *local* neighborhood interactions scale up and interact with *global* systemic experiences, or how *diversity* of members injects sufficient creativity into a system to enable dynamic behavior, while at the same time harmonizing with *redundant* features essential for systemic coherence. For instance, in this third study, I described how "similar curriculum ... experience [becomes] an obvious source of redundancy, as the members of the community know enough collectively about the subject to begin [shared] conversation[s]" (Ricks, 2007, p. 83). Curriculum, therefore, becomes an important anchor that grounds educational complex system formation.

More particularly, the three teachers were selected from an initial pool of 18 potential mathematics teachers recommended by local mathematics education experts as demonstrating regular, robust, student-to-student mathematics dialogue in their classrooms. One of the three teachers selected had served as her "district's mathematics specialist for 3 years while the district attempted to implement a mathematics curriculum program funded by the National Science Foundation and recognized by the National Council of Teachers of Mathematics (NCTM) as embracing a problem-solving approach" (Ricks, 2007, p. 36). But "because of backlash from parents, the district made the innovative curriculum optional, and many teachers had returned to the [original, more traditional] curriculum" (Ricks, 2007, p. 36), but not the district mathematics specialist. Another of the three selected teachers "had worked with a regional education services agency to help local schools in the district implement the current [state-mandated] mathematics curriculum" (Ricks, 2007, p. 38). There were several results from this study:

- Varied types of teacher dialogue correlate with reform-oriented approaches for curriculum implementation;
- Teachers modified their teaching approaches depending on their perceptions of student ability, for instance, one teacher taught her prealgebra and algebra class with identical, reform-oriented pedagogy, while a second teacher taught her prealgebra classes using more traditional pedagogy but used the reform-oriented pedagogy in her algebra class; and
- Class-based mathematical emergence contributed to the "shared public enacted curriculum" (Ricks, 2007, p. 59), and teachers personally psychologized available "curriculum resources" (p. 77)

to select and/or generate tasks maintaining students' high cognitive demand (Stein et al., 2000).

Study Four: Comparing U.S. and Chinese Teachers' Enactment of Identical, Intended Curriculum

The fourth research study (Yuan & Ricks, 2011) investigated how a Chinese teacher and a U.S. teacher designed and taught an identical, intended curriculum to classrooms of Chinese students. This study helped to disentangle curriculum from the student body taught. In particular, we examined how a U.S. teacher might design curriculum without the constraints of U.S. students and the cultural baggage teachers perceive that U.S. students have. The unusual aspect of a U.S. teacher instructing a class of *Chinese* students facilitated a unique methodological possibility. Often, U.S. teachers blame U.S. students' perceived behavioral issues, cultural complacency, lackadaisical motivation, and procedural mindsets as hurdles when implementing reform-oriented curriculum (i.e., lessons with fewer, more challenging tasks; more student dialogue and board-work; open-ended whole class discussion). This study demonstrated the U.S. teacher's embedded U.S. cultural norms still triumphed despite the change to teaching a Chinese student body. Additionally, the study revealed differences in the structure of both Chinese and U.S. lesson plans and enactment of the lessons, as predicted by prior research on country-based cultural scripts (Stigler & Hiebert, 1999). For example, in Yuan and Ricks (2011) the U.S. teacher-to-student dialogue ratio was almost four times that of the Chinese teacher, the U.S. teacher exhibited haphazard board work with continual board erasure, the U.S. teacher never had a student come to the board, and the U.S. teacher excessively scaffolded tasks in ways that undercut students' cognitive difficulty (Stein et al., 2000).

Study Five: Shanghai (China) Teaching Research Groups

The fifth study investigated Shanghai curriculum reform implementation through a form of Chinese lesson study involving teaching research groups (Yang & Ricks, 2011, 2012). Yang and Ricks (2011) stated: "The teaching research system has gradually evolved ... into a powerful school-based form of professional development for implementing curriculum reform [including] helping teachers [enact] ecological pedagogy; shifting attention from textbooks ... creating a study atmosphere ... and ... sharing teaching experience" (p. 42). In particular, we described how crucial incident analysis during the teaching research group [TRG] lesson study prepared teachers to address typical student misconceptions of mathematics curriculum topics. We explained how "the TRG is the basic unit in [a broader teaching research organizational] network; its main responsibility is conducting research on teaching to solve the practical problems facing teachers" (Yang & Ricks, 2011, p. 42) as they strive to enact mandates of an intended government curriculum.

Study Six: East Asian Mathematics Education

The sixth study is an ongoing conglomeration of research (Lu & Ricks, 2012; Ricks, 2010a, 2010b, 2014, 2017; Ricks et al., 2009) that examines factors affecting the interpretation, implementation, and innovation of mathematics curriculum in six mainland Chinese cities (Shanghai, Beijing, Tianjing, Hangzhou, Guilin, and Changshu) and in one South Korean city (Seoul) through videography-based research of public/private school mathematics lessons, various forms of East Asian lesson study, and interviews with educators and educational administrators. The data were collected over several years of visits to China and Korea, and data analysis is ongoing. For me, these research visits enabled firsthand juxtapositional life-experience (a critical component of constant comparative methodology), enhancing my ability to compare and contrast U.S. curricular challenges with those occurring in East Asian nations.

Study Seven: Multi-Year Iterative Lesson Study Implementations

The seventh study investigated 15 years of iterative modifications to lesson study implementation in secondary mathematics methods courses to align its enacted curriculum with its intended curricular goals. Peterson et al. (2019) stated: "Our two main methods course goals ... were for secondary [teacher candidates] to have rich conversations about (1) the mathematics of their lessons, and (2) how students think about that mathematics" (p. 528). Although Study One focused on lesson study's impact on public school curriculum enactment during a single semester's implementation, Study Seven examined the year-by-year changes made to the overall lesson-study process in the methods course to achieve the methods course's curricular goals. The lesson study implementation described in Study One formed part of the second year's iteration of lesson study in Study Seven. Over the course of 15 years of implementation, subtle but significant modifications to the lesson-study process improved enactment of the methods course's curricular goals, facilitated by (a) the methods instructor visiting to Japan for two months to study the Japanese student-teaching process (Corey et al., 2010; Peterson, 2010), (b) changing the textbooks referenced for lesson planning in the methods course, (c) introducing knowledgeable others' expertise at key stages of the lesson-study process, (d) encouraging diversity through exposure to varied lesson-study experiences, and (e) honoring the integrity of the iterative nature of lesson study implementation.

Study Eight: Teaching in Mathematics Education

The eighth research study is a longitudinal auto-ethnographic study, starting in 2007, of my personal struggles to teach reform-based curriculum in university elementary mathematics methods courses, including integrating modified East Asian lesson study practices and field-based experience in the adjacent university laboratory school. The lessons learned about curriculum research from the other seven studies inform this work. Additionally, through firsthand experience, this study has brought into clearer focus the role entrenched bureaucracies play in mathematics curriculum reform efforts. Part of this study documents the frustrations and hurdles our School of Education's Elementary Education Program faces as we balance shifting local and state bureaucratic requirements with our desire to honor educational research findings, particularly related to diversity, equity, and inclusion (Ricks, 2009, 2017).

LESSONS LEARNED

Disentangling mathematics curriculum from enmeshing contexts to facilitate curriculum research is problematic. I discuss the lessons learned while conducting curriculum research as four overarching themes. These themes stretch across these research studies and aid in the untangling of this intermeshing: (1) utilizing the benefits of complexity theory; (2) using juxtaposition for data contrast, including collaboration with bilingual colleagues; (3) casting wide nets, including enhancement with technology; and (4) recognizing the increasing effect of bureaucracies on all aspects of curriculum and curriculum research.

Lesson 1: The Benefits of Complexity Theory

Complexity science is an emerging conglomeration of perspectives that investigates holistic, self-organizing phenomena arising through the systemic interaction of agents (Davis & Simmt, 2003). The first and most important lesson I have learned about curriculum research over the course of two decades of curricular scholarship is: *The paradigmatic power of complexity theory illuminates curricular research by thinking about all aspects of education as interacting, nested, complex systems of agents, which are themselves other complex systems*. The first two studies were limited by lack of this applicable paradigm. In the third study, I utilized my initial understandings of complexity theory with great reward; with each subsequent study, my improvement as a mathematics curriculum researcher correlated directly with my increasing development as a complexivist researcher.

More particularly, I use complexity science theory to recast mathematics curriculum as a systemic phenomenon that emerges through the mutual interaction of many sub-systems (Rogers, 2017). Complexity theory has allowed me to understand how curricular issues (a) emerge as enmeshed systems, (b) exhibit fuzzy, ill-defined boundaries, (c) are iterative and fractal (self-similar) in nature, (d) are chaotic and unpredictable, (e) are dynamic and adaptable, (f) are intelligent and self-organizing, and (g) are legitimate living entities that self-reproduce (Davis & Sumara, 2001; Ricks, 2009, 2017; Stigler & Hiebert, 1999). I consider these various complex systems to be legitimate lifeforms. Just as bacteriologists (Shapiro, 1998) decades ago crossed a paradigmatic Rubicon by considering the *colony* of bacteria to be the primary focal organism for study rather than the individual bacterium, I believe that educational scholarship has much to learn by similarly thinking of *holistic* educational complex systems—especially curricular systems—as living organisms in their own right. For instance, curricular complex systems exhibit immune responses to perceived threats (e.g., the U.S. parental backlash against *Common Core State Standards*[1] adoption in many U.S. states) and may be collectively smarter than the curriculum researchers attempting to study them (Surowiecki, 2004). Many curriculum researchers report similar findings about the power of applying complex paradigms to matters of curriculum research (de Freitas et al., 2019; Remillard & Reinke, 2018); I urge other curriculum researchers to consider the same.

Specifically, my favorite definition of curriculum merges complexity theory with Dewey's (1923) conception of curriculum as the product of teachers' psychologizing human-race organized subject matter. I view *curriculum as emergent psychologization of subject-matter by some level of educational complex system*. For example, officially mandated government curricula are psychologized subject matter at the national, regional, or state level. Intended curriculum for teachers is the product of district, school, departmental, or individual teacher psychologization of the psychologized curricular material descending from above. Individuals are also complex systems, made up of smaller, nested complex systems; for instance, the brain is a complex system of neurons, interlocked with the complex system of the immune system. Enacted curriculum at the classroom level is class-, small-group, or individual student psychologizing of subject-matter from higher levels, depending on the pedagogical approach used.

Complexity theory helps illuminate the benefits of various endeavors. For instance, a reform curriculum is helpful by incorporating better, baseline, psychologized subject-matter from higher educational systems. Reform pedagogies are powerful by enabling stronger, class-based, curricular psychologizing, which enhances individual psychologizing of mathematics subject-matter. Certain forms of professional development, such as lesson study, are useful because they enhance collective teacher psychologizing of pedagogical issues, including increasing teachers' capacity to psychologize the intended curriculum. Finally, complexity theory helps explain how reform efforts might be improved by enabling better, participatory, curricular psychologizing by parents, policymakers, and/or special interest groups.

Complex perspectives have proven their potency at reshaping my curricular research perspectives on a theoretical level. On a practical level, complexity theory has also influenced my curriculum research *praxis*, especially through the use of *patient longitudinal iteration*. My curriculum research has become a nested, spiralized phenomenon utilizing process reflection (Ricks, 2011). All lessons learned in previous studies form a personal *research curriculum* ready for application in the next research study; over time, a deep reservoir of experience-based praxis nourishes my new attempts at curricular research. A slow-but-steady, cumulative progression of curricular-research expertise builds over many years. For example, initial work on using modified lesson study in U.S. teacher training (Study One) was re-examined with subsequent iterations of methods-based, lesson-study use in Study Seven over the dozen-plus intervening years. Study Six, comprising curricular-based research across multiple East Asian cities, is another example of long-term, cumulative commitment, unfolding over many years during many separate trips. Although this study was summarized as a unitary research endeavor, in reality it consisted of numerous sub-studies, the boundaries of each blurring into the next investigation. A third example of harnessing lessons learned during my studies on curriculum occurs in the ongoing Study Eight's rolling of lessons learned (including during Study Eight) into each subsequent semester's syllabus.

As part of this patient longitudinal iteration, I have learned to be *more wise and expansive in my requests to research subjects for the use of research data*, such as long-term, secure storage and analysis, including sharing of data with other researchers who may eventually join my research teams, and the use of data for professional development purposes in educational settings (e.g., methods course curriculum, inservice training, research conferences). Such iterative use highlights how curriculum research itself can become one of the curricular components for current and future curriculum researchers—another layer of iteration. As curriculum researchers, we are in the process of creating our own domain of activity.

Lesson 2: Enhancing Contrast Through Juxtaposition

A second critical lesson about curriculum research I have learned over the course of these eight studies is: *Juxtaposition can be designed into research to enhance contrast.* This contrast helps disentangle elements of curriculum research from embedding contexts. I learned the value of juxtaposition as a design element for curriculum research starting in Study One; to encourage rich discussions, we purposefully created lesson study groups from mixes of teacher candidates expressing conceptual or procedural tendencies in journal prompts. Juxtapositional design was also manifest in Study Two as participating teachers experienced the reform-oriented pedagogy of intended, state-mandated curriculum as it was actually intended (i.e., reform-oriented). The participating teachers struggled to harmonize their own positive experiences while doing open-ended investigations with the anticipated enactment of such curriculum in the upcoming school year, revealing that between intended and enacted curriculum there exists a tenuous anticipatory domain over which teachers fret. I further used juxtapositional design in Study Three to enhance contrasting elements through a unique combination of parameters: (a) two teachers from a western U.S. state, one from an eastern U.S. state; (b) two teachers taught both algebra and pre-algebra classes; (c) one of the teachers from the western state taught in an affluent district, the other in a poor district.

Culture is another enmeshing complex system that confounds attempts at curriculum research. Culture affects the paradigmatic perspectives of all curriculum researchers, introducing biases and blind spots that taint any research endeavor because culture remains invisible until juxtaposed against another culture. For example, to remedy this dilemma, Study Four had both a Chinese and a U.S. teacher transform identical intended curriculum into actual lessons to enable comparison of enacted curriculum at a level not possible if either had been studied in isolation. Studying Chinese and Korean mathematics lessons (Studies Four, Five, and Six) across seven separate cities, and both Japanese and Chinese professional development principally during lesson study in Studies One, Five, and Six has helped make sense of U.S. curricular dynamics in new ways. Juxtaposing the research sites across the six Chinese cities allowed subtle differences to emerge between educational cultures and visions; for instance, Beijing, being a historical city manifests more traditional Chinese culture, whereas newer metropolises like Shanghai vibe with the pulse of internationalization, capitalism, and entrepreneurship. Visiting Korea after China brought out still more cross-national differences. Further, juxtaposing Japanese and Chinese lesson study allowed greater comparing and contrasting in my quest to adapt this valuable form of professional development to U.S. contexts. Ultimately, I feel that Japanese lesson study is a powerful form of professional development because of its integrity to Dewey's (1933) and Schön's (1983) concepts of process reflection (Ricks, 2011). The proximity provided by personal visits to other countries' schools is another aspect of juxtapositional research that I have found valuable; being there, on the ground, in immediate contact with curriculum innovators, actual teachers and students during mathematics lessons, or in the moment during lesson study debriefing meetings, has yielded priceless experience whose worth is difficult to emphasize. I also utilized the presence afforded by proximity to obtain various international curriculum materials (textbooks, lesson plans, professional development materials) and form potential future networking contacts that would not have been possible otherwise.

Another aspect of cross-cultural juxtapositional design cannot be overstated: the coordination of research endeavors with bilingual colleagues and subsequent in-depth collaboration. Language barriers continue to be a significant hurdle to methodological designs in many internationalized studies, especially for researchers like me who only speak one native tongue. I was fortunate to have bilingual cooperation in several studies (Study Four, Five, and Six). Two of the three teachers in Study Three were also bilingual (English, Spanish). Although language itself was not an explicit focus of any research study described in this chapter, the presence of bilingual colleagues can help in the quest for curriculum excellence. Additionally, collaboration (whether with bilingual colleagues or not) magnifies research effort. Working as a group makes all aspects of curriculum research easier as we harness the power of our "collective brain" (Leikin, 2004, p. 239).

Finally, I have utilized juxtapositional design as an organizing principle (Cuoco et al., 2010) to iteratively develop the curriculum of my own university mathematics methods classes (Ricks, 2009), creating what I call a *juxtapositional curriculum* that attempts to use international cultural conditions as contrasting leverage to dislodge or temporarily relieve some of the pinching pressure of current U.S. teaching norms. For example, I use Stigler and Hiebert's (1999) *The Teaching Gap* as a principal text, along with videos of international lessons. I pose challenging mathematics tasks used internationally to create teacher candidate experiences with Japanese "structured problem solving" (Stevenson & Nerison-Low, 1998; Stigler & Hiebert, 1999) or Chinese "one problem with variation" (Wang & Murphy, 2004) pedagogies. Comparing and contrasting international pedagogical approaches troubles my students' entrenched procedural expectations about their own anticipated future pedagogy. For several semesters we also established international pen pal partners between my methods students and the methods students of a Chinese colleague. When teaching the accompanying methods course lab (a field-based component pairing teacher candidates with a public-school mentor teacher), I have chosen classes in struggling, local public schools with high free/reduced lunch, students-of-color ratios instead of more settled schools in affluent neighborhoods. Roughly half of my methods students have never attended public school themselves, so these juxtapositional experiences are eye-opening, and help deepen the urgency for my students to self-examine the underlying assumptions of what they believe quality teaching to be. Juxtapositional curriculum appears useful towards closing various teaching "gaps" described by Stigler and Hiebert (1999) that plague U.S. education. Purposeful juxtaposition in design introduces clarifying contrast that illuminates confounding aspects being considered. In summary, leveraging the power of juxtapositional design (e.g., data contrast, personal proximity, bilingual cooperation, collaboration) is a promising new tool for curriculum researchers.

Lesson 3: Casting Wide Nets to Capture Special Cases

Considered as a complex system itself, curriculum is never a monolithic entity, irrespective of the level of analysis. Complex system emergence (Johnson, 2001) can be occasioned, but never forced; behaviors are chaotic, unpredictable, oscillating between equilibriums, forever dynamic. The third major theme I have learned is to metaphorically *cast a wide net to catch the exotic fish* to deal with these dilemmas. The wider the net cast, the greater the chance of catching those very special moments in complex system emergence that cannot be planned for in advance. An obvious application of this principle occurred during videography of mathematics lessons and lesson study meetings, filming U.S., Chinese, and Korean mathematics lessons with multiple cameras to provide a greater field of view of classrooms, increase redundancy in case technology fails, and improve audio capture. Multiple camera angles with clever camera placement and multiple sound capture facilitate the study of mathematics curriculum implementation. Affordable, compact, wearable cameras have been effective to capture whole-class dynamics with built-in, wide-angle (fish-eye) lenses (Study Six). Their small size makes them easily portable when traveling through airport customs, in taxis, and between research sites; they are conveniently inconspicuous (i.e., not distracting to instruction) when deployed around classrooms from varying vantage points, such as mounted on hand-sized tripods placed on surrounding shelves or windowsills. They have been a valuable addition to the larger, person-controlled cameras with built-in zoom lenses that focus on precise details of classroom action. Intense

Disentangling Mathematics Curriculum From Contextual Enmeshment 593

zooming to focus on something that seems important to the videographer at that moment immediately blinds the videorecording to other potentially important contextual classroom dynamics unless other perspectives are covered by these rolling (recording) cameras.

It is easier to metaphorically cast a wide fishing net a few times to find the exotic fish—hauling in many fish, sorting through them, casting back into the sea the unneeded fish—than it is to continually re-cast smaller nets again and again, hoping perhaps to catch a single exotic fish. For example, gaining access to research sites requires tremendous effort, amiable participants, and a good deal of luck; once in, it is better to utilize the site as best as possible, instead of thinking later to organize another research site. I study mathematics curriculum through a complex systems framework, and the behavior of complex systems is unpredictable, dynamic, and often subtle. So, I have found that collecting larger data sets enables capture of those special moments that can be analyzed in greater detail. The wider the net cast, the more fish collected, the greater the probability of catching the exotic fish. For example, in Study One, I studied two separate lesson study groups, and selected only one of the groups for fine-grained analysis (the group with the richer data set). In Study Three, I initially studied 18 expert mathematics teachers, later honing the dataset to only three teachers for the full, multi-week video study. I filmed each teacher for the equivalent of six weeks of instructional time in eight of their classes. From this large dataset, I selected several excellent lesson episodes for the fine-grained analysis required for case-study research; I had sorted through the wide net's catch, and finally found my exotic fish.

I have learned that part of casting a wide net is: *Utilize potential resources at your disposal, including the magnifying effect of technology*. We sometimes fish by hand, although boats with more powerful engines rapidly reach richer fishing ranges, sonar finds deeply hidden schools of fish, and mechanized machinery hauls in larger, heavier nets. All eight of my studies have harnessed technology in data collection and/or analysis, including laptops to facilitate participant-teacher reflection on curricular issues through electronic journaling and digital software for task solutions (in Study Two), electronic communication across continents (in Studies Four through Seven), and videotaping/audiotaping of classroom dialogue, lesson study, and/or interviews (in all eight studies). In Studies Three, Six, and Eight, I utilized multiple camera angles and audiorecording locations to maximize the chance of capturing the highest-quality data. Toward the end of data collection in Study Three, I was using five separate cameras around the periphery of the teacher's classroom, fed with audio through a mixer from two, high-quality, omni-directional, ceiling-hung microphones, to capture classroom dynamics that could be combined with table-mounted microphones to capture selected, small-group work. Such precision, especially to capture hard to hear student conversations, has provided an unprecedented look into classroom curricular dynamics. The continuing expansion of technological tools that can be rapidly employed in curriculum research proffers exciting new possibilities that promise to nourish our continued efforts. For example, I am currently utilizing artificial intelligence to analyze classroom dialogue patterns from large datasets, something others are also experimenting with; this exciting new technology can easily be extended into the area of textbook or other written curricular analysis, where large amounts of curriculum can be analyzed rapidly. Other newer technologies, like comfortable virtual reality, affordable eye-tracking hardware/software, and fMRI (functional magnetic resonance imaging) of brain activity, extend our fishing range as curriculum researchers into new, uncharted, unfished seas.

Lesson 4: Bureaucratic Dangers

The fourth and final lesson I am learning when doing mathematics curriculum research is: *Entrenched educational bureaucracies entangle curriculum and disrupt attempts at curriculum research*, usually in negative ways by destabilizing the democratic agency of system participants and calcifying the dynamism needed for responsive curricular innovation. Bureaucracies can act as demagogic pedagogues, deflecting the potency of positive curricular momentum through proceduralization, standardization, quantization, and eventual corporatization. My newer research (Study Eight, and beyond) investigates the role of power imbalances, principally in the form of bureaucratic overreach, to disrupt these useful curriculum endeavors. Bureaucracy often magnifies the power and personality of particular people in inequitable and undemocratic ways. Bureaucracies can also act as gatekeepers that frustrate researchers' access to curriculum research sites. Rarely are there meaningful and accessible checks and balances on bureaucratic power. I believe that mathematics curriculum research should give the influential topic of bureaucratic power (especially overreach) more attention in the years to come, especially as contemporary trends point to ever-tighter administrative control of curriculum at all systemic levels. In particular, researchers might examine how productive efforts at mathematics curriculum reforms are being undermined, derailed, or diluted by bureaucratic agendas, and also, perhaps to learn of ways to navigate the terrain of researchers and policymakers working together.

CONCLUSION

In this chapter, I detailed some significant issues facing our scholarly community of curriculum researchers, and how some cross-cutting themes I learned while doing mathematics curriculum research may illuminate potential solution paths through this landscape. In particular, I highlighted eight specific studies that were conducted to understand more about how curriculum reform can be achieved, and what barriers prevent its improvement. Over these eight studies, I learned about the paradigmatic potential of complexity theory to recast curricular endeavors as part of nested complex systems, the power of juxtapositional design in curriculum research methodology, specific strategies for casting wide nets to catch the exotic fish, and the growing influence and potential dangers of bureaucratic power. Each lesson illuminates ways in which curriculum can begin to be disentangled from contextual enmeshment. I close by reemphasizing the need for our community of mathematics curriculum researchers to understand how bureaucratic entrenchment can stifle both curriculum improvement and further research on curriculum.

REFERENCES

Corey, D. L., Peterson, B. E., Lewis, B. M., & Bukarau, J. (2010). Are there any places that students use their head? Principles of high-quality Japanese mathematics instruction. *Journal for Research in Mathematics Education, 41*(5), 434–478.

Cuoco, A., Goldenberg, E. P., & Mark, J. (2010). Contemporary curriculum issues: Organizing a curriculum around mathematical habits of mind. *Mathematics Teacher, 103*(9), 682–688.

Davis, B., & Simmt, E. (2003). Understanding learning systems. *Journal for Research in Mathematics Education, 34*, 137–167.

Davis, B., & Sumara, D. (2001). Learning communities: Understanding the workplace as a complex system. *New Directions for Adult and Continuing Education, 92*, 85–95.

de Freitas, E., Ferrara, F., & Ferrari, G. (2019). The coordinated movements of collaborative mathematical tasks: The role of affect in transindividual sympathy. *ZDM Mathematics Education, 51*, 305–318. https://doi.org/10.1007/s11858-018-1007-4

Dewey, J. (1923). *The child and the curriculum.* University of Chicago Press.

Dewey, J. (1933). *How we think: A restatement of the relation of reflective thinking to the educative process.* Heath.

Glaser, B. G. (1965). The constant comparative method of qualitative analysis. *Social Problems, 12*, 436–445.

Heck, D., Chval, K., Weiss, I., & Ziebarth, S. W. (Eds.). (2012). *Approaches to studying the enacted mathematics curriculum.* Information Age Publishing.

Hill, H., Rowan, B., & Ball, D. (2005). Effects of teachers' mathematics knowledge for teaching on student achievement. *American Education Research Journal, 42*(2), 371–406.

Johnson, S. (2001). *Emergence: The connected lives of finance, brains, cities, and software.* Scribner.

Kilpatrick, J., Mesa, V., & Sloane, F. (2006, November 9-11). *Algebra teaching and learning viewed internationally* [Paper presentation]. Paper presented at the 2nd International Education Association's International Research Conference, Brookings Institution.

Lather, P. (2004). Scientific research in education: A critical perspective. *British Educational Research Journal, 30*(6), 759–772.

Leikin, R. (2004). The wholes that are greater than the sum of their parts: Employing cooperative learning in mathematics teachers' education. *Journal of Mathematical Behavior, 23*, 223–256.

Lewis, C. (2002). *Lesson study: A handbook of teacher-led instructional change.* Research for Better Schools.

Li, Y., & Lappan, G. (Eds.). (2014). *Mathematics curriculum in school education.* Springer.

Lu, L., & Ricks, T. E. (2012). Chinese mathematics teaching reform in four years time: The DJP model. In L. R. Van Zoest, J.-J. Lo, & J. L. Kratky (Eds.), *Proceedings of the 34th annual meeting of the North American Chapter of the International Group for the Psychology of Mathematics Education* (p. 571). Western Michigan University.

National Governors Association Center for Best Practices & Council of Chief State School Officers. (2010). *Common core state standards for mathematics.* Retrieved May 2, 2022, from http://thecorestandards.org/

Nipper, K., Ricks, T. E., Kilpatrick, J., Mayhew, L., Thomas, S., Kwon, N. Y., Klerlein, J. T., & Hembree, D. (2011). Teacher tensions: Expectations in a professional development institute. *Journal of Mathematics Teacher Education, 14*, 375–392. https://doi.org/10.1007/s10857-011-9180-1

Peterson, B. E. (2010). Mathematics student teaching in Japan: Where's the management. In G. Anthony & B. Grevhold (Eds.), *Teachers of mathematics: Recruitment and retention, professional development and identity* (pp. 135–144). Kristiansand.

Peterson, B. E., Teuscher, D., & Ricks, T. E. (2019). Lesson study in a mathematics methods course: Overcoming cultural barriers. In R. Huang, A. Takahashi, & J. Pedro da Ponte (Eds.), *Theory and practice of lesson study in mathematics, advances in mathematics education* (pp. 527–548). Springer Nature. https://doi.org/10.1007/978-3-030-04031-4_26

Remillard, J. T., & Heck, D. J. (2014). Conceptualizing the enacted curriculum in mathematics education. In D. R. Thompson & Z. Usiskin (Eds.), *Enacted mathematics curriculum: A conceptual framework and research needs* (pp. 121–148). Information Age Publishing.

Remillard, J., & Reinke, L. (2018). Mathematics curriculum in the United States: New challenges and opportunities. In D. R. Thompson, M. A. Huntley, & C. Suurtamm (Eds.), *International perspectives on mathematics curriculum* (pp. 133–164). Information Age Publishing.

Ricks, T. E. (2003). *An investigation of reflective processes during lesson study by mathematics preservice teachers* [Unpublished Master's thesis, Brigham Young University].

Ricks, T. E. (2007). *The mathematics class as a complex system* [Unpublished Doctoral dissertation, University of Georgia].

Ricks, T. E. (2009). Juxtapositional pedagogies in mathematics methods courses. In S. L. Swars, D. W. Stinson, & S. Lemons-Smith (Eds.), *Proceedings of the 31st annual meeting of the North American Chapter of the International Group for the Psychology of Mathematics Education* (Vol. 5, pp. 1244–1252). Georgia State University.

Ricks, T. E. (2010a). China, microbes, and complexity: How to improve mathematics education. In H. Zhang (Ed.), *Complexity, Chinese culture, and curriculum reform* (pp. 112–122). East China Normal University Press.

Ricks, T. E. (2010b). Juxtaposing Chinese and American mathematics education commitments. In P. Brosnan, D. B. Erchick, & L. Flevares (Eds.), *Proceedings of the 32nd Annual Meeting of the North American Chapter of the International Group for the Psychology of Mathematics Education* (Vol. 6, pp. 354–362). The Ohio State University.

Ricks, T. E. (2011). Process reflection during Japanese lesson study experiences by prospective secondary mathematics teachers. *Journal of Mathematics Teacher Education, 14*, 251–267. https://doi.org/10.1007/s10857-010-9155-7

Ricks, T. E. (2014). Why always greener on the other side? The complexity of Chinese and US mathematics education. In B. Sriraman, J. Cai, K. H. Lee, F. Lianghuo, Y. Shimuzu, L. C. Sam, & K. Subramanium (Eds.), *The first sourcebook on Asian research in mathematics education: China, Korea, Singapore, Japan, Malaysia and India* (pp. 759–784). Information Age Publishing.

Ricks, T. E. (2016). Analysis of singular-plural dialogue of mathematics teachers. In M. B. Wood, E. E. Turner, M. Civil, & J. A. Eli (Eds.), *Proceedings of the 38th Annual Meeting of the North American Chapter of the International Group for the Psychology of Mathematics Education* (p. 1424). The University of Arizona.

Ricks, T. E. (2017). Reflective capabilities of mathematics education systems in China, Japan, and the United States. In J.-W. Son, T. Watanabe, & J.-J. Lo (Eds.), *What matters? Research trends in international comparative studies in mathematics education* (pp. 237–258). Springer.

Ricks, T. E., Lu, L., & Fleener, M. J. (2009). Understanding students' thinking from an interrelationship perspective. *Journal of Mathematics Education, 17*(6), 66–71.

Rogers, G. W. (2017). *"I'm Not Good at Math": Mathematical illiteracy and innummeracy [sic] in the United States* [Unpublished Doctoral dissertation, Georgia Southern University]. Digital Commons@Georgia Southern. https://digitalcommons.georgiasouthern.edu/etd/1597

Schön, D. A. (1983). *The reflective practitioner: How professionals think in action*. Basic Books.

Shapiro, J. A. (1998). Thinking about bacterial populations as multicellular organisms. *Annual Reviews of Microbiology, 52*, 81–104.

Stein, M. K., Smith, M. S., Henningsen, M. A., & Silver, E. A. (2000). *Implementing standards-based mathematics instruction: A casebook for professional development*. Teachers College Press.

Stevenson, H. W., & Nerison-Low, R. (1998). *To sum it up: Case studies of education in Germany, Japan, and the United States*. National Institute on Student Achievement, Curriculum, and Assessment. Office of Educational Research and Improvement. U.S. Department of Education.

Stigler, J. W., & Hiebert, J. (1999). *The teaching gap*. Free Press.

Surowiecki, J. (2004). *The wisdom of crowds: Why the many are smarter than the few and how collective wisdom shapes business, economics, societies, and nations*. First Anchor Books.

Suurtamm, C., Huntley, M. A., & Thompson, D. R. (2018). What might be learned from examining curricular perspectives across countries? In D. R. Thompson, M. A. Huntley, & C. Suurtamm (Eds.), *International perspectives on mathematics curriculum* (pp. 1–8). Information Age Publishing.

Thompson, D. R., Huntley, M. A., & Suurtamm, C. (Eds.) (2018). *International perspectives on mathematics curriculum*. Information Age Publishing.

Thompson, D. R., & Usiskin, Z. (Eds.). (2014). *Enacted mathematics curriculum: Conceptual framework and research needs*. Information Age Publishing.

Tienken, C. H. (2008). Rankings of national achievement test performance and economic strength: Correlation or conjecture? *International Journal of Educational Policy & Leadership, 3*(4), 1–15.

Tierney, R. J. (2001). An ethical chasm: Jurisdiction, jurisprudence, and the literacy profession. *Journal of Adolescent and Adult Literacy, 45*(4), 260–276.

Wang, T., & Murphy, J. (2004). An examination of coherence in a Chinese mathematics classroom. In L. Fan, N.-Y. Wong, J. Cai, & S. Li (Eds.), *How Chinese learn mathematics: Perspectives from insiders* (pp. 107–123). World Scientific Publishing Company.

Yang, Y., & Ricks, T. E. (2011). How crucial incidents analysis support Chinese lesson study. *International Journal for Lesson and Learning Studies, 1*(1), 41–48. https://doi.org/10.1108/20468251211179696

Yang, Y., & Ricks, T. E. (2012). Chinese lesson study: Developing classroom instruction through collaborations in school-based teaching research activities. In Y. Li & R. Huang (Eds.), *How Chinese teach mathematics and improve teaching* (pp. 51–65). Routledge.

Yuan, Z., & Ricks, T. E. (2011). A comparative study on Chinese and U.S mathematics teachers' classroom teaching based on mathematical problems: A case of Geometric Probability Model. *Fujian Education, 827*(46), 34–36.

ENDNOTE

1. National Governors Association Center for Best Practices & Council of Chief State School Officers (2010).

CHAPTER 28

EVERYTHING I REALLY NEED TO KNOW ABOUT CURRICULUM, I LEARNED FROM CSMC

Jill A. Newton
Purdue University

My introduction to mathematics curricular research started in 2004 as a graduate assistant in the Center for the Study of Mathematics Curriculum (CSMC) at Michigan State University. Since then, I have collaborated on investigations of written, enacted, and assessed curricula and explored relationships among these curricular levels. In this chapter, I highlight four lessons learned from CSMC and subsequent research: (1) results from research matter and the rich learning experiences are never wasted; (2) the importance on the research team of diverse and dissenting perspectives; (3) the power of theoretical, conceptual, and analytic frameworks for curriculum research; and (4) the complexities of curriculum research. I share reflections related to significant collaborations, my changing perception of the nature of research, and embracing challenges and complexities. I also pose questions for myself and the field about the need to foreground attention to social contexts of mathematics education in curriculum scholarship, emphasizing the importance of the stories we tell and their potential impact.

Keywords: curriculum framework; curriculum research; mathematics curriculum

The title of this chapter was inspired by the popular set of essays, *All I Really Need to Know I Learned in Kindergarten* (Fulghum, 1986). As I reflected on my curriculum research experiences, I could trace nearly all lessons learned to my work as a graduate student in the Center for the Study of Mathematics Curriculum (CSMC). I begin this chapter with background information about CSMC and my work on the project, followed by a summary of lessons learned through these experiences and subsequent research projects. I include what I learned, how I learned it, and who I learned it from, acknowledging that those of us currently conducting mathematics curriculum research are standing on the shoulders of giants.

INTRODUCTION

In Spring 2004, I visited Michigan State University (MSU) to meet with Cassandra Book, a friend of a friend and an Associate Dean in the College of Education, about the possibility of working with student teachers at MSU. I had been teaching mathematics overseas for nine years; my dad had been recently diagnosed with cancer and I was headed back to the U.S. hoping to do something different yet related to mathematics education. Cass suggested that I consider a PhD in mathematics education; I had a BS in mathematics from MSU and a MA in international education from George Washington University (U.S.). As a first-generation college student with no awareness that such a doctoral degree concentration existed, I was both intimidated and intrigued. Cass made a few calls and set up opportunities for me to talk with several professors. Fast forward to my last meeting that afternoon. I left Elizabeth (Betty) Phillips's office with a large stack of *Connected Mathematics Project* (*CMP*) books, having been invited to be a CSMC graduate student fellow.

CSMC was a U.S. National Science Foundation (NSF)-funded Center for Learning and Teaching (CLT) that focused on scholarly inquiry and professional development related to issues of K–12 mathematics curriculum; developing leadership capacity related to curricular design, analysis, implementation, and evaluation; advancing a mathematics curricular research agenda; and creating an organizational structure to encourage and facilitate communication and productive collaboration among mathematics curriculum scholars. The original CMSC team was led by Barbara Reys (PI, University of Missouri), along with Co-PIs Christian Hirsch (Western Michigan University), Glenda Lappan (MSU), and Robert (Bob) Reys (University of Missouri). Just prior to the launching of CSMC, curriculum had been identified as one of six principles for school mathematics in *Principles and Standards for School Mathematics* (*PSSM*) (National Council of Teachers of Mathematics [NCTM], 2000). Barbara Reys had served on the writing team for this document, so CSMC's focus on curriculum was a natural fit for her.

As a mathematics education researcher, I was "raised" in the context of curriculum research. From the beginning, I read about teachers' use of curricular materials (e.g., Drake et al., 2001; Remillard & Bryans, 2004), *Standards*-based (NCTM, 1989) curriculum in what we called "The Senk and Thompson book" (Senk & Thompson, 2003), the complexity of evaluating curriculum research (e.g., National Research Council [NRC], 2004), and ways to frame curriculum research in what we called the "curriculum handbook chapter" (Stein et al., 2007).[1] At MSU I worked in an office in which I had regular contact with numerous people who were writing and field testing *Connected Mathematics* (*CMP*) (Lappan et al., 2006), a NSF-funded, *Standards*-based middle-grades mathematics curriculum. At that time, two of the lead *CMP* textbook writers were at MSU (Glenda Lappan and Betty Phillips),[2] together with other writers (Christopher Danielson, Yvonne Grant, Jacqueline Stewart) and *CMP* pilot teachers (e.g., Mary Bouck, Kathy Dole, Terry Keusch). My interactions with these people at MSU provided a front-row seat to authentic theory-practice curricular connections, and reinforced the notion that to improve teaching and learning in mathematics classrooms it is essential for curriculum researchers to work hand-in-hand with diverse teams of people.

On my first day as a graduate student at CSMC in Fall 2004, Glenda met with Gregory Larnell (another new CSMC research assistant) and me to explain the goals of a CSMC study to analyze the current grades K–8[3] standards for mathematics from 42 states. We were charged to begin with the algebra standards and each handed a CD-ROM with data. Greg and I still look back on this experience in amazement as we walked to our shared office, wondering how we would tackle this daunting assignment. Our admiration for Betty and Glenda began that day as they entrusted us with this important and challenging work. What is most notable about this extensive study, completed by many members of the CSMC research team, is that almost immediately upon publication of two volumes of these analyses, the research seemingly became immaterial because in 2010 the *Common Core State Standards for Mathematics* (*CCSSM*) was published (National Governors Association Center for Best Practices [NGA] and the Council of Chief State School Officers [CCSSO]). Most states now no longer use the standards we painstakingly analyzed for several years. Many lessons were learned during this investigation. The most important for me was: *The learning that results from research matters regardless of unpredictable future events; the rich learning experiences offered to the research team and others (e.g., participants, audience) are never wasted.* That which was learned may be different than expected, but everyone involved learned something and that learning will inform future research and practice.

Since this first encounter with mathematics curricular research, I have collaborated on many investigations of written, enacted, and assessed curricula (e.g., Giorgio-Doherty et al., 2021; Mintos et al., 2018; Newton, 2011; Stehr et al., 2016) and explored relationships among these curricular levels (e.g., Newton, 2012; Newton & Kasten, 2013). In the following sections, I describe three additional lessons learned from CSMC and subsequent research experiences: (1) the importance of having a research team with members having diverse and dissenting perspectives; (2) the power of theoretical, conceptual, and analytic frameworks for curriculum research; and (3) the complexities of curriculum research.

THE IMPORTANCE OF BEING PART OF A RESEARCH TEAM WITH MEMBERS HAVING DIVERSE AND DISSENTING PERSPECTIVES

From my perspective, one of the greatest strengths of CSMC was the diverse community of people involved in the project. Beyond the leadership team, the CSMC community was comprised of additional senior researchers, research associates (i.e., faculty members at institutions not officially part of CSMC), graduate students, curriculum developers, K–12 school administrators, and teachers. This diversity facilitated a complex collaborative and mentoring network that informed and strengthened all project activities. We collaborated and met with nearly every subset of team members across the years of the project, providing various opportunities to learn from one another. Because members of the CSMC team had a range of expertise (e.g., curriculum development, curricular implementation, curricular evaluation), they offered unique insights into curriculum research related to, among other things, assumptions about practice, curricular coherence, and curricular adoption across the U.S. As you can imagine, the more novice among us potentially learned the most; however, the leaders often made explicit—through their words and actions—that the range of voices and generations of individuals on the project were assets and all input and questions were valued. At CSMC conferences and project meetings, there were countless opportunities as graduate students progressed through their doctoral program to present and discuss our research with those more senior than us and those who were newer to the work. During graduate school, I often presented and published research with other CSMC graduate students (e.g., Sarah Kasten, MSU; Shannon Dingman, University of Missouri) and faculty (e.g., Glenda Lappan, MSU; Bob Reys, University of Missouri). Many of these collaborations and extensions of them continue today.

Valuing School Personnel

Through the example of the leadership team, we learned to value the insights of school personnel to inform our research. This included teachers and administrators, who were represented at all three research sites and at all CSMC events. For example, at MSU, we worked closely with Valerie

Mills, a school administrator at Oakland Intermediate School District. Valerie's contributions were invaluable. She was well versed in best practices for teaching mathematics, knew the research well, and was a strong advocate for teachers in her district. We also engaged with teachers and learned alongside them in professional development settings. At MSU, I collaborated with Betty, Glenda, and Sandy Wilcox on a long-term professional development experience in local schools. This was my first opportunity to facilitate teacher communities, and I was in expert hands.

We began our teacher collaboration with an examination of the mathematics curriculum used in the district, finding that teachers had not had the opportunity to look across the different curricula to interrogate the coherence among them. I remember one activity, led by Sandy, in which each grade-level team of teachers developed a poster related to the attention given to fractions in their grade level. They were asked to answer the following questions: What do you assume that students know about fractions when they come to your classroom? What do students learn about fractions in your classroom and how do they learn it? What do you expect students to know about fractions when they leave your classroom? These posters were then displayed in sequence from K–8. We were all amazed by many things, including that several fraction concepts and procedures were repeated multiple times and some important concepts and procedures were absent. The collaborative learning from this and other activities was enlightening for me and demonstrated the importance of teachers engaging in investigations and analysis of their curriculum as equal partners in research. The professional development and research teams also learned from the teachers, many of whom had been in mathematics classrooms and school leadership positions for many years. For example, teachers highlighted the realities of teaching students at a range of proficiencies on any topic; in the fraction example, reteaching fraction conceptualizations and operations was necessary to ensure that struggling students had opportunities to develop their knowledge.

Valuing Dissenting Perspectives

When I began writing this section, I had not yet added the word "dissenting" to the title. However, when thinking about the CSMC team, my mind went immediately to three CSMC team members: (1) Jere Confrey, who was the Chair of the Committee for a Review of the Evaluation Data on the Effectiveness of NSF-Supported and Commercially Generated Mathematics Curriculum Materials (NRC, 2004) at the time (now Professor Emeritus, North Carolina State University); (2) Zalman Usiskin, who was the Director of the University of Chicago School Mathematics Project (https://ucsmp.uchicago.edu/) at the time (now Professor Emeritus, University of Chicago); and (3) Iris Weiss who was the President of a research and evaluation firm (http://www.horizon-research.com/) at the time (now Retired President, Horizon Research, Inc.). Jere, Zal, and Iris brought strong opinions to the curriculum research enterprise based on many years of experience and often asked challenging questions about various aspects of the CSMC research agenda, framework, and particular studies. For example, I remember heated discussions about the seemingly impossible research evidence requirements established by the NRC (NRC, 2004) and the What Works Clearinghouse (https://ies.ed.gov/ncee/wwc/) to establish a curriculum as "effective." Jere emphasized the critical need for broad, rigorous research standards, and the curriculum developers in the crowd (e.g., Betty, Glenda, Zal) argued that no existing curriculum would meet the standards.

Another topic debated among CSMC personnel related to developing curricula that are appealing to and usable by teachers. A curriculum that sits on a shelf, no matter how theoretically "perfect," serves no purpose. These negotiations inevitably led to rich discussions about how to develop curricula and how to conduct research to establish curricular effectiveness to inform research and school audiences. As a novice researcher, I greatly appreciated the opportunity to see a team of scholars negotiate complex ideas, sometimes in seemingly contentious ways, and then go to dinner and laugh, tell stories, and catch up on everyone's families. They recognized they were smarter together and that the multiple perspectives and experiences represented were critical to the quality and impact of CSMC curriculum activities.

Valuing Belongingness

CSMC's organizational structure also included Research Associates (e.g., Óscar Chávez, Jeff Choppin, Corey Drake, Mary Ann Huntley, Janine Remillard). These curriculum researchers, at various stages of their careers, were faculty at institutions that were not necessarily part of CSMC. They provided senior faculty access to innovative ideas and insights and contributed to a tiered mentoring system. For CSMC graduate students, the CSMC Research Associates provided invaluable support as we navigated our doctoral programs and established our own research agendas. The impact has been ongoing as this implicit but significant mentorship continues today.

Being a member of the CSMC community meant that you never, even as a graduate student and novice researcher, felt alone at a mathematics education conference; there were always other CSMC people around and you felt part of something bigger than yourself. Moreover, we had opportunities to engage with graduate students and faculty from other

NSF-funded mathematics education CLTs with structures similar to CSMC (e.g., Center for the Mathematics Education of Latinos [CEMELA], Diversity in Mathematics Education [DIME]). The numerous opportunities for team members to present individually and collectively at annual CSMC and national conferences enhanced the sense of belonging, both in the CSMC research community and the larger mathematics education research community. These experiences allowed us to hit the ground running as new faculty, as we had developed a network of curriculum researchers, strong skills for collaborating, and an understanding of important factors to consider for curriculum development, evaluation, and research.

Developing Curriculum Research Communities

Although I have never led a project as large as CSMC, I took these lessons to heart and tried to model them in my work. My first NSF-funded grant, *Preparing Teachers of Algebra* (*PTA*), was a collaborative grant with MSU. I do not remember exactly how the idea for *PTA* arose, but there were certainly CSMC connections. Sharon Senk was the PI; she mentored Yukiko Maeda and me as Co-PIs throughout the project. Sharon had worked with Yukiko on a study at MSU and Yukiko brought quantitative research expertise to the leadership team. We had internal and external advisory boards with individuals who provided diverse perspectives. We hired multiple graduate students at both universities and, in the spirit of CSMC, provided opportunities for them to collaborate on research projects and present and publish with faculty and one another. In fact, 10 years later, the last of these collaborations is in the final stages. This involves Hyunyi Jung, a Purdue graduate student at the time and now on the faculty at the University of Florida, and Eryn (Stehr) Maher, an MSU graduate student at the time and now on the faculty at Georgia Southern University.

We also hired, through several initiatives at Purdue, multiple undergraduate researchers (UGRs) to work on *PTA*. These UGRs brought important skills, experiences, and insights to our research. For example, two of the UGRs were skilled with technologies that enhanced our capacity to disseminate our findings, increasing broader impact. Several of the students were in the secondary mathematics teacher education (SMTE) program at the time. We were studying SMTE programs and their current experiences were invaluable in preparing survey and interview questions. All UGRs had opportunities to present work with other UGRs, graduate students, and faculty. *PTA* was modeled after CSMC in the layers of mentoring—Sharon as a senior researcher, Yukiko and I as junior faculty researchers, graduate students as more novice researchers, and UGRs as brand-new researchers, each group and individual bringing a unique set of valuable experiences and insights.

In two other curriculum projects, the lessons learned from CSMC and *PTA* about the value of diverse voices continued. First, I collaborated with a team of faculty and teachers from around the state of Indiana on a collaborative professional development experience, *Creating Algebra Teacher Communities for Hoosiers (CATCH)*. Faculty from four universities worked with each other and teachers from high-needs schools near their universities to design and enact four summers of algebra learning opportunities for teachers. These experiences were just as educative for faculty as for the teachers. I immediately invited two graduate students, Andrew Hoffman and Brooke Max, to co-facilitate the sessions at our site, recognizing that their creativity, energy, and insights would be invaluable. We visited the school district in which we were interested in working, Frankfort Community Schools, bringing just a pad of paper and our ideas, as it was essential that the goals and curriculum of the project were co-developed with the teachers. We first met with the 10 middle- and high-school mathematics teachers, and one of them mentioned that they often co-taught their mathematics classes with special education teachers. This seemed like a perfect opportunity to add additional voices with different areas of expertise to the project; so, we invited the five special education teachers to our next meeting, who then joined the *CATCH* team. We designed opportunities for the faculty and teachers from the four sites to meet at the Indiana Council of Teachers of Mathematics (ICTM) meeting each year. We presented our work together there and at the annual Indiana STEM Conference. Our personal and professional collaborations with these teachers have continued as, for example, they talk to our student teachers each year about how to develop positive co-teaching relationships.

In another NSF-funded research grant, *Co-developing a Curriculum Coherence Toolkit with Teachers (C3T2)*, a collaboration with Corey Drake (CSMC Research Associate, MSU), Amy Olson (Duquesne University), and Marcy Wood (University of Arizona), we again sought a diverse set of partners. Corey and I are qualitative curriculum researchers; Amy is a quantitative researcher with a background in assessment and psychology; Marcy is a qualitative researcher with extensive elementary education experience. Our advisory board included curriculum researchers, a K–12 curriculum specialist, and a researcher who focuses on creating ambitious learning opportunities in historically underserved communities. Throughout the project, we funded graduate students with diverse areas of experience and expertise. As always, I invited UGRs to join our research team. They presented and published research in venues designed for UGRs (e.g., *Purdue Undergraduate Research Journal*) and in more traditional mathematics education conferences and journals with other *C3T2* team members.

In summary, *I learned from CSMC and subsequent curriculum research experiences how to develop a research team*. I learned the importance of including researchers at different levels of their careers (e.g., senior researchers, junior faculty, graduate students, and UGRs), school personnel, and individuals with a range of other expertise (e.g., quantitative research, curriculum development, teaching particular grade levels). I learned to seek out those with diverse and even dissenting perspectives because considering alternative approaches and explanations will strengthen the research. I learned to support those on the research team, providing opportunities for them to participate in all aspects of the research enterprise. Equally important, I learned to encourage research team members to support, mentor, and collaborate with one another, creating leadership and presentation/publication opportunities for them.

THE POWER OF THEORETICAL, CONCEPTUAL, AND ANALYTIC FRAMEWORKS FOR CURRICULUM RESEARCH

I remember hearing about frameworks for the first time in a doctoral seminar at MSU and looking around the room, wondering if others were as simultaneously confused and blown away as me. Of course, I was aware that everyone has their own perspectives on the world, but naming these lenses and explicitly using them to analyze data were new to me. I also remember a talk given by Denise Spangler in which she expertly described such frameworks. I have often recommended this talk, and the associated papers (Mewborn, 2005; Spangler & Williams, 2019) to my graduate students. Over time, I came to enjoy exploring, selecting, and utilizing conceptual and analytic frameworks in my curriculum research. At the same PME-NA conference in which Denise gave her "frameworks talk," I attended a discussion group (Choppin et al., 2005) focused on mathematics classroom discourse, that is, what curriculum researchers might refer to as enacted curriculum. In this session, Mary Truxaw, Megan Staples, Beth Herbel-Eisenmann, and Jennifer Seymour demonstrated the diverse set of stories that can be told from data as each analyzed the same set of classroom transcripts, a written form of enacted curriculum, using different analytic frameworks. I found this fascinating, as it opened the possibility of telling many equally valid and important stories and problematized for me the notion of research as an objective enterprise. In particular, the use of frameworks in curriculum research highlights important aspects of mathematical content and processes of standards, cognitive demand of tasks, and alignment between standards and assessments. In the sections that follow, I describe my learning how to use and develop theoretical, conceptual, and analytic frameworks for curriculum research.

Conducting Analysis of the K–8 State Standards

In the CSMC standards analyses in which we explored the grades K–8 standards from 42 states, we collaboratively selected and utilized multiple frameworks. First, Gregory Larnell and I used NCTM's (2000) four algebra descriptors as the initial framework to explore the K–8 algebra standards: *Understand patterns, relations, and functions*; *Represent and analyze mathematical situations and structures using algebraic symbols*; *Use mathematical models to represent and understand quantitative relationships*; *Analyze change in various contexts*. We reviewed each grade level expectation (GLE) in the state standards documents and identified more than 3000 GLEs associated with algebra, as defined by the NCTM descriptors. Our framework evolved over time, as is often the case, and we organized our data into three primary strands (*Patterns*; *Functions*; and *Expressions, Equations, and Inequalities* [EEI]) and two secondary strands (*Properties*; *Relationships Between Operations*). In Newton et al. (2006), we described aspects of these GLEs, including how they built across grade levels, how they varied in level of specificity, and which topics were present in most of the states to form a pseudo common curriculum.

At this same time, I conducted a parallel analysis of the grades K–8 geometry GLEs (Newton, 2011). Sharon Senk suggested I review the work of Pierre van Hiele and Dina van Hiele-Geldof (e.g., van Hiele, 1988; van Hiele & van Hiele-Geldof, 1958), who explored children's geometric thinking and the teaching associated with the development of this thinking. Sharon had used the levels described in van Hiele's framework in a previous study of secondary mathematics students' geometric thinking (e.g., Senk, 1989). The van Hiele levels offered a useful lens through which to investigate the GLEs, revealing significant attention across the states at Levels 1 (Visualization) and 2 (Analysis) in elementary and middle school and then a jump to Level 4 (Formal Deduction) in high school. Notable was the lack of attention to Level 3 (Informal Deduction), perhaps offering some explanation for students' challenges with proof.

As part of this same set of analyses, Leslie Dietiker, Aladar Horvath, and I used the process components (i.e., Formulate questions, Collect data, Analyze data, Interpret results) from the *Guidelines for Assessment and Instruction in Statistics Education* (*GAISE*) Framework (Franklin et al., 2007) to investigate the statistics GLEs (1711 in total) in the 42 state standards (Newton et al., 2011). We coded each GLE for one or more of the statistical process components. The GAISE framework allowed us to notice that students were expected to spend much more time analyzing data and interpreting results than formulating questions and collecting data. That is, they were usually given data and asked to do things with it. Using this framework, we identified Type I and Type II expectations within each component, sometimes in the same GLE. Type I expectations required students to be able to do the process (e.g., Pose questions and gather data about themselves and their surroundings; MO, Kindergarten). In Type II expectations, students evaluated the process in some way (e.g., Recognize and analyze faulty interpretation or representation of data; MD, Grade 7). Type I expectations were much more common than Type II expectations, particularly in some states.

Finally, Sarah Kasten and I analyzed the K–8 measurement standards, specifically 1601 GLEs that addressed one-, two-, and three-dimensional measurement (Kasten & Newton, 2011). We searched for an appropriate framework to use. However, when we piloted possible frameworks in a review of the GLEs, no one framework seemed to answer our questions. So, we elected to use a set of frameworks, including measurement growth point descriptors (Clarke et al., 2003) and three sets of "big ideas" in spatial measurement (Lehrer, 2003; Lehrer et al., 2003; Stephan & Clements, 2003). We adapted and combined these conceptual frameworks to describe the state curricula, both within the individual states and in comparisons across states. In further exploration of measurement standards, Sarah and I used two popular curriculum frameworks (i.e., Surveys of Enacted Curriculum, SEC; Webb Alignment Tool, WAT) to measure the alignment between standards and assessments in three grade levels in three states (Newton & Kasten, 2013). We found affordances and limitations of both models. In particular, the WAT Depth-of-Knowledge criteria coupled with the alignment index and tile charts of the SEC provided complementary perspectives on the alignment between standards and assessments.

Selecting Curriculum Research Frameworks

In *PTA*, Sharon, Yukiko, and I were determined to develop a curricular framework that would capture the breadth and depth of school algebra as we investigated how teachers were being prepared to teach algebra in SMTE programs. As a research team, we consulted every document we could find that addressed these topics, including recommendations from professional organizations (e.g., *Mathematical Education of Teachers* from the Conference Board of Mathematical Sciences, 2001; *InTASC Model Core Teaching Standards* from the Council of Chief State School Officers, 2011; *PSSM* from NCTM, 2000), books and book chapters (e.g., Chazan, 2000; Driscoll, 1999; Usiskin, 1988), research articles (e.g., McCrory et al., 2012; Moses et al., 1999), and the *Common Core State Standards for Mathematics* (*CCSSM*) (NGA & CCSSO, 2010). We divided the team into work groups and entered the framework into a spreadsheet that resulted in 486 lines in four sections (i.e., Algebra Content, Equity, *CCSSM*, Teaching and Learning). I remember the day we first shared this preliminary work with our advisory board. There were audible gasps! We knew this framework would be unwieldy, but we wanted to start with a comprehensive approach. We eventually organized the 486 lines into seven themes, which we used for our survey and interviews: (1) Nature and Structure of Algebra, (2) Functions, (3) Contexts and Modeling, (4) Reasoning and Proof, (5) Connections, (6) Tools and Technology, and (7) Equity. In reading this, all this work may sound like a waste of time; however, as I mentioned in my first lesson at the beginning of the chapter, such work is never wasted. By the time we began our data collection, our research team had learned valuable information about algebra and conducted a systematic review that would serve us well throughout the project … and to be honest, it was fun.

The most challenging and rewarding framework I have used in my curriculum research is Anna Sfard's (2008) Commognitive framework. Her book and the associated research articles (e.g., Ben Yehuda et al., 2005; Sfard, 1998, 2007) made my head hurt, but her framework inspired me with its eloquence and challenged me with its complexity. Of course, I decided to give it a whirl for my dissertation (really, Jill?). I was lucky that Anna was a visiting professor at MSU for several years while I was there. The Commognition theory made perfect sense to me because it combined communication and cognition in ways I had never considered. I had struggled to make sense of learning that happens inside the mind and learning that happens with others. Commognition not only addressed that dilemma, but also developed the idea of learning as changes in discourse practices, which blew my mind! I remember Anna talking in a seminar with someone who used Systemic Functional Linguistics (SFL) in her work, asking why language and meaning making needed to be separated; she suggested they could be one and the same. In fact, the changes in discourse were the learning and the evidence of learning simultaneously. So, I read and learned about the key constructs of the theory, including mathematical words, visual mediators, routines, and endorsed narratives. I decided to investigate the relationship between the written and enacted curriculum in a sixth-grade mathematics unit on fractions using these Commognitive constructs to tell the story. I describe the findings in Newton (2012). It was fascinating work, and the framework revealed important aspects of the written curriculum and how one amazing teacher implemented these lessons. Although I felt I was "in over my head" the whole time I was doing this project, in hindsight it was an important experience for me to challenge myself intellectually and have someone push me to think hard about ideas that were unfamiliar and intimidating to me. When my graduate students read about Commognition they are wowed, and I am still wowed too.

In summary, *I learned from CSMC and subsequent curriculum research experiences how to select and utilize theoretical, conceptual, and analytic frameworks.* I learned how to review research literature, policy documents, and publications of professional organizations in search of potential frameworks to answer particular research questions, and when possible, to answer the questions in several ways to provide a rich description of the data or phenomenon. When the "right" framework was not available, I learned to modify, merge, or create frameworks for particular purposes. I learned there are multiple, equally valid stories to be told from a set of data and that all research, including curriculum research, is subjective in some way as we inevitably bring our own lenses to the research enterprise.

THE COMPLEXITIES OF CURRICULUM RESEARCH

The complexities of curriculum research are endless; they occur in every aspect of the process described in the GAISE framework, including formulating questions, collecting, and analyzing data, and interpreting results. Further, every other aspect of curriculum, beyond curricular research, is also complex: development, selection, evaluation, use, and impact. As people who engage with curriculum, we are taking a complex and exciting journey. It was challenging to think about the specific lessons learned related to these complexities. Perhaps the lessons are to expect the complexities and even to embrace them, knowing that they make both the process and the product more interesting and potentially more impactful for the research team and audience.

When conducting the original CSMC state standards analyses, we analyzed data and reported findings in what we saw as a systematic way, including selecting an appropriate framework, coding the data, categorizing, and counting various aspects of the data, and noticing features of the outcomes of those analyses. For example, the team at the University of Missouri used innovative charts to map topics as they developed across the K–8 state standards—when were they introduced, developed, and mastered (see Reys et al., 2006). At the time, the magnitude of the complexities did not really occur to me because our process felt messy, but at the same time, relatively straightforward. My colleague, Gregory Larnell, brought a more critical perspective to our work and helped me learn to do this by analyzing the verbs used in the GLEs. He intended to write something based on a framework akin to Bloom's taxonomy, in which he compared the cognitive level of state standards based on verb analysis. However, he found that verb use is much more complex than anticipated and a simple method for determining the cognitive level of a GLE based on the verb(s) was doomed given the complexities of sentence structure; the context presented in the GLE, and the meaning attributed to the verbs (see Larnell & Smith, 2011). As described in the previous section, we make decisions about what we will attend to in the data and there can be consequences to these decisions. We need, therefore, to be ever mindful of the stories we tell and the impact of those stories.

Establishing Curricular Efficacy

I have not conducted research in which I compared the efficacy of different curricula. However, I am aware of the complexities of this research genre. I "grew up" as a researcher in CSMC hearing and reading about the importance of such studies and was privy to many discussions about what it would take, in the sense of curricular research, to establish such efficacy. This was at the time when many NSF-supported, *Standards*-based curricula were finding their place in the market. Researchers were trying to establish the efficacy of these materials for students given their attention to conceptual understanding, curricular coherence, and engaging tasks. At the same time, others were working against these efforts, making a case for the continued use of more traditional mathematics curriculum and pedagogies. Here we were bumping up against the math wars. (See Schoenfeld, 2004 for details about the "math wars.") The closest I came to this type of research was a collaboration in 2017 on a research companion chapter about curriculum research with Yvonne Grant in which we aimed to support the curriculum principle for school mathematics described in *Principles to Actions* (NCTM, 2014). In one section of the chapter, we described six studies that endeavored to establish the efficacy of particular curricula; we described the curriculum studied, instruments used, and findings from each study. I will not go into the details of the analysis (see pp. 117–120 in Newton & Grant, 2017 for more information), but what we found was a complex picture of curricular effectiveness, including many factors that may contribute to results, variation of results on different types of items and with different groups of students, and a range of suggested ways to measure effectiveness. Although overall student-centered curricula carried the day, the answer was far from simple—more research is certainly needed, but also great care must be taken when adopting and implementing curriculum. Additionally, the context must be carefully considered. Highlighting the complexity of curricular implementation and research, I once heard Alan Schoenfeld say at a CSMC meeting that without *Standards*-based curriculum-focused professional learning experiences for teachers, it is possible that a more traditional curriculum could be more effective.

Conducting Curriculum Research

During our work on *PTA*, several members of the research team wrote an article in which we highlighted some of the methodological complexities and lessons learned as we administered a national survey to collect information about how U.S. secondary mathematics teacher education (SMTE) program curricula were preparing algebra teachers (Maeda et al., 2014). First, we benefitted from the combination of conducting

a large-scale national survey to paint a broad picture of the phenomenon (i.e., the preparation of algebra teachers in SMTE programs) and the potential for quantitative data analysis. In addition, we conducted case studies at five universities in which we were able to use what we learned in the national survey to investigate the phenomenon in particular contexts, using qualitative methods to write rich descriptions of SMTE program curricula. In *PTA*, these case studies were purposefully selected SMTE programs at universities that varied across size, demographics, geographic location, and program requirements and structure. Of course, using both quantitative and qualitative methods introduced complexity to enact the process and disseminate the product, but the benefits greatly outweighed the costs.

Second, we conducted several rounds of think-aloud pilot surveys prior to survey distribution, which proved to be helpful as we refined our language and the structure of our questions and gauged survey respondent fatigue. As with all surveys, we needed to balance what we wanted to learn with how much time respondents (all busy people) would spend completing the survey.

Third, we discovered how challenging it can be to obtain a representative sample of a target population, resulting in the need to change our sampling frame. In the end, we used the Carnegie classification of universities to determine universities to survey. However, this framing was imperfect because the classification of a university is not always indicative of the nature and size of the SMTE program. We never anticipated the complexities we would confront when we sought to identify and contact individuals at universities that housed SMTE programs because no national list of SMTE programs exists. One year and a team of five research assistants later, our sample was complete. Finally, we learned the importance of defining terminology (e.g., opportunity to learn [OTL]) in the survey and interviews. In an effort to clarify our definitions, we included the following on the first page of our survey:

- *Secondary mathematics is the mathematics taught in grades 5 to 12.*
- *Secondary mathematics teacher is a teacher who teaches secondary mathematics.*
- *Initial certification program is a program that prepares candidates for their first teaching certificate. The program may be offered at the undergraduate or graduate level.*
- *Secondary mathematics teacher education program is a college- or university-based program in which an individual completes necessary coursework in order to be eligible to teach secondary mathematics. In cases in which the program is a post-baccalaureate credential or master's program, the program leads to initial certification.*

It is tempting but problematic to assume that educational constructs have the same meaning across stakeholder groups, research fields, or individuals. One of the first books I read that introduced me to the complexities of curriculum research called for mixed methods approaches (National Research Council, 2010), and I have certainly seen the benefits of this approach in my own work. The authors of that book also emphasized that curriculum research dilemmas typically have no easy solutions; rather, they recommended a reasoned approach using multiple sources of data and methods of analysis to move toward answers to complex questions we need to ask about curriculum.

Different versions of these same lessons played out in my NSF-funded project, *C3T2*. In this project, we began with a national survey. We learned an important lesson from advice given to us by people at NSF, as we originally planned to survey and investigate grades K–12 teachers' use of mathematics curriculum. At the suggestion of NSF personnel, we focused on the curricular use of elementary teachers, and then ultimately surveyed and interviewed teachers in grades 3–5. As we began to collect our case study data, we realized the importance of this advice. Even within this narrow range of grade levels, it was revealed that teachers were using up to 14 different mathematics curricular materials which we needed to investigate; a larger set of grade levels would likely have produced an unwieldy set of materials.

As in *PTA*, the national survey was instrumental in helping us develop interview protocols. Once again, we piloted all protocols (survey and three interviews) in think-aloud settings so teachers could share how they were thinking about the questions we were asking them in writing on the survey and orally in the interviews. We learned many things, but the most often mentioned issues concerned language. We made assumptions that the way we thought and talked about particular aspects of curriculum, learning, and teaching would be similar to those of the teachers. However, discourses are audience- and context-specific, and the pilot teachers helped us use better language to clarify our meaning for our teacher participants. Most of the researchers on the team had not taught in grades K–12 classrooms in a while; we changed given our new roles and because school discourse is always evolving. As a result of this pilot work, we struck a balance between asking open-ended questions to provide teachers with opportunities to create their own narratives and offering multiple choices for the purpose of answering some of our research questions.

One particularly painful example is the time needed to develop a definition for *curricular materials*. This seems like a relatively straightforward term. It was at the heart of our study, making it very important, but we first had to agree on its meaning as a research team and then work with teachers to increase the chances that we would all be on the same page when using the term. Here is where we landed, but conversations have continued even after the survey and protocols were finalized.

> *We are defining math curricular materials as the following: Any materials used by teachers for the purposes of planning, teaching, and/or assessment.*
> - *This includes (a) "packaged curriculum;" (b) individual lesson plans, activities, and materials; and (c) electronic and online resources and apps.*
> - *Examples include textbooks and teacher guides purchased by your school (such as Everyday Mathematics or Houghton Mifflin Math), online curricula (such as EngageNY), online resources (such as IXL, Zearn, or BrainPop), materials you downloaded from websites (such as TeachersPayTeachers or Pinterest), or materials you create yourself.*

Of course, COVID turned everything on its head in all types of research, including curriculum research. We had just finalized our national teacher survey when everything came to a standstill. Once we regrouped in 2020, we revised the entire survey because we hypothesized that COVID had significantly changed teachers' curriculum use to more electronic sources. Because we did not administer the original survey, we will never know the extent of the shift. But, teachers' curricular use was much more complex than we anticipated; the trickle-down effect of this was that our curriculum research was also more complex. (See Chapter 18 in this volume for more information about the lessons learned as a result.)

In summary, *I learned from CSMC and subsequent curriculum research experiences that every aspect of curriculum (e.g., development, implementation) is complex and every stage of research (e.g., formulating questions, analyzing data) is complex, guaranteeing that curriculum research is bound to be complex.* Despite these complexities, there is interesting and important research to be conducted and a large collaborative network of curriculum researchers to do it. I learned that challenges are inherent in both qualitative and quantitative methods, that definitions and language matter, and that even the best-laid plans can, and often do, go awry. I learned that a critical approach to curriculum research decisions is essential, and we need to be vigilant about the stories we tell and the potential impact of those stories.

DISCUSSION

I am sure I have learned more in the writing of this chapter than anyone will learn from reading it. In our academic lives, reflection is a luxury, and this chapter has given me an excuse to do just that, to sit with memories and ask myself questions (some uncomfortable) about research decisions, approaches, and collaborations. I set out to describe the lessons I have learned in nearly 20 years of conducting mathematics curriculum research and landed on four: (1) the results from research matter and the rich learning experiences are never wasted; (2) the importance on the research team of diverse and dissenting perspectives; (3) the power of theoretical, conceptual, and analytic frameworks for curriculum research; and (4) the complexities of curriculum research. Over the past 10 years, I have also conducted research related to study abroad curricula. I have not described this work explicitly in the chapter; however, many of the lessons are common and there is overlap in the research. For example, one social justice framework that I have used extensively in study abroad curriculum research, developed by Nancy Fraser (e.g., 2000, 2005), was adapted for the educational context by Courtney Cazden (2012), who has also conducted mathematics education research.

Our field is undergoing a reckoning of sorts, with scholars among us demanding attention to social contexts, to important and authentic action toward equity, and to recognizing that mathematics is not available to everyone (e.g., Gutiérrez, 2013; Martin et al., 2010; Stinson, 2004). While I was conducting my CSMC research, I took a class with Danny Martin (via distance when that was not really a thing). Some MSU graduate students had been given a copy of the syllabus for his course, "Social Contexts in Mathematics Education." We desperately wanted to take the course, but he was at the University of Illinois Chicago (UIC). He generously agreed that we could use his syllabus and meet with him during a visit at UIC. We were thrilled, and Helen Featherstone at MSU agreed to be the faculty advisor for the course. One uncomfortable question I asked myself in the process of writing this chapter was: "Why, while I was taking Danny's course and conducting curriculum research, I never asked myself questions like those being asked in the course readings: Who is this mathematics curriculum for? Who does and does not benefit from the curriculum? How should mathematics curriculum research attend to issues of social contexts and social justice? Are curricula offering humanizing learning opportunities for students?"

In 2000, Sarah Lubienski reported findings from a study in which she investigated whether problem-solving curricula might have the unintended consequence of further disadvantaging students of lower socioeconomic status (SES), in part, because of language demands. This seemed like an important question to ask but was on delicate terrain as the NSF-funded *Standards*-based (i.e., problem-solving) curricula were trying to make their case for wider acceptance. I always appreciated that Sarah posed the question. I think it raised awareness, putting this possibility on the curriculum community's radar as the intention of these curricula always was to provide more learners with opportunities for access to engaging and successful learning experiences. I wondered why I was not asking this question as well, especially given that I grew up in a working-class family similar to the students about whom Sarah was concerned. In fact, although my research has had aspects of equity-type contexts and questions embedded in it, I have not foregrounded these in my mathematics education curriculum research. I do foreground issues of equity and social justice in my teaching of undergraduate and graduate students and other research, but less so in my mathematics curriculum research. This is yet another lesson learned and more for me (and us) to think about: *How can I, and the mathematics curriculum research community, foreground equity and social justice in its agenda?*

Other takeaways from writing this chapter and reflecting on my curriculum research experiences were to highlight theory-practice connections and ensure those involved in mathematics teaching practice are at the table in curriculum research, to be bold and innovative when selecting and using frameworks, and that we are "smarter together" (e.g., Featherstone et al., 2011). I credit CSMC for my current awareness that all research, and scholarship, more generally, is relational. What is important is learning from one another throughout the research process and never losing sight of the ultimate goal of improving mathematics teaching and learning. Given the new curricular landscape resulting from, in part, the COVID pandemic and the proliferation of electronic curriculum resources, there is much to be investigated and learned to help those in the classroom who, given the current context for students and teachers, need our research and support more than ever. Among other things, this chapter turned out to be a love letter of sorts to all of those affiliated with CSMC from whom I have learned so much—the leadership team, advisory board, research associates, graduate students, school administrators, and teachers. The lessons described are a small part of the extensive CSMC legacy. My hope is that these lessons provide useful insights for future curriculum researchers.

REFERENCES

Ben-Yehuda, M., Lavy, I., Linchevski, L., & Sfard, A. (2005). Doing wrong with words: What bars students' access to arithmetical discourses. *Journal for Research in Mathematics Education*, *36*(3), 176–247. https://www.jstor.org/stable/30034835

Cazden, C. B. (2012). A framework for social justice in education. *International Journal of Educational Psychology*, *1*(3), 178–198. http://dx.doi.org/10.4471/ijep.2012.11

Chazan, D. (2000). *Beyond formulas in mathematics and teaching: Dynamics of the high school algebra classroom*. Teachers College Press.

Choppin, J., Ares, N., Herbel-Eisenmann, B., Hoffmann, A., Seymour, J., Staples, M., Truxaw, M., Wagner, D., Casa, T., & DeFranco, T. (2005). Discussion group on mathematics classroom discourse. In G. M. Lloyd, M. Wilson, J. L.

M. Wilkins, & S. L. Behm (Eds.), *Proceedings of the 27th annual meeting of the North American Chapter of the International Group for the Psychology of Mathematics Education* (pp. 106–109). Virginia Polytechnic University. http://www.pmena.org/pmenaproceedings/PMENA%2027%202005%20Proceedings.pdf

Clarke, D., Cheeseman, J., McDonough, A., & Clarke, B. (2003). Assessing and developing measurement in young children. In D. H. Clements & G. Bright (Eds.), *Learning and teaching measurement* (pp. 68–79). National Council of Teachers of Mathematics.

Conference Board of the Mathematical Sciences. (2001). *The mathematical education of teachers*. American Mathematical Society and Mathematical Association of America.

Council of Chief State School Officers [CCSSO]. (2011). *InTASC model core teaching standards: A resource for state dialogue*.

Drake, C., Spillane, J. P., & Hufferd-Ackles, K. (2001). Storied identities: Teacher learning and subject-matter context. *Journal of Curriculum Studies*, *33*(1), 1–23. https://doi.org/10.1080/00220270119765

Driscoll, M. (1999). *Fostering algebraic thinking: A guide for teachers, grades 6–10*. Heinemann.

Featherstone, H., Crespo, S., Jilk, L., Oslund, J., Parks, A., & Wood, M. (2011). *Smarter together! Collaboration and equity in the elementary math classroom*. National Council of Teachers of Mathematics.

Franklin, C., Kader, G., Mewborn, D., Moreno, J., Peck, R., Perry, M., & Scheaffer, R. (2007). *Guidelines for assessment and instruction in statistics education (GAISE) report: A pre-K–12 curriculum framework*. American Statistical Association.

Fraser, N. (2000). Rethinking recognition. *New Left Review*, *3*, 107–120. https://newleftreview.org/issues/ii3/articles/nancy-fraser-rethinking-recognition

Fraser, N. (2005). Reframing global justice. *New Left Review*, *36*, 69–88.

Fulghum, R. (1986). *All I really need to know I learned in Kindergarten: Uncommon thoughts on common things*. Ivy Books.

Giorgio-Doherty, K., Baniahmadi, M., Newton, J., Olson, A. M., Ferguson, K., Sammons, K., Wood, M. B., & Drake, C. (2021). COVID and curriculum: Elementary teachers report on the challenges of teaching and learning mathematics remotely. *Journal of Multicultural Affairs*, *6*(2). https://scholarworks.sfasu.edu/jma/vol6/iss2/3/

Gutiérrez, R. (2013). The sociopolitical turn in mathematics education. *Journal for Research in Mathematics Education*, *44*(1), 37–68. https://doi.org/10.5951/jresematheduc.44.1.0037

Jones, D. L., & Jacobbe, T. (2014). An analysis of the statistical content in textbooks for prospective elementary teachers. *Journal of Statistics Education*, *22*(3). https://doi.org/10.1080/10691898.2014.11889713

Kasten, S., & Newton, J. (2011). Analysis of K-8 measurement grade level expectations. In J. P. Smith, III (Ed.), *Variability is the rule: A companion analysis of K–8 state mathematics standards* (pp. 13–40). Information Age Publishing.

Lappan, G., Fey, J. T., Fitzgerald, W. M., Friel, S. N., & Phillips, E. D. (2006). *Connected mathematics 2*. Pearson Prentice Hall.

Larnell, G. V., & Smith, J. P., III. (2011). Verbs and cognitive demand in K–8 geometry and measurement grade level expectations. In J. P. Smith, III (Ed.), *Variability is the rule: A companion analysis of K–8 state mathematics standards* (pp. 95–118). Information Age Publishing.

Lehrer, R. (2003). Developing understanding of measurement. In J. Kilpatrick, W. G. Martin, & D. Shifter (Eds.), *A research companion to Principles and Standards for School Mathematics* (pp. 179–192). National Council of Teachers of Mathematics.

Lehrer, R., Jaslow, L., & Curtis, C. (2003). Developing an understanding of measurement in the elementary grades. In D. H. Clements & G. Bright (Eds.), *Learning and teaching measurement* (pp. 100–121). National Council of Teachers of Mathematics.

Lubienski, S. T. (2000). Problem solving as a means toward "mathematics for all": An exploratory look through a class lens. *Journal for Research in Mathematics Education*, *31*(4), 454–482. https://doi.org/10.2307/749653

Maeda, Y., Alexander, V. G., Newton, J., & Senk S. L. (2014). Development of a national survey for secondary mathematics teacher education programs: Challenges and lessons learned. *Assessment Update*, *26*(3), 5–6, 12–13.

Martin, D., Gholson, M. L., & Leonard, J. (2010). Mathematics as gatekeeper: Power and privilege in the production of knowledge. *Journal of Urban Mathematics Education*, *3*(2), 12–24.

McCrory, R., Floden, R., Ferrini-Mundy, J., Reckase, M. D., & Senk, S. L. (2012). Knowledge of algebra for teaching: A framework of knowledge and practices. *Journal for Research in Mathematics Education*, *43*(5), 584–615. https://doi.org/10.5951/jresematheduc.43.5.0584

Mewborn, D. S. (2005). Frameworks for research in mathematics teacher education. In G. M. Lloyd, M. Wilson, J. L. M. Wilkins, & S. L. Behm (Eds.), *Proceedings of the 27th annual meeting of the North American Chapter of the International Group for the Psychology of Mathematics Education* (pp. 31–39). Virginia Polytechnic University. http://www.pmena.org/pmenaproceedings/PMENA%2027%20 2005%20Proceedings.pdf

Mintos, A., Hoffman, A., Kersey, E., Newton, J., & Smith, D. (2018). Learning about issues of equity in secondary mathematics teacher education programs. *Journal of Mathematics Teacher Education*, *22*(5), 433–458. https://doi.org/10.1007/s10857-018-9398-2

Moses, R. P., Kamii, M., Swap, S. M., & Howard, J. (1999). The Algebra Project: Organizing in the spirit of Ella. *Harvard Educational Review*, *59*(4), 423–443.

National Council of Teachers of Mathematics [NCTM]. (1989). *Curriculum and evaluation standards for school mathematics*.

National Council of Teachers of Mathematics [NCTM]. (2000). *Principles and standards for school mathematics*.

National Council of Teachers of Mathematics [NCTM]. (2014). *Principles to action: Ensuring mathematical success for all*.

National Governors Association Center for Best Practices & Council of Chief State School Officers [NGA & CCSSO]. (2010). *Common core state standards for mathematics*. https://www.thecorestandards.org/Math/

National Research Council. (2004). *On evaluating curricular effectiveness: Judging the quality of K–12 mathematics evaluations*. National Academies Press.

National Research Council. (2010). *Preparing teachers: Building evidence for sound policy*. Committee on the Study of Teacher Preparation Programs in the United States. National Academies Press.

Newton, J. (2011). An examination of K–8 geometry state standards through the lens of van Hiele's levels of geometric thinking. In J. P. Smith, III (Ed.), *Variability is the rule: A companion analysis of K–8 state mathematics standards* (pp. 71–94). Information Age Publishing.

Newton, J. (2012). Investigating the mathematical equivalence of written and enacted middle school Standards-based curricula: Focus on rational numbers. *International Journal of Educational Research, 51–52*, 66–85. https://doi.org/10.1016/j.ijer.2012.01.001

Newton, J., & Grant, Y. (2017). "The right stuff": Curriculum to support the vision of NCTM and *CCSSM*. In D. Spangler & J. Wanko (Eds.), *Enhancing classroom practice with research behind Principles to Actions* (pp. 113–128). National Council of Teachers of Mathematics.

Newton, J., Horvath, A., & Dietiker, L. (2011). The statistical process: A view across K–8 state standards. In J. P. Smith, III (Ed.), *Variability is the rule: A companion analysis of K–8 state mathematics standards* (pp. 119–159). Information Age Publishing.

Newton, J., & Kasten, S. E. (2013). Two models for evaluating alignment of state standards and assessments: Competing or complementary perspectives? *Journal for Research in Mathematics Education, 44*(3), 550–581. https://doi.org/10.5951/jresematheduc.44.3.0550

Newton, J., Larnell, G., & Lappan, G. (2006). Analysis of K–8 algebra grade-level learning expectations. In B. J. Reys (Ed.), *The intended mathematics curriculum as represented in state-level curriculum standards: Consensus or confusion?* (pp. 59–88). Information Age Publishing.

Remillard, J. T., & Bryans, M. B. (2004). Teachers' orientations toward mathematics curriculum materials: Implications for teacher learning. *Journal for Research in Mathematics Education, 35*(5), 352–388. https://doi.org/10.2307/30034820

Remillard, J. T., & Heck, D. J. (2014). Conceptualizing the curriculum enactment process in mathematics education. *ZDM Mathematics Education, 46*(5), 705–718. https://doi.org/10.1007/s11858-014-0600-4

Reys, B., Dingman, S., Olson, T., Sutter, A., Teuscher, D., & Chval, K. (2006). Analysis of K–8 number and operation grade-level learning expectations. In B. J. Reys (Ed.), *The intended mathematics curriculum as represented in state-level curriculum standards: Consensus or confusion?* (pp. 15–58). Information Age Publishing.

Schoenfeld, A. H. (2004). The math wars. *Educational Policy, 18*(1), 253–286. https://doi.org/10.1177/0895904803260042

Senk, S. L. (1989). Van Hiele levels and achievement in writing geometry proofs. *Journal for Research in Mathematics Education, 20*(3), 309–321. https://doi.org/10.2307/749519

Senk, S. L., & Thompson, D. R. (Eds.). (2003). *Standards-based school mathematics curricula: What are they? What do students learn?* Lawrence Erlbaum.

Sfard, A. (1998). On two metaphors for learning and the dangers of choosing just one. *Educational Researcher, 27*(2), 4–13. https://doi.org/10.3102/0013189X027002004

Sfard, A. (2007). When the rules of discourse change, but nobody tells you: Making sense of mathematics learning from a commognitive standpoint. *The Journal of the Learning Sciences, 16*(4), 567–615. https://doi.org/10.1080/10508400701525253

Sfard, A. (2008). *Thinking as communicating: Human development, the growth of discourses, and mathematizing.* Cambridge University.

Spangler, D. A., & Williams, S. R. (2019). The role of theoretical frameworks in mathematics education research. In K. R. Leatham (Ed.), *Designing, conducting, and publishing quality research in mathematics education* (pp. 3–16). Springer.

Stehr, E. M., Jung, H., Newton, J., & Senk, S. L. (2016). Supporting preservice teachers' use of connections and technology in algebra teaching and learning. In M. B. Wood, E. E. Turner, M. Civil, & J. A. Eli (Eds.), *Proceedings of the 38th annual meeting of the North American Chapter of the International Group for the Psychology of Mathematics Education* (pp. 913–916). University of Arizona.

Stein, M. K., Remillard, J. T., & Smith, M. S. (2007). How curriculum influences student learning. In F. K. Lester, Jr. (Ed.), *Second handbook of research on mathematics teaching and learning* (Vol. I, pp. 319–369). National Council of Teachers of Mathematics.

Stephan, M., & Clements, D. H. (2003). Linear and area measurement in prekindergarten to grade 2. In D. H. Clements & G. Bright (Eds.), *Learning and teaching measurement* (pp. 3–15). National Council of Teachers of Mathematics.

Stinson, D. (2004). Mathematics as "gate-keeper"(?): Three theoretical perspectives that aim toward empowering all children with a key to the gate. *The Mathematics Educator, 14*(1), 8–18.

Usiskin, Z. (1988). Conceptions of school algebra and uses of variables. In A. F. Coxford & A. P. Shulte (Eds.), *The ideas of algebra, K–12* (pp. 8–19). National Council of Teachers of Mathematics.

Van Hiele, P. M. (1988). The child's thought and geometry. In D. Fuys, D. Geddes, & R. Tischler (Eds.), *English translation of selected writings of Dina van Hiele-Geldof and Pierre M. van Hiele* (pp. 243–252). Brooklyn College.

Van Hiele, P. M., & van Hiele-Geldof, D. (1958). A method of initiation into geometry at secondary school. In H. Freudenthal (Ed.), *Report on methods of initiation into geometry* (pp. 67–80). J. B. Wolters.

ENDNOTES

1. The CSMC team later expanded the Stein et al. framework to include factors that influence curriculum. The expanded framework developed by the CSMC team can be found in Jones and Jacobbe (2014) and in Remillard and Heck (2014).

2. The other primary authors of *CMP* were Jim Fey, Bill Fitzgerald, and Susan Friel.

3. In the United States, it is common for elementary school to include grades K–5 (ages 5–11), middle grades to include grades 6–8 (ages 11–14), and high school to include grades 9–12 (ages 14–18).

ABOUT THE EDITORS AND AUTHORS

EDITORS

Mary Ann Huntley is Senior Lecturer of Mathematics in the Department of Mathematics at Cornell University (U.S.), where she teaches mathematics courses and leads outreach activities for Grades K–12 students and teachers. Since earning a PhD in Curriculum and Instruction (mathematics education) at the University of Maryland, Mary Ann has consulted on a variety of national and international research and evaluation projects, with clients including the World Bank and U.S. Department of Education. Her research involves examining middle- and high-school mathematics curriculum from various perspectives, including the intended, enacted, and achieved curriculum. She previously served as a program officer in Instructional Materials Development at the U.S. National Science Foundation, was awarded a National Academy of Education Spencer Postdoctoral Fellowship, and received the American Association of Colleges of Teacher Education Outstanding Dissertation Award. Mary Ann is a co-editor of the book series *Research in Mathematics Education*, published by Information Age Publishing.

Christine Suurtamm is Professor Emeritus of Mathematics Education at the University of Ottawa, Canada. Her research focuses on the complexity of mathematics teachers' classroom practice, particularly on formative assessment practices as opportunities to attend to students' mathematical thinking. She is also Director of the Pi Lab, a research facility funded by the Canada Foundation for Innovation. She has been Principal Investigator on several large-scale projects in mathematics teaching and learning and has published in mathematics education and assessment research journals as well as practice-based journals. Most recently, she was the Research Advisor to the Ontario Ministry of Education during the reform of both the elementary (Grades 1–8) and Grade 9 mathematics curriculum. Her contributions to mathematics education and her excellence in research and teaching have been recognized through multiple university and national awards. She is a co-editor of the series *Research in Mathematics Education*, published by Information Age Publishing.

Denisse R. Thompson, PhD, is Professor Emerita of Mathematics Education at the University of South Florida (U.S.), having retired in 2015 after 24.5 years on the faculty. Her research interests include curriculum development and evaluation, with over 30 years of involvement with the University of Chicago School Mathematics Project. She is also interested in mathematical literacy, the use of children's literature in the teaching of mathematics, and in issues related to assessment in mathematics education. She has published in the *Journal for Research in Mathematics Education*, the *Journal of Mathematics Teacher Education*, *ZDM Mathematics Education*, *Investigations in Mathematics Learning*, *International Journal of Science and Mathematics Education*, *Journal of Educational Measurement*, and the practice-based journals of the National Council of Teachers of Mathematics. She is a co-editor of the series *Research in Mathematics Education*, published by Information Age Publishing.

AUTHORS

Mona Baniahmadi is a Research Associate in the Texas Advanced Computing Center (TACC) at the University of Texas, Austin (U.S.). At TACC, she evaluates STEM[1] education research programs; she also evaluates and supervises internal programs and external projects to assess their impact across K–12 and college settings. Before joining TACC, Mona worked as a Postdoctoral Researcher in the College of Education at the University of Texas, Austin, collaborating on research focused on mathematics and elementary teacher professional learning. Mona received her EdD in Educational Leadership from Duquesne University, Pittsburgh. She has conducted research on computer science education, elementary mathematics education, science and mathematics curriculum design, and teacher professional learning programs. Mona is a former elementary and secondary teacher, higher education instructor, mentor, textbook author, and curriculum director with over 10 years of experience in developing and implementing innovative teaching methods and instructional strategies to provide equal learning opportunities for students in classrooms.

Maria Blanton is a Senior Scientist at TERC, Cambridge, Massachusetts (U.S.). Her primary research interests are in developing early algebra learning progressions and designing effective interventions to support children's early algebraic thinking. She has led numerous federally funded research projects and has published her work in leading research journals in mathematics education. She is co-editor of the research volumes *Algebra in the Early Grades* (2008, Taylor/Francis) and *Teaching and Learning Proof Across the Grades* (2009) and author or co-author of the practitioner-oriented books *Algebra and the Elementary Classroom* (2008), *Developing Essential Understanding of Algebraic Thinking for Teaching Mathematics in Grades 3–5* (2011), and *Teaching with Mathematical Argument* (2018). She was the 2020 recipient of the Award for Interdisciplinary Excellence in Mathematics Education through Texas A&M University.

Douglas H. Clements, Distinguished University Professor, Kennedy Endowed Chair in Early Childhood Learning, and co-Executive Director of the Marsico Institute for Early Learning and Literacy at the University of Denver, Colorado (U.S.), is a major scholar in the field of early childhood mathematics education, whose work has relevance to the research field, to the classroom, and to the educational policy arena. Formerly a preschool and kindergarten teacher, he has published over 200 refereed research studies, 30 books, 107 chapters, and 300 additional works, and has directed more than 40 funded projects. At the national level, his contributions have led to the development of new mathematics curricula, teaching approaches, teacher training initiatives, and models of "scaling up" interventions. Dr. Clements has served on six national research committees and authored reports for the National Academies of Sciences, Engineering, and Medicine (NASEM), including the National Research Council and Institute of Medicine, the President's National Mathematics Advisory Panel, and the Common Core State Standards Committee (National Governors Association and Council of Chief State School Officers).

Priscila Dias Corrêa is an Associate Professor at the University of Windsor, in Ontario, Canada. She is passionate about mathematics education and strives to offer high quality and meaningful mathematics learning experiences to her students. She holds a PhD in Mathematics Education from the University of Alberta, Canada; a MSc in Electrical Engineering from the Pontifical Catholic University of Rio de Janeiro, Brazil; a BEd in

Mathematics from the Federal University of Rio de Janeiro, Brazil; and a BEng in Electrical Engineering from the Federal Center for Technological Education Celso Suckow da Fonseca, Brazil. Among her research areas of interest are mathematical proficiency, mathematics curriculum, mathematics assessment, racialized mathematics learning experiences, mathematics modeling, and computational thinking. Priscila brings her research into her undergraduate and graduate teaching, aiming to enhance and advance her students' teaching practices and mathematics knowledge for teaching.

Lara Dick is Associate Professor of Mathematics Education in the Department of Mathematics at Bucknell University, Pennsylvania (U.S.). She earned a doctoral degree in mathematics education and a master's degree in applied mathematics from North Carolina State University (U.S.). Lara teaches combined elementary content and methods courses, secondary mathematics methods courses, and mathematics courses such as statistics and calculus. Lara's research is focused on engaging preservice teachers with noticing their students' mathematical thinking through various frames, including lesson study, technology-enhanced tasks, and cognitive demand of tasks.

Shannon Dingman is Professor of Mathematics Education in the Department of Mathematical Sciences at the University of Arkansas (U.S.). He earned a PhD in Curriculum and Instruction from the University of Missouri, where he was a fellow of the National Science Foundation-funded Center for the Study of Mathematics Curriculum (CSMC). His research agenda focuses on mathematics curriculum, namely, the analysis of mathematics standards and mathematics textbooks. He is currently studying the curricular reasoning of mathematics teachers. He also studies student quantitative reasoning and has co-authored the textbook *Case Studies for Quantitative Reasoning: A Casebook of Media Articles*.

Kristin Doherty is a PhD candidate at Michigan State University (U.S.) in the Department of Teacher Education. Her primary research focus is on how elementary teachers' curricular use and pedagogy can support students in the process of mathematical justification, in ways that promote students' epistemic agency and autonomy. She is also interested in how practice-based teacher education can support teachers to facilitate mathematics discussions that are responsive to students. Kristin's research is informed by her experience as an elementary and middle-school mathematics teacher. Prior to her doctoral studies, Kristin worked at a small charter school as the mathematics coordinator for grades K–8 as well as the school's middle-school mathematics teacher. Kristin's teaching experience also includes teaching all subjects in fifth and sixth grades at a Montessori school. Currently, she teaches elementary mathematics methods courses at Michigan State University.

Thérèse Dooley is Emeritus Associate Professor in Mathematics Education at the Institute of Education, Dublin City University (DCU), Ireland. Her research interests include the role of discussion in the development of mathematical ideas, mathematical task design, Lesson Study, and politeness theory. She recently collaborated with others at DCU to conduct a review of research in literacy and numeracy for early years, primary, and post-primary learners. She was a member of the National Council for Curriculum and Assessment's Early Childhood and Primary Mathematics Development Group (2016–2024).

Corey Drake is Director of Professional Learning at The Math Learning Center (MLC) in Oregon (U.S.). Prior to joining MLC, she was Professor of Teacher Education and Mathematics Education and the Director of Teacher Preparation at Michigan State University. Her research focuses on teacher learning from and about curriculum materials, specifically related to providing more equitable mathematics experiences for elementary students. Her work has been funded by the National Science Foundation and the Spencer Foundation, and published in venues including *Educational Researcher*, the *Journal of Teacher Education*, the *Journal of Mathematics Teacher Education*, and *Teaching Children Mathematics*.

Elizabeth Dunphy is Emeritus Associate Professor in Early Childhood Education at the Institute of Education, Dublin City University, Ireland. Her research interests include eliciting children's perspectives, young children's learning, early childhood mathematics, and pedagogy and assessment in early childhood education. She served as Chairperson of the National Council for Curriculum and Assessment's Early Childhood and Primary Mathematics Development Group (2016–2024).

Alden J. (AJ) Edson is Research Assistant Professor of Mathematics Education at Michigan State University (U.S.). His research focuses on secondary school mathematics curriculum design and development using design-based research methodologies. AJ is particularly interested in studying the enactment of curriculum materials in a digital world and the affordances of innovative mathematics curriculum materials as a context for teacher learning. He initiates and implements research and development grants related to the Connected Mathematics Project (CMP) and assists with other activities to support CMP. His experience with other curriculum materials that are funded by the U.S. National Science Foundation (e.g.,

Core-Plus Mathematics, *Transition to College Mathematics and Statistics*) involves collaborating with teams of mathematicians and statisticians, mathematics and statistics educators, and school mathematics teachers, with the goal of providing students and teachers with problem-based, inquiry-oriented materials.

Nicole L. Fonger is Associate Professor of Mathematics and Mathematics Education at Syracuse University (U.S.). She earned a PhD in Mathematics Education at Western Michigan University (U.S.). Nicole's research agenda focuses on supporting students' meaningful learning of algebra in K–12 school settings. She conducts empirical and theoretical research, design studies, and youth participatory action research. As a community-engaged researcher, she builds relationships with teachers and youth, focusing on historically responsive literacy and social justice mathematics to support marginalized students. Among her honors, she is a Service, Teaching and Research (STaR) Fellow of the U.S. Association of Mathematics Teacher Educators, and she received the U.S. National Council of Teachers of Mathematics outstanding publication award for linking research and practice.

Doris Fulwider has many years of teaching experience in a primary high-ability classroom. She earned her Gifted and Talented teaching license and two master's degrees from Purdue University (U.S.), including one in curriculum and instruction with a concentration in mathematics education. Doris recently retired from classroom teaching and now teaches elementary mathematics methods courses, serves as a research assistant, and is a doctoral student at Purdue University. Doris is passionately interested in facilitating and nurturing sense-making in the elementary mathematics classroom, supporting teachers, and engaging families.

Karen C. Fuson is Professor Emerita of Learning Sciences, School of Education and Social Policy, Northwestern University (U.S.). She was a member of the National Research Council's (NRC) Committees that wrote *Adding It Up*, *Mathematics Learning in Early Childhood*, and *How Students Learn*. Professor Fuson was on the feedback team for the *Common Core State Standards-Mathematics*, worked on the learning progressions for these standards, and advised PARCC (Partnership for Assessment of Readiness for College and Careers) and Smarter Balance on their mathematics test design and items. She has published over 80 research articles on mathematics teaching and learning and given many conference presentations. She is the author of the PK through Grade 6 math program *Math Expressions* published by Houghton Mifflin Harcourt which is based on research in the

NRC reports and other research, on aspects of international mathematics programs, and on results from her research team.

Angela Murphy Gardiner is a Senior Research Associate at TERC, Cambridge, Massachusetts (U.S.). A former elementary classroom teacher, she has extensive expertise in designing and implementing early algebra interventions in elementary grades as part of regular classroom instruction. Her work also includes designing and leading in-person and virtual teacher professional development around early algebraic thinking. She has published widely (e.g., *Journal for Research in Mathematics Education* and *Teaching Children Mathematics*) and has led national presentations that focus on transforming early algebra research for practitioner audiences (e.g., National Council of Teachers of Mathematics annual conferences).

Raz Harel is the Head of the Mathematics Department at David Yellin College of Education, Israel. His research centers on students' justification processes, automated feedback systems, and example-based online assessments of mathematical reasoning. Dr. Harel also has experience in designing software packages, interactive textbooks, and projects related to mathematical inquiry learning.

Laurel Hendrickson is a PhD student in the Department of Teaching, Learning, and Sociocultural Studies (TLS) at the University of Arizona (U.S.). She has taught English and Spanish as second and foreign languages for 19 years in public and private schools in the United States, Mexico City, and Bucharest (Romania) in grades K–12, in community colleges, and in community settings. Her educational background includes a bachelor's degree in Spanish, and a MEd in Curriculum and Instruction for Secondary Spanish from Portland State University (U.S.). Laurel writes curriculum for all grades and is moved by the determination to improve that many language learners demonstrate in her classes. Her focus in the TLS program at the University of Arizona is in motivation, linguistics, and the development of language cognition in students.

Julie Houle is a pedagogical consultant in elementary mathematics for the Lester B. Pearson School Board in Montreal, Canada. She was a classroom teacher at the elementary level for 13 years and moved into her role as a consultant in 2018. Following attendance at a workshop on early numeracy led by Helena Osana, Julie was inspired to create a toolkit based on research in children's numeracy to provide teachers with accessible and concrete activities to support their classroom teaching. Through a funded partnership with Concordia University in Montreal, Julie designed and constructed the *Numeracy Kit for Kindergarten 5-year-olds* (*NyKK-5*). Since its creation, Julie has been involved in research with colleagues across Canada to validate the kit and investigate its impact on teachers' classroom practice.

Nan Jiang is a PhD student at the University of Arizona (U.S.) in the Department of Teaching, Learning, and Sociocultural Studies (TLS). Her primary research focus is on literacy development of learners with developing English proficiency (e.g., children and English language learners) using a cultural-historical perspective. Prior to her doctoral studies, Nan worked as an English teacher in two Chinese universities as well as a part-time English teacher in several Chinese educational services. Her educational background includes a bachelor's degree in business English from Jinan University in China, and a MEd in English Language Learners from Vanderbilt University in Nashville, Tennessee (U.S.).

Dustin L. Jones is Professor of Mathematics Education in the Department of Mathematics and Statistics and Associate Dean of the College of Science and Engineering Technology at Sam Houston State University in Huntsville, Texas (U.S.). His research interests include mathematics textbooks, statistics education, and mathematics teacher preparation. His professional mission is to help people understand mathematics better. He has taught mathematics courses at the elementary, secondary, and collegiate levels. He has served on the board of the Association of Mathematics Teacher Educators and is a co-host of the Teaching Math Teaching Podcast (http://www.teachingmathteachingpodcast.com).

Lisa Kasmer retired from the Department of Mathematics at Grand Valley State University in Michigan (U.S.) where was was Professor of Mathematics Education. Her research interest included how mathematics teachers use their curricular reasoning to make sense of and subsequently enact and reflect on their lessons. Lisa also directed a faculty-led study abroad program to Tanzania each year, which was an opportunity for preservice mathematics teachers to teach in local Tanzanian schools. While they learned about teaching mathematics in a culture much different than their own, the preservice teachers considered issues such as ethnomathematics, culturally relevant pedagogy, and different methods of instructional delivery.

Hangil Kim is a mathematics teacher at Chungnam High School in Daejeon, Korea, and earned his doctoral degree in STEM Education at the University of Texas at Austin (U.S.). His research interests concern the engagement of students in proving and reasoning as well as teachers' views about the use of examples in proving-related activities, and ways in which such activities promote students' learning to prove. His work has appeared in research journals in mathematics education (e.g., *Journal of Educational Research of Mathematics*, *Research in Mathematics Education*) and in the edited volume, *Conceptions and Consequences of Mathematical Argumentation, Justification, and Proof*. He has also served as a referee for *Research in Mathematics Education*.

Kelsey Klein is a Senior Research Associate at Education Development Center (EDC) in the U.S. She has worked on multiple large-scale projects studying the implementation and impact of high-school mathematics and computer science curricula, and has expertise in research design, survey and assessment development, and quantitative methodologies. She was the quantitative data analyst for the *Supporting Success in Algebra* project, funded by the U.S. National Science Foundation, working with co-PI Laura O'Dwyer. In 2021, she received her PhD in Measurement, Evaluation, Statistics and Assessment from Boston College. Dr. Klein's dissertation focused on the development of a valid and reliable measure of mathematics anxiety for high-school students utilizing a unique item format and guided by Rasch measurement theory.

Eric Knuth is a Program Director in the Division for Research on Learning at the U.S. National Science Foundation, where he oversees research that aims to improve STEM teaching and learning. Prior to joining the National Science Foundation, he was a mathematics education professor for five years at the University of Texas at Austin (U.S.) and for 18 years at the University of Wisconsin-Madison (U.S.). His research interests concern the engagement of students in core mathematical practices, including practices related to algebraic thinking and proving. His work has appeared in the premier research journals in mathematics education, in the learning sciences, and in education, more generally. He also served as chair of the American Educational Research Association's Special Interest Group for Research in Mathematics Education as well as a member of the National Council of Teachers of Mathematics Research Committee.

Anne Lafay is a speech-language pathologist and neuropsychologist currently at the University Savoie Mont Blanc in Chambéry, France, as an Assistant Professor in the Department of Psychology and a member of the Psychology and Neurocognition Lab. She completed her doctorate at Laval University in Quebec, Canada, on the numerical cognitive deficits in children with mathematics learning disorders. She held a postdoctoral fellowship at Concordia University in Montreal, Canada, in Helena Osana's Mathematics Teaching and Learning Lab. Lafay has considerable experience working with children and adolescents with language and learning disorders in reading and mathematics, and researches mathematics development and learning disabilities with a focus on understanding children's difficulties, identifying and assessing children's learning needs, and developing effective interventions.

Boram Lee is a doctoral candidate in the STEM Education program at the University of Texas at Austin (U.S.). Her research interests lie in analyzing the effects of mathematics video games on improving elementary school students' mathematical reasoning, including algebraic thinking. She focuses on theorizing design frameworks for mathematics video games, developing mathematics video games using the Unity 3D game engine, and analyzing students' gameplay. She is particularly interested in the development of algebraic thinking, both in instruction and through video games. Before joining this graduate program, she was teaching elementary students in Seoul, Korea, while running diverse math activity clubs.

Kristy Litster is Associate Professor in the Department of Teacher Education at Valdosta State University (U.S.), and the program coordinator for the Valdosta State University Elementary Education Master's Program. Her specialization is in curriculum and instruction, with a concentration in mathematics education and leadership. Dr. Litster's research interests are focused on relationships between instructional practices and student-enacted levels of cognitive demand in elementary mathematics. She also works with preservice and in-service teachers to develop high cognitively demanding mathematics tasks to engage, enhance, and extend student learning in both face-to-face and digital learning environments.

Lorraine M. Males is the Julie & Henry Bauermeister Associate Professor in Education and Human Sciences, Mathematics Education in the Department of Teaching, Learning, and Teacher Education at the University of Nebraska-Lincoln (U.S.). She is a former middle- and high-school mathematics and computer science teacher. Her work focuses on supporting secondary mathematics teachers and graduate students in learning to use resources to support learning of all students.

June Mark is a Managing Project Director at Education Development Center (EDC) in the U.S. Her research and development work in mathematics and computer science education focuses on instructional materials development, teacher professional learning, and curriculum implementation. Her work is dedicated to ensuring all students receive high-quality mathematics and computer science learning and focuses on increasing access and equity for students who are historically underrepresented in the STEM fields. She is the lead author of *Transition to Algebra* and co-author of the professional book *Making Sense of Algebra: Developing Students' Mathematical Habits of Mind*. She is currently co-PI of *Supporting Success in Algebra*, PI of *Equitable Computer Science Implementation in All NYC Schools* (a research-practice partnership with the New York City Public Schools), and co-PI of *Mathematics through Programming in the Elementary Grades*, all funded by the U.S. National Science Foundation. She is also project director of the U.S. Department of Education-funded *Improving Equity in AP Computer Science Principles: Scaling Beauty and Joy of Computing*.

Josh Markle is Assistant Professor in Mathematics Education at the University of Alberta, Canada. His research focuses on spatial reasoning and embodiment in teaching and learning mathematics.

Kim Markworth is a mathematics educator with three decades of experience at the elementary, middle, and university levels. She taught elementary and middle-school mathematics for 12 years before becoming a mathematics teacher educator at Western Washington University in Bellingham, Washington (U.S.). From 2020 to 2023, she served as the Director of Content Development at The Math Learning Center, where she led a team in developing *Bridges in Mathematics (Third Edition)*. Kim is passionate about designing problem-solving experiences that engage all students in challenging mathematics. She has authored two books, *Problem Solving in All Seasons, Grades K–2* and *Grades 3–5*, and designed *Concept Quests*, collections of rich tasks complementing mathematics curricula in grades 2–8.

Lynn M. McGarvey is Professor of Mathematics Education at the University of Alberta, Canada. Her research focuses primarily on mathematical understanding and generalizations of children within the context of geometric, spatial, and algebraic reasoning. She is also interested in the historical and international development of mathematics education curriculum.

Nitsa Movshovitz-Hadar is Emerita Professor at the Technion Israel Institute of Technology. She is a mathematics educator who loves mathematics and cares to share her passion with others. She completed her PhD at Berkeley (U.S.) in 1975 with Leon Henkin as her thesis advisor, and upon her return to Israel, joined the academic staff at Technion. Since then, she has initiated many R&D projects aimed at enlivening mathematics teaching and learning and has published numerous papers in professional mathematics education journals. She also served as the content designer of DraMath, a video series in mathematics produced by the Israel Educational TV, and as director of MadaTech, the Israel National Science Museum. Since 2012, she has been on the advisory board of MoMath, the U.S. National Mathematics Museum in New York City.

GwiSoo Na serves as a Professor in the Department of Mathematics Education at the Cheongju National University of Education in Korea. Prior to joining the Cheongju National University of Education, she was a research fellow for four years at the Korea Institute for Curriculum and Evaluation. Her research interests concern the development of mathematics curriculum and the engagement of students in mathematical proving and reasoning. Her work has appeared in the premier research journals in mathematics education and listed as the most-cited-paper in the *Journal of Educational Research in Mathematics* (*JERM*), published by the Korea Society of Educational Studies in Mathematics. She serves as an advisor of Science-Mathematics-Information Education Expert Committee of the Korea Foundation for the Advancement of Science and Creativity. She also served as an advisor of the Science-Mathematics-Information Education Convergence Committee of the Korea Ministry of Education as well as the editor-in-chief of *JERM*.

Jill A. Newton is Professor of Mathematics Education at Purdue University (U.S.). She earned a PhD in Curriculum, Teaching, and Educational Policy at Michigan State University, a MA in International Education at George Washington University (U.S.), and a BS in Mathematics at Michigan State University. She began her career teaching mathematics as a Peace Corps Volunteer in Papua New Guinea and has taught in the Democratic Republic of Congo, Bulgaria, Tanzania, the United States, and Venezuela. At Purdue University, she teaches graduate and undergraduate mathematics education courses and courses offered in association with her Tanzania study-abroad program. Her research focuses on K–12 mathematics curriculum, secondary mathematics teacher education programs, and study-abroad programs. She has been awarded research grants from the National Science Foundation and Spencer Foundation, serving as Principal Investigator on *Codeveloping a Curriculum Coherence Toolkit with Teachers* and Co-PI on *Excellence in STEM Teaching in Indiana*.

Laura O'Dwyer is Department Chair and Professor in the Measurement, Evaluation, Statistics and Assessment Department at Boston College (U.S.). She is a quantitative methodologist with extensive experience designing and conducting large-scale randomized controlled trials and quasi-experimental studies to examine the impact of educational interventions. She is affiliated with the Center for the Study of Testing, Evaluation and Educational Policy (CSTEEP) at Boston College, and her research has been funded by the U.S. National Science Foundation, the Institute of Education Sciences (IES), and the U.S. Department of Education.

Amy M. Olson is Associate Professor in the Department of Educational Foundations and Leadership at Duquesne University (U.S.). As a scholar, she examines the equity implications of teachers' beliefs about academic content and learning (both their own and their students), often in the context of mathematics classrooms. At the heart of her work is an exploration of what it means to be successful in school. Amy considers herself a teacher-scholar and is intentional about engaging with research that informs her pedagogy. She teaches in undergraduate and graduate teacher education and educational leadership programs with the goals of supporting and sustaining the work of justice-oriented educators.

Travis A. Olson is Professor of Mathematics Education in the Department of Teaching and Learning at the University of Nevada, Las Vegas (U.S.), where he teaches undergraduate, masters, and doctoral-level mathematics education courses. His research involves curriculum analyses, curriculum use in mathematics classrooms, and student and teacher understandings of mathematical constructs around teaching and learning. He is a writing team member of the Association of Mathematics Teacher Educators' (AMTE) national-level *Standards for Preparing Teachers of Mathematics*, and has served in various roles for the AMTE, National Council of Supervisors of Mathematics, Research Council on Mathematics Learning, and National Council of Teachers of Mathematics.

Arielle Orsini is a PhD student in the Department of Education at Concordia University in Montreal, Canada. She has experience as an instructor for the Mathematics Teaching course at Concordia University in the Early Childhood and Elementary Education (BA) program. She is also a member of the Mathematics Teaching and Learning Lab, led by Helena Osana. Arielle has been a part of the Early Numeracy Project since its inception and became the lead research assistant for the *Numeracy Kit for Kindergarten 5-Year-Olds* (*NyKK-5*) project. Her research focuses on the affordances of manipulatives and using manipulatives for play-based mathematics education.

Helena P. Osana is Professor of Education at Concordia University in Montreal, Canada, and Principal Investigator of the Mathematics Teaching and Learning Lab. Osana's training is in educational psychology with research emphases in cognition, mathematics education, and quantitative analytic methods, particularly as applied to classroom contexts. Supported by numerous grants from external agencies, she studies children's numeracy and problem-solving strategies, as well as the pedagogical factors that support their mathematics learning. Osana is a leading expert on how children learn mathematics with instructional representations, including images, diagrams, and manipulatives. She also investigates the role of children's individual differences, such as prior knowledge and linguistic factors, to provide nuanced accounts of how mathematics instruction impacts learning. Osana is actively involved in mathematics teacher education at the preservice and inservice levels and conducts research on approaches to professional development for preschool and elementary teachers.

Mary Beth Piecham is a Research Scientist at Education Development Center (EDC) in the U.S. She has significant experience leading large-scale curriculum studies, with expertise in data management, research design, qualitative and mixed-methods analyses, district data sharing, and protection of human subjects. She was the project manager for the U.S. National Science Foundation (NSF)-funded *Supporting Success in Algebra* project, a study of the implementation and impact of *Transition to Algebra*. Previous projects at EDC include her work as co-PI for the NSF-funded *Mathematical Practices Implementation* study of the impact of a high-school curriculum on teachers' knowledge and practice, and as project director for the *Focus on Mathematics Phase II*, a study of mathematical habits of mind secondary teachers use in their own practice.

Kelsey Quaisley is a postdoctoral scholar in the Mathematics Department at Oregon State University (U.S.). Her research interests include understanding and elevating the experiences of instructors and students, as well as supporting newer mathematics instructors at the tertiary level and prospective instructors of mathematics at the primary and secondary levels. Her dissertation was a narrative inquiry into the experiences of a new instructor of mathematics content courses for prospective elementary teachers.

Leticia Rangel is a retired professor at the Federal University of Rio de Janeiro (UFRJ), Brazil. She believes in the power of education as an instrument for promoting equity. The training of mathematics teachers has been her goal since joining as a mathematics teacher at the UFRJ Elementary and Secondary School, in 1993. She holds a PhD in Systems and Computing Engineering, focused on mathematics teacher training (2015), from the Alberto Luiz Coimbra Institute for Graduate Studies and Research in Engineering at the Federal University of Rio de Janeiro (COPPE/UFRJ), and a master's in mathematics from the Institute of Mathematics at the Federal University of Rio de Janeiro (IM/UFRJ). Leticia has been researching in the field of Mathematics Education, with an emphasis on training teachers who teach mathematics and on mathematics knowledge for teaching.

Janine Remillard is Professor of Education at the University of Pennsylvania's Graduate School of Education (U.S.), where she serves as the chair of the Learning, Teaching, and Literacies Division. Her teaching and research focus on mathematics teacher learning and classroom practices, teacher education, and the role of tools in supporting teacher learning. She has a strong commitment to teaching and learning in urban contexts and making mathematics humanizing for teachers and students. She co-founded the Community Based Mathematics Project, a hub for developing and sharing locally relevant and social justice-oriented mathematics lessons that teachers can modify to fit the needs of their classrooms. Remillard is active in the mathematics education community in the U.S. and internationally and is involved in cross-cultural research on teachers' use of mathematics curriculum resources. Her most recent book, *Elementary Mathematics Curriculum Materials: Designs for Student Learning and Teacher Enactment*, was published in 2020.

Thomas E. Ricks is Associate Professor in the School of Education at Louisiana State University in Baton Rouge, Louisiana (U.S.). His research interests interweave mathematics and science education teaching and learning, teacher preparation, curricular issues, complexity science, and international cross-cultural comparisons, particularly between the U.S. and China. His recent work describes educational systems at various nested levels as legitimate and intelligent lifeforms, raising awareness about expanding curricular reform efforts that incorporate these educational systems.

Nicole Rigelman is Professor of Mathematics Education in the Department of Curriculum and Instruction at Portland State University (U.S.) and the Education Program Officer for the Math Learning Center in Oregon (U.S.). She is a former middle-school mathematics teacher and K–12 district mathematics specialist. At the university level, she has taught mathematics methods and content courses for preservice teachers. Rigelman currently serves as coordinator of Portland State University's Mathematics Instructional Leader Program, where she supports content, pedagogy, and leadership development for practicing teachers and teacher leaders. She is passionate about developing partnerships with schools and supporting instruction focused on deepening each and every student's mathematical thinking and discourse as well as increasing agency and enthusiasm for problem solving. Rigelman has authored many articles and book chapters and was co-author for *Catalyzing Change in Early Childhood and Elementary Mathematics: Initiating Critical Conversations*.

Ingrid Ristroph recently received her PhD in STEM Education at the University of Texas at Austin (U.S.). Her research interests are investigating equitable mathematics teaching practices with the end goal of improving the engagement of students in mathematics classrooms. She is interested in teacher practices fostering algebraic reasoning, supporting student discourse, and instructional practices that attend to issues of authority in the classroom. Previously, Ingrid worked at the Charles A. Dana Center, where she developed secondary mathematics courses and led professional development for inservice teachers. Currently, she works with Austin Independent School District as an Academic Specialist supporting secondary mathematics teachers.

Bima Sapkota is Assistant Professor in the School of Mathematical and Statistical Sciences at the University of Texas Rio Grande Valley (U.S.). She earned her PhD in mathematics education from Purdue University and her master's in mathematics education from Tribhuvan University, Nepal. Her primary research areas are practice-based teacher education, discipline-specific teaching and learning, elementary teachers' curricular decisions, and social justice in education. She investigates how mathematics preservice teachers develop and conceptualize discipline-specific practices when they engage in practice-based instructional activities. She teaches elementary, middle school, and secondary mathematics methods and content courses as well research methods courses.

Julie Sarama is Distinguished University Professor and Kennedy Endowed Chair in Innovative Learning Technologies at the University of Denver, Colorado (U.S.). She has taught high-school mathematics, computer science, middle-school gifted mathematics, and early childhood mathematics. She has directed over 30 projects funded by the U.S. National Science Foundation (NSF), the U.S. Department of Education's Institute of Education Sciences (IES), the National Institute of Health (NIH), and the Office of Special Education Programs (OSEP). She has also developed and programmed over 50 award-winning educational software products. She conducts research on young children's development of mathematical concepts and competencies, implementation and scale-up of educational reform, professional development models and their influence on student learning, and implementation and effects of software environments (including those she has created) in mathematics classrooms. These studies have been published in more than 115 refereed articles, 10 books, 160 chapters, and over 100 additional publications.

Amanda Gantt Sawyer is Associate Professor of Mathematics Education in the Middle, Secondary, and Mathematics Education Department in the College of Education at James Madison University, Virginia (U.S.). She earned a doctoral degree in mathematics education from the University of Georgia (U.S.), a master's degree in applied mathematics from the University of South Carolina (U.S.), and a master's degree in elementary education from the University of South Carolina. Her research focuses on mathematics teachers' selection, critical curation, and implementation of mathematics resources from virtual resource pools, as well as teachers' views on the nature of mathematics, teaching of mathematics, and learning mathematics.

Emily Saxton is Director of Research and Evaluation at The Math Learning Center (MLC) in Oregon (U.S.). She holds a PhD in applied psychology with a specialization in developmental psychology. She has worked as a mixed methods researcher in a variety of professional settings, including RMC Research, Human Services Research Institute, Portland Metro STEM Partnership, and Portland State University. Her research interests center on understanding academic and motivational development in K–12 school settings, developing measurement systems, and addressing pressing questions for STEM education. As a researcher working in applied settings, she prioritizes communicating actionable and intuitive findings with key stakeholders. Her work has been published in a variety of scholarly journals, including the *International Journal of Behavioral Development*, the *International Journal of Science Education*, and *Studies in Educational Evaluation*.

Ruti Segal is a Senior Lecturer and the Chair of the master's program in Primary School Mathematics and Science and Head of the Department of Mathematical Education in Primary School (BEd) at the Oranim College of Education, Tivon, Israel. She teaches undergraduate and graduate students and supervises graduate students' research. She taught high school mathematics for 26 years and served as the Senior Assistant to the Superintendent of School Mathematics in Israel. Dr. Segal's research focuses on mathematical knowledge for teaching, particularly as a part of professional development within a community of practice, techno-pedagogical content knowledge, and high school mathematics professional development through interweaving Mathematical-News-Snapshots into the curriculum. As an academic advisor for Vidactica—a Video and Didactic (Vidact) project conducted at the Weizmann Institute of Science, she focused on how self-video based on curiosity-driven discourse can enhance teachers' professional growth. In addition, she currently leads an intercollegiate research group that deals with the integration of technologies in teaching mathematics and science.

Sharon L. Senk, PhD, is Professor Emerita from Michigan State University (U.S.), where she had appointments in the Department of Mathematics and the Program in Mathematics Education. Earlier, she served as Assistant Professor at Syracuse University (U.S.), was a visiting Research Associate at the University of Chicago (U.S.), and taught mathematics in secondary schools in Illinois, Massachusetts, and Michigan. Her major research interests are the teaching and learning of algebra and geometry, mathematics curriculum development, assessment in mathematics, and the preparation of mathematic teachers, with publications appearing in many books and journals, including *Investigations in Mathematics Learning, Journal of Educational Measurement*, the *Journal of Mathematical Behavior, Journal for Research in Mathematics Education, Journal of Teacher Education, Mathematics Teacher, School Science and Mathematics*, and *ZDM Mathematics Education*.

Mary C. Shafer is Associate Professor Emerita in the Department of Mathematical Sciences at Northern Illinois University (NIU) (U.S.), where for 21 years she taught mathematics education courses at the graduate and undergraduate levels and co-directed three Mathematics and Science Partnership grants funded by the U.S. Department of Education. Prior to her work at NIU, she was an Associate Researcher and Research Co-Director with Thomas A. Romberg at the Wisconsin Center for Education Research, University of Wisconsin-Madison (U.S.) on *A Longitudinal/Cross-Sectional Study of the Impact of* Mathematics in Context *on Student Mathematical Performance*, which was funded by the (U.S.) National Science Foundation. Her research interests include curriculum research and teachers' pedagogical decisions and classroom assessment practices.

Atara Shriki is Associate Professor in the Mathematics Education Department at the Kibbutzim College of Education, Israel, where she teaches undergraduate and graduate students, and supervises graduate students. Her research interests concern the professional development of mathematics teachers, the nurturing of mathematical creativity, STEAM[2] education, the history of mathematics, and aspects related to online teaching and learning mathematics. She also writes children's books designed to introduce big mathematical ideas to young children.

Boaz Silverman is a curriculum developer and teacher educator at the Center for Educational Technology in Tel Aviv, Israel, focusing on context-rich curriculum materials and on providing multiple types of explanations and justifications. He has a PhD in Mathematics Education from the Weizmann Institute of Science, Israel. His research interests include the role of context in the learning of mathematics and the contribution of textbooks to teaching and learning mathematics.

Deborah Spencer is a Managing Project Director at Education Development Center (EDC) in the U.S. She has over 30 years of experience leading mathematics education initiatives that support the use of high-quality curricula and equitable instructional practices, with a focus on underserved students PK–12. She is principal investigator for the (U.S.) National Science Foundation (NSF)-funded *Supporting Success in Algebra* study, a study of the implementation and impact of *Transition to Algebra*, an innovative curriculum for students underprepared for algebra. She is co-PI for the NSF-funded *Mathematics through Programming in the Elementary Grades,* a project creating programming microworlds that aim to integrate computational thinking with elementary mathematics to increase all children's access to critical content in both domains. She is also a senior advisor to EDC's *Young Mathematicians* project, which works in partnership with preschool families and HeadStart centers to support playful engagement with mathematics for young children at school and at home.

Ana C. Stephens is a Researcher at the Wisconsin Center for Education Research at the University of Wisconsin–Madison (U.S.). Her work addresses the development of students' algebraic reasoning in the elementary and middle school years, with a focus on levels of sophistication in students' strategies for algebraic problem solving and students' developing understandings of mathematical equivalence. This work has led to numerous presentations (e.g., PME-NA, AERA) and publications in various journals, such as *Journal of Mathematics Teacher Education*, *Journal for Research in Mathematics Education*, *Journal of Mathematical Behavior*, and *Mathematical Thinking and Learning*. Stephens is also interested in connecting research and teaching and leads mathematics methods courses for elementary and middle school teachers.

Rena Stroud is Associate Professor of Education at Merrimack College, Massachusetts (U.S.). As a quantitative methodologist, her areas of expertise include study design, basic ANOVA and regression techniques, hierarchical linear modeling, structural equation modeling, and inter-rater reliability estimation. She has presented at numerous conferences on education, child development, and cognition (e.g., American Educational Research Association, National Association for Research in Science Teaching), and has published widely (e.g., *Journal for Research in Mathematics Education*, *Mathematical Thinking and Learning*, *American Educational Research Journal*).

Despina Stylianou is Professor of Mathematics Education at City College of New York-CUNY (U.S.). Her research in mathematical cognition and argumentation in Grades K–8 explores the habits of mind critical to learning mathematics, particularly around algebraic concepts and practices. Her work has led to numerous publications and presentations (e.g., *Journal for Research in Mathematics Education*, *Educational Studies in Mathematics*, *Journal for Mathematics Teacher Education*). Her research has been supported by several national and state grants, including an Early Career Award by the National Science Foundation.

Alison Tellos is a PhD student in the Department of Education at Concordia University in Montreal, Canada. Alison has been a research assistant on the *Numeracy Kit for Kindergarten 5-Year-Olds* (*NyKK-5*) since October 2020. She has experience providing professional development workshops to kindergarten teachers on early numeracy activities to incorporate into their current classroom practices. Alison's research focuses on how LEGO® bricks can support children's learning in mathematics and on supporting teachers in effective ways to use LEGO® bricks in the classroom. Her doctoral research involves the use of LEGO® bricks with children to gain a deep understanding of their conceptual understanding in mathematics and their learning through instruction.

Maria S. Terrell earned her PhD at the University of Virginia (U.S.). She is a research mathematician with experience and interest in pre-college and college-level curriculum innovation and reform. Maria served as principal or co-principal investigator on numerous grants from the U.S. National Science Foundation, including projects supporting development of materials for high-school mathematics outreach, instructional innovation in calculus and undergraduate geometry, and materials to assess and evaluate innovation in engineering mathematics instruction. During her 30 years at Cornell University, she served as Visiting Associate Professor of Mathematics, Assistant Dean in the College of Arts and Sciences, and Senior Lecturer and Director of Teaching Assistant Programs in the Department of Mathematics. She has co-authored articles in the geometry of rigid structures; together with Peter Lax, she has co-authored two calculus books. Maria is currently retired.

Dawn Teuscher is Professor of Mathematics Education in the College of Computational, Mathematical, and Physical Sciences at Brigham Young University in Utah (U.S.). She earned a PhD in Curriculum and Instruction from the University of Missouri, where she was a fellow for the National Science Foundation-funded Center for the Study of Mathematics Curriculum (CSMC). Her research agenda focuses on mathematics curriculum, specifically the analyses of K–12 mathematics standards and middle-school mathematics textbooks. Currently, she studies mathematics teachers' curricular reasoning to determine how teachers reason about their curricular decisions as they plan and enact mathematics lessons.

Jennifer S. Thom is Associate Professor of Mathematics Education and Curriculum Studies, Faculty of Education at the University of Victoria, Canada. Her current research interests include spatial reasoning in classroom mathematics, the nature of mathematical understanding, STEM, and deaf/hard of hearing (D/HH) learners' mathematics.

Zalman Usiskin is Professor Emeritus of Education at the University of Chicago (U.S.), where he was an active faculty member from 1969 through 2007. He began his career as a high-school mathematics teacher. His work has focused on the teaching and learning of arithmetic, algebra, and geometry, with particular attention to applications of mathematics at all levels, the use of transformations and related concepts in geometry, algebra, and statistics, and the use of calculator and computer technology. From 1971 through 1979, he authored or co-authored texts for each of the four years of high-school mathematics. In 1983, he helped initiate the University of Chicago School Mathematics Project (UCSMP). He served as its overall director from 1987 until 2019 and led the development of all editions of the UCSMP materials for middle and secondary schools.

Marcy B. Wood is Professor of Mathematics Education in the Department of Teaching, Learning, and Sociocultural Studies at the University of Arizona (U.S.). Her research interests arise from her experiences as a third- and fourth-grade teacher in Albuquerque, New Mexico. She is curious about the mathematical learning of elementary students, specifically students who have not yet been successful in mathematics. Her research uses the tools of discourse and identity to examine mathematical learning and to assist teachers in supporting the mathematical learning of their students. She also draws upon sociocultural theories, cognitive theories, and conceptual metaphors as structures for understanding mathematical learning.

Michal Yerushalmy is Professor Emeritus of the Department of Mathematics Education at the University of Haifa, Israel. Yerushalmy was previously the Vice-President and Dean of Research of the University of Haifa (2011–2016). She served as the chair of the Department of Education, and established and chaired the Department of Mathematics Education. Yerushalmy studies mathematical learning and teaching with a focus on design and implementation of technology-based environments. Yerushalmy authored and designed numerous software packages and interactive textbooks. She co-authored, with J. L. Schwartz, the *Geometric Supposer* and designed the *VisualMath* algebra and calculus curriculum. Yerushalmy's research and development focus is on ways to use technology to make mathematical inquiry learning everywhere available, and on ways to support learning and teaching with online formative assessment (The STEP project).

Julie Zeringue is a Research Scientist at Education Development Center (U.S.). She has worked on multiple large-scale projects looking at curriculum implementation and how school, district, and state contexts affect use. She also has experience in qualitative analysis and district recruitment. Her research interests include exploring students' conceptions of beginning algebraic concepts, and understanding what factors encourage or limit student enrollment in and access to AP Computer Science Principles (AP CSP). Dr. Zeringue served as a co-editor of *Classics in Mathematics Research* (2004) with Tom Carpenter and John Dossey. More recently, she has co-published articles on mathematics textbook selection, algebraic thinking, and factors influencing students' inclination to take the AP CSP course.

ENDNOTES

1. STEM represents Science, Technology, Engineering, and Mathematics.
2. STEAM represents Science, Technology, Engineering, the Arts, and Mathematics.

Printed in the USA
CPSIA information can be obtained
at www.ICGtesting.com
CBHW062102290924
15091CB00003BA/60